# High Levels of Natural Radiation and Radon Areas:
# Radiation Dose and Health Effects

# High Levels of Natural Radiation and Radon Areas: Radiation Dose and Health Effects

Proceedings of the 6th International Conference on High Levels of Natural Radiation and Radon Areas, held in Osaka, Japan between 6 and 10 September 2004

*Editors:*

**Tsutomu Sugahara**
The Japan Health Foundation
Kyoto
JAPAN

**Yasuhito Sasaki**
National Institute of Radiological Sciences
Chiba
JAPAN

**Hiroshige Morishima**
Kinki University
Atomic Energy Research Institute
Osaka
JAPAN

**Isamu Hayata**
National Institute of Radiological Sciences
Chiba
JAPAN

**Mehdi Sohrabi**
National Radiation Protection Department
Atomic Energy Organization
Tehran
IRAN

**Suminori Akiba**
Kagoshima University
Kagoshima
JAPAN

ELSEVIER

2005

ELSEVIER B.V.  
Radarweg 29  
P.O. Box 211, 1000 AE Amsterdam  
The Netherlands

ELSEVIER Inc.  
525 B Street, Suite 1900  
San Diego, CA 92101-4495  
USA

ELSEVIER Ltd  
The Boulevard, Langford Lane  
Kidlington, Oxford OX5 1GB  
UK

ELSEVIER Ltd  
84 Theobalds Road  
London WC1X 8RR  
UK

© 2005 Elsevier B.V. All rights reserved.

This work is protected under copyright by Elsevier, and the following terms and conditions apply to its use:

Photocopying  
Single photocopies of single chapters may be made for personal use as allowed by national copyright laws. Permission of the Publisher and payment of a fee is required for all other photocopying, including multiple or systematic copying, copying for advertising or promotional purposes, resale, and all forms of document delivery. Special rates are available for educational institutions that wish to make photocopies for non-profit educational classroom use.

Permissions may be sought directly from Elsevier's Health Science Rights Department, Elsevier Inc., 625 Walnut Street, Philadelphia, PA 19106, USA: phone (+1) 215 238 7869, fax: (+1) 215 238 2239, E-mail: healthpermissions@elsevier.com. You may also complete your request on-line via the Elsevier homepage (http://www.elsevier.com), by selecting 'Customer Support' and then 'Obtaining Permissions'.

In the USA, users may clear permissions and make payments through the Copyright Clearance Center, Inc., 222 Rosewood Drive, Danvers, MA 01923, USA; phone: (+1) (978) 7508400, fax: (+1) (978) 7504744, and in the UK through the Copyright Licensing Agency Rapid Clearance Service (CLARCS), 90 Tottenham Court Road, London W1P 0LP, UK; phone: (+44) 207 631 5555; fax: (+44) 207 631 5500. Other countries may have a local reprographic rights agency for payments.

Derivative Works  
Tables of contents may be reproduced for internal circulation, but permission of Elsevier is required for external resale or distribution of such material.  
Permission of the Publisher is required for all other derivative works, including compilations and translations.

Electronic Storage or Usage  
Permission of the Publisher is required to store or use electronically any material contained in this work, including any chapter or part of a chapter.

Except as outlined above, no part of this work may be reproduced, stored in a retrieval system or transmitted in any form or by any means, electronic, mechanical, photocopying, recording or otherwise, without prior written permission of the Publisher. Address permissions requests to: Elsevier Rights Department, at the fax and e-mail addresses noted above.

Notice  
No responsibility is assumed by the Publisher for any injury and/or damage to persons or property as a matter of products liability, negligence or otherwise, or from any use or operation of any methods, products, instructions or ideas contained in the material herein. Because of rapid advances in the medical sciences, in particular, independent verification of diagnoses and drug dosages should be made.

First edition 2005

Library of Congress Cataloging in Publication Data  
A catalog record from the Library of Congress has been applied for.

British Library Cataloguing in Publication Data  
A catalogue record from the British Library has been applied for.

International Congress Series No. 1276  
ISBN: 0-444-51431-7  
ISSN: 0531-5131

∞ The paper used in this publication meets the requirements of ANSI/NISO Z39.48-1992 (Permanence of Paper).  
Printed in The Netherlands.

Working together to grow  
libraries in developing countries

www.elsevier.com | www.bookaid.org | www.sabre.org

ELSEVIER    BOOK AID International    Sabre Foundation

www.ics-elsevier.com

# Preface

Naturally occurring background radiation is a topic that has evoked curiosity and concern of the scientist and layman alike in recent years due to the shift in focus of health effects, from acute high level exposure of radiation to chronic low level exposure. Many places around the world have high levels of natural background radiation due to elevated levels of primordial radionuclides in the soil and their decay products like Radon and Thoron in the environment. Of late, Technologically Enhanced Naturally Occurring Radioactive Material (TENORM) has also contributed to the burden of background radiation. A large amount of data on the health effects of radiation has been collected from several scientific studies conducted in these areas.

With a view to disseminating the information acquired and expanding the knowledge base, the first International Conference on areas of high natural radioactivity was held in Pocos de Caldas, Brazil in 1975. Subsequently, four more conferences were held in Mumbai, India (1981), Ramsar, Iran (1990), Beijing, China (1996), and Munich, Germany (2000). The 6th International Conference on High Levels of Natural Radiation and Radon Areas was held from September 6 to 10, 2004 at the November Hall of Kinki University, Osaka, Japan, in the hope of providing opportunities for scientists from different disciplines from all over the world to meet and discuss the various scientific issues that would provide an impetus for research pertaining to the area of natural radiation including TENORM.

The conference program was composed of ten special lectures, four workshops, three panel discussions, thirty-two oral presentations, one hundred and eleven poster presentations, summary presentations, and a general discussion. Rich in content, it was compiled with thoughtful consideration given to the effective discussion of aspects of dosimetry, cytogenetics, biology, epidemiology, and radiation protection regarding high levels of natural radiation and radon. Measurements of natural radiation and radon, biological studies such as effects on the biosphere, molecular mechanisms and chromosome aberration, and health risk estimation such as cancer were discussed on the basis of research conducted in areas of potentially high radiation such as; Yangjiang, Guangdong in China, Karunagappally, Kerala in India, Ramsar in Iran, Guarapari and Araxá in Brazil, etc.

0531-5131/ © 2004 Elsevier B.V. All rights reserved.
doi:10.1016/j.ics.2004.12.014

The conference was truly international, with two hundred and five participants including eighty overseas participants from thirty-two countries. The organizers would like to thank all the participants and those who have worked so hard for its success, and hope on this occasion of publishing the proceedings, that this volume will serve as both an excellent review and valuable reference for the future study of high level natural radiation and radon areas.

Hiroshige Morishima
Chairman of the Organizing Committee

## Conference Officers and Committees
International Scientific Committee
Chairman         Tsutomu Sugahara (The Japan Health Foundation, Japan)
Vice Chairman    Yasuhito Sasaki (National Institute of Radiological Sciences, Japan)
Members          Burton Bennett (Radiation Effects Research Foundation, Japan)
                 Werner Burkart (IAEA)
                 Elisabeth Cardis (WHO/IARC)
                 Zhanat Carr (WHO)
                 Norman Gentner (UNSCEAR)
                 Abel J. González (IAEA)
                 Gerry M. Kendall (National Radiological Protection Board, UK)
                 Mehdi Sohrabi (ICHLNRRA)
                 Sentaro Takahashi (National Institute of Radiological Sciences, Japan)
                 Luxin Wei (National Institute for Radiological Protection, China)

## Organizing Committee
Chairman         Hiroshige Morishima (Kinki University, Japan)
Vice Chairman    Isamu Hayata (National Institute of Radiological Sciences, Japan)
Members          Kenzo Fujimoto (National Institute of Radiological Sciences, Japan)
                 Naoto Fujinami (Kyoto Prefectural Institute of Hygienic & Environmental Sciences, Japan)
                 Kazunobu Fujitaka (National Institute of Radiological Sciences, Japan)
                 Manabu Fukumoto (Tohoku University, Japan)
                 Sadaaki Furuta (Japan Nuclear Cycle Development Institute, Japan)
                 Shunichi Hisamatsu (Institute for Environmental Sciences, Japan)
                 Takao Iida (Nagoya University, Japan)
                 Michiaki Kai (Oita University of Nursing and Health Sciences, Japan)
                 Masahito Kaneko (Radiation Effects Association, Japan)
                 Taeko Koga (Kinki University, Japan)
                 Toshisou Kosako (The University of Tokyo, Japan)
                 Shin-ichiro Miyazaki (The Kansai Electric Power Co., Inc, Japan)
                 Mitsuoki Morimyo (National Institute of Radiological Sciences, Japan)
                 Toshi Nagaoka (Japan Atomic Energy Research Institute, Japan)
                 Sei-ichi Nakamura (Health Research Foundation, Japan)
                 Ohtsura Niwa (Kyoto University, Japan)
                 Toshiaki Ogiu (National Institute of Radiological Sciences, Japan)
                 Kazuo Sakai (Central Research Institute of Electric Power Industry, Japan)
                 Masao Sasaki (Professor Emeritus of Kyoto University, Japan)
                 Atsuhiko Takeda (Health Research Foundation, Japan)
                 Zufan Tao (National Institute for Radiological Protection, China)
                 Hideo Tatsuzaki (National Institute of Radiological Sciences, Japan)
                 Kanji Torizuka (Health Research Foundation, Japan)
                 Tadashi Tsujimoto (Electron Science Institute, Japan)
                 Masami Watanabe (Nagasaki University, Japan)
                 Yuji Yamada (National Institute of Radiological Sciences, Japan)

## Program Committee
Chairman         Suminori Akiba (Kagoshima University, Japan)
Members          Naoto Fujinami (Kyoto Prefectural Institute of Hygienic & Environmental Sciences, Japan)
                 Manabu Fukumoto (Tohoku University, Japan)
                 Masahide Furukawa (National Institute of Radiological Sciences, Japan)
                 Seiji Kodama (Nagasaki University, Japan)
                 Raghu R.K. Nair (Regional Cancer Centre, Kerala, India)
                 Yukiko Shimizu (Radiation Effects Research Foundation, Japan)
                 Quanfu Sun (National Institute for Radiological Protection, China)
                 Shinji Tokonami (National Institute of Radiological Sciences, Japan)

## Acknowledgements

"The 6th International Conference on High Levels of Natural Radiation and Radon Areas - Radiation Dose and Health Effects -" was organized by Kinki University and the National Institute of Radiological Sciences, in cooperation with UNSCEAR, IAEA, and WHO. It has concluded successfully, and the organizers' sincere thanks are due to:

- The Program committee, who were responsible for the entire program, from compiling the invited sessions and contributed oral/poster sessions, to the arrangement of the committee meeting of the ICHLNRRA.
- The foundations, institutions, organizations, companies and individuals who supported the conference financially or physically.
  - The Health Research Foundation which has carried out research, sometimes on a cooperative basis in China, India, Iran and Brazil, and contributed for a long time as a kind of "center" for high level natural radiation research in Japan.
  - Japan Vascular Disease Research Foundation
  - Japan Nuclear Cycle Development Institute
  - The Commemorative Organization for the Japan World Exposition ('70)
  - Inoue Foundation for Science
  - The Kao Foundation For Arts And Science
  - Osaka Convention & Tourism Bureau
  - CHIYODA TECHNOL CORPORATION
  - THE KANSAI ELECTRIC POWER COMPANY
  - TOKYO ELECTRIC POWER COMPANY
  - PONY INDUSTRY CO., LTD.
  - Fuji Electric Systems Co., Ltd.
  - ALOKA CO., LTD.
  - NAGASE LANDAUER, LTD.
  - Tsutomu Sugahara, M.D.

# Contents

| | |
|---|---|
| **Preface** | v |
| **Conference Officers and Committees** | vii |
| **Acknowledgements** | viii |
| **Invited papers** | 1 |

**What did we learn from radiobiological and genetic studies in HBRAs?**

Chromosome study in high background radiation area in Southern China
*C. Wang, W. Zhang, M. Minamihisamatsu, H. Morishima, Y. Yuan, T. Jiang,
D. Chen, L. Wei, T. Sugahara and I. Hayata*    3

Investigations on the health effects of human population residing in the high-level
natural radiation areas in Kerala in the southwest coast of India
*M.V. Thampi, V.D. Cheriyan, G. Jaikrishan, B. Das, C.J. Kurien,
E.N. Ramachandran, C.V. Karuppasamy, B. Ravikumar, D.C. Soren,
U. Vijayan, P.K.M. Koya, V.J. Andrews, V. Anilkumar, A. Mitra,
M. Madhusoodhanan, K.V. Aravindan and M. Seshadri*    8

New findings in the very high natural radiation area of Ramsar, Iran
*M. Ghiassi-Nejad, M.M. Beitollahi, N. Fallahian and M. Saghirzadeh*    13

**Factors for the induction of DNA rearrangements leading to malignant disease**

Chromosomal mutations by low dose radiation vs. those by other mutagenic factors
*I. Hayata*    17

Extranuclear target and low dose radiation risk assessment
*T.K. Hei, H. Zhou and M. Suzuki*    21

Molecular dissection of in vivo DNA rearrangements induced by radiation and
chemical mutagens
*T. Nohmi and K.-i. Masumura*    25

**Other topics on cytogenetics and biology**

Chromosome aberrations in peripheral lymphocytes of individuals living in
dwellings with an increased level of indoor radon concentrations
*G. Stephan, U. Oestreicher and R. Lehmann*    29

Cytogenetic studies of populations exposed to Chernobyl fallout
D.C. Lloyd   33

**Can we estimate individual radiation doses of HBRA residents?**

Results of Yangjiang study in China and an experience of Ramsar survey in Iran
N. Fujinami, T. Koga, H. Morishima and T. Sugahara   37

Individual dose estimation—our experience with the Karunagappally study in Kerala, India
K. Raghu Ram Nair, S. Akiba, V.S. Binu, P. Jayalekshmi, P. Gangadharan, M. Krishnan Nair and B. Rajan   41

Indoor radon, thoron, and thoron daughter concentrations in Korea
Y.J. Kim, H.Y. Lee, C.S. Kim, B.U. Chang, B.H. Rho, C.K. Kim and S. Tokonami   46

The Spanish experience on HBRA
L.S. Quindos, P.L. Fernández, C. Sainz, J. Gomez, J.L. Matarranz and E. Suarez Mahou   50

**Radon exposure and its potential risk**

Lung cancer risk due to radon in dwellings—evaluation of the epidemiological knowledge
H.E. Wichmann, A. Schaffrath Rosario, I.M. Heid, M. Kreuzer, J. Heinrich and L. Kreienbrock   54

Thoron in the living environments of Japan
H. Yonehara, S. Tokonami, W. Zhuo, T. Ishikawa, K. Fukutsu and Y. Yamada   58

Occupational exposure to radon-experience and approach to regulatory control
E. Ettenhuber   62

Radon retrospective measurements
C. Samuelsson   66

**Significance of radon and thoron measurements on dose estimation**

Thoron versus radon: measurement and dosimetry
N.H. Harley, P. Chittaporn, R. Medora, R. Merrill and W. Wanitsooksumbut   72

Rn–Tn discriminative measurements and their dose estimates in Chinese loess plateau
Y. Yamada, S. Tokonami, W. Zhuo, H. Yonehara, T. Ishikawa, M. Furukawa, K. Fukutsu, Q. Sun, C. Hou, S. Zhang and S. Akiba   76

Radon dosimetry and its implication for risk
A. Birchall and J.W. Marsh     81

Reference fields and calibration techniques for Rn-220 measuring instruments
E. Gargioni and D. Arnold     85

**Other topics on dosimetry**

Dosimetric considerations for environmental radiation and NORM
S.L. Simon     89

Measurements to determine the radiological impact of uranium and thorium in soils in the darling scarp
L.F. Toussaint     93

**What did we learn from epidemiological studies in HBRAs?**

Cancer and non-cancer epidemiological study in the high background radiation area of Yangjiang, China
J. Zou, Z. Tao, Q. Sun, S. Akiba, Y. Zha, T. Sugahara and L. Wei     97

What did we learn from epidemiological studies in high background radiation area in India
P. Jayalekshmi, P. Gangadharan, V.S. Binu, R.R.K. Nair, M.K. Nair, B. Rajan and S. Akiba     102

Mortality and morbidity from cancer in the population exposed to high level of natural radiation area in Ramsar, Iran
A. Mosavi-Jarrahi, M. Mohagheghi, S. Akiba, B. Yazdizadeh, N. Motamedi and A. Shabestani Monfared     106

Pattern of cancer mortality in some Brazilian HBRAs
L.H.S. Veiga and S. Koifman     110

WHO network and international collaborative project on residential radon risk
Z. Carr and M. Repacholi     114

Commentary on information that can be drawn from studies of areas with high levels of natural radiation
C. Elisabeth     118

**Cosmic rays**

High-level doses brought by cosmic rays
K. Fujitaka     124

Factors affecting cosmic ray exposures in civil aviation
G.M. Kendall     129

## Depleted uranium

Properties, use and health effects of depleted uranium
*W. Burkart, P.R. Danesi and J.H. Hendry* — 133

Public health and environmental aspects of DU
*J.P. McLaughlin* — 137

A comparison of human exposure to natural radiation and DU in parts of the Balkan region
*Z.S. Zunic, K. Fujimoto and I.V. Yarmoshenko* — 141

## Summary

Summary of HBRA epidemiological studies
*Q. Sun and Z. Carr* — 147

Summary of dosimetry (radon and thoron) studies
*S. Tokonami* — 151

Summary of biological studies
*S. Kodama* — 155

Overall summary and comments
*M. Kai* — 159

## Contribution of HBRA studies to health risk assessments and regulatory control

Contribution of high natural background radiation area studies to an evolved system of radiological protection
*M. Kaneko* — 162

Problems with HBRA studies in health risk assessment
*G.M. Kendall* — 166

New public dose assessment from internal and external exposures in low- and elevated-level natural radiation areas of Ramsar, Iran
*M. Sohrabi and M. Babapouran* — 169

## Oral presentations — 175

## Cytogenetics and biology

Changes of chromosome aberration rate and micronucleus frequency along with accumulated doses in continuously irradiated mice with a low dose rate of $\gamma$-rays
*K. Tanaka, A. Kohda, K. Ichinohe and T. Matsumoto* — 177

Dose and dose rate effects of low dose ionizing radiation on activation of p53 in immortalized murine cells
T. Sugihara, J. Magae, R. Wadhwa, S.C. Kaul, Y. Kawakami, T. Matsumoto, K. Tanaka and Y. Oghiso     179

Histologic assessment of regenerating rat liver under low-dose rate radiation exposure
D.V. Guryev     181

Increased radiosensitivity of splenic T lymphocytes in pregnant mice
Y. Igari, K. Igari, H. Kakihara, F. Kato, A. Ootsuyama and T. Norimura     183

Effects of low dose-rate irradiation on the glucose metabolism in type II diabetes model mice
T. Nomura and K. Sakai     185

Adaptive response, split dose and survival of mice
A. Bhan Tiku and R.K. Kale     187

Heavy ion induced transformation of telomerase immortalized human small airway epithelial cells
C.Q. Piao, Y.L. Zhao, M. Suzuki, A. Balajee and T.K. Hei     189

Mechanisms of liver carcinogenesis by chronic exposure to alpha-particles form internally deposited Thorotrast
L. Wang, D. Liu, T. Shimizu and M. Fukumoto     192

Mangiferin, a glucosylxanthone, protects against the radiation-induced micronuclei formation in the cultured human peripheral blood lymphocytes
G.C. Jagetia and V.A. Venkatesha     195

Association of glutathion *S*-transferase and chromosomal aberrations as a means to determine occupational exposure
A. Movafagh, F. Maleki, S.G. Mohammadzadeh and S. Fadaei     197

Hormone levels associated with immune responses among inhabitants in HLNRAs of Ramsar-Iran
F. Zakeri and A. Kariminia     199

Radioadaptive responses induced in lymphocytes of the inhabitants in Ramsar, Iran
S.M.J. Mortazavi, A. Shabestani-Monfared, M. Ghiassi-Nejad and H. Mozdarani     201

## Dosimetry

Elevated radon concentrations in a Pleistocenic cave operating as a show cave
C. Papastefanou, M. Manolopoulou, S. Stoulos, A. Ioannidou and
E. Gerasopoulos   204

Radon in groundwater: analysis of causes using GIS and multivariate statistics: A case study in the Stockholm county
K. Skeppström and B. Olofsson   206

Current studies on radon gas in Thailand
P. Wanabongse, S. Tokonami and S. Bovornkitti   208

Natural radioactivity in the high background radiation area at Erasama beach placer deposit of Orissa, India
D. Sengupta, A.K. Mohanty, S.K. Das and S.K. Saha   210

High radon areas in Norway
T. Strand, C. Lunder Jensen, K. Ånestad, L. Ruden and G. Beate Ramberg   212

Levels of indoor radon, thoron, and their progeny in Himalaya
R.C. Ramola   215

Long-term measurements of radon progeny concentrations with LR 115 SSNTDs
K.N. Yu, D. Nikezic, F.M.F. Ng, B.M.F. Lau and J.K.C. Leung   217

Convenient methods for evaluation of indoor thoron progeny concentrations
W. Zhuo and S. Tokonami   219

Radon exhalation from a ground surface during a cold snow season
H. Yamazawa, T. Miyazaki, J. Moriizumi, T. Iida, S. Takeda, S. Nagara,
K. Sato and T. Tokizawa   221

The migration of U-238 in the system "soil–plant" and its effect on plant growth
A.A. Kasianenko, G.A. Kulieva, A.N. Ratnikov, T.L. Jigareva and V.A. Kalchenko   223

$^{210}$Po and $^{210}$Pb content in environmental and human body samples in the Ramsar area, Iran
H. Samavat, M.R.D. Seaward, S.M.R. Aghamiri and A. Shabestani Monfared   225

High resolution analyses of temporal variations of airborne radionuclides
K. Komura, Y. Yamaguchi, N. Muguntha Manikandan, Y. Murata, M. Inoue
and T. Iida   227

Naturally accumulated radiation doses and dating of archaeologically burnt
materials using luminescence from white minerals
T. Hashimoto, Y. Nukata and T. Yawata   231

**Epidemiology and radiation protection**

Case-control study of residential radon and childhood leukemia in Japan:
results from preliminary analyses
S. Yoshinaga, S. Tokonami, S. Akiba, H. Nitta, M. Kabuto and JCCSG   233

The risk of lung cancer in HBR area in India—a case control study
V.S. Binu, P. Gangadharan, P. Jayalekshmi, R.R.K. Nair, M.K. Nair, B. Rajan
and S. Akiba   236

Improvement of the radon situation at former uranium mining and milling sites
in East Germany
P. Schmidt and J. Regner   238

Investigation and reduction of personnel exposure levels in Bavarian water
supply facilities
S. Körner, M. Trautmannsheimer and K. Hübel   240

The ENVIRAD project: a way to control and to teach how to protect from
high indoor radon level
A.M. Esposito, M. Ambrosio, E. Balzano, L. Gialanella, M. Pugliese, V. Roca,
M. Romano, C. Sabbarese and G. Venoso   242

Radiation hazard and protection for the nuclear weapon terrorism
J. Takada   245

**Posters**   247

**Cytogenetics and biology**

Effects of radon and thermal therapy on osteoarthritis
K. Yamaoka, F. Mitsunobu, K. Hanamoto, T. Kataoka and Y. Tanizaki   249

Induction of micronuclei in rat tracheal epithelial cells following radon exposure
at air–liquid interface culture
K. Fukutsu, Y. Yamada, W. Zhuo, S. Tokonami and A. Koizumi   251

Simplification of analysis for comparison of radioactive characteristics and
its application to some minerals for radon therapy
*K. Hanamoto and K. Yamaoka* — 253

Bystander cellular effects in normal human fibroblasts irradiated with
low-density carbon ions
*M. Suzuki, C. Tsuruoka, N. Yasuda, K. Matsumoto and K. Fujitaka* — 256

Effects of low dose-rate long-term gamma-ray irradiation on DNA damage in
mouse spleen
*K. Otsuka and K. Sakai* — 258

Detection of radiation-induced mutations in the liver after partial hepatectomy
using HITEC transgenic mice
*A. Ootsuyama, F. Kato and T. Norimura* — 260

Role of *p53* gene in genetic instability induced by ionizing radiation
*K. Igari, Y. Igari, H. Kakihara, F. Kato, A. Ootsuyama and T. Norimura* — 262

Adaptive response of bone marrow stem cells induced by low-dose rate
irradiation in C57BL/6 mice
*K. Shiraishi, A. Tachibana, M. Yonezawa and S. Kodama* — 264

The life saving role of radioadaptive responses in long-term interplanetary
space journeys
*S.M.J. Mortazavi, J.R. Cameron and A. Niroomand-Rad* — 266

Biosorption of chromium(III) by new bacterial strain (NRC-BT-2)
*M. Rabbani, H. Ghafourian, S. Sadeghi and Y. Nazeri* — 268

Uptake and removal of nickel by new bacterial strain (NRC-BT-1)
*H. Ghafourian, M. Rabbani, Y. Nazeri and S. Sadeghi* — 270

Biosorption of $Cs^+$ by new bacterial strain (NRC-BT-2)
*Y. Nazeri, S. Sadeghi, M. Rabbani and H. Ghafourian* — 272

The effect of low dose-rate γ-radiation on the chemiluminescence of blood
serum at chronic inflammation in rats
*N. Klimenko, M. Onyshchenko, N. Dikij and E. Medvedeva* — 274

**Measurements**

An intercomparison exercise for thoron gas measurement
*T. Ishikawa, A. Cavallo and S. Tokonami* — 276

Particle size measurement of radon decay products using MOUDI and GSA
*S. Tokonami, K. Fukutsu, Y. Yamada and Y. Yatabe* — 278

Practicality of the thoron calibration chamber system at NIRS, Japan
Y. Kobayashi, S. Tokonami, H. Takahashi, W. Zhuo and H. Yonehara — 281

Measurements of radon, thoron and their progeny in a dwelling in Gifu prefecture, Japan
C. Németh, S. Tokonami, T. Ishikawa, H. Takahashi, W. Zhuo and M. Shimo — 283

Soil radon flux and outdoor radon concentrations in East Asia
W. Zhuo, M. Furukawa, Q. Guo and Y. Shin Kim — 285

Estimation of radon-222 exhalation rate and control of radon-222 concentration in ventilated underground space
J. Moriizumi, M. Mori, E. Sasao, H. Yamazawa and T. Iida — 287

Measurements of concentrations and its ratio of radon decay products in rainwater by gamma-ray spectrometry with a low background germanium detector
M. Takeyasu, T. Iida, T. Tsujimoto and K. Yamasaki — 289

Radon exhalation rate monitoring in/around the closed uranium mine sites in Japan
Y. Ishimori and Y. Maruo — 291

Radon-222 concentrations and the environmental radiation dose rates at Misasa spa districts, Japan
M. Inagaki, S. Nojiri, T. Takemura, T. Koga, H. Morishima, Y. Tanaka, M. Mifune and I. Kobayashi — 293

Distribution characteristics of natural radionuclides at some spa districts in Japan
T. Koga, T. Takemura, S. Nojiri, H. Morishima, M. Inagaki, Y. Tanaka and M. Mifune — 295

Optimization of measuring methods on size distribution of naturally occurring radioactive aerosols
K. Yamasaki, Y. Oki, Y. Yamada, S. Tokonami and T. Iida — 297

Measurement of radon concentration in water using direct dpm method of liquid scintillation counter
Y. Yasuoka, T. Ishii, T. Sanada, W. Nitta, Y. Ishimori, Y. Kataoka, T. Kubo, H. Suda, S. Tokonami, T. Ishikawa and M. Shinogi — 299

Airborne and waterborne radon concentrations in houses with the use of groundwater
T. Ishikawa, S. Yoshinaga and S. Tokonami — 301

Emanating power of $^{220}$Rn from a powdery radiation source
T. Iimoto, T. Nagai, N. Sugiura and T. Kosako — 303

The environment dose measurement using the OSL dosimeter
*I. Kobayashi, H. Sekiguchi, F. Tatsuta, H. Komori, T. Koga and H. Morishima*    305

Eolian dust may be an effective expander of HBRA: a case of China–Ryukyu connection
*M. Furukawa, W. Zhuo, S. Tokonami, N. Akata and Q. Guo*    307

Radon and thoron in Yongding Hakka Earth Building in Fujian Province of China
*M. Wei, W. Zhuo, Y. Luo, M. Furukawa, X. Chen, Y. Yamada, D. Lin and F. Jian*    309

Natural radiation levels in Fu'an city in Fujian Province of China
*Y. Luo, W. Zhuo, M. Wei, S. Tokonami, W. Wang, Y. Yamada, J. Chen and M. Chen*    311

A pilot survey on indoor radon and thoron progeny in Yangjiang, China
*G. Qiuju, C. Bo and S. Quanfu*    313

High background radiation valley formed by Peitou Hot Spring
*C.-J. Chen, P.-H. Lin and C.-C. Huang*    315

Natural radionuclide distribution in soil samples around Kudankulam Nuclear Power Plant Site (Radhapuram taluk of Tirunelveli district, India)
*G.M. Brahmanandhan, D. Khanna, J. Malathi and S. Selvasekarapandian*    317

Measurement of activity concentrations of $^{40}$K, $^{238}$U and $^{232}$Th in soil samples of Agastheeswaram taluk, Kanyakumari district, India
*D. Khanna, J. Malathi, G.M. Brahmanandhan and S. Selvasekarapandian*    319

Study of primordial radionuclide distribution in sand samples of Agastheeswaram taluk of Kanyakumari district, India
*J. Malathi, G.M. Brahmanandhan, D. Khanna and S. Selvasekarapandian*    321

Primordial radionuclides concentrations in the beach sands of East Coast region of Tamilnadu, India
*K.S. Lakshmi, S. Selvasekarapandian, D. Khanna and V. Meenakshisundaram*    323

Measurement of primordial radionuclide distribution in the soil samples along the East coast of Tamilnadu, India
*V. Meenakshisundaram, K.S. Lakshmi, J. Malathi and S. Selvasekarapandian*    325

Indoor gamma dose measurement along the East coast of Tamilnadu, India using TLD
*S. Selvasekarapandian, K.S. Lakshmi, G.M. Brahmanandhan and V. Meenakshisundaram*    327

Study of background radiation from soil samples of Udumalpet Taluk of
Coimbatore district, India
R. Sarida, G.M. Brahmanandhan, J. Malathi, D. Khanna,
S. Selvasekarapandian, R. Amutha, V. Meenakshisundaram
and V. Gajendran      329

Study of background radiation from soil samples of Pollachi taluk, Tamilnadu,
India
R. Amutha, G.M. Brahmanandhan, J. Malathi, D. Khanna,
S. Selvasekarapandian, R. Sarida, V. Meenakshisundaram
and V. Gajendran      331

Enrichment of natural radionuclides in monazite areas of coastal Kerala
Y. Narayana, P.K. Shetty and K. Siddappa      333

Thoron ($^{220}$Rn) levels in dwellings around normal and high background
areas in India
T.V. Ramachandran, K.P. Eappen, R.N. Nair and Y.S. Mayya      335

Background radiation exposure levels: Indian scenario
T.V. Ramachandran, K.P. Eappen and Y.S. Mayya      337

Radiological impact of utilization of phosphogypsum and fly ash in building
construction in India
V.K. Shukla, T.V. Ramachandran, S. Chinnaesakki, S.J. Sartandel and
A.A. Shanbhag      339

Indoor and outdoor radon levels and their diurnal variations in the environs
of southwest coast of India
N. Karunakara, H.M. Somashekarappa, K.M. Rajashekara and K. Siddappa      341

Study of indoor gamma radiation in Coimbatore City, Tamilnadu, India
J. Malathi, A.K. Andal Vanmathi, A. Paramesvaran, R. Vijayshankar
and S. Selvasekarapandian      344

Natural radioactivity in South West Coast of India
N. Karunakara, H.M. Somashekarappa and K. Siddappa      346

Transportation of radionuclides from Western Ghats to Arabian sea through
some major rivers of South India
K.M. Rajashekara, Y. Narayana, N. Karunakara and K. Siddappa      348

Application of well-type NaI(Tl) detector for indoor radon measurements
D. Al-Azmi      350

Assessment of exposure to the population of Russia from radon
A.M. Marenny, S.M. Shinkarev, A.V. Penezev, M.N. Savkin and M. Hoshi — 352

Altai Region of Russia—a high radon potential area
V.P. Borisov, I.P. Saldan and A.P. Strokov — 354

Assessment of exposure to the population of Moscow from natural sources of radiation
A.M. Marenny, S.M. Shinkarev, A.V. Penezev, A.V. Frolova, Yu.A. Morozov, S.E. Okhrimenko, M.N. Savkin and M. Hoshi — 356

Comparison of contemporary and retrospective radon concentration measurement in dwellings in Poland
J. Jankowski, J. Skubalski, J. Olszewski, P. Szalanski and A. Zak — 358

Radon on underground tourist routes in Poland
J. Olszewski, W. Chruścielewski and J. Jankowski — 360

Radioactivity in vine cellars in Hungary and Slovenia
P. Szerbin, J. Vaupotic, I. Csige, I. Kobal, I. Hunyadi, L. Juhász and E. Baradács — 362

Tenorm's around coal fired power plant tailings ponds in Hungary
P. Szerbin, L. Juhász, I. Csige, A. Várhegyi, J. Vincze, T. Szabó and F.-J. Maringer — 365

Evaluation of the technologically enhanced naturally occurring radioactive material in Hungary
L. Juhász, P. Szerbin and I. Czoch — 367

Inaccuracies in assessing doses from radon in workplaces
C. Németh, S. Tokonami, J. Somlai, T. Kovács, N. Kávási, Z. Gorjánácz, A. Várhegyi and J. Hakl — 369

Naturally occurring alpha emitting radionuclides in drinking water (Hungary) and assessment of dose contribution due to them
T. Kovács, E. Bodrogi, J. Somlai, V. Jobbágy, P. Dombovári and C. Németh — 371

Public exposure to radon and thoron progeny in Romania
O. Iacob and C. Grecea — 373

Radon exposure in Slovenian kindergartens and schools
J. Vaupotič and I. Kobal — 375

Blower door method and measurement technology in radon diagnosis
A. Froňka and L. Moučka — 377

Radon programme in the Czech Republic—experience and further research
A. Froňka, J. Hůlka and J. Thomas                                                                  379

Effective dose calculation using radon daughters and aerosol particles measurement
in Bozkov Dolomite Cave
L. Thinova, Z. Berka, E. Brandejsova, V. Zdimal and D. Milka                                       381

Measurement of radon daughters in water and in air using the detection unit
"YAPMARE" with a YAP:Ce scintillation detector
L. Thinova, A. Kunka, P. Maly, F. de Notaristefani, K. Blazek, T. Trojek and
L. Moucka                                                                                          383

A radon survey in some regions of Turkey
Y. Yarar, T. Günaydı and E. Kam                                                                    385

Environmental radioactivity concentrations of Tekirdağ
Y. Yarar and E. Kam                                                                                387

Radiological impact in an area of elevated natural radioactivity background:
the case of the island of Ikaria–Aegean Sea, Greece
T. Georgia and F. Heleny                                                                           390

Beyond track etch monitoring—the gap between passive and active devices
has been bridged!
T. Streil, S. Feige and V. Oeser                                                                   392

MyRIAM: an active electronic online personal inhalation dose meter for detection
of $\alpha$- and $\beta$-radiation on aerosols
T. Streil and V. Oeser                                                                             394

Estimation of the radon dose in buildings by measuring the exhalation rate
from building materials
V. Steiner, K. Kovler, A. Perevalov and H. Kelm                                                    397

Retrospective radon assessments in a high radon dwelling in Ireland
J.P. McLaughlin, K. Kelleher, H. Jiménez-Nápoles and L. León-Vintró                                399

Radiation exposure from high-level radiation area and related mining and
processing activities of Jos Plateau, central Nigeria
I.I. Funtua and S.B. Elegba                                                                        401

Radon concentrations in caves of Parque Estadual do Alto Ribeira (PETAR),
SP, Brazil: preliminary results
S. Albergi, B.R.S. Pecequilo and M.P. Campos                                                       403

Distribution pattern of natural radionuclides in Lake Nasser bottom sediments
A.E. Khater, Y.Y. Ebaid and S.A. El-Mongy 405

Natural radioactivity contents in tobacco
N. Abd El-Aziz, A.E.M. Khater and H.A. Al-Sewaidan 407

$^{235}$U–γ emission contribution to the 186 keV energy transition of $^{226}$Ra in environmental samples activity calculations
Y.Y. Ebaid, S.A. El-Mongy and K.A. Allam 409

Determination of radium isotopes in mineral water12 samples by α-spectrometry
G. Jia, G. Torri, P. Innocenzi, R. Ocone and A. Di Lullo 412

Variation of terrestrial gamma radiation in Toki, Japan—comparison between gamma-ray spectrometry using Ge semiconductor and ICP-MS measurement
Y. Fujikawa, M. Fukui, T. Baba, T. Yoshimoto, E. Ikeda, M. Saito, H. Yamanishi and T. Uda 415

Spectrometry characteristics of photon fields and atmospheric radionuclide deposits monitoring in one part of Southern Bohemia
J. Kluson, L. Thinova, T. Cechak and T. Trojek 418

GPS-based handheld device for measuring environmental gamma radiation and mapping contaminated areas
J. Paridaens 420

Development of a remote radiation monitoring system using unmanned helicopter
S. Okuyama, T. Torii, Y. Nawa, I. Kinoshita, A. Suzuki, M. Shibuya and N. Miyazaki 422

Thermoluminescence mechanism on $SiO_2$ phosphor
M. Takami, M. Ohta and H. Yasuda 424

Radon anomaly related to the 1995 Kobe earthquake in Japan
Y. Yasuoka, T. Ishii, S. Tokonami, T. Ishikawa, Y. Narazaki and M. Shinogi 426

Generation of runaway electrons induced by radon progeny products in thunderstorm electric fields and the initiation of lightning discharges
T. Torii, T. Nozaki, T. Sugita, T. Nishijima and Z.-I. Kawasaki 428

**Epidemiology and radiation protection**

Residential radon and childhood leukemia: a metaanalysis of published studies
S. Yoshinaga, S. Tokonami and S. Akiba 430

New approach to dose optimization for members of the public
*T. Hattori and K. Sakai*   432

A study on the necessity of boarding control for international flights
*H. Yasuda and K. Fujitaka*   434

Cancer risk due to exposure to high levels of natural radon in the inhabitants of Ramsar, Iran
*S.M.J. Mortazavi, M. Ghiassi-Nejad and M. Rezaiean*   436

Living in high natural background radiation areas in Ramsar, Iran. Is it dangerous for health?
*A.S. Monfared, F. Jalali, H. Mozdarani, M. Hajiahmadi and H. Samavat*   438

The need for considering social, economic, and psychological factors in warning the general public from the possible risks due to residing in HLNRAs
*S.M.J. Mortazavi, A. Abbasi, R. Asadi and A. Hemmati*   440

Lung cancer risk due to radon exposure for 10 or 20 years
*J. Chen*   442

Radiation doses to cardiovascular system due to absorbed radon
*J. Chen and R.B. Richardson*   444

**Award of honor**   447

**Concluding remarks**   450

**Author index**   455

**Keyword index**   461

# Invited papers

ELSEVIER

www.ics-elsevier.com

# Chromosome study in high background radiation area in Southern China

Chunyan Wang[a], Wei Zhang[a], Masako Minamihisamatsu[b], Hiroshige Morishima[c], Yongling Yuan[d], Tao Jiang[a], Deqing Chen[a], Luxin Wei[a], Tsutomu Sugahara[e], Isamu Hayata[b],*

[a]National Institute for Radiological Protection, Chinese Center for Disease Control and Prevention, Beijing 100088, China
[b]National Institute of Radiological Sciences, Chiba 263-8555, Japan
[c]Kinki University, Osaka 577-8502, Japan
[d]Labor Hygiene Institute of Hunan Province, Changsha 410007, China
[e]Health Research Foundation, Kyoto 606-8225, Japan

**Abstract.** We have studied the chromosomes of lymphocytes of the residents in a high background radiation area (HBRA) and a control area (CA) in the southern China. The level of natural radiation in HBRA is three to five times higher than that in control area. Unstable type aberrations (dicentrics and rings) were analyzed in 22 individuals in HBRA and 17 individuals in CA. Stable type aberrations (translocations) were examined in 28 elderly individuals and 6 children in HBRA and 24 elderly individuals and 8 children in CA. There was statistically significant difference in the increasing rates between those two areas in adult. In case of stable aberrations, the frequencies were much higher than those of the unstable aberrations both in HBRA and in control. Statistically, there was no difference in the frequencies between HBRA and CA, but the frequencies in children were lower than those in elderly individuals in both groups. Individual variation was small in children while that in adults was large. The amount of chromosome aberrations induced by radiation is not significant when it is compared with that induced by other mutagenic factors such as smoking in the Southern China. © 2004 Elsevier B.V. All rights reserved.

*Keywords:* Natural radiation; Chromosome aberration; Dicentric and ring; Translocation; Human lymphocyte

---

* Corresponding author. Tel./fax: +81 43 206 3080.
  E-mail address: hayata@nirs.go.jp (I. Hayata).

0531-5131/ © 2004 Elsevier B.V. All rights reserved.
doi:10.1016/j.ics.2004.11.021

## 1. Introduction

In order to know the quantitative risk of low dose radiation on human health, a China–Japan collaborative cytogenetic study in the lymphocytes of the residents in a high background radiation area (HBRA) and its control area (CA) in Guangdong province, China, has been performed since 1992 [1–5]. The level of natural radiation in HBRA is three to five times higher than that in CA. The increase is caused by radionuclides such as Th-232 and U-238 decay products in the soil and building materials. The residents are mostly farmers and their families live in those rural areas over several generations. The genetic and cultural backgrounds of the residents in both areas are very similar.

In the present paper, we review our cytogenetic studies that have been made since 1992 and discuss the effect of low dose radiation on human health.

## 2. Materials and methods

For the study of dicentirc and ring chromosomes (Dic+Ring) [1], 22 individuals were from eight families in HBRA. Among them, 8 were in the grandparental, 6 were in the parental, and 8 were in child generations. Seventeen individuals were from five families in CA. Among them, 6 were in the grandparental, 6 were in the parental, and 5 were in child generations. For the study of translocations, 28 elderly individuals and 6 children in HBRA and 24 elderly individuals and 8 children in CA were analyzed [4,5]. By means of a questionnaire survey obtained prior to blood sampling, they stated no history of significant medical exposure except for the occasional chest X-ray examinations that contribute minor amount to the cumulative dose of the individuals.

Individual dose was measured with electric pocket dosimeters (Aloka PDM-101) for 24 h and/or thermoluminescence dosimeters (National UD-200S) for 2 months [6].

Chromosome preparations were made according to the method for the study of the effect of low dose radiation [1,2]. For the analysis of Dic+Ring, metaphases were stained with Giemsa's solution and examined as described before [1]. For the detection of Translocations, chromosomes 1, 2 and 4 were painted and metaphases were analyzed as reported before [2].

For the statistical analysis, Mann–Whitney $U$-test, a variance test of the homogeneity of the Poisson distribution and the Spearman rank correlation test were applied.

## 3. Results and discussion

Regarding Dic+Ring, a total of 101,394 cells and in average 2,600 cells per subject were analyzed [1]. Frequencies of Dic+Ring per 1000 cells were 0.49–5.45 in HBRA and 0–2.86 in CA. The frequencies per age (increasing rates per year) of Dic+Ring in the residents excluding children were varied from 0.029 to 0.092 (mean=0.056) in 1000 cells per year in HBRA, while 0 to 0.052 (mean=0.030) in 1000 cells per year in CA. There was statistically significant difference in the increasing rates between those two areas in adult ($P<0.01$, Mann–Whitney $U$-test).

In the case of translocations, a total of 317,915 cells and in average 4,800 cells per subject were analyzed [4,5]. Frequencies of translocations in relation to the dose and to the age are plotted in Fig. 1. The mean frequencies of translocations in children were 3.8±1.1 (from 2.2 to 5.3, $\chi^2=1.6$ with 5DF) in HBRA and 3.2±2.0 (from 0 to 5.8, $\chi^2=8.8$ with 7DF) in

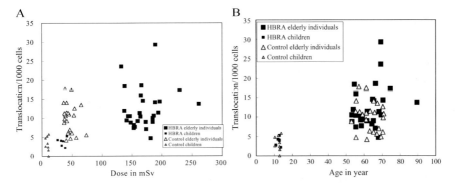

Fig. 1. Frequencies of translocations in relation to dose (A) and age (B) in 28 adults and 6 children in HBRA and 24 adults and 8 children in CA.

controls. There were two outliers in the elderly group. One outlier was due to medical exposure by fluoroscopy. The other had a high value of unknown reason. The mean frequencies of translocations in elderly individuals were 11.3±3.6 (from 4.7 to 18.6, $\chi^2$=28.67 with 25DF, when two outliers were excluded) in HBRA and 10.0±3.8 (from 4.2 to 17.8, $\chi^2$=33.21 with 23DF) in CA. Whether two outliers were excluded or not, no significant difference was found in the frequencies of translocations between HBRA and CA ($P$>0.05, Mann–Whitney $U$-test) in both children and elderly individuals. On the other hand, correlation between age groups and translocation frequencies was significant at the 1% level ($r_s$=0.486 with 62DF, Spearman rank correlation test) in both HBRA and CA.

Since increased frequencies of translocations as a result of smoking were reported in many papers [7–10], we investigated the effect of smoking. As shown in Fig. 2, when the adult in HBRA (excluding two outliers) and CA were classified into four subgroups such as HBRA non-smokers, HBRA smokers, CA non-smokers and CA smokers, a significant difference was found in the frequencies of translocations between CA smokers and CA non-smokers±$P$<0.05, Mann–Whitney $U$-test) [5]. Tendency of difference (A near $T$-value of 0.05 level) was found in a comparison of HBRA smokers and CA nonsmokers. But there was no difference in the frequencies among other subgroups. These results indicate that smoking had more effect on the yield of translocations than radiation in those areas.

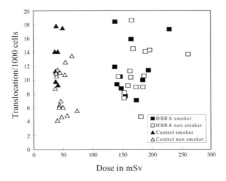

Fig. 2. Frequencies of translocations in relation to dose in 15 nonsmokers and 11 smokers in HBRA and 16 nonsmokers and 7 smokers in CA.

The mean values of the frequencies of translocations were 11.7, 11.0, 13.4 and 8.4 in HBRA smoker, HBRA nonsmoker, CA smoker and CA nonsmoker, respectively. The mean value of the frequency of translocations in HBRA nonsmokers was higher than that in CA nonsmokers. The mean value in HBRA smoker was lower than that in CA smokers. There is a possibility that the effect of smoking is slightly suppressed by the activated enzymes induced by the elevated level of radiation in HBRA for prevention or repair of DNA damages.

Increase of Dic+Ring was observed in HBRA where the dose rate was lower than one cGy per year. This dose rate is the condition that less than one track of radiation passes through a cell in 2 months. Therefore, there seems to be no threshold dose for the induction of chromosome aberration.

Dicentrics and translocations are induced by radiation in about equal frequency [11]. Dicentrics are highly specific indicators of radiation exposures. They are unstable and gradually eliminated from the body. On the other hand, translocations are the indicator of total effect of all kinds of clastogens. Translocations are stable aberrations and accumulate in the body. They may relate to the risk of malignant and congenital diseases [12]. The present results indicate that the elevated level of natural radiation increases the chromosome aberrations. However, the amount of the increased chromosome aberrations is not significant when it is compared with the total amount of those of other mutagenic factors such as smoking, chemicals and metabolic factors. It does not seem possible to detect significant increase in the congenital and malignant diseases caused by the DNA rearrangements that is attributable to the radiation three to five times higher than normal level of natural radiation.

## References

[1] T. Jiang, et al., Dose–effect relationship of dicentric and ring chromosomes in lymphocytes of individuals living in the high background radiation areas in China, J. Radiat. Res. 41 (2000) 63–68 (suppl.).
[2] I. Hayata, et al., Chromosome translocation in residents of the high background radiation areas in Southern China, J. Radiat. Res. 41 (2000) 69–74 (Suppl.).
[3] I. Hayata, et al., Chromosome translocation in residents of high background radiation area in China, in: W. Burkart, M. Sohrabi, A. Bayer (Eds.), High Levels of Natural Radiation and Radon Areas: Radiation Dose and Health Effects, International Congress Series, Excerpta Medica, vol. 1225, Elsevier, Amsterdam, 2002, pp. 199–205.
[4] W. Zhang, et al., Imperceptible effect of radiation based on stable type chromosome aberrations accumulated in the lymphocytes of residents in the high background radiation areas in China, J. Radiat. Res. 44 (2003) 69–74.
[5] W. Zhang, et al., Effect of smoking on chromosomes compared with that of radiation in the residents of a high-background radiation area in China, J. Radiat. Res. 45 (2004) 441–446.
[6] H. Morishima, et al., Study of the indirect method of personal dose assessment for the inhabitants in HBRA of China, in: L. Wei, T. Sugahara, Z. Tao (Eds.), High Levels of Natural Radiation: Radiation Dose and Health Effects, International Congress Series, Excerpta Medica, vol. 1136, Elsevier, Amsterdam, 1997, pp. 223–233.
[7] D.H. Moore II, et al., A study of the effects of exposure on cleanup workers at the Chernobyl nuclear reactor accident using multiple end points, Radiat. Res. 148 (1997) 463–475.
[8] S. Pressl, A. Edwards, G. Stephan, The influence of age, sex, and smoking habits on the background level of FISH-detected translocations, Mutal. Res. 442 (1999) 89–95.
[9] E.J. Tawn, C.A. Whitehouse, Stable chromosome aberration frequencies in men occupationally exposed to radiation, J. Radiol. Prot. 23 (2003) 269–278.

[10] R.V. Burim, et al., Clastogenic effect of ethanol in chronic and abstinent alcoholics, Mutat. Res. 560 (2004) 187–198.
[11] W. Zhang, I. Hayata, Preferential reduction of dicentrics in reciprocal exchanges due to the conbination of the size of broken chromosome segments by radiation, J. Hum. Genet. 48 (2003) 531–534.
[12] L. Hagmar, et al., Chromosomal aberrations in Lymphocytes predict human cancer: a report from the European Study Group on Cytogenetic Biomarkers and Health (ESCH), Cancer Res. 58 (1998) 4117–4121.

# Investigations on the health effects of human population residing in the high-level natural radiation areas in Kerala in the southwest coast of India

M.V. Thampi[a,*], V.D. Cheriyan[a], G. Jaikrishan[b], B. Das[a],
C.J. Kurien[a], E.N. Ramachandran[a], C.V. Karuppasamy[a],
B. Ravikumar[a], D.C. Soren[a], Usha Vijayan[a], P.K.M. Koya[a],
V.J. Andrews[a], V. Anilkumar[a], A. Mitra[a], M. Madhusoodhanan[c],
K.V. Aravindan[b], M. Seshadri[b]

[a]*Low Level Radiation Research Laboratory, Department of Atomic Energy, Low Level Radiation Studies Section, Bhabha Atomic Research Centre, 691 001, Kollam, Kerala, India*
[b]*Low Level Radiation Studies Section, Bhabha Atomic Research Centre, Mumbai, India*
[c]*Peadiatrics Unit, Government Victoria Hospital, Kollam, India*

**Abstract.** Monitoring of newborns at birth for clinically and cytogenetically observable malformations as well as a comprehensive Health Audit Survey are being carried out in the high-level natural radiation (HLNR) area of Kerala, a southwest state of India. A total of 92,689 newborns were monitored from August 1995 to June 2004 and overall incidence of stillbirths and malformations was 0.51% and 2.03%. Multiple logistic regression analyses of congenital malformation and stillbirths do not suggest any correlation with the radiation levels. Health Audit Survey was carried out in three Panchayats, and the pattern of both birth defects and late onset diseases was similar. The karyotype analysis of 23,844 cord blood samples was carried out. The overall incidence of constitutional karyotype anomalies was $4.86 \pm 0.45$ per 1000. © 2004 Published by Elsevier B.V.

*Keywords:* Malformation; Cytogenetics; Natural Radiation; Kerala

---

* Corresponding author. Tel.: +91 474 2740 449; fax: +91 474 2749 533.
*E-mail address:* mvgthampi@yahoo.co.in (M.V. Thampi).

0531-5131/ © 2004 Published by Elsevier B.V.
doi:10.1016/j.ics.2004.11.152

## 1. Introduction

Investigations on the health effects of high-level natural radiation (HLNR) on human population are being carried out in the densely populated monazite bearing HLNR area of Kerala, a southwest state of India on the Arabian coast. The average per capita dose received by the population in the 55-km-long and 0.5-km-wide belt is about four times the normal background radiation level, and the dose rate varies from 1.0 to 45.0 mGy/year [1,2]. The details of the HLNR area and methodology of studies being carried out by the Department of Atomic Energy, Government of India are described elsewhere [3–5]. As part of the programme, hospital-based monitoring of newborns at birth for clinically observable malformations, cytogenetic analysis and a comprehensive house-to-house Health Audit Survey are being carried out. The prevalence of various end-points studied are compared between HLNR and Normal Level Natural Radiation (NLNR) areas as well as at different background dose levels and forms the subject matter of this paper.

## 2. Materials and methods

The newborns from four government hospitals in the HLNR and NLNR areas were clinically screened for identifiable malformation at birth by trained doctors. Cord blood samples were used for lymphocyte microculture technique, and Giemsa stained/G-banded slides were analysed for karyotype anomalies. The detailed methodology of these studies are described elsewhere [3–5]. The mean radiation dose at the area of residence of the couples is used for dose–response analysis and classification of HLNR (>1.5 mGy/year) and NLNR ($\leq$ 1.5 mGy/year) areas. The house-to-house Health Audit Survey was conducted in collaboration with the departments of Health and Social welfare, Government of Kerala. The basic data collection was carried out by Anganwadi Workers, having good rapport and routine interaction with the population. They were trained to elicit information on sociodemography, life style, reproductive history of married women, congenital malformation/late onset chronic diseases, etc. and record it in the precoded proforma.

## 3. Results

A total of 92,689 (60,544 from HLNRA; 32,145 from NLNRA) newborns (91,368 singletons, 648 twins; seven triplets, one quadruplet) were monitored for malformations from August 1995 to June 2004. About 85% of mothers was in the age group of 20–29 years at the time of delivery. The overall incidence of stillbirths was 0.51% and that of malformations was 2.03%, the system involved being musculoskeletal (29%) followed by genitourinary (18%), ear (10%), gastrointestinal (8%), cardiovascular (7%), central nervous (6%) and others (22)%. Multiple logistic regression analyses of malformation and stillbirths showed dependence on maternal age, gravida status, ethnicity, gender of the baby and consanguinity but do not suggest any correlation with radiation levels (Table 1). The frequency of Down syndrome also were similar in HLNR (7.3/10,000) and NLNR (5.9/10,000) areas ($P$>0.2), with an overall incidence of 1 in 1471. The relatively lower incidence of stillbirths, malformations, DS and lower birthweight

Table 1
Multiple logistic regression analysis of congenital malformations and stillbirth

| Characteristic | No. of newborns | % | Malformed no. | F[a] | OR[b] | 95% CI | Stillbirth no. | F[a] | OR[b] | 95% CI |
|---|---|---|---|---|---|---|---|---|---|---|
| *Maternal Age* | | | | | | | | | | |
| 15–19 | 6075 | 06.6 | 104 | 17.1 | 1.00 | | 39 | 6.4 | 1.00 | |
| 20–21 | 14,556 | 15.7 | 295 | 20.3 | 1.23 | 0.98–1.54 | 62 | 4.3 | 0.66* | 0.44–0.98 |
| 22–23 | 21,335 | 23.0 | 413 | 19.4 | 1.20 | 0.96–1.50 | 82 | 3.8 | 0.57* | 0.38–0.85 |
| 24–25 | 19,999 | 21.6 | 392 | 19.6 | 1.24 | 0.99–1.56 | 106 | 5.3 | 0.76 | 0.51–1.12 |
| 26–27 | 13,984 | 15.0 | 273 | 19.5 | 1.25 | 0.98–1.59 | 81 | 5.8 | 0.81 | 0.53–1.22 |
| 28–29 | 8593 | 09.3 | 182 | 21.2 | 1.37* | 1.06–1.78 | 40 | 4.7 | 0.62 | 0.39–1.00 |
| ≥30 | 8138 | 08.8 | 210 | 25.8 | 1.68* | 1.30–2.17 | 60 | 7.4 | 0.93 | 0.59–1.46 |
| *Gravida* | | | | | | | | | | |
| 1 | 42,197 | 45.5 | 864 | 20.5 | 1.00 | | 193 | 04.6 | 1.00 | |
| 2 | 42,184 | 45.5 | 841 | 19.9 | 0.92 | 0.83–1.02 | 198 | 04.7 | 1.04 | 0.83–1.29 |
| 3 | 7207 | 07.8 | 132 | 18.3 | 0.80* | 0.66–0.98 | 61 | 08.5 | 1.78* | 1.30–2.45 |
| ≥4 | 1092 | 01.2 | 32 | 29.3 | 1.21 | 0.83–1.75 | 18 | 16.5 | 3.24* | 1.93–5.43 |
| *Gender* | | | | | | | | | | |
| Male | 47,492 | 51.2 | 1166 | 24.6 | 1.00 | | 266 | 5.6 | 1.00 | |
| Female | 45,188 | 48.8 | 703 | 15.6 | 0.63* | 0.57–0.69 | 204 | 4.5 | 0.81* | 0.67–0.97 |
| *Consanguinity* | | | | | | | | | | |
| Absent | 90,552 | 97.7 | 1809 | 20.00 | 1.00 | | 447 | 04.9 | 1.00 | |
| Present | 2128 | 02.3 | 60 | 28.2 | 1.43* | 1.10–1.85 | 23 | 10.8 | 2.04* | 1.34–3.12 |
| *Ethnicity* | | | | | | | | | | |
| Nair | 14,688 | 15.9 | 267 | 18.2 | 1.00 | | 60 | 4.1 | 1.00 | |
| Ezhava | 20,588 | 22.2 | 442 | 21.5 | 1.18* | 1.01–1.38 | 100 | 4.9 | 1.20 | 0.87–1.65 |
| Viswakarma | 4729 | 05.1 | 96 | 20.3 | 1.11 | 0.88–1.41 | 21 | 4.4 | 1.05 | 0.64–1.75 |
| Other Hindu | 21,880 | 23.6 | 453 | 20.7 | 1.15 | 0.98–1.34 | 133 | 6.1 | 1.44* | 1.06–1.96 |
| Christian | 11,790 | 12.7 | 226 | 19.2 | 1.06 | 0.89–1.27 | 58 | 4.9 | 1.19 | 0.82–1.71 |
| Muslim | 19,005 | 20.5 | 385 | 20.3 | 1.16 | 0.99–1.37 | 98 | 5.2 | 1.18 | 0.84–1.64 |
| *Radiation dose (mGy/year)* | | | | | | | | | | |
| ≤1.50 | 32,144 | 34.7 | 678 | 21.1 | 1.00 | | 163 | 5.1 | 1.00 | |
| 1.51–3.00 | 51,778 | 55.9 | 1023 | 19.8 | 0.94 | 0.85–1.03 | 272 | 5.3 | 1.04 | 0.86–1.27 |
| 3.01–6.00 | 5507 | 05.9 | 107 | 19.4 | 0.92 | 0.75–1.13 | 18 | 3.3 | 0.64 | 0.39–1.05 |
| ≥6.01 | 3251 | 03.5 | 61 | 18.8 | 0.89 | 0.68–1.17 | 17 | 5.2 | 0.97 | 0.58–1.62 |
| Total | 92,680 | 100.0 | 1869 | 20.2 | | | 470 | 05.1 | | |

[a] F—Frequency/1000 newborns.
[b] Odds ratio. Coefficients obtained from logistic regression analysis, including all variables indicated in the model as a categorical variable. Reference category is indicated with an OR of 1.00. Nine intersex cases not included in analysis.
* Significant at 5% level.

(<2500 g, 8.2% among live singleton newborns) is commensurate with the younger maternal age [6].

Health Audit Survey has been conducted in three Panchayats (99 Anganwadies, about 23,000 households) in the study area, and the data are being processed. Percentage of individuals reported to have congenital malformation and late onset chronic diseases (arthritis, asthma, blindness, cataract, deafness, diabetes, dumbness, epilepsy, heart disease, hypertension, hypotension, leukemia, other cancers, neurological disorders, psychiatric disoders, thyroid diseases, vitiligo, etc.) in different age groups based on 10,800 households of 46 Anganwadies is depicted in Fig. 1. Although the pattern of

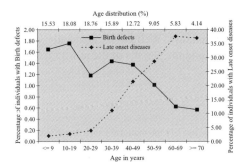

Fig. 1. Age-specific prevalence of birth defects and late onset diseases.

background radiation levels vary widely in the three Panchayats, the pattern of both birth defects and late onset diseases was similar in the three Panchayats.

The karyotype analysis of 23,844 (8004 from NLNRA and 15,840 from HLNRA) cord blood samples was carried out. The overall incidence of karyotype anomalies was 4.86±0.45 per 1000 (4.80±0.55 in HLNR area and 5.00±0.79 in NLNR area) and is comparable with published figures [7]. Both structural and numerical chromosomal anomalies were comparable between HLNR and NLNR areas and did not seem to be associated with the background dose levels. Numerical anomalies were slightly higher among males (2.76±0.47 vs. 2.34±0.45), whereas structural aberrations were higher among females (1.71±0.37 v/s 2.95±0.51).

## 4. Discussion

Screening newborns for congenital malformation is a practical approach to assess transmission of genetic damage, if any, caused by chronic exposure to HLNR. The studies carried out do not suggest an increase of any of the end-points in HLNR area so far. The natural radiation dose rates of the area may not be high enough to induce genetic damage to be measured/discerned by screening congenital malformations. Apart from the multifactorial origin of congenital malformations, the intricacies of dominant and recessive mutations which could have occurred in the past generation(s) and the uncertainties involved in the dose estimation owing to mobility of the population and differential dose exposure profile of parents make matters more complex. The studies are being expanded to assess the impact of HLNR on chronic diseases through the Health Audit Survey. This will make it possible to gauge the impact of HLNR on some selected disease conditions as the etiopathogenesis of diseases are not similar and the pathway of other mutagens and radiation may also be different.

## Acknowledgements

The authors wish to thank K.P. Bhaskaran, R. Prabhakaran and P.S. Mohan for their excellent technical assistance in tissue culture. The contribution of V. Yesodharan, Y. Sundaran, K.K. Thomas and M.P. George in the conduct of Health Audit Survey is gratefully acknowledged. Thanks are also due to the medical and paramedical staff of the state department of Health and Family Welfare, Anganwadi workers and Supervisors of department of Social Welfare and the parents of the newborns for their consent for drawing cord blood samples.

## References

[1] D.S. Bharatwal, G.H. Vaze, Proc. 2nd UN International Conf. Peaceful Uses of Atomic Energy, Geneva, vol. 23, 1958, p. 156.
[2] C.M. Sunta, In: M. Sohrabi, J.U. Ahmed, S.A. Durani (Eds.), Proc. International Conference on High Levels of Natural Radiation, Tehran, Iran, vol. 71, Atomic Energy Organization of Iran, 1993, p. 71.
[3] G. Jaikrishan, et al. Radiation Research 152 (1999) S149–S153.
[4] V.D. Cheriyan, et al. Radiation Research 152 (1999) S154–S158.
[5] M.V. Thampi, et al. In: W. Burkart, M. Sohrabi, A. Bayer (Eds.), High Levels of Natural Radiation and Radon Areas: Radiation Dose and Health Effects, International Congress Series, vol. 1225, 2002, pp. 207–211.
[6] C.A. Hecht, E.B. Hook, Am. J. Med. Genet. 62 (1196) 376–385.
[7] UNCEAR, Sources, Effects and Risks of Ionization Radiation, Report to the General Assembly, United Nations, New York, 1993.

# New findings in the very high natural radiation area of Ramsar, Iran

M. Ghiassi-Nejad[a,b,*], M.M. Beitollahi[a], N. Fallahian[a], M. Saghirzadeh[b]

[a]National Radiation Protection Department, Iranian Nuclear Regulatory Authority, Tehran, Iran
[b]Department of Biophysics, Tarbiat Modarres University, P.O. Box 14155-4838, Tehran, Iran

**Abstract.** Ramsar, a northern coastal city of Iran, has some high levels of natural radiation areas (HLNRAs) as well as about 50 hot springs with low and high radium contents usually used for medical treatment purposes. Different radiological measurements, as well as cytogenetical, immunological and preliminary biological and epidemiological studies have already been performed in this region. The purpose of this study is to review new findings in very high background radiation areas of Ramsar, in order to identify information relevant to exposures in these areas and the potential and known impacts of such exposures on public health, and also to make recommendations for further studies. At present, there is no reliable radio-epidemiological data regarding the incidence of cancer in HLNRAs of Ramsar. In this regard, more extensive researches are invited, in particular on epidemiology and radiobiology as well as on internal dose assessment. © 2004 Elsevier B.V. All rights reserved.

*Keywords:* Natural radiation; Exposure; Ramsar

## 1. Introduction

Ramsar region, a northern coastal city of Iran, has been the subject of concern as a highly radioactive region for the past 40 years. The hot springs, which are used by the inhabitants and visitors as spas, are the main sources for the distribution of natural radionuclides (especially for $^{226}$Ra and its decay products), in turn leading to the creation of high levels of natural radiation areas (HLNRAs) in the region [1].

The present study briefly reviews the investigations that have already been carried out in HLNRAs of Ramsar, with the emphasis on new findings and related results, to mainly

---

\* Corresponding author. Tel.: +98 21 8011001; fax: +98 21 8006544.
*E-mail address:* ghiassi@mailcity.com (M. Ghiassi-Nejad).

identify the potential and known impacts of exposures in HLNRAs of this region on public health, and also to make recommendations for further studies.

## 2. Studies on HLNRAs of Ramsar

The most studies carried out in such areas include a survey of environmental indoor and outdoor gamma exposure; external personal dosimetry of inhabitants; determination of $^{226}$Ra in soil, beans, food stuff, public water supplies, hot springs, etc.; determination of $^{222}$Rn levels in rooms and different locations of houses, schools and Ramsar Hotels in different seasons; cytogenetic and preliminary biological studies of inhabitants compared to that of a control area; radiological studies of a house with a high potential for internal and external exposure; studies of building construction materials; air particle size distributions; vegetable-to-soil concentration ratios (CRs) for $^{226}$Ra; exposure to $^{226}$Ra from consumption of vegetables and biosorption of $^{226}$Ra via bacterial strains [1–14]. The geographical location of Ramsar is illustrated in Fig. 1.

## 3. Results and discussion

People in some areas of Ramsar receive an annual effective dose from background radiation of up to 260 mSv year$^{-1}$, substantially higher than that is permitted for radiation workers [8]. However, the preliminary results of the studies of the people living in high background radiation areas of this region show no observable detrimental effect [9]. Furthermore, the frequency of chromosome aberrations (CA) in the lymphocytes of the inhabitants of very high background radiation areas of Ramsar after exposure to a challenge dose of 1.5 Gy gamma rays was significantly lower than that of the inhabitants of a normal background radiation area [8], suggesting enhanced repair of induced DNA damage resulting from exposure to elevated levels of background radiation [9]. These findings suggest that exposure to high levels of natural radiation does not require remedial or corrective actions [9].

The results of studies on the long-term effects of high level natural radioactivity on some immunological and cytogenetical parameters in Ramsar inhabitants showed a

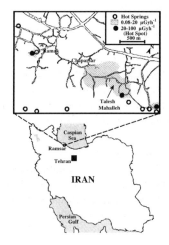

Fig. 1. Geographical location of Ramsar region.

significant increase of CD69 expression on TCD4+ stimulated cells ($P<0.004$) and a significant increase of total serum IgE ($P<0.05$), and also higher incidence of stable and unstable chromosomal aberrations in the HLNRA group compared to the control group with normal background radiation ($P<0.05$). Other humoral immune parameters did not show significant differences between the two groups [10].

Some of the local physicians strongly believe that the population living in these areas does not reveal increased solid cancer or leukemia incidences [8].

Some findings on the hematological alterations of prolonged exposure of housewives in high background radiation areas of Ramsar also showed no harmful effects.

In general, the radiobiological and epidemiological studies in these areas have not as yet shown any evidence of significant increase in health detriment compared with that in normal background areas [11]; however, more epidemiological studies are required to achieve the radiation effects on human beings, especially the risk estimate.

Other radiological and radiobiological investigations have been performed in this region, the results of which are as follows:

The uptake values of some bacterial strains isolated from hot springs and soil of HLNRAs of Ramsar (A-C, T-A, A-E and CH-G) at a radioactivity level of $4.3\pm0.2$ Bq ml$^{-1}$ were $202\pm53$, $171\pm49$, $123\pm11$ and $53\pm6$ Bq g$^{-1}$, respectively. Biosorption was the only component in the total $^{226}$Ra uptake and metabolism was not involved in the uptake process [12].

In addition, minimum and maximum CR were measured in leafy and root vegetables with average values of $1.6\times10^{-2}$ and $4.0\times10^{-3}$, respectively [13]. Besides, the CR values measured in edible vegetables in a highly radioactive region of Ramsar are similar to those values in low background radiation areas [14]. The mean effective dose resulting from $^{226}$Ra due to consumption of edible vegetables by adults in the critical group in Talesh Mahalleh (a district of the city of Ramsar) has also been estimated to be 72.3 μSv year$^{-1}$ that is about 12 times greater than the average of effective dose resulting from this radionuclide due to combined intake of all foods and drinking water in normal background areas [13].

## 4. Conclusions and recommendations

The conclusions from the results of this survey are:

(a) Radiation hormesis and adaptive response induced by low doses should be done in the region. In addition, dosimetric data are required to facilitate further radiological and epidemiological researches.
(b) More research on the effects of high natural radiation on plants at the chromosomal level is needed to evaluate if there is any difference between the plants in HLNRA and normal areas.
(c) Further studies are needed to clarify stimulus effects of radiation and more susceptibility of people to the late effects of radiation exposure and allergic diseases. In this regard, more extensive researches are needed, in particular on epidemiology and radiobiology as well as on internal dose assessment.
(d) Long-term monitoring of health status of residents, aero-biological studies and cytokine assays are recommended in these areas. In addition, evaluation of a cancer registry in the region is required.

(e) Some bacterial strains might be appropriate candidates for designing biosorption systems.
(f) There is no significant difference between the CR values of edible vegetables in a highly radioactive region and the corresponding values in the low background radiation areas.
(g) Uptake of $^{226}$Ra via roots of vegetables is directed toward the leaves.

## References

[1] M. Ghiassi-Nejad, et al., Radiological parameters of a house with high levels of natural radiation Ramsar, Iran, in: W. Burkart, M. Sohrabi, A. Bayer (Eds.), Proceedings of the 5th International Conference on High Levels of Natural Radiation and Radon Areas: Radiation Dose and Health Effects, Munich, Germany, 4–7 September 2000, Elsevier Science Publications, The Netherlands, 2002, pp. 33–37.
[2] B. Khademi, A. Sekhavat, H. Parnianpour, Area of high natural background radiation in the northern part of Iran, in: T.L. Cullen, E. Penna Franca (Eds.), International Symposium on Areas of High Natural Radioactivity. Pocos de Caldas, Brasil, 16–20 June 1975, Pontificia Universidade Católica, Brasil, 1977, p. 186.
[3] B. Khademi, A.A. Alemi, A. Nasseri, Transfer of radium from soil to plants in an area of high natural radioactivity in Ramsar, Iran, in: Th.F. Gessel, W.M. Lowder (Eds.), Proceedings of Symposium on Natural Radiation Environment, Houston, TX, 23–28 April 1978, US Department of Energy, USA, 1980, pp. 600–610.
[4] M. Sohrabi, Recent radiological studies in high level natural radiation areas of Ramsar, in: M. Sohrabi, J.U. Ahmed, S.A. Durrani (Eds.), Proceedings of International Conference on High Levels of Natural Radiation, Ramsar, Iran, 3–7 November 1990, IAEA Publication Series, IAEA, Vienna, 1993, pp. 39–47.
[5] M. Sohrabi, et al., Determination of $^{222}$Rn levels in houses, schools and Hotels of Ramsar by AEOI passive radon diffusion dosimeters, in: M. Sohrabi, J.U. Ahmed, S.A. Durrani (Eds.), Proceedings of International Conference on High Levels of Natural Radiation, Ramsar, Iran, 3–7 November 1990, IAEA Publication Series, IAEA, Vienna, 1993, pp. 365–375.
[6] M. Sohrabi, World high level natural radiation and/or radon-prone areas with special regard to dwellings, in: L. Wei, T. Sugahara, Z. Tao (Eds.), Proceedings of 4th International Conference on High Levels of Natural Radiation, Beijing, China, 21–25 October 1996, Elsevier, Amsterdam, 1997, pp. 57–68.
[7] M. Sohrabi, A.R. Esmaili, New public dose assessment of elevated natural radiation areas of Ramsar (Iran) for epidemiological studies, in: W. Burkart, M. Sohrabi, A. Bayer (Eds.), Proceedings of the 5th International Conference on High Levels of Natural Radiation and Radon Areas: Radiation Dose and Health Effects, Munich, Germany, 4–7 September 2000, Elsevier Science Publications, The Netherlands, 2002, pp. 15–24.
[8] M. Ghiassi-Nejad, et al., Very high background radiation areas of Ramsar, Iran: preliminary biological studies, Health Phys. 82 (1) (2002) 87–93.
[9] P.A. Karam, et al., ICRP evolutionary recommendations and the reluctance of the members of the public to carry out remedial work against radon in some high-level natural radiation areas, in: T. Sugahara, O. Nikaido, O. Niwa (Eds.), Radiation and Homeostasis, Elsevier Sciences Publications, Amsterdam, 2002, pp. 35–37.
[10] M. Ghiassi-Nejad, et al., Long-term immune and cytogenetic effects of high level natural radiation on Ramsar inhabitants in Iran, J. Environ. Radioact. 74 (2004) 107–116.
[11] M. Sohrabi, The state-of-the-art on worldwide studies in some environments with elevated naturally occurring radioactive materials (NORM), Appl. Radiat. Isotopes 49 (3) (1998) 169–188.
[12] D. Satvatmanesh, et al., Biosorption of $^{226}$Ra in high level natural radiation areas of Ramsar, Iran, J. Radioanal. Nucl. Chem. 258 (3) (2003) 483–486.
[13] M. Ghiassi-Nejad, et al., Exposure to $^{226}$Ra from consumption of vegetables in the high level natural radiation area of Ramsar, J. Environ. Radioact. 66 (2003) 215–225.
[14] M.M. Beitollahi, et al., Vegetable-to-soil concentration ratio (CR) for $^{226}$Ra in a highly radioactive region, Book of Contributed Papers of International Conference on the Protection of the Environment from the Effects of Ionizing Radiation, Stockholm, Sweden, 6–10 October, IAEA, Vienna, 2003, pp. 155–158.

www.ics-elsevier.com

# Chromosomal mutations by low dose radiation vs. those by other mutagenic factors

Isamu Hayata*

*National Institute of Radiological Sciences, 263-8555, Japan*

**Abstract.** In order to estimate the genotoxic risk due to low dose radiation, it is essential to distinguish the effect of radiation from those of chemicals. In this presentation, the difference between radiation-induced chromosomal aberrations and those induced by chemical mutagens, and the characteristics of dicentrics and translocations are described. Then, the significance of the chromosome aberrations in the normal living circumstances is discussed. It is shown that dicentrics and translocations are very sensitive and useful biomarkers in the study of the effect of low dose radiation in comparison with the effect of other clastogens. It is suggested that any effect of radiation such as malignant diseases attributable to the alteration of DNA would not be revealed by the epidemiological study at the excess dose level of 20 cSv in the normal living circumstances. © 2004 Elsevier B.V. All rights reserved.

*Keywords:* Chromosome aberration; Radiation; Chemical mutagen; Clastogen; Health effect

## 1. Introduction

In order to estimate the genotoxic risk due to low dose radiation, it is essential to distinguish the effect of radiation from those of chemicals. In this presentation, first, I will describe the difference between radiation-induced chromosomal aberrations and those induced by chemical mutagens, and the characteristics of dicentrics and translocations. Then, I will discuss the significance of chromosome aberrations in the peripheral lymphocytes in the normal living circumstances.

---

\* Tel.: +81 43 206 3080; fax: +81 43 206 3080.
  *E-mail address:* hayata@nirs.go.jp.

0531-5131/ © 2004 Elsevier B.V. All rights reserved.
doi:10.1016/j.ics.2004.11.022

## 2. Difference between chromosomal aberrations caused by ionizing radiation and those by chemical mutagens

Chromosomal aberrations are caused by direct DNA breakage, replication of a damaged DNA template, inhibition of DNA synthesis and other mechanisms such as inhibition of enzymes relating to chromosome formation [1]. Ionizing radiation induces direct DNA breaks that result in chromosome type aberrations involving both chromatids of chromosomes in cells in the $G_0/G_1$ ($G_0$: non cycling $G_1$ and $G_1$: period before DNA synthesis) phase of the cell cycle, and chromatid type aberrations involving only one of the two chromatids as well as chromosome type aberrations in cells in the $S/G_2$ (S: DNA synthesis period and $G_2$: period after DNA synthesis) phase [2]. Ionizing radiation does induce a DNA single-strand break (SSB) in $G_0/G_1$ that leads to chromatid type aberrations. But DNA SSB induced by ionizing radiation is quickly repaired [3] or becomes a double-strand break (clustered SSB) leading to chromosome type aberrations [4]. In consequence, all radiation-induced aberrations in $G_0/G_1$ phase become chromosome type. Ionizing radiation is S phase independent clastogen.

Although ionizing radiation thus induces chromosomal aberrations attributable to DNA breaks, chemicals cause a variety of chromosomal aberrations attributable to coiling deficiency, over condensation, aneuploid, polyploid, as well as structural aberrations. The structural aberrations caused by the chemicals (except for radiation mimetic agents such as bleomycin [5,6] and restriction enzymes [7,8]) are generally chromatid type and formed only after passing S phase. Some chemicals induce DNA double-strand breaks leading to chromosome type aberrations by interfering with spontaneous endogenous processes related to oxidative metabolism, to errors during DNA replication and to various forms of site-specific DNA recombination [9]. Those are indirect effects of chemicals on DNA breaks. Chemicals are generally S phase-dependent clastogens.

The peripheral lymphocytes are normally in the $G_0$ (non-cycling $G_1$) stage of the cell cycle and they replicate DNA only after being stimulated in the cell culture [2]. When they are exposed to either ionizing radiation or chemicals in vivo, their resultant chromosome aberrations are different, i.e., the former induces chromosome type aberrations and the latter induces chromatid type aberrations (Fig. 1). Thus, they show morphologically different features at the metaphase stage in the first cell division cycle after exposure to those clastogens. But, the chromosome type aberrations and chromatid type aberrations become indistinguishable in the second (and later) cell divisions, because the chromatid type rearrangement is copied on both chromatids (derived chromosome type aberration).

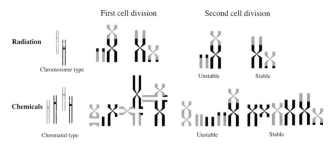

Fig. 1. Chromosomal aberrations induced by radiation and chemicals in the peripheral lymphocytes ($G_0/G_1$ phase).

## 3. Characteristics of dicentrics and translocations

Dicentrics and translocations originated from reciprocal rearrangements are induced in the peripheral lymphocytes in about equal frequency by radiation [10]. The former is unstable to be eliminated through cell division, while the latter is stable to be accumulated in the body. Both of them are induced by chemicals also. There are three kinds of dicentrics due to the difference in the formation mechanism. The first type is the dicentric formed by tandem fusion (telomere fusion), which does not accompany a fragment. The second type is the dicentric derived from a chromatid type aberration, which is rarely seen in the first metaphase in vitro in the peripheral lymphocytes, because it has to pass through at least one cell cycle in vivo during which more than 50% of them are theoretically eliminated. The second type also dose not accompany a fragment. The last type is the dicentric caused by the reciprocal exchange of a chromosome type rearrangement. This type accompanies a fragment. Therefore, a dicentric with a fragment observed in the first metaphase in the cultured lymphocyte is a highly specific indicator of radiation exposures.

On the other hand, translocations are stable type and are not eliminated through cell division. It is not possible to distinguish radiation-induced translocation from chemical-induced translocation. They are indicators of the accumulated effects of all kinds of clastogens.

We performed our study in a high background radiation area based on these morphological characteristics [11–15]. It is proved that those chromosome aberrations are very sensitive and useful biomarkers in the study of the effect of low dose radiation in comparison with the effect of other clastogens.

## 4. Significance of chromosome aberrations in the peripheral lymphocytes in the normal living circumstances

Induction yield of dicentrics per cSv with X or gamma rays is reported to be 0.3 per 1000 cells in the human lymphocytes [16]. Accordingly, the yield of translocations would be about the same with this rate. If the average dose of exposure per year in human body is assumed to be 0.24 cSv [17], the accumulate dose becomes 15.6 cSv and 4.68 translocations per 1000 cells would be induced by radiation in the body in 65 years. It

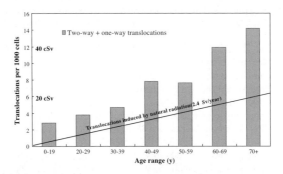

Fig. 2. Background frequency of translocations with age (after Sorokine-Durm et al. [18]). Linear line shows the calculated frequency of translocations induced by natural radiation, when induction yield of translocations per cSv is 0.3 per 1000 cells and the average dose of exposure per year in human body is assumed to be 0.24 cSv.

is reported that, on an average, about 12 translocations per 1000 lymphocytes in healthy individuals whose ages were around 65 years were detected, although large individual variations were found [12,18] (Fig. 2).

This value indicates that significant amount of chromosomal mutations due to the clastogens other than radiation are induced at the various rates among individuals in human. It is reported that increase of the structural chromosomal aberrations in the peripheral lymphocytes relates to the increase of the risk of malignant diseases [19]. The background frequency of 12 per 1000 cells is equivalent to the frequency to be induced by the radiation at the dose of about 40 cSv, that is, more than two times of the radiation to be received at the age of 65 years old in the natural living circumstances. Therefore, any effect of radiation such as malignant diseases attributable to the alteration of DNA would not be revealed by the epidemiological study at the excess dose level of 20 cSv in an ordinary life.

## References

[1] R.J. Albertini, et al., IPCS guidelines for the monitoring of genotoxic effects of carcinogens in humans, Mutat. Res. 463 (2000) 111–172.
[2] IAEA Technical reports series No. 405, Cytogenetic analysis for radiation dose assessment, A manual, 2001.
[3] M.A. Bender, et al., Mechanism of chromosomal aberration production: III. Chemicals and ionizing radiation, Mutat. Res. 23 (1974) 197–212.
[4] M. Gulston, et al., Processing of clustered DNA damage generates additional double strand breaks in mammalian cells post-irradiation, Nucleic Acids Res. 32 (2004) 1602–1609.
[5] B.K. Vig, R. Lewis, Genetic toxiccology of bleomycin, Mutat. Res. 55 (1978) 121–145.
[6] L.F. Povik, M.J.F. Austin, Genotoxicity of bleomycin, Mutat. Res. 257 (1991) 127–143.
[7] R.A. Winergar, R.J. Preston, The induction of chromosome aberrations by restriction endonucleases that produce blunt-end or cohesive-end double-strand breaks, Mutat. Res. 19 (1988) 141–149.
[8] M. Jasin, Chromosome breaks and genomic instability, Cancer Investig. 18 (2000) 76–78.
[9] G. Lliakis, et al., Mechanisms of DNA double strand break repair and chromosome aberration formation, Cytogenet. Genome Res. 104 (2004) 14–20.
[10] W. Zhang, I. Hayata, Preferential reduction of dicentrics in reciprocal exchanges due to the combination of the size of broken chromosome segments by radiation, J. Hum. Genet. 48 (2003) 531–534.
[11] L. Wei, T. Sugahara, An introductory overview of the epidemiological study on the population at the high background radiation areas in Yangjiang China, J. Radiat. Res. 41 (2000) 1–7 (suppl.).
[12] T. Jiang, et al., Dose–effect relationship of dicentric and ring chromosomes in lymphocytes of individuals living in the high background radiation areas in China, J. Radiat. Res. (Suppl. 41) (2000) 63–68.
[13] W. Zhang, et al., Imperceptible effect of radiation based on stable type chromosome aberrations accumulated in the lymphocytes of residents in the high background radiation area in China, J. Radiat. Res. 44 (2003) 69–74.
[14] I. Hayata, et al., Effect of high-level natural radiation on chromosomes of residents in southern China, Cytogenet. Genome Res. 104 (2004) 237–239.
[15] C. Wang, et al., Chromosome study in high background area in southern China, 1276 (2005) 3–7.
[16] D.C. Lloyd, et al., Chromosomal aberrations in human lymphocytes induced in vitro by very low doses of X-rays, Int. J. Radiat. Biol. 61 (1992) 335–343.
[17] UNSCEAR 1988 Report: Sources, effects and risks of ionizing radiation, united Nations Scientific Committee on the Effects of Atomic Radiation, 1988. Report to the general assembly with annexes. United Nations sales publication E88. IX. 7. United Nations, New York.
[18] I. Sorokine-Durm, C. Whitehouse, A. Edwards, The variability of radiation yields amongst control populations, Radiat. Prot. Dosim. 88 (2000) 93–99.
[19] L. Hagmar, et al., Chromosomal aberrations in lymphocytes predict human cancer: a report from the European Health (ESCH), Cancer Res. 58 (1988) 4117–4121.

# Extranuclear target and low dose radiation risk assessment

Tom K. Hei[a,*], Hongning Zhou[a], Masao Suzuki[b]

[a]Center for Radiological Research, College of Physicians and Surgeons, Columbia University, 630 West 168th Street, New York, NY 10032, USA
[b]National Institute of Radiological Sciences, Chiba, Japan

**Abstract.** The classical dogma of radiation biology stated that the genotoxic effects of ionizing radiation such as mutations and carcinogenesis are due mainly to direct damage to the nucleus. Recent evidence accumulated from both in vitro and in vivo studies, however, suggested that extranuclear target and the bystander effects may play an important role in mediating the radiobiological effects of both high and low LET radiations. These data indicate that the relevant target for various radiobiological endpoints is larger than an individual cell and suggest a need to reconsider the validity of the linear extrapolation model in making risk estimate for low dose radiation exposure, an issue that is the subject of considerable debate. © 2004 Published by Elsevier B.V.

*Keywords:* Bystander effect; Extranuclear target; Mutation; Radiation risk

## 1. Introduction

Ever since X-rays were shown to induce mutation in *Drosophila* more than 70 years ago, prevailing dogma considered the genotoxic effects of ionizing radiation such as mutations and carcinogenesis as being due mostly to direct damage to the nucleus. As such, generations of students in radiation biology have been taught that such heritable biological effects are the consequence of a direct radiation–nuclear interaction. In fact, evidence suggesting that this simple statement is not strictly true has been around for decades. For example, Kotval and Gray [1] had shown that α-particles which passed close,

---

* Corresponding author. Tel.: +1 212 305 8462; fax: +1 212 305 6850.
E-mail address: tkh1@columbia.edu (T.K. Hei).

0531-5131/ © 2004 Published by Elsevier B.V.
doi:10.1016/j.ics.2004.11.020

but not through, the chromatid thread had a significant probability of producing chromatid and isochromatid breaks or chromatid exchanges, the modern day definition of a bystander effect derived mainly from the work based on micro-dosimetric principles conducted more than a decade ago [2].

While circumstantial evidence in support of a by-stander effect appears to be consistent, direct proof of such extranuclear/extracellular effects are most convincingly demonstrated using charged particle microbeams. Using the Columbia University microbeam to target an exact fraction of cells in a population and irradiated their nuclei with a lethal dose of alpha particles, Zhou et al. [3] showed that the measured mutant fraction was three times higher than the expected background value. Since all directly hit cells were reproductively dead, the mutant fraction obtained must come from the non-irradiated cells in the vicinity of the hit ones.

## 2. Bystander mutations can be induced by a single alpha particle

Since bronchial epithelial cells exposed to environmental radon in homes rarely have more than one particle traversal at any one time, it is important to ascertain whether this bystander effect can be demonstrated at low doses of alpha particle, a dose as low as a single traversal per cell. Consistent with our previous finding, traversal of the nucleus with a single alpha particle was only slightly cytotoxic to $A_L$ cells resulting in a surviving fraction of ~0.79+0.05 [4]. Furthermore, the yield of $CD59^-$ mutants induced in populations of $A_L$ cells in which 100% of the cells had received exactly one alpha particle through the nucleus was not significantly different from the mutant fraction obtained when only 20% of the cells were hit with a single alpha particle [5].

## 3. Is bystander mutagenesis an artifact of the hamster hybrid cells

To demonstrate that the bystander genotoxic effect observed thus far with the $A_L$ cells is not an artifact of the human hamster hybrid cells and the findings are relevant to human lung cancer incidence as a result of exposure to environmental radon, it is necessary to show that primary human bronchial epithelial cells exhibit a similar bystander genotoxic response as well. Since mutagenesis studies in primary epithelial cells are difficult to conduct due to the limited life span of the cultures, a highly sensitive genotoxic endpoint, that of G2 phase premature chromosome condensation technique, to measure chromatid breaks in bystander normal human bronchial epithelial (NHBE) cells was used. Commercially available NHBE cells were plated on microbeam dishes and irradiated as described above. The incidence of chromatid-type breaks induced in these cells where a single alpha particle was delivered to the nuclei in either 20 or 100% of the cultures was determined. As shown in Fig. 1, in a population in which every NHBE cell was irradiated, 90% of the cells contained three or more chromatid breaks (upper right hand panel). Assuming no interaction between the irradiated and non-irradiated cells, 72% of the cells were expected to contain no breaks when 20% of the cells were hit with one particle through the nucleus (lower left hand panel). In actuality, only 38% of the cells in this population showed no chromatid breaks (lower right hand panel). Furthermore, the profile of chromatid breaks was very different from that in which 100% of the cells in the population were hit. The proportion of NHBE cells with multiple chromatid breaks that

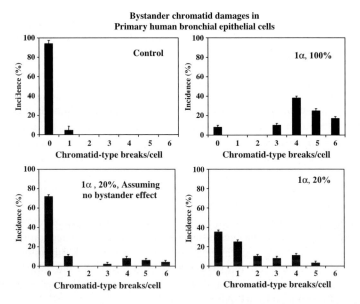

Fig. 1. Induction of chromatid-type breaks per cell from populations of commercially available primary human bronchial epithelial (NHBE) cells in which 0%, 20% or 100% of cells were traversed by exactly one alpha particle through the nuclei. See text for detail.

was so prevalent in directly hit cells were much smaller in the bystander population. These data clearly illustrated that the bystander genotoxic response can also be demonstrated in NHBE cells and is not an artifact that pertains to the $A_L$ cells.

## 4. Possible mechanisms of the bystander effect

The mechanism of radiation induced genotoxicity is not clear and is likely to be complex and involves multiple pathways. Table 1 lists the possible process involved. It is likely that a multiple signaling cascade involving both an initiating event and downstream signaling steps are necessary to mediate the bystander process. Since gap junctional communication plays an important role in mediating the bystander signaling, a variety of possible signaling molecules with molecular size less than 1000 Da, such as reactive oxygen species, reactive nitrogen species, long lived organic radicals, protein hydro-

Table 1
Characteristics of the bystander effect

- Multiple pathways likely
- P53 gene function is NOT essential for the process
- In confluent cultures, gap junction mediated cell-to-cell communication is essential
- In sparse cultures, secreted signaling molecules, particularly reactive oxygen and reactive nitrogen species are involved
- It is not clear if ROS are only the initiating signaling event that triggers other down stream, more stable secondary signaling molecules

peroxides, growth factors, cytokines including $TGF_\beta$ and prostaglandins, may be involved in the process. Studies are currently underway in many laboratories in identifying the true nature of the signaling molecules.

## 5. Relevance to low dose radiation risk assessment

Accurate risk assessment of human exposure to ionizing radiations traditionally has been compromised, in that reliable data are available only for relatively high doses. Cancer risk estimates over the dose range from 0.5 to 2.5 Sv are available from the epidemiological study of the A-bomb survivors. Risks at lower doses can only be inferred by an extrapolation from the high dose risks. The question is, what is the risk of radiation exposures above the level of natural background but below the lowest dose for which risks are known from the A-bomb survivors? Both ICRP and NCRP recommend a linear no-threshold extrapolation, but this has generated a great deal of controversy and is a much debated issue, since it involves issues of major societal and economic concern. The bystander observations imply that the relevant target for various radiobiological endpoints is larger than an individual cell and suggest a need to reconsider the validity of the linear extrapolation model in making risk estimate for low dose radiation exposure, an issue that is the subject of considerable debate.

## Acknowledgements

The authors would like to thank Ms. Sarah Baker for critical reading of the manuscript. Work was supported by funding from the National Institute of Health grants CA 49062, ES 12888, NIH Resource Center Grant RR 11623, Environmental Center grant ES 10349 and from the US Department of Energy DEFG-ER63441.

## References

[1] J.P. Kotval, L.H. Gray, Structural changes produced in microspores of *Tradescantia* by α-particles, J. Genet. 48 (1947) 135–154.
[2] H. Nagasawa, J. Little, Induction of sister chromatid exchanges by extremely low doses of alpha particles, Cancer Res. 52 (1992) 6394–6396.
[3] H.N. Zhou, et al., Induction of a bystander mutagenic effect of alpha particles in mammalian cells, Proc. Natl. Acad. Sci. U. S. A. 97 (2000) 2099–2104.
[4] T.K. Hei, et al., Mutagenic effects of a single and an exact number of alpha particles in mammalian cells, Proc. Natl. Acad. Sci. U. S. A. 94 (1997) 3765–3770.
[5] H.N. Zhou, et al., Radiation risk to low fluences of alpha particles may be greater than we thought, Proc. Natl. Acad. Sci. U. S. A. 98 (2001) 14410–14415.

# Molecular dissection of in vivo DNA rearrangements induced by radiation and chemical mutagens

Takehiko Nohmi*, Ken-ichi Masumura

*Division of Genetics and Mutagenesis, National Institute of Health Sciences, 1-18-1 Kamiyoga, Setagaya-ku, Tokyo 158-8501, Japan*

**Abstract.** Cellular DNA is continuously exposed to radiation and chemicals. To analyze the deletion mutations in a whole body system, *gpt* delta mice has been established. In this mouse model, deletions in lambda phage DNA integrated in the chromosome are preferentially selected as Spi⁻ (*s*ensitive to *P*2 *i*nterference) phages, which can then be subjected for molecular analysis. Here, we report the sequence characteristics of deletions induced by ionizing radiations in liver, ultraviolet light B (UVB) in epidermis, mitomycin C (MMC) in bone marrow and heterocyclic amine 2-amino-1-methyl-6-phenylimidazo[4,5-*b*] pyridine (PhIP) in colon. About half of the large deletions occur between short direct-repeat sequences and the remainder had flush ends, suggesting that they are generated during the repair of double-strand breaks in DNA. Radiations, UVB and MMC efficiently induced the large deletions whereas PhIP mainly generated one base deletions in runs of guanine bases. The most predominant mutations in untreated mice were one base deletions in runs of adenine bases. Possible mechanisms of the intra-chromosomal deletion mutations are discussed. © 2004 Elsevier B.V. All rights reserved.

*Keywords:* Deletion; Point mutation; Genome instability; Transgenic mouse; Transgenic rat

## 1. Introduction

Humans are exposed to various DNA-damaging agents, which induce oxidation, methylation and deamination of DNA, adduct formation and strand sessions. Of the various DNA lesions, the most detrimental is probably double-strand breaks (DSBs) in DNA. DSBs can be induced by ionizing radiations (IRs), anti-cancer therapeutic agents

---

\* Corresponding author. Tel.: +81 3 3700 9873; fax: +81 3 3707 6950.
*E-mail address:* nohmi@nihs.go.jp (T. Nohmi).

0531-5131/ © 2004 Elsevier B.V. All rights reserved.
doi:10.1016/j.ics.2004.11.017

such as mitomycin C (MMC) or when replicative DNA polymerases encounter single-stranded breaks or DNA lesions [1]. DSBs in DNA trigger a set of cellular responses such as delay of cell cycle and apoptosis. However, if not properly repaired or eliminated, DSBs enhance the frequency of illegitimate recombination, i.e., deletions, amplifications and translocations of chromosomes, leading to genome instability and tumorigenesis. Human cells are estimated to suffer more than 2000 imperfect repair sites in the chromosome at the end of 70 years of life because of the spontaneous DSBs [2].

To gain insights into the mechanisms of deletion mutations in vivo, we have established a novel transgenic mouse, named *gpt* delta [3,4]. The mice carry tandem repeats of lambda EG10 DNA in the chromosome 17, which are retrievable as phage particles by in vitro packaging reactions. The rescued phages are then subjected to Spi$^-$ (*s*ensitive to *i*nterference) selection. This selection takes advantage of the restricted growth of wild-type lambda phage in P2 lysogens. Only mutant lambda phages that are deficient in the functions of both the *gam* and *redBA* genes can grow well in P2 lysogens and display the Spi$^-$ phenotype. Simultaneous inactivation of both the *gam* and *redBA* genes is usually induced by deletions in the region. Using the shuttle vector mutation assay, deletions can be analyzed at a molecular level in various organs of mice. In addition, 6-thioguanine (6-TG) selection has been incorporated for the detection of point mutations, i.e., base substitutions and frameshifts. Here, we report the characteristics of deletion mutations induced by radiation and chemicals. The mutation spectra of *gpt* mutations have been reported previously [5].

## 2. Molecular nature of Spi$^-$ mutations

We have analyzed more than 400 independent Spi$^-$ mutants at a sequence level. They were rescued from the liver, spleen, bone marrow, epidermis, colon and kidney of *gpt* delta mice and the mutations were induced by IRs (heavy ion, X-ray and gamma-ray radiations), ultraviolet light B (UVB), MMC and 2-amino-1-methyl-6-phenylimidazo[4,5-*b*] pyridine (PhIP). Mutants rescued from untreated mice were also analyzed. We categorized the mutants into five classes (classes I to V) based on the deletion sizes and the sequence characteristics of the junctions (Fig. 1).

Class I mutants include large deletions exceeding 1 kb. The maximum length of the deletions is about 10 kb. Almost half of Class I mutants has short homologous sequences at the junctions of mutants. Thus, we subdivided Class I mutants into Class I-A and I-B. Mutants of Class I-A are large deletions that exhibit short homologous sequences of 1 to 12 bp at the junctions. Class I-B mutants are large deletions without short homologous sequences. Intriguingly, about 10% of Class I-A and I-B mutants have insertion sequences in the junctions. These extra nucleotides are often called "filler DNA" and the genetic rearrangements that arise by illegitimate recombination in mammalian cells have such sequences at about 10% [6]. The length of the insertion sequence is usually 1 or 2 bp but the maximum insertion was 14 bp. Class I-A and I-B mutants delete regions of both the *gam* and *redBA* genes or a region in the *gam* gene and the upstream. They are induced by IRs, UVB and MMC.

Class II include middle-size deletions of more than 1 bp but less than 1 kb. Some have short homologous sequences but others not as in the case of Class I mutants. Mutations of

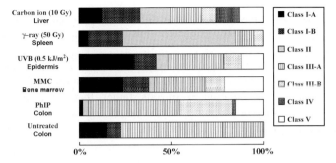

Fig. 1. Schematic representation of various classes of Spi⁻ mutations. Class I-A, large deletions with short homologous sequences at the junctions; Class I-B, large deletions without short homologous sequences; Class II, middle size (more than 1 bp but less than 1 kb) deletions; Class III-A, single bp deletions at run sequences; Class III-B, single bp deletions at non-run sequences; Class IV, complex mutations; Class V, miscellaneous mutations. The references of each treatment are as follows: carbon ion [9]; gamma-ray [7]; UVB [10]; MMC [11]; PhIP[8]; untreated [8].

more than 20 bp but less than 600 bp are frequently identified in the spleen of mice irradiated with gamma-ray at a dose of 50 Gy.

Class III mutants are 1 bp deletions in the *gam* gene. These small deletions are not supposed to induce Spi⁻ mutations. However, translation of the *gam* and *redBA* genes is probably linked, and the *gam* gene is first transcribed so that the 1 bp deletions in the *gam* gene may interfere with the start of translation of the downstream *redBA* genes, thereby functionally inactivating not only *gam* but also *redBA* [7]. We subdivided Class III mutants into Class III-A and III-B: the former mutations occur at run sequences such as AAAAAA and the latter at non-repetitive sequences. Class III-A mutants occurring at A:T repetitive sequences are the most predominant type of mutations in untreated mice. PhIP induces Class III-A deletions. However, most of them occur at G:C repetitive sequences or beside run sequences [8].

Class IV is a complex type of mutations where the exact junctions could not be identified because of the genome rearrangements. This class of mutants is frequently observed in p53 knockout mice irradiated with heavy ion [9]. Class V includes miscellaneous mutations.

## 3. Possible mechanisms of deletion mutations in vivo

Deletions with short (Class I-A) or no (Class I-B) homologous sequences at their junctions have been observed in a number of mutant genes implicated in human diseases including cancer. Indeed, about 40% of large deletions in human disorders are characterized by the presence of very short sequence homologies at the breakpoints [12]. On the basis of the sequence characteristics observed in the junctions of Spi⁻ mutants, we suggest that non-homologous end joining (NHEJ) repair plays an important role in the generation of intra-chromosomal deletions such as Spi⁻ Class I mutants. This pathway involves the DNA end-binding heterodimer Ku70/Ku80, DNA-PK$_{CS}$, XRCC4 and DNA ligase IV. Although some of these proteins play an essential role in the maintenance of genome stability and suppression of tumorigenesis, NHEJ repair pathway has the potential to induce deletion mutations. If two incompatible ends are generated by IRs, they first have to be converted to ligatable ends by enzymatic processing which often causes deletions.

Genome rearrangements associated with oxidative stress are important in carcinogenesis. In this regard, *gpt* delta rat [13] could be important because most of carcinogenesis studies are undertaken in rats rather than in mice. The genome rearrangements can directly be examined in the target organs of carcinogenesis in *gpt* delta rats kept under various nutritional conditions or exposed to radiation and chemicals.

## Acknowledgments

Part of this study was financially supported by the Budget for Nuclear Research of the Ministry of Education, Culture, Sports, Science and Technology Japan. This work was also supported by Grants-in-aid for Cancer Research from the Ministry of Health, Labor and Welfare, Japan, and for Basic Research from the Japan Health Science Foundation.

## References

[1] G.C. Smith, S.P. Jackson, The DNA-dependent protein kinase, Genes Dev. 13 (1999) 916–934.
[2] M.R. Lieber, et al., Mechanism and regulation of human non-homologous DNA end-joining, Nat. Rev., Mol. Cell Biol. 4 (2003) 712–720.
[3] T. Nohmi, T. Suzuki, K. Masumura, Recent advances in the protocols of transgenic mouse mutation assays, Mutat. Res. 455 (2000) 191–215.
[4] T. Nohmi, et al., A new transgenic mouse mutagenesis test system using Spi$^-$ and 6-thioguanine selections, Environ. Mol. Mutagen. 28 (1996) 465–470.
[5] T. Nohmi, K. Masumura, gpt delta transgenic mouse: a novel approach for molecular dissection of deletion mutations in vivo, Adv. Biophys. 38 (2004) 97–121.
[6] D.B. Roth, et al., Oligonucleotide capture during end joining in mammalian cells, Nucleic Acids Res. 19 (1991) 7201–7205.
[7] T. Nohmi, et al., Spi$^-$ selection: an efficient method to detect gamma-ray-induced deletions in transgenic mice, Environ. Mol. Mutagen. 34 (1999) 9–15.
[8] K. Masumura, K. Matsui, M. Yamada, Characterization of mutations induced by 2-amino-1-methyl-6-phenylimidazo[4,5-*b*] pyridine in the colon of gpt delta transgenic mouse: novel G:C deletions beside runs of identical bases, Carcinogenesis 21 (2000) 2049–2056.
[9] F. Yatagai, et al., Heavy-ion-induced mutations in the gpt delta transgenic mouse: effect of p53 gene knockout, Environ. Mol. Mutagen. 40 (2002) 216–225.
[10] M. Horiguchi, et al., Molecular nature of ultraviolet B light-induced deletions in the murine epidermis, Cancer Res. 61 (2001) 3913–3918.
[11] A. Takeiri, et al., Molecular characterization of mitomycin C-induced large deletions and tandem-base substitutions in the bone marrow of gpt delta transgenic mice, Chem. Res. Toxicol. 16 (2003) 171–179.
[12] T. Morris, J. Thacker, Formation of large deletions by illegitimate recombination in the HPRT gene of primary human fibroblasts, Proc. Natl. Acad. Sci. U. S. A. 90 (1993) 1392–1396.
[13] H. Hayashi, et al., Novel transgenic rat for in vivo genotoxicity assays using 6-thioguanine and Spi$^-$ selection, Environ. Mol. Mutagen. 41 (2003) 253–259.

www.ics-elsevier.com

# Chromosome aberrations in peripheral lymphocytes of individuals living in dwellings with an increased level of indoor radon concentrations

Günther Stephan[a],*, Ursula Oestreicher[a], Rainer Lehmann[b]

[a]*Federal Office for Radiation Protection, Department of Radiation Protection and Health, Ingolstädter Landstrasse. 1, D-85764 Oberschleissheim, Germany*
[b]*Federal Office for Radiation Protection, Department of Radiation Protection and Health, Berlin, Germany*

**Abstract.** Sixty one individuals living in dwellings with increased indoor radon concentrations (80–13,000 Bq/m$^3$) were investigated cytogenetically. The frequency of dicentric chromosomes (dic) and centric rings (cr) showed a dependency on the time integrated radon concentration from ≤1000 to 10,000 Bq/m$^3$ · y. At values higher than 10,000 Bq/m$^3$ · y, the frequency remained on the same level. The frequency of translocations was determined in 16 persons living in houses with radon concentrations >5000 Bq/m$^3$. In persons older than 40 years, a significantly increased frequency was observed in comparison to the age-matched controls ($p<0.05$). Since most of the translocations were found in stable cells, it is concluded that translocations are also induced in blood-forming tissue and are transmitted to peripheral blood. There is good agreement between physically calculated doses, according to ICRP and BEIR VI, and biologically calculated doses for blood and bone marrow.
© 2004 Elsevier B.V. All rights reserved.

*Keywords:* Cytogenetic analyses; Indoor radon; Dicentric chromosomes; Translocations

## 1. Introduction

From miners, it is well documented that exposure to radon and its decay products results in an increased frequency of lung cancer [1,2]. For this reason, indoor radon exposure with regard to radiation protection of the general public has become a matter of interest worldwide. Epidemiological studies on indoor radon and lung cancer showed consistently a small increase in lung cancer with increasing radon exposure [3,4] that is comparable to the

---

* Corresponding author. Tel.: +49 1888 333 2210; fax: +49 1888 333 2205.
*E-mail address:* GStephan@bfs.de (G. Stephan).

0531-5131/ © 2004 Elsevier B.V. All rights reserved.
doi:10.1016/j.ics.2004.11.018

results from uranium miners studies [2]. Moreover, there are some reports dealing with an indoor radon related risk for leukaemia [5,6].

The main target tissue of radon and its decay products is the respiratory tract. Part of the inhaled radon is transmitted into blood and is then transported to other tissues including the blood-forming stem cells. In accordance with BEIR VI [2], domestic radon concentration of 20 Bq/m$^3$ results in an annual lung dose of 500 μGy, whereas blood receives an annual dose of only about 1 μGy. The annual radon and thoron derived dose to active marrow for an adult was calculated to be 90 and 30 μSv, respectively [7].

We have performed cytogenetic investigations in individuals living in dwellings with increased levels of indoor radon concentrations up to >5000 Bq/m$^3$. The chromosome analyses in peripheral blood lymphocytes from these individuals may contribute to the question whether significant doses are received by peripheral lymphocytes and target cells in the bone marrow.

## 2. Individuals, material and methods

The Federal Office for Radiation Protection (BfS) carried out measurements of domestic radon concentrations by means of the passive α track method. On the basis of this measurements, 61 individuals (29 females and 32 males) living in dwellings with radon concentrations ranging between 80 and 13,000 Bq/m$^3$ were selected. At the time of blood sampling, the persons were aged between 6 and 80 years (mean age 44.6 years; 8 children and 53 adults). Data of confounding factors such as medical radiation exposure, intake of medical drugs, and smoking habits were recorded in questionnaires.

From all individuals, structural chromosome aberrations were analysed by using FPG-staining. For quantitative calibrations, dicentric chromosomes (dic) and centric rings (cr) were used. Additionally, the FISH-technique was applied to analyse translocations in 16 persons (8 females and 8 males) out of 6 houses of the highest exposure group (>5000 Bq/m$^3$).

The culture technique, chromosome preparations and FPG staining have been described elsewhere [8], also the procedures for fluorescence in situ hybridisation (FISH) [9]. In cells carrying translocations, unpainted chromosomes were also analysed in order to distinguish between stable and unstable cells.

## 3. Results and discussion

From 61 radon exposed persons, 60,056 FPG stained cells were analysed. To avoid smoking as a confounding factor, persons smoking more than 20 cigarettes per day were excluded from the statistical analysis. The mean frequency of dic+cr (2.51±0.21/1000 cells) of these persons was significantly higher than that of our control group (1.03±0.17/1000 cells). The members of the control group (mean age 43.0 years) lived in the Munich area where the indoor radon concentration does not exceed the German average of 50 Bq/m$^3$. Published cytogenetic data show contradictory results [10,11,12]. The reason for this may be that radon concentration in dwellings of the investigated persons and/or the number of individuals including the numbers of scored cells were too low.

Because radiation exposure of the persons is determined by indoor radon concentration (Bq/m$^3$) and residence time (years: y), the frequency of dic+cr was related to the time integrated radon concentration (Bq/m$^3$ · y). Up to about 2000 Bq/m$^3$ · y the frequency of dic+cr did not differ from our control group. After this, it becomes clear that the frequency of dic+cr increases with increasing time integrated radon concentration (Fig. 1). When the time integrated concentration exceeded 10,000 Bq/m$^3$ · y, the frequency of dic+cr remained on the same level: 10,000–100,000 Bq/m$^3$ · y, 2.71±0.39 dic+cr/1000 cells; 100,000–800,000 Bq/m$^3$ · y, 2.92±0.49 dic+cr/1000 cells.

For individuals with a time integrated radon concentration of >150,000 Bq/m$^3$ · y, the mean indoor radon concentration was 7,050 Bq/m$^3$ and the mean time of residence was 47 years. According to BEIR VI, the resulting blood dose is about 17 mGy. The mean frequency of dic+cr of the same individuals is 3.25±0.51 per 1000 cells. The dose relationship obtained by irradiation with Americium-241 α-particles ($y=0.27\pm0.02$ D) [13] was used to convert this aberration frequency into dose. The resulting biologically estimated dose of 12±3 mGy is in good agreement with physical dose calculation.

From 16 persons (mean age 48.6 years) who lived in dwellings with radon concentrations, >5000 Bq/m$^3$ 23,315 cells were analysed for translocations with the FISH-technique. The mean frequency of translocations showed a slight but not significant increase in comparison with our control group [9]. If only the persons older than 40 years of age were taken into consideration, the mean frequency of translocations (13.38±2.22 per 1000 cells) is significantly increased in comparison with the age-matched control (7.50±0.72 per 1000 cells). Since the average residence time of older persons in radon houses is longer (41.6 years) than that of the persons aged under 40 years (13.9 years), they received a higher dose, which may be the reason for the increased translocation frequency. In published studies, no relation between radon concentrations and translocation frequencies was found [14,15]. In these studies, however, radon concentrations in dwellings were much lower than >5000 Bq/m$^3$.

Translocations only in stable cells must be considered to answer the question whether the observed translocations are induced directly in the peripheral lymphocytes, or if they may be partly induced in the blood-forming tissue and are then transmitted to peripheral blood. From in vitro exposure of lymphocytes to α-particles from an Americium-241 source, we know that in the dose range between 0.02 and 0.1 Gy, the frequency of induced translocations is about two times higher than that for dicentrics [16]. If the observed translocation frequency in the study group would be exclusively induced in peripheral blood, the frequency ratio of translocations to dicentrics should also be about 2. In the study group, however, the frequency of translocations in stable cells was about nine times higher than that of dicentrics. If the excess translocations would be induced in the blood-forming tissue and reach the

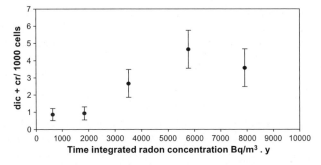

Fig. 1. Dependency of dic+cr/1000 cells on the time integrated radon concentration up to 10,000 Bq/m$^3$ · y.

peripheral blood they should occur in stable cells, since these cells are able to proliferate: 84 translocations out of 91 translocations were observed in stable cells.

According to ICRP 65, the dose for the red bone marrow after inhalation of radon without decay products is $2.00 \times 10^{-10}$ Sv/(Bq h m$^{-3}$) [1]. The mean residence time in radon houses of the investigated persons is about 42 years, and the mean radon concentration in this group is about 8000 Bq/m$^3$. Assuming, for example, that subjects stayed in their houses about 12 h a day, they received a calculated bone marrow dose of about 300 mSv, which corresponds to 15 mGy, based on an RBE of 20 for α-radiation. According to BEIR VI, the estimated absorbed dose to red bone marrow for radon concentration of 20 Bq/m$^3$ is between 0.5 and 6 μGy y$^{-1}$. Therefore, the calculated dose for the above-mentioned persons lies between 8 and 100 mGy. The dose–response curve for α-particles and translocations only in stable cells was found to be: $y=(21.52\pm2.36)\times10^{-2}$ D [16]. When the observed translocation frequency in stable cells is used in combination with the dose–response curve, a dose of $62\pm18$ mGy can be calculated. There is quite a good correlation between the physically calculated dose to bone marrow and the biologically estimated dose.

## References

[1] ICRP, International Commission on Radiological Protection, Protection against Radon-222 at home and at work, Publication, vol. 65, Pergamon Press, Oxford, 1994.
[2] BEIR VI, National Research Council, Committee on Health Risk of Exposure to Radon: Health Effects of Exposure to Radon, National Academy Press, Washington, DC, 1999.
[3] S.C. Darby, D.C. Hill, Health effects of residential radon: a European perspective at the end of 2002, Radiat. Prot. Dosim. 104 (2003) 321–329.
[4] J.H. Lubin, Studies of radon and lung cancer in North America and China, Radiat. Prot. Dosim. 104 (2003) 315–319.
[5] A.K.M.M. Haque, A.E. Kirk, Environmental radon and cancer risk, Radiat. Prot. Dosim. 45 (1992) 639–642.
[6] J.P. Eatough, D.L. Henshaw, Radon and monocytic leukaemia in England, J. Epidemiol. Community Health 47 (1993) 506–507.
[7] R.B. Richardson, J.P. Eatough, D.L. Henshaw, Dose to red bone marrow from natural radon and thoron exposure, Br. J. Radiol. 64 (1991) 608–624.
[8] G. Stephan, S. Pressl, Chromosomal aberrations in peripheral lymphocytes from healthy subjects as detected in first cell division, Mutat. Res. 446 (1999) 231–237.
[9] G. Stephan, S. Pressl, Chromosome aberrations in human lymphocytes analysed by fluorescence in situ hybridization after in vivo irradiaton, and in radiation workers, 11 years after an accidental radiation exposure, Int. J. Radiat. Biol. 71 (1997) 293–299.
[10] H.J. Albering, et al., Indoor radon exposure and cytogenetic damage, Lancet 340 (1992) 739.
[11] M. Bauchinger, et al., Chromosome aberrations in peripheral lymphocytes from occupants of houses with elevated indoor radon concentrations, Mutat. Res. 310 (1994) 135–142.
[12] A. Maes, A. Poffijn, L. Verschaeve, Case report: karyotypic and chromosome aberration analysis of subjects exposed to indoor radon, Health Phys. 71 (1996) 641–643.
[13] E. Schmid, et al., Analyses of chromosome aberrations in human peripheral lymphocytes induced by in vitro α-particle irradiation, Radiat. Environ. Biophys. 35 (1996) 179–184.
[14] M. Bauchinger, et al., Quantification of FISH-painted chromosome aberrations after domestic radon exposure, Int. J. Radiat. Biol. 70 (1996) 657–663.
[15] C. Lindholm, et al., Domestic radon exposure and the frequency of stable or unstable chromosomal aberrations in lymphocytes, Int. J. Radiat. Biol. 75 (1991) 921–928.
[16] J.F. Barquinero, G. Stephan, E. Schmid, Effect of americium-241 α-particles on the dose–response of chromosome aberrations in human lymphocytes analysed by fluorescence in situ hybridization, Int. J. Radiat. Biol. 80 (2004) 155–164.

ELSEVIER

www.ics-elsevier.com

# Cytogenetic studies of populations exposed to Chernobyl fallout

## D.C. Lloyd*

*National Radiological Protection Board, Radiation Effect Dept., Chilton, Didcot, Oxfordshire, OX11 0RQ, UK*

**Abstract.** This review considers chromosome aberration studies, mainly using the dicentric assay, of persons continuously resident in areas of Chernobyl fallout and others, evacuated from highly contaminated places, and relocated sometimes to areas of lesser contamination. It has been possible to use cytogenetics to estimate group mean doses that could be used as representative of dose to the wider communities from which the study subjects were drawn. This is an important supplement to physical reconstruction of doses as an input to epidemiology. There are indications from the cytogenetics that in some instances, modelling committed doses from environmental monitoring data have underestimated the population dose. © 2004 Published by Elsevier B.V.

*Keywords:* Chernobyl accident; Chromosomal aberrations; Biodosimetry

## 1. Introduction

The 1986 Chernobyl accident irradiated a large number of people mostly because of the deposition of fission products over wide tracts of land. The immediate emergency medical/public health response was, of course, to attend to subjects close by the reactor who were heavily exposed and to organise evacuation of communities within a 30-km radius as the extent of the contamination became apparent.

In the context of the subject of the present meeting, the accident resulted in large numbers of people who have continued to be exposed to radiation doses from long-lived fission products, notably radiocaesium and strontium. Based on environmental sampling and modelling, the various exposure pathways to humans, the wide-scale population doses are low but nevertheless appreciably above the natural background that existed prior to the

---

* Tel.: +44 1235 822700; fax: +44 1235 833891.
 *E-mail address:* david.lloyd@nrpb.org.

0531-5131/ © 2004 Published by Elsevier B.V.
doi:10.1016/j.ics.2004.11.036

accident. These people comprise populations, particularly in Belarus, who have remained resident in their home communities and others who were evacuated from more heavily polluted places. These latter were often moved to more lightly contaminated areas because of the difficulty in finding locations that had been completely spared from the fallout.

Direct medical consequences of the radiation exposures have been deterministic injuries to the highly exposed individuals and an increase in thyroid cancers in children. Reports also exist of increased frequencies of other cancers and a range of nonmalignant diseases [1]. However, the extent to which these can be directly linked to irradiation is debatable. The socioeconomic distress caused by the breakup of the Soviet Union, together with the stress and deprivation prevalent in communities 'blighted' by Chernobyl fallout are important factors that are very likely to have impacted on public health.

## 2. Cytogenetic biomarkers of population exposure

The analysis for chromosomal aberrations, notably the dicentric in cultured blood lymphocytes, is a long-established method for investigating radiation exposure and is routinely used as a biodosemeter [2]. The Chernobyl accident prompted a number of cytogenetic studies, and naturally, in the immediate postaccident period, effort was concentrated, to good effect, on the patients with acute radiation sickness. The next most important group were the 'liquidators', people drafted in over the ensuing years for recovery and clean-up operations. The cytogenetic surveying of representatives from this population showed elevated chromosomal aberration levels, but, with a few notable exceptions [3], the dicentric frequencies indicated that their doses generally did not exceed the maximum limit of 250 mGy that had been set for this unique operation [4].

Cytogenetic surveys of the general populations in contaminated areas generally assumed lower priority and thus tended to commence later. This is unfortunate because the dicentric is an unstable aberration and is removed from the lymphocyte pool as cells are naturally replaced. For persons with normal haematology, a reasonable 'rule of thumb' is that the dicentric is replaced with a 3-year half-life. Thus, in a blood sample taken from a person continuously exposed to an elevated background, the dicentric frequency will reflect an average dose over about 4–5 years. Surveys initiated much later than this are likely to underestimate the cumulative dose, and certainly today, one could only contemplate using a retrospective biodosemeter, such as the fluorescence in situ hybridisation (FISH) translocation technique. The following review is of necessity selective and draws upon some studies, which, because of their timing and/or size, could be considered most informative.

About 5 years postaccident, Sevan'kaev et al. [5] sampled a large number of young people continuously resident in contaminated communities of Belarus and Russia, who had been <12 years in 1986, and included some who were conceived and/or born after the accident. One finding from this survey, and others subsequently [6], was a number of children with 'rogue' cells; highly aberrant metaphases that had previously been reported from surveys of such diverse groups as primitive tribes, deep-sea divers, and A-bomb survivors. In the Chernobyl context, the possibility was raised that they could be associated with intense localised sources, such as radioiodine in a thyroid or inhaled 'hot particles'. Both possibilities were convincingly ruled out [7] based on biophysical and microdosimetric arguments, and it was concluded that a viral origin was most probable [8].

The overall conclusion of the survey was that, with the exception of one village, the mean dicentric frequencies were not elevated compared with a control group. The exception was the most heavily contaminated community where the average dose to date, modelled from environmental monitoring, was 100 mSv. The dicentrics indicated an average dose of 110 mGy and, in adults from the same village, 130 mGy. Given the assumptions and approximations in all calculations, this close level of agreement is probably fortuititous. Enough children who had been exposed in utero were included in the study to permit the conclusion of no significant difference in the level of induced aberrations in their lymphocytes compared with those exposed only postnatally.

Approximately 10 years after the accident, two laboratories, in Belarus and Ukraine, belatedly revealed that within a few days to weeks after the accident, and well within the time window for the dicentric assay, they had collected blood from residents of contaminated villages and/or resettled evacuees from the exclusion zone. In both instances, the evacuees went to clean locations, unlike the group discussed above. These studies were eventually published [9,10].

The Belarus study again concentrated on children where a resident group's dicentric frequency indicated a mean dose of 230 mGy and that for a group of evacuees 410 mGy. Both of these values are markedly higher than doses of around 10–20 mSv [11] that were calculated for citizens of Pripyat city and surrounding rural parts of the exclusion zone prior to evacuation. The discrepancy might be due to incorporated radionuclides, particularly those with the shorter half-lives where, apart perhaps from some isotopes of iodine, there are insufficient data to allow reliable estimates to be made of probable doses. Environmentally based dose calculations made by later surveys were from essentially the longer-lived nuclides. The unexpectedly higher doses detected cytogenetically were therefore probably received during the first 2 weeks. Among the evacuee children, an interesting age-dependent trend was noted, such that higher dicentric yields were found in the younger (6–10 years) children compared with the older subgroup (11–15 years). On further investigation, it was found that this was entirely due to the older girls and this led to speculation that some behavioural or lifestyle characteristic of older girls compared with other children served to lessen their radiation dose during the critical 2 weeks.

The Ukranian study examined adult and child evacuees and reported that citizens of Prypiat who left on day 3 had a mean dicentric yield indicative of 320 mGy, whilst a group from nearby rural areas who left later, day 10, had an average dose of 490 mGy. Again, these are values much higher than the 10–20 mSv that was calculated from external $\gamma$ dose rate measurements. Again, the incorporation of $\gamma$-emitting fission products early on was considered the most likely reason for the discrepancy.

With the passage of time, as the dicentric assay becomes less meaningful, the cytogenetic emphasis should move towards FISH, but to date, there have been few studies on long-term residents. Salomaa [6] took samples in 1993 as one of the last villages in Russia was being evacuated and concluded a mean dose of ~60 mGy, which compared well with probably <100 mSv calculated from measurements of fallout. Around the same time, another FISH study [12] was made of inhabitants of four Belarus villages and concluded mean group doses of 180–400 mGy, but no comparisons were made with levels of fallout. Finally, ~13–15 years later, the Prypiat evacuee group described above for whom a high early mean dose of 320 mGy was estimated was resampled, and FISH then indicated ~200 mGy [13]. Given the statistical uncertainties, these two values are compatible.

## 3. Conclusions

The studies of people exposed protractedly to Chernobyl fallout are relevant to the topic of this meeting on populations living in areas of elevated background radiation. With the limited cytogenetic resources in the former Soviet Union, which at that time only had the dicentric assay available, these people attracted less urgency. Nevertheless, a number of studies were initiated covering a wide geographical catchment and time span. In a number of instances, the results have indicated higher average doses to the cohorts than were calculated by physical reconstruction. The purpose of cytogenetic studies on such cohorts, who clearly are not at risk of deterministic injury, is retrospective biodosimetry as an input to epidemiology studies for stochastic disease in the affected populations. Cytogenetics raises important issues because, particularly regarding the early postaccident period, it is suggesting that higher doses may have been received than have been previously assumed or calculated from sometimes uncertain physical data or models.

## References

[1] V.G. Bebeschko, O.A. Bobyliova, Medical consequences of the Chernobyl nuclear power plant accident: experience of 15-year studies, Int. Congr. Ser. 1234 (2002) 267–279.
[2] IAEA, Cytogenetic analysis for radiation dose assessment, a manual, Tech Rept Ser 405, Vienna: IAEA, 2001.
[3] A.V. Sevan'kaev, et al., High exposures to radiation received by workers inside the Chernobyl sarcophagus, Radiat. Prot. Dosim. 59 (2) (1995) 85–91.
[4] A.V. Sevan'kaev, et al., A survey of chromosomal aberrations in lymphocytes of Chernobyl liquidators, Radiat. Prot. Dosim. 58 (2) (1995) 85–91.
[5] A.V. Sevan'kaev, et al., Chromosomal aberrations in lymphocytes of residents of areas contaminated by radioactive discharges from the Chernobyl accident, Radiat. Prot. Dosim. 58 (4) (1995) 247–254.
[6] S. Salomaa, et al., Unstable and stable chromosomal aberrations in lymphocytes of people exposed to Chernobyl fallout in Bryansk, Russia. Int. J. Radiat. Biol. 71 (1) (1997) 51–59.
[7] A.V. Sevan'kaev, et al., 'Rogue' cells observed in children exposed to radiation from the Chernobyl accident, Int. J. Radiat. Biol. 63 (3) (1993) 361–367.
[8] J.V. Neel, et al., 'Rogue' lymphocytes among Ukranians not exposed to radioactive fall-out from the Chernobyl accident, Proc. Natl. Acad. Sci. U. S. A. 89 (1992) 6973–6977.
[9] L.S. Mikhalevich, et al., Estimates made by dicentric analysis for some Belarussian children irradiated by the Chernoby accident, Radiat. Prot. Dosim. 87 (2) (2000) 109–114.
[10] N.A. Maznik, et al., Chromosomal dosimetry for some groups of evacuees from Prypiat and Ukranian liquidators at Chernobyl, Radiat. Prot. Dosim. 74 (1/2) (1997) 5–11.
[11] I.A. Likhtarev, V.V. Chumack, V.S. Repin, Retrospective reconstruction of individual and collective external gamma doses of population evacuated after the Chernobyl accident, Health Phys. 66 (6) (1994) 643–652.
[12] F. Darroudi, A.T. Natarajan, Biological dosimetric studies in the Chernobyl accident, on populations living in the contaminated areas (Gomel regions) and in Estonian clean-up workers, using FISH technique, in: A. Karaglou, G. Deamet, G.N. Kelly, H.G. Menzel (Eds.), The Radiological Consequences of the Chernobyl Accident, CEC, Luxembourg, 1996, pp. 1067–1072.
[13] A. Edwards, et al., Biological estimates of dose to inhabitants of Belarus and Ukraine following the Chernobyl accident, Radiat. Prot. Dosim. 111 (2) (2004) 211–219.

# Results of Yangjiang study in China and an experience of Ramsar survey in Iran

Naoto Fujinami[a,*], Taeko Koga[b], Hiroshige Morishima[b], Tsutomu Sugahara[c]

[a]*Kyoto Prefectural Institute of Hygienic and Environmental Sciences, 395 Murakami-cho, Fushimi-ku, Kyoto, 612-8369, Japan*
[b]*Kinki University Atomic Energy Research Institute, Osaka, 577-8502, Japan*
[c]*Health Research Foundation, Kyoto, 606-8225, Japan*

**Abstract.** While we attempted to assess the annual effective doses of inhabitants living in high levels of natural radiation areas (HLNRAs) in Yangjiang, China, the number of subjects was too large to determine directly each dose of all inhabitants with personal dosimeters. Individual dose rates were therefore estimated from ambient radiation dose rates and occupancy factors, based on the result of preliminary examination, which had indicated a good correlation between dose values obtained through this estimation and personal dosimetry. In Ramsar, Iran, studding was observed around springs with a high level of natural radiation, and outdoor dose rates varied largely and irregularly, even within a narrow area. Owing to the uneven distribution of natural radiation sources in building materials, indoor dose rates were so grossly different, even among rooms, that it appeared difficult to accurately estimate individual dose rates of inhabitants in these areas by the abovementioned indirect method. Consequently, it would be reasonable that direct and indirect methods are applied properly according to objects of study and some conditions, such as level of individual doses and number of inhabitants. © 2004 Elsevier B.V. All rights reserved.

*Keywords:* High levels of natural radiation area; NaI(Tl) scintillation survey meter; Electronic personal dosimeter; Thermoluminescence dosimeter; Environmental radiation; Individual dose

---

\* Corresponding author. Tel.: +81 75 621 4169; fax: +81 75 612 3357.
*E-mail address:* fujinami@mbox.kyoto-inet.or.jp (N. Fujinami).

0531-5131/ © 2004 Elsevier B.V. All rights reserved.
doi:10.1016/j.ics.2004.11.082

## 1. Introduction

Some epidemiological and cytogenetic studies on inhabitants living in high levels of natural radiation areas (HLNRAs) have been performed to examine the health effects of exposure to low-dose radiation. As a part of international collaborative studies in this field, we have carried out radiological surveys in some HLNRAs. The present report deals with methods for assessing individual radiation doses of inhabitants in HLNRAs, on the basis of results of Yangjiang study in China and an experience of Ramsar survey in Iran.

## 2. Materials and methods

Measurements of absorbed dose rate in air were made at 1 m above the ground and at the earth's surface using an Aloka TCS-166 NaI(Tl) scintillation survey meter. The individual dose rates of inhabitants in HLNRAs were determined with Aloka PDM-101 electronic personal dosimeters and/or Panasonic UD-200S thermoluminescence dosimeters.

## 3. Results and discussion

China–Japan collaborative epidemiological studies on inhabitants living in HLNRAs have been performed in Yangjiang of Guangdong Province, China, since 1991 [1–4]. In these studies, Yangdong and Yangxi Counties in Yangjiang were selected as HLNRAs, and Enping County as a control area, taking into account the environmental radiation doses, where we undertook surveys of annual effective doses that these inhabitants received from natural radiation sources. Geographical averages for field radiation doses could not be used because of a large difference in radiation dose among houses. Although personal dosimetry, as well as indoor and outdoor surveys, appeared to be required, the number of dosimetric subjects was too large to determine directly each dose of all inhabitants with personal dosimeters. For the above reason, an indirect method was devised for estimating individual dose rates from ambient radiation dose rates obtained by environmental survey and occupancy factors, which were fractions of time spent in a certain place [2–4]. To examine this estimation method for its validity, we determined actual dose rates with personal dosimeters for appropriately selected families and compared them with those estimated. Fig. 1 indicates the relationship of the evaluated annual effective doses due to external terrestrial radiation between the two methods. This result means that a good correlation existed between the dose rate values obtained through our estimation and personal dosimetry. This fact enabled us to estimate individual dose rates from ambient radiation dose rates and occupancy factors. Using a similar indirect method for estimating individual doses, Sun et al. [5] estimated the excess relative risk of solid cancer mortality in these HLNRAs, Yangjiang.

In Ramsar, Iran, we surveyed radiation dose rates around several hot springs and inside six dwellings of inhabitants living in HLNRAs in September 1999 and November 2000 [6,7], in cooperation with the Iranian Nuclear Regulatory Authority. Studding was observed around springs with a high level of natural radiation. Outdoor dose rates varied largely and irregularly, even in a narrow area, while these radiation levels were not static but dynamic. As is shown in Fig. 2, there was also a large difference within indoor dose rates among rooms, as well as among houses, because of the nonuniform distribution of

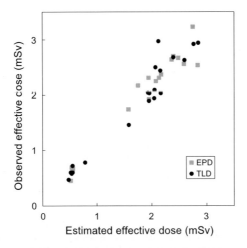

Fig. 1. Correlation of annual effective doses between those determined with personal dosimeters and those estimated from ambient radiation dose rates and occupancy factors.

travertine, which was contained in building materials. These results are consistent with those published by Sohrabi [8,9] and Sohrabi and Esmaili [10]. It appeared, therefore, too difficult to accurately estimate the individual dose rates of the inhabitants in these areas by an indirect method such as described previously, that a direct method with a personal dosimeter should be applied, especially to subjects of cytogenetic study, who are expected

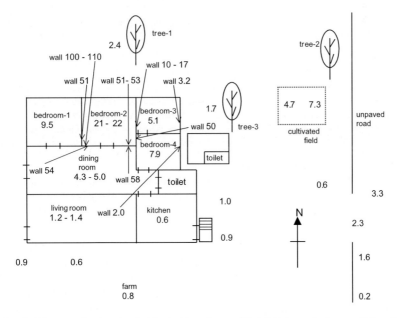

Fig. 2. Absorbed dose rate in air at 1 m height, inside and around Taleshi's house in Ramsar. Values for "walls" indicate absorbed dose rate on the surface of these (unit of measurment: μGy/h).

to have received high doses. ESR dosimetry using teeth would be necessary to determine individual cumulative doses. In addition, dosimetric assessment of inhalation exposure of inhabitants living in such houses would require measurements of radon concentration in each room, since these concentrations were very different among rooms [6,7].

## 4. Conclusions

Although it is generally desirable to determine directly the individual doses of all inhabitants in HLNRAs with personal dosimeters, in cases where it is difficult, it would be reasonable that direct and indirect methods are applied properly according to the objects of study and some conditions, such as level of individual doses and number of inhabitants.

## References

[1] L. Wei, T. Sugahara, An introductory overview of the epidemiological study on the population at the high background radiation areas in Yangjiang, China, J. Radiat. Res. 41 (2000) 1–7 (Suppl.).
[2] H. Morishima, et al., Dose measurement, its distribution and individual external dose assessments of inhabitants in the high background radiation areas in China, J. Radiat. Res. 41 (2000) 9–23 (Suppl.).
[3] Y. Yuan, et al., Recent advances in dosimetry investigation in the high background radiation area in Yangjiang, China, in: L. Wei, T. Sugahara, Z. Tao (Eds.), High Levels of Natural Radiation 1996 Radiation Dose and Health Effects. ICS-1136, Elsevier, Amsterdam, 1997, pp. 223–233.
[4] H. Morishima, et al., Study of the indirect method of personal dose assessment for the inhabitants in HBRA of China, in: L. Wei, T. Sugahara, Z. Tao (Eds.), High Levels of Natural Radiation 1996 Radiation Dose and Health Effects. ICS-1136, Elsevier, Amsterdam, 1997, pp. 235–240.
[5] Q. Sun, et al., Excess relative risk of solid cancer mortality after prolonged exposure to naturally occurring high background radiation in Yangjiang, China, J. Radiat. Res. 41 (2000) 43–52 (Suppl.).
[6] N. Fujinami, et al., Radiological survey performed in Guarapari in Brazil and Ramsar in Iran, in: J. Peter, G. Schneider, A. Bayer (Eds.), High Levels of Natural Radiation Radon Areas: Radiation Dose and Health Effects, vol. 2, Bundesamt fur Strahlenschutz, Munich, 2002, pp. 11–14. ISSN 0937-4469.
[7] M. Ghiassi-Nejad, et al., Radiological parameters for a house with high levels of natural radiation, Ramsar Iran, in: W. Burkart, M. Sohrabi, A. Bayer (Eds.), High Levels of Natural Radiation and Radon Areas: Radiation Dose and Health Effects. ICS-1225, Elsevier, Amsterdam, 2002, pp. 33–37.
[8] M. Sohrabi, Recent radiological studies of high level natural radiation areas of Ramsar, in: M. Sohrabi, J.U. Ahmed, S.A. Durrani (Eds.), Proceedings of International Conference on High Levels of Natural Radiation, Ramsar, Iran, 3–7 November 1990, IAEA Publication Series, IAEA, Vienna, 1993, pp. 39–47.
[9] M. Sohrabi, et al., Determination of Rn-222 levels in houses, schools and hotels of Ramsar by AEOI passive radon diffusion dosimeters, in: M. Sohrabi, J.U. Ahmed, S.A. Durrani (Eds.), Proceedings of International Conference on High Levels of Natural Radiation, Ramsar, Iran, 3–7 November 1990. IAEA Publication Series, IAEA, Vienna, 1993, pp. 365–375.
[10] M. Sohrabi, A.R. Esmaili, New public dose assessment of elevated natural radiation areas in Ramsar (Iran) for epidemiological studies, in: W. Burkart, M. Sohrabi, A. Bayer (Eds.), High Levels of Natural Radiation and Radon Areas: Radiation Dose and Health Effects. ICS-1225, Elsevier, Amsterdam, 2002, pp. 15–24.

www.ics-elsevier.com

# Individual dose estimation—our experience with the Karunagappally study in Kerala, India

K. Raghu Ram Nair[a],*, S. Akiba[b], V.S. Binu[c], P. Jayalekshmi[c], P. Gangadharan[c], M. Krishnan Nair[a], B. Rajan[a]

[a]*Radiation Physics Division, Regional Cancer Centre, Trivandrum, Kerala 695011, India*
[b]*Kagoshima University, Kagoshima, Japan*
[c]*Natural Background Radiation Registry (NBRR), RCC, India*

**Abstract.** A radiation cohort consisting of four panchayats and a control area consisting of two panchayats was identified in the high background radiation area of Karunagappally after measuring air kerma values of indoor and outdoor radiation levels in 75,052 houses in the area. Radioactivity in the soil was estimated from 181 samples collected from seven panchayats. Thoron-in-breath analysis was done in 87 subjects living in low, medium and high radiation level areas. Occupancy factor of 7711 subjects in different age groups was also estimated. Annual effective dose of individuals was calculated assuming the cosmic ray component of 0.227 and 0.252 mGy/year indoor and outdoor, respectively. Risk analysis will be attempted after measuring the individual dose of a sample population in the area. © 2004 Elsevier B.V. All rights reserved.

*Keywords:* Terrestrial radiation; Radiation level measurement; Environmental radiation dosimetry; Annual effective dose; Individual dose estimation

## 1. Introduction

Karunagappally (77°E, 9°N) is a Taluk (an administrative unit) 192 km$^2$ in area in the southwest coast of India in the state of Kerala with a population of 389,000. It is further sub-divided into 12 panchayats, each having an area in the range of 7–25 km$^2$. The coastal part of this area is around 25 km long and the soil content in this part has 1% monazite, which contains 8–10% $^{232}$Th [1]. For the first time a comprehensive survey of

* Corresponding author. Tel.: +91 471252290; fax: +91 4712447454.
*E-mail address:* raghurkn@rcctvm.org (K. Raghu Ram Nair).

0531-5131/ © 2004 Elsevier B.V. All rights reserved.
doi:10.1016/j.ics.2004.10.030

the entire taluk (administrative unit) was undertaken by the Regional Cancer Centre in 1990. The following radiation measurements were carried out in a 7-year period till 1997 along with collection of socio demographic profile and cancer incidence in the population.

## 2. Radiation level measurement: scintillometer survey and correlation with TLD

Seventy-five thousand fifty-two (75,052) houses in the taluk were covered using scintillometers. The reading was taken in the centre of the main room of the house at a height of 1 m above the ground. Another reading outside the house at a distance of 3 m away from the wall of the house and at a height of 1 m above the ground was also taken. The scintillometer reading was correlated with TLD measurement using natural calcium fluoride as the phosphor. About 50 mg of the powder was dispensed inside a brass capsule having 3 mm diameter, 8 mm length and a wall thickness of 1.5 mm. Two such capsules were placed inside a plastic locket with a wall-thickness of 2 mm. The locket was placed inside the house taped below the cot or table at a height of 50–100 cm above the ground in a room having maximum occupancy. The dose was measured in quarterly integrating cycles for the whole year at the same place. This was done in 800 randomly selected houses in the taluk. A scintillometer reading was taken at the same height at the same house every quarter. The mean of the four spot readings ($\mu R/h$) was converted to annual dose (mGy/year) and correlated with the TLD dose.

The final outdoor and indoor reading of all the houses in the taluk was obtained by converting the scintillometer spot reading to TLD equivalent annual dose (mGy/year). Outdoor radiation level was less than 2 mGy/year in 37,167 (49.5%) houses. Of the remaining houses, 26,669 (35.5%) are in the range of 2–4.9 mGy/year, 7973 (10.6%) houses are in the range of 5–9.9 mGy/year and 1849 (2.5%) houses are in the range of 10–14.9 mGy/year. There are 1394 (1.9%) houses at 15 mGy/year and above of which 670 (0.9%) are above 20 mGy/year.

Ward-wise dose distribution from low, medium and high radiation level areas in the taluk show that the distribution is not uniform but shows high and low values in all the three areas. This introduces uncertainties in individual dose estimation.

In the higher radiation level areas, the indoor radiation level was lower than outdoor. This shows that the construction material of the walls offers a limited protection to radiation; concrete and brick walls offering maximum and thatched walls minimum.

## 3. Soil analysis

The 192 km$^2$ area was divided into 1 km$^2$ grids. Soil samples were collected from the four corners and the centre of each grid using a 30-cm-long auger having a diameter of 6 cm. The collected samples were mixed thoroughly, dried, sieved and transferred into 275 ml cylindrical plastic containers. The containers were sealed airtight and kept for 1 month to attain radioactive equilibrium. $^{232}$Thorium (Th), $^{238}$Uranium (U) and $^{40}$Potassium (K) content of the samples was analysed using a 10 cm dia Na I (Tl) detector and 4 K MCA. From the concentration of Th, U and K present in the soil sample, the absorbed dose rate in air due to the three radionuclides and their daughters

Table 1
Thoron-in-breath analysis of 87 subjects living in low, medium and high radiation areas

| Indoor radiation level, mGy/year | No. of subjects | Mean Th burden, Bq |
|---|---|---|
| <1.2 (mean 0.72) | 16 | 6.3 |
| 1.2<3.8 (mean 2.04) | 33 | 8.9 |
| >3.8 (mean 15.64) | 38 | 10.91 |
| $p=0.0094$ | | - |

was calculated using the concentration-to-dose conversion factors for the three radionuclides in nGy/h/Bq/kg [2].

A total of 181 soil samples from seven panchayats were analysed. The concentration of Th, U and K was calculated in Bq/kg and compared with the scintillometer reading at 1 m above the ground level at the site of sample collection. A correlation ($r=0.9524$) was obtained between the two.

Th is the dominant fraction in all the samples irrespective of radiation levels. The contribution of K to the total dose is only 3% for low level areas. For the high background areas there is no contribution from K at all. The contribution of dose from K in the soil in the seven panchayats of Karunagappally taluk is only 1%.

The soil activity in terms of Bq/kg was analysed to assess its composition. In the low level areas, the soil contained 24%K, 18% U and 58% Th; in the medium level areas the composition was 12%, 15% and 73%, respectively, and in the high background areas the levels were 3% K, 15% U and 82% Th. The overall composition showed 12% K, 15% U and 73% Th.

## 4. Measurement of Thoron-in-breath

Subjects living in the coastal Panchayats where Th content in the soil is very high can ingest Th through food and water. A part of the Thoron emanating from this Th migrates to the lung via blood and come out in the exhaled breath. So a breath analysis for Thoron will give information regarding the ingested level of Th in the subjects.

Table 2
Fraction of time spent/day indoor and outdoor for different age groups (3783M+3928F=7711)

| Age, years | Male | | Female | | Age, years | Male | | Female | |
|---|---|---|---|---|---|---|---|---|---|
| | In | Out | In | Out | | In | Out | In | Out |
| 0– | 0.87 | 0.13 | 0.89 | 0.11 | 40– | 0.50 | 0.50 | 0.66 | 0.34 |
| 1– | 0.76 | 0.24 | 0.76 | 0.24 | 45– | 0.50 | 0.50 | 0.67 | 0.33 |
| 5– | 0.63 | 0.37 | 0.64 | 0.36 | 50– | 0.52 | 0.48 | 0.68 | 0.32 |
| 10– | 0.59 | 0.41 | 0.62 | 0.38 | 55– | 0.55 | 0.45 | 0.68 | 0.32 |
| 15– | 0.57 | 0.43 | 0.64 | 0.36 | 60– | 0.60 | 0.40 | 0.68 | 0.32 |
| 20– | 0.53 | 0.47 | 0.67 | 0.33 | 65– | 0.61 | 0.39 | 0.70 | 0.30 |
| 25– | 0.51 | 0.49 | 0.69 | 0.31 | 70– | 0.65 | 0.35 | 0.70 | 0.30 |
| 30– | 0.50 | 0.50 | 0.67 | 0.33 | 75+ | 0.71 | 0.29 | 0.77 | 0.23 |
| 35– | 0.50 | 0.50 | 0.66 | 0.34 | All | 0.57 | 0.43 | 0.68 | 0.32 |

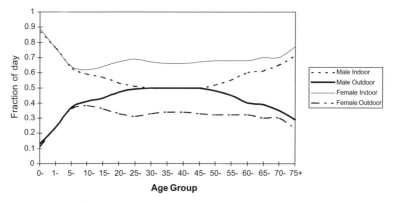

Fig. 1. Fraction of time spent/day indoor and outdoor for different age groups (3783M+3928F=7711).

Subjects from high, medium and low background radiation areas were included in the study. Two persons each, preferably husband and wife from a house, were selected for the study. The measurement was done by the Double Filter (DF) system [3,4].

Eighty-seven subjects were studied using this technique. The results show that the subjects from high and medium indoor radiation level areas show a significant increase ($p=0.0094$ ANOVA) in the Th body burden as compared to low indoor radiation level area (Table 1). This implies that ingestion of Th takes place in subjects living in high radiation level areas and the need for calculating the ingested dose while computing annual and cumulative doses.

## 5. Identification of radiation cohort and calculation of the annual effective dose

From the air kerma measurements carried out, four panchayats, viz. Chavara, Neendakara, Alappad and Panmana, were identified as high background radiation areas. This constitutes the radiation cohort. Oachira and Thevalakkara Panchayats were designated as Control Area. A survey was carried out to assess the house occupancy factor by interviewing all inmates in all age groups of every 50th house in the area. A total of 3783 males and 3928 females (Total 7711) were interviewed for the study. The fraction of time spent indoor and outdoor was estimated for both sexes in all age groups. Assuming the air kerma values for the cosmic ray component as 0.227 mGy/

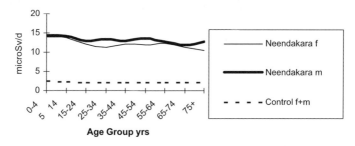

Fig. 2. Variation in the dose rate for male and female in the high background panchayat of Neendakara in comparison to the control area.

Table 3
No. of subjects in radiation cohort and control panchayats in different age groups

| Age, years | Radiation cohort | Control | Total | % | Age, years | Radiation cohort | Control | Total | % |
|---|---|---|---|---|---|---|---|---|---|
| 0–4 | 9859 | 4405 | 14,264 | 8.4 | 45–54 | 8925 | 5189 | 14,114 | 8.3 |
| 5–14 | 24,379 | 11,390 | 35,769 | 21.2 | 55–64 | 7593 | 4158 | 11,751 | 6.9 |
| 15–24 | 23,158 | 11,500 | 34,658 | 20.4 | 65–74 | 5101 | 2833 | 7934 | 4.7 |
| 25–34 | 18,542 | 8352 | 26,894 | 15.9 | 75+ | 1827 | 1304 | 3131 | 1.8 |
| 35–44 | 14,012 | 7103 | 21,115 | 12.4 | Total | 113,396 | 56,234 | 169,630 | 100- |

year for indoor and 0.252 mGy/year for outdoor, the annual effective dose was calculated using the formula:

$$\text{Annual effective dose (mSv)} = \{[\text{indoor dose/year} - 0.227] \times \text{OFindoor}$$
$$+ [\text{outdoor dose/year (median) of panchayat}$$
$$- 0.252] \times \text{OFoutdoor}\} \times \text{CF}$$

OF—Occupancy factor; CF—absorbed dose to effective dose conversion factor [2].

Table 2 shows the occupancy factor calculated for different age groups for both sexes. Fig. 1 shows the variation in the time spent/day indoor and outdoor for both sexes in different age groups. Fig. 2 shows the variation in dose rate/day in males and females in the high background panchayat of Neendakara in the radiation cohort in comparison to that in the control area. Males are exposed to more doses than females as the former spends more time outdoors. Table 3 shows the number of subjects in different age groups in the radiation cohort and control areas. More than 70% of the subjects are in the age group of 5–35 years.

The calculated effective dose has to be verified with measured individual dose before attempting risk analysis estimates.

## References

[1] WHO Technical Report Series 166,1959.
[2] UNSCEAR, 1993.
[3] P. Kotrappa, et al., Modified double filter system for measuring Radon–Thoron in the environment and in exhaled breath, Advances in Radiation Protection Monitoring IAEA-SM-229/31, Vienna, 1979.
[4] Y.S. Mayya, et al., Measurement of 220Rn in exhaled breath of Th plant workers, Health Phys. 51.6 (1986) 737–744.

# Indoor radon, thoron, and thoron daughter concentrations in Korea

Y.J. Kim[a,*], H.Y. Lee[a], C.S. Kim[a], B.U. Chang[a], B.H. Rho[a], C.K. Kim[b], S. Tokonami[c]

[a]*Korea Institute of Nuclear Safety, 19 Gusong-dong, Yusong, Taejon, South Korea*
[b]*International Atomic Energy Agency, Austria*
[c]*National Institute of Radiological Sciences, Japan*

**Abstract.** A nationwide survey for radon ($^{222}$Rn), thoron ($^{220}$Rn), and thoron daughters has been conducted. The arithmetic mean of annual radon concentration in Korean dwellings was $53.4 \pm 57.5$ Bq/m$^3$. The indoor radon concentration showed a lognormal distribution with a geometric mean and its standard deviation of $43.3 \pm 1.8$ Bq/m$^3$. The arithmetic mean of annual thoron and its progeny concentrations in Korean dwellings were $45.2 \pm 110.1$ and $0.99 \pm 1.00$ Bq/m$^3$, respectively. The equilibrium factor between thoron and its progeny was 0.022. The arithmetic mean indoor gamma dose rate was $212 \pm 52$ nGy/h. The radon concentrations in the traditional and modern style houses were about two times higher than those in apartments. The average annual effective doses to the general public from radon, thoron, and gamma dose rate were 1.35, 0.23, and 1.03 mSv/y, respectively. © 2004 Elsevier B.V. All rights reserved.

*Keywords:* Radon; Thoron; Thoron daughter; Annual effective dose

## 1. Introduction

The presence of radon ($^{222}$Rn) and thoron ($^{220}$Rn) in indoor environment constitutes a major health hazard for man [1]. In the viewpoint of public health protection from radon and thoron, many surveys [2–4] have been performed to estimate indoor radon, thoron, and their daughter concentrations in many countries. In Korea, some studies [5–7] on the

---

* Corresponding author. Tel.: +82 42 868 0464; fax: +82 42 868 0556.
  *E-mail address:* k337kyj@kins.re.kr (Y.J. Kim).

0531-5131/ © 2004 Elsevier B.V. All rights reserved.
doi:10.1016/j.ics.2004.11.161

concentrations of radon in homes have been carried out. However, there has been no study on indoor thoron and thoron daughter concentration.

In this study, a nationwide survey for radon ($^{222}$Rn), thoron ($^{220}$Rn), and thoron daughters has been conducted to provide data on the annual average indoor radon, thoron, and thoron daughter concentration in Korean dwellings and to estimate the effective dose to the general public.

## 2. Materials and methods

For the nationwide survey of indoor radon, thoron, and thoron daughter concentrations, six cities and nine provinces were selected as its measurement areas, covering representatively the Republic of Korea, taking into consideration the administration boundary. The sample size for the nationwide radon survey was 2603 dwellings during from 1999 to 2000 [7] and 970 dwellings during from 2002 to 2004. For thoron and thoron daughters survey, 450 dwellings were surveyed from 2002 to 2004.

A typical traditional Korean house has walls built of mud block coated with cement and has a ceiling of hardboard and the roof covered with roof tiles made of mud. On the other hand, modern style houses have walls built with concrete block, the roof and ceiling built with concrete slab.

The radon survey in Korea was carried out from December 1999 to November 2000, covering four seasons [7]. A passive nuclear track detector (RadTrak®, Landauer, USA) was placed in the bedroom of target houses for four 3-month periods. To reduce the effect of thoron on the measured radon concentrations, the detector was installed at a position over 1 m from the wall.

Table 1
Indoor radon concentrations in dwellings of Korea (unit: Bq/m$^3$)

| Survey boundary | Radon | | Thoron | | Thoron daughter | |
|---|---|---|---|---|---|---|
| | Arithmetic mean | Geometric Mean | Arithmetic mean | Geometric Mean | Arithmetic mean | Geometric Mean |
| Seoul | 45.1±66.5 | 36.0±1.7 | 51.0±159.1 | 10.7±6.8 | 0.95±0.81 | 0.72 |
| Busan | 38.6±22.8 | 34.3±1.6 | – | – | – | – |
| Taegu | 41.4±22.4 | 36.9±1.6 | – | – | – | – |
| Gwangju | 57.0±31.2 | 50.2±1.7 | – | – | – | – |
| Taejon | 52.9±37.6 | 45.5±1.7 | 32.2±38.7 | 13.8±5.3 | 0.74±0.69 | 0.56 |
| Incheon | 38.2±20.2 | 34.3±1.6 | 31.2±59.3 | 9.2±6.0 | 1.04±1.81 | 0.70 |
| Gyeonggi | 51.6±36.8 | 43.9±1.7 | 34.7±50.5 | 10.3±6.2 | 0.91±0.85 | 0.69 |
| Gangwon | 72.5±56.6 | 59.0±1.9 | 46.4±79.6 | 14.8±6.7 | 0.87±0.86 | 0.66 |
| Chungbuk | 80.5±146.0 | 56.0±2.0 | 64.5±192.6 | 15.5±6.8 | 1.12±1.20 | 0.80 |
| Chungnam | 74.8±72.2 | 59.2±1.9 | 49.3±134.6 | 14.7±5.8 | 1.01±0.92 | 0.81 |
| Jeonbuk | 68.8±82.9 | 53.7±1.9 | 53.9±78.7 | 20.2±5.9 | 1.28±1.01 | 0.99 |
| Jeonnam | 61.3±42.4 | 51.5±1.8 | – | – | – | – |
| Gyeongbuk | 49.5±31.1 | 42.6±1.7 | – | – | – | – |
| Gyeongnam | 52.2±41.9 | 42.8±1.8 | – | – | – | – |
| Jeju | 58.9±48.8 | 47.3±1.9 | – | – | – | – |
| Nationwide mean | 53.4±57.5 | 43.3±1.8 | 45.2±110.1 | 12.7±6.3 | 0.99±1.00 | 0.73 |

"–" are under surveying.

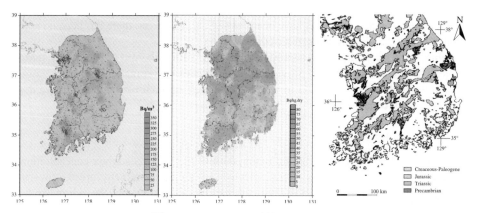

Fig. 1. Distribution maps of indoor $^{222}$Rn concentration (left), $^{226}$Ra concentration in surface soil (center), and old granite in Korea (right).

The survey for radon, thoron, and thoron daughters on three cites and five provinces was carried out from June 2002 to March 2004, and the survey on the remaining three cites and four provinces will be finished until December 2004 with Radopot® (Hungary) and thoron progeny monitor (Japan), which were placed in the bedroom of each home for four 3-month periods. To measure background gamma radiation, three thermoluminescent dosimeters (TLDs) were also placed at the same place.

## 3. Results and discussions

The indoor radon concentration showed a lognormal distribution, and the arithmetic annual mean radon concentration was 53.4 Bq/m$^3$, with 1.7% above 200 Bq/m$^3$ of the recommended value in ICRP 65 (1993).

The regional distributions of indoor radon, thoron, and thoron daughter concentrations were summarized in Table 1. Some areas of the country, particularly Chungbuk, Gangwon,

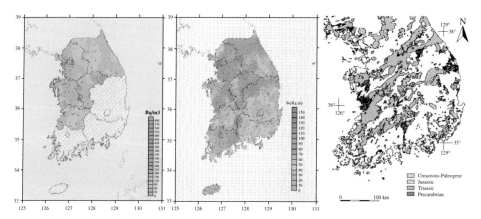

Fig. 2. Distribution maps of indoor $^{220}$Rn (Tn) concentration (left), $^{232}$Th concentration in surface soil (center), and old granite in Korea (right). The blank areas are under surveying.

Table 2
Indoor radon, thoron, and thoron daughter concentrations with type of dwellings, and average annual effective doses

| Type of dwellings | Annual mean concentration (Bq/m$^3$) | | | Effective dose (mSv/y) | | |
|---|---|---|---|---|---|---|
| | Radon | Thoron | Thoron daughters | Radon | Thoron | Gamma dose rate |
| Traditional style | 68.6±78.6 | 56.2±154.1 | 1.06±1.03 | 1.73 | 0.28 | 1.02 |
| Modern style | 71.4±73.6 | 44.1±101.4 | 0.92±0.97 | 1.80 | 0.22 | 1.10 |
| Apartment | 36.0±16.1 | 33.7±40.9 | 1.04±1.05 | 0.91 | 0.17 | 1.91 |
| Nationwide mean | 53.4±57.5 | 45.2±110.1 | 0.99±1.00 | 1.35 | 0.23 | 1.03 |

and Jeonnam provinces, showed relatively high radon, thoron, and thoron daughter concentrations.

Figs. 1 and 2 illustrate distribution maps of indoor radon and thoron concentrations. The locations of high indoor radon and thoron concentration are in accordance with the distribution maps of $^{226}$Ra and $^{232}$Th concentrations in surface soil, respectively, and granite.

Table 2 gives indoor radon, thoron, and thoron daughter concentrations with type of dwellings and average annual effective doses. The Korean traditional style and modern style houses showed relatively high concentration of indoor radon and thoron, compared with apartment. This result may be due to the differences of building materials and structure. According to preliminary experiments, $^{226}$Ra and $^{232}$Th concentrations in mud was much higher than that in other building materials. Most of the apartment complex are high-rise buildings and have a parking lot in the underground, where the ventilation of air is good. Consequently, lower radon and thoron concentrations in apartments seem to be explained by their building materials and ventilation rates.

The annual effective doses aroused from indoor radon, thoron, and gamma dose rate were 1.35, 0.23, and 1.03, respectively. The inhalation of radon and thoron accounts for about 60% of the total natural radiation dose to the public.

## References

[1] H.S. Virk, S. Navjeet, Indoor radon/thoron survey report from Hamirpur and Una districts, Himachal Pradesh, India, Appl. Radiat. Isot. 52 (2000) 137–141.
[2] M. Doi, et al., Spatial distribution of thoron and radon concentrations in the indoor air of a traditional Japanese wooden house, Health Phys. 66 (1994) 43–49.
[3] Nation Radiological Protection Board, Radon in dwellings in England: 1997 Review, NRPB-R293:1997.
[4] United Nations Scientific Committee on the Effects of Atomic Radiation (UNSCEAR), Sources and effects of ionizing radiation, United Nations, New York, 2000.
[5] H.Y. Lee, et al., Risk analysis of population exposure to the alpha-radiation (IV), Rep. Natl. Inst. Health 31 (2) (1994) 543 (in Korean).
[6] J.H. Kim, et al., A nationwide survey of indoor radon measurement in Korea (II), Annu. Rep. KFDA 1 (1997) 313–320 (in Korean).
[7] C.K. Kim, et al., Nationwide survey of radon levels in Korea, Health phys. 84 (2003) 354–361.

www.ics-elsevier.com

# The Spanish experience on HBRA

L.S. Quindos[a,*], P.L. Fernández[a], C. Sainz[a], J. Gomez[a], J.L. Matarranz[b], E. Suarez Mahou[b]

[a] Department of Medical Physics, Faculty of Medicine, University of Cantabria, C/ Cardenal Herrera Oria s/n 39011 Santander, Spain
[b] Nuclear Safety Council, c/ Justo Dorado 11, 28040 Madrid, Spain

**Abstract.** During the last two decades, the Department of Medical Physics of the University of Cantabria has been involved in the development of a Radiation Protection Programme throughout Spain. In the framework of this Programme, over 5000 measurements of indoor radon were carried out nationwide. A geometric mean radon concentration in air of 45 Bq m$^{-3}$ with a standard deviation of 2.7 and a range of variation from 10 to 15,400 Bq m$^{-3}$ were found. After that, several regional surveys were conducted to determine exposure to natural sources of radiation for people living in the vicinity of the old Spanish Uranium Mines and the Spanish Nuclear Power Stations as well as for those living in the populated areas of Sierra de Guadarrama close to Madrid city and Villar de la Yegua town. This paper summarizes the main results obtained from the measurements performed paying special attention to those concerning the High Background Radiation Areas of the country. © 2004 Elsevier B.V. All rights reserved.

*Keywords:* Dose; Radon; Radiation; Survey

## 1. Introduction

Environmental radioactivity, in special radon measurements, and radiation protection have been the subjects on which the Department of Applied and Medical Physics of the University of Cantabria (Santander, Spain) has focussed his research efforts during the last 20 years. From a nationwide point of view, it could be said that the Spanish Radon Programme began in 1988 with the development of a national survey in Spanish houses [1]. The data coming from the 2000 measurements carried out in this survey represented a valuable basis to face rigorously the radon issue in Spain.

* Corresponding author. Tel.: +34942201974; fax: +34942201991.
*E-mail address:* quindosl@unican.es (L.S. Quindos).

0531-5131/ © 2004 Elsevier B.V. All rights reserved.
doi:10.1016/j.ics.2004.09.051

In addition, since 1991 the Spanish Nuclear Safety Council, together with the National Uranium Company and some Universities have developed the so-called MARNA project [2]. This project is a nationwide study with the aim of estimating potential radon emission from external gamma dose rates and radium calculations taking into consideration geological parameters and empirical correlations found between outdoor external gamma dose rates and $^{226}$Ra concentration in soil.

Bearing in mind the information obtained from both studies, several regional surveys were conducted to get a more detailed knowledge of the exposure to natural sources of radiation for people living in specific areas. Not only indoor radon but also external gamma dose rate and radioactivity in soils were measured in the vicinity of the old Spanish Uranium Mines and of the Spanish Nuclear Power Stations, as well as in populated areas as Villar de la Yegua Town and the named Sierra de Guadarrama.

## 2. Methods

Exposure to natural sources of radiation to people living in the mentioned areas was determined from indoor radon measurements, external gamma dose rate and radioactivity in soils measurements. In all the surveys described in this paper, the following procedures were carried out.

Indoor radon measurements were performed by using track etched detectors CR-39 exposed for a 6-month period in order to evaluate average radon concentration values. In all the measurements, a seasonal correction factor was assumed in order to make the results obtained over a 6-month period representative of the actual mean annual indoor radon concentration [3].

External gamma radiation was measured with a Mini-Instruments Environmental Monitor type 6-80 with an energy compensated Geiger Muller tube MC-70, specially designed to measure environmental levels of gamma.

The radioactivity in soil determinations were focused on measuring the activity concentrations of $^{226}$Ra, $^{232}$Th and $^{40}$K. All the soil samples were dried in an oven at 100 °C for 24 h, reweighed to determine the water content, sieved to remove stones and pebbles and crushed to pass through a 1-mm mesh sieve. Finally, the prepared sample was packed in a sealable 250 mL PVC can and left for at least 4 weeks before counting by gamma espectrometry in order to ensure that radioactive equilibrium between $^{226}$Ra, $^{222}$Rn, and short-lived radon progeny was reached. Gamma espectrometry measurements were made using a low background HPGe detector with a relative efficiency of 20% and a resolution of 1.86 keV at 1.33 MeV.

## 3. Description of surveys and results

### 3.1. Surroundings of the Spanish Nuclear Power Plants

During 1998 and 1999, financially supported by the Spanish Nuclear Safety Council, regional surveys were conducted to evaluate natural radiation exposure of the people living in the vicinity of the Spanish nuclear power stations. There are six facilities working in the country and the population of these regions is about 200,000.

A remarkable result was found in the surroundings of Almaraz nuclear power plant in the province of Caceres where the highest mean annual effective dose to the population was found [4]. The estimated value, of 4.07 mSv year$^{-1}$, is 1.6 times higher than the national average value. The reason of this significant difference in dose value with the other nuclear power stations was the high radon concentrations found in homes. Thus, in order to perform a more accurate assessment of the dose coming from radon in this area, a new and more extensive survey on indoor radon was carried out in the named Campo Arañuelo region around the Almaraz nuclear power station. This study revealed the presence of a high radon level area called La Vera in the northern side of the Campo Arañuelo region. It was found in La Vera a 9% of houses with indoor radon concentrations higher than 400 Bq m$^{-3}$. In addition, the new dose assesment gave a value of 6 mSv per year in La Vera, with a maximum of 25 mSv per year estimated in Jarandilla, a town belonging to the former area where 30% of houses had radon concentrations over 400 Bq m$^{-3}$.

## 3.2. Vicinity of the Spanish old Uranium Mines

From 2000 to 2001 and under the sponsorship of the Spanish Nuclear Safety Council, the surveys in the six uranium-mining areas in the country were carried out. The exploitation period ranged from 1950 to 1980, and between 1987 and 1996, a general decommissioning plan was carried out. One of the main objectives of the plan was to reduce and control radon flow and contamination of water. The population of these areas is over 400,000 inhabitants.

The highest geometric mean radon concentration and annual effective dose for natural sources, of 111 Bq m$^{-3}$ and 5.1 mSv year$^{-1}$, respectively, were found in the surroundings of Albala uranium mine [5]. Estimated mean annual effective doses for the six areas studied ranged between 3.2 and 5.1 mSv per year, which is between 1.2 and 2 times higher than the national average value. A 14% of houses over 400 Bq m$^{-3}$ were found in the vicinity of the Albala uranium mine.

## 3.3. Sierra de Guadarrama

The area called Sierra de Guadarrama situated in the North of the province of Madrid have been subject of another regional survey. The first national study showed high percentages of houses with radon concentrations higher than 200 and 400 Bq m$^{-3}$ (European Union recommendation concerning radon concentrations in new and old houses, respectively) [6]. Due to the prevalence of granitic rocks in the soil composition of this area and the considerable residential growth (the population of this region has been increased from about 500,000 people in 1990 to 1.5 million in 2000) a regional survey is now ongoing from 2002.

Until now, the indoor radon measurements indicate that the 14% of houses have levels above 400 Bq m$^{-3}$ and 30% above 200 Bq m$^{-3}$. In addition, the geometric mean radon concentration is 180 Bq m$^{-3}$ which is about four times higher than the national average value.

## 3.4. Villar de la Yegua town

Perhaps the most important high background radiation area in Spain is the Villar de la Yegua town where the highest indoor radon concentration, up to 25,000 Bq m$^{-3}$, have been found, and effective doses coming from natural sources as high as 40 mSv per year has been estimated. Several surveys have been carried out from 1988 to now, confirming Villar de la Yegua as a high radon level area [7]. The main results concerning radon concentration show a geometric mean of 818 Bq m$^{-3}$, 18.2 times higher than the national value, and percentages of houses with concentrations above 400 and 1000 Bq m$^{-3}$, of 75% and 25%, respectively.

## 4. Conclusions

The results of the presented surveys will be used as a data base source for the development of even more specific studies in the regions where high radon levels have been detected in order to get a better knowledge of the sources of radon in houses and decide if countermeasures for radon remediation in dwellings should be taken.

The comparison between the MARNA Project's predicted map of potential indoor radon concentration and those obtained from the abovementioned surveys shows a good correlation [8]. This agreement has been found on a national basis (scale 1:1,000,000) showing the interest of maps as those from MARNA Project when data as shown are available, minimizing costs and efforts in the development of national radon programme.

Nevertheless, it is interesting to test the correlation between predicted and measured average indoor radon concentration for lower scales (1:50,000). To this end, a collaborative project between the research groups MARNA (Nuclear Safety Council) and RADON (University of Cantabria) is now ongoing in a geographical area for which enough geological information exists and present-day measurements of radon concentrations in the air are available.

## References

[1] L.S. Quindós, P.L. Fernández, J. Soto, National survey on indoor radon in Spain, Environ. Int. 17 (1991) 449–453.
[2] E. Suarez, J.A. Fernández, Project MARNA: natural gamma radiation map, Rev. Soc. Nucl. Esp. (1997) 58–65.
[3] J. Miles, C. Howarth, Memorandum: validation scheme for laboratories making measurements of radon in dwellings: 2000 revision. National Radiological Preotection Board. NRPB-M1140. (2000). Chilton, Didcot, Oxfordshire OX11.
[4] L.S. Quindós, et al., Natural radiation exposure in the vicinity of Spanish nuclear power stations, Health Phys. 85 (5) (2003) 594–598.
[5] L.S. Quindós, et al., Population dose in the vicinity of old Spanish uranium mines, Sci. Total Environ. 329 (2004) 283–288.
[6] European Union, Council Directive 90/143/EC of 21 February 1990 on the protection of the public against indoor exposure to radon, Off. J. Eur. Communities (1990).
[7] International Commission on Radiological Protection, Protection against radon-222 at home and at work, ICRP Publication, vol. 65, Pergamon Press, Oxford, 1994.
[8] L.S. Quindós, et al., Natural gamma radiation map (MARNA) and indoor radon levels in Spain, Environ. Int. 29 (2004) 1091–1096.

www.ics-elsevier.com

# Lung cancer risk due to radon in dwellings— evaluation of the epidemiological knowledge

H.E. Wichmann[a,b,*], A. Schaffrath Rosario[a], I.M. Heid[a], M. Kreuzer[c], J. Heinrich[a], L. Kreienbrock[d]

[a]*GSF National Research Center, Institute of Epidemiology, Neuherberg, Germany*
[b]*Ludwig-Maximilians-University, Chair of Epidemiology, Munich, Germany*
[c]*BfS-Federal Office for Radiation Protection, Neuherberg, Germany*
[d]*Hannover School of Veterinary Medicine, Institute for Biometry, Epidemiology and Information Processing, Hannover, Germany*

**Abstract.** Studies on underground miners have shown an increased risk of lung cancer after exposure to high doses of radon and radon daughters. However, in the range of typical indoor radon exposures, there are still uncertainties on the strength of the association between radon exposure and the risk of lung cancer. Since 1990, population-based epidemiological studies have been performed in the USA, Canada, Sweden, Finland, Germany, Czech Republic, Austria, France, Italy, Spain, and China. Some of them did not find an effect, others have shown a statistically significant association of radon exposure and lung cancer risk. The reason for this heterogeneity is found in the different study designs, different ways of measuring radon, different ranges of radon concentrations, varying information on confounding factors, but most importantly limited sample size in most of the individual studies. Therefore, pooling studies are performed now in North America and Europe, which are close to be finished. © 2004 Elsevier B.V. All rights reserved.

*Keywords:* Radon; Lung cancer; Epidemiology; Case-control study

## 1. Introduction

Based on studies of uranium miners, it is well established that the radioactive gas Rn222 and its progeny are human carcinogens [1,2]. The literature on individual case-control

---

* Corresponding author. GSF National Research Center, Institute of Epidemiology, Ingolstaedter Landstr. 1, D-85764 Neuherberg, Germany. Tel. +49 89 3187 4066; fax +49 89 3187 4499.
  *E-mail address:* wichmann@gsf.de (H.E. Wichmann).

0531-5131/ © 2004 Elsevier B.V. All rights reserved.
doi:10.1016/j.ics.2004.10.010

studies published until 2003 [3–24] shows a broad spectrum of results. Meanwhile further studies have been published or are in press [25–27]. The reason for the broad spectrum of results is found in different study designs, different radon measurements, different ranges of radon concentrations, but most important small sample size in most of the studies. Combined analysis may overcome at least the problem of sample size [9,15,25].

In the following, the most important findings of the pooled analysis of the two German studies are summarized [27]. This pooling is unique because the original studies had an identical study design.

## 2. Methods

The studies were performed in West Germany from 1990 to 1996 [11,22] and in East Germany from 1990 to 1997 [12,23]. The pooled analysis uses an extended database containing 2963 cases and 4232 controls. Incident lung cancer cases with a histologically or cytologically confirmed diagnosis were identified via hospitals. Controls were selected from the German registration offices and frequency-matched to the cases by sex, age and region. All cases and controls were interviewed face-to-face. Radon measurements were carried out for 1 year using alpha-track detectors in the living room and the bedroom in houses occupied in the last 5–35 years.

Odds ratios (OR) and asymptotic 95% confidence intervals (CI) were calculated via conditional logistic regression. Linear trends per 100 $Bq/m^3$ were estimated and expressed as excess odds ratios (EOR). All risk estimates were adjusted for smoking and asbestos exposure, the two main confounders.

## 3. Results

A statistically significant elevated OR was found for radon concentrations above 140 $Bq/m^3$ with an OR of 1.40 (CI 1.03 to 1.89) (see Table 1). The overall excess odds ratio (EOR) per 100 $Bq/m^3$ was 0.10 (CI −0.02 to 0.30).

Table 1
Lung cancer risk due to indoor radon exposure, based on time-weighted average radon exposure in the last 5–35 years, pooled German case-control data (2963 cases, 4232 controls), 1990–1997[27]

|  | No. of cases/controls | Adjusted relative risk* | 95% Confidence interval |
|---|---|---|---|
| All subjects |  | Odds ratio |  |
| <50 $Bq/m^3$ | 1532/2177 | 1.00 | Reference/1 |
| 50–79 $Bq/m^3$ | 997/1498 | 0.97 | 0.85, 1.11 |
| 80–139 $Bq/m^3$ | 314/410 | 1.06 | 0.87, 1.30 |
| ≥140 $Bq/m^3$ | 120/147 | 1.40 | 1.03, 1.89 |
| Trend estimate per 100 $Bq/m^3$ |  | Excess odds ratio |  |
| All subjects | 2963/4232 | 0.10 | −0.02, 0.30 |

\* Categorical relative risks estimated by odds ratio in a conditional logistic regression, stratifying for age, sex and region, and adjusting for smoking by log (packyears+1), three categories for ex-smoking and a binary indicator for smoking of other tobacco products, and for a binary indicator of occupational asbestos exposure. Trend estimate is the excess odds ratio of lung cancer per 100 $Bq/m^3$ increase in radon exposure, estimated in a conditional linear risk model using the same stratification and adjustment variables.

If the analysis was restricted to less mobile participants (which occupied only one home during the 5–35 years before interview), the EOR per 100 Bq/m$^3$ increased to 0.14 (CI −0.03 to 0.55). The increased risk nearly completely depended on the analysis of the bedroom measurements with EOR=0.13 (CI 0.00 to 0.33), whereas the analysis of the living room measurements showed a negligible risk. If the study was analysed by histological subtype, the strongest risk was found for small cell carcinoma (EOR=0.29, CI 0.04 to 0.78). For squamous cell carcinoma and other histologic types, the risk estimate was lower (EOR=0.10, upper CI limit 0.46, and EOR=0.17, CI −0.08 to 1.14), and for adenocarcinoma, no increase in risk was observed (EOR=0.00, upper CI limit 0.25).

## 4. Discussion

An excess relative risk per 100 Bq/m$^3$ of EOR=0.10 was found in this study which is very close to the result of the meta-analysis of eight studies with EOR=0.09 [15], the result of the pooled analysis of seven North American studies EOR=0.11 [25]. Furthermore, all these results from indoor studies fit nicely to the predicted EOR=0.12 based on data of low exposed miners [1]. This impressively demonstrates that the heterogeneity found in many smaller individual studies mainly was due to lack of sufficient power.

The observed increase in risk estimates when restricting the analysis to subjects living only in one home in the last 5–35 years or by using bedroom measurements can be explained by a reduced error in exposure assessment. A reduction of measurement error can also possibly be achieved by analysing glass objects in the household via surface monitors [28,29]. Here an even higher EOR was found [4]. However, it is still open how these measurements are influenced by the presence of tobacco smoke [30].

With respect to histological subtypes, the German analysis reproduces what has been found in other indoor studies [8,25] as well as in studies on uranium miners [1,24,31,32] with a higher risk for small cell carcinoma. This is especially true in comparison to adenocarcinoma, for which no increase in risk was found.

In conclusion, the German pooling study supports the result of the North American pooling study [25] qualitatively and quantitatively. The European pooling study in which more than 7000 cases and more than 14,000 controls will be included [9] is on the way. Finally, a worldwide pooling project will follow with about 12,000 cases and 21,000 controls, which is expected to be finished in 2006.

## References

[1] BEIR VI, National Research Council, Health effects of exposure to radon, BEIR VI. Committee on Health Risks of exposure to radon (BEIR VI), Board on Radiation Effects Research, Commission on Life Science, National Academy Press, Washington, DC, 1999.
[2] IARC, IARC Monographs On The Evaluation Of Carcinogenic Risks To Humans, Ionizing Radiation, Part 2: Some Internally Deposited Radionuclides, vol 78, IARC Press, Lyon, France, 2001.
[3] M.C. Alavanja, et al., Residential radon exposure and lung cancer among nonsmoking women, J. Nat. Cancer Inst. 86 (1994) 1829–1837.
[4] M.C. Alavanja, et al., Residential radon exposure and risk of lung cancer in Missouri, Am. J. Public Health 89 (1999) 1042–1048.
[5] A. Auvinen, et al., Indoor radon exposure and risk of lung cancer: a nested case-control study in Finland, J. Nat. Cancer Inst. 88 (1996) 966–972 (Erratum. J. Nat. Cancer Inst. 1996; 90:401–402.).

[6]  J.M. Barros-Dios, et al., Exposure to residential radon and lung cancer in Spain: a population-based case-control study, Am. J. Epidemiol. 156 (2002) 548–555.
[7]  W.J. Blot, et al., Indoor radon and lung cancer in China, J. Nat. Cancer Inst. 82 (1990) 10–25.
[8]  S.C. Darby, et al., Risk of lung cancer associated with residential radon exposure in south-west England: a case-control study, Br. J. Cancer 78 (1998) 394–408.
[9]  S.C. Darby, D.C. Hill, On behalf of the European Collaborative Group on Residential Radon and Lung Cancer, Health effects of residential radon: a European perspective at the end of 2002, Rad. Prot. Dosimetry 104 (2003) 321–329.
[10] R.W. Field, et al., Residential radon gas exposure and lung cancer: the Iowa Radon Lung Cancer Study, Am. J. Epidemiol. 151 (2000) 1091–1102.
[11] L. Kreienbrock, et al., Case-control study on lung cancer and residential radon in West Germany, Am. J. Epidemiol. 153 (2001) 42–52.
[12] M. Kreuzer, et al., Residential radon and risk of lung cancer in Eastern Germany, Epidemiology 14 (2003) 559–568.
[13] F. Lagarde, et al., Residential radon and lung cancer among never-smokers in Sweden, Epidemiology 12 (2001) 396–404.
[14] E.G. Letourneau, et al., Case-control study of residential radon and lung cancer in Winnipeg, Manitoba, Canada, Am. J. Epidemiol. 140 (1994) 310–322.
[15] J.H. Lubin, J.D. Boice, Lung cancer risk from residential radon: meta-analysis of eight epidemiologic studies, J. Nat. Cancer Inst. 89 (1997) 49–57.
[16] G. Pershagen, et al., Residential radon exposure and lung cancer in women, Health Phys. 63 (1992) 179–186.
[17] G. Pershagen, et al., Residential radon exposure and lung cancer in Sweden, New Engl. J. Med. 330 (1994) 159–164.
[18] F.E. Pisa, et al., Residential radon and risk of lung cancer in an Italian alpine area, Arch. Environ. Health 56 (2001) 208–215.
[19] J.B. Schoenberg, et al., Case-control study of residential radon and lung cancer among New Jersey women, Cancer Res. 50 (1990) 6250–6254.
[20] L. Tomasek, et al., Study of lung cancer and residential radon in the Czech Republic, Centr. Europ. J. Publ. Health 3 (2001) 150–153.
[21] Z. Wang, et al., Residential radon and lung cancer risk in a high-exposure area of Gansu province, China, Am. J. Epidemiol. 155 (2002) 554–564.
[22] H.E. Wichmann, et al., Lungenkrebsrisiko Durch Radon In Der Bundesrepublik Deutschland (West), Ecomed Verlag, Landsberg/Lech, Germany, 1998, [in German].
[23] H.E. Wichmann, et al., Lungenkrebsrisiko Durch Radon In Der Bundesrepublik Deutschland (Ost), Ecomed Verlag, Landsberg/Lech, Germany, 1999, [in German].
[24] S.X. Yao, et al., Exposure to radon progeny, tobacco use and lung cancer in a case-control study in southern China, Radiat. Res 138 (1994) 326–336.
[25] D. Krewski, et al., Risk of lung cancer in North America associated with residential radon. Epidemiology, in press.
[26] J.H. Lubin, et al., Risk of lung cancer and residential radon in China: pooled results of two studies, Int. J. Cancer 109 (2004) 132–137.
[27] H.E. Wichmann, et al., Increased lung cancer risk due to residential radon in a pooled and extended analysis of studies in Germany. Health Physics, in press.
[28] D.J. Steck, et al., Po210 implanted in glass surfaces by long term exposure to indoor radon, Health physics 83 (2002) 261–271.
[29] F. Bochicchio, J.P. McLaughlin, C. Walsh, Comparison of radon exposure assessment results: Po210 surface activity on glass objects vs. contemporary air radon concentration, Radiat. Meas. 36 (2003) 211–215.
[30] C.R. Weinberg, Potential bias in epidemiologic studies that rely on glass-based retrospective assessment of radon, Environ. Health Perspect. 103 (1995) 1042–1046.
[31] G. Saccomanno, et al., A comparison between the localization of lung tumors in uranium miners nd in non-miners from 1947 to 1991, Cancer 77 (1996) 1278–1283.
[32] M. Kreuzer, et al., Histopathologic findings of lung carcinoma in German uranium miners, Cancer 89 (2000) 2613–2621.

www.ics-elsevier.com

# Thoron in the living environments of Japan

H. Yonehara*, S. Tokonami, W. Zhuo, T. Ishikawa,
K. Fukutsu, Y. Yamada

*Radon Research Group, Research Center for Radiation Safety, National Institute of Radiological Sciences, 4-9-1, Anagawa, Inage-ku, Chiba 263-8555, Japan*

**Abstract.** High thoron concentrations have been investigated in various living environments with radon–thoron discriminative measurements; however, behavior of the thoron and its decay products and their effects on human health have not been clearly elucidated yet. In Japan, some kinds of building materials rich in natural radionuclides of thorium series, such as clay and soil plaster, are commonly used for construction and/or surface finishing of wall in traditional Japanese style houses. Measurements with different types of monitor for thoron and its decay products were carried out to investigate general characteristics of thoron in Japanese living environments. From the result of the measurements, thoron concentrations ranged from almost 0 to more than thousands Bq m$^{-3}$. Extremely high concentrations can be found in the places near wall surface, and it decreases drastically as distance from the surface of the wall increases. Thoron concentration was found to be changed drastically with time. The equilibrium equivalent thoron concentrations (EETC) ranged from nearly 0 to about 6 Bq m$^{-3}$. Correlation between the thoron concentration and EETC was not good. This report summarized the results of an investigation of thoron in living environments of Japan. © 2004 Elsevier B.V. All rights reserved.

*Keywords:* Thoron; Radon; Indoor; Japan

## 1. Introduction

Results of measurements of thoron ($^{220}$Rn) concentrations in living environments in various countries have been reported. High concentrations at some places were investigated in the measurements. In some cases, thoron was found as a factor which could interfere with the measurement of radon ($^{222}$Rn) concentration. Behavior of the thoron and its decay products and their effects on human health have not been clearly

---

* Corresponding author. Tel.: +81 43 206 3099; fax: +81 43 206 4097.
  *E-mail address:* yonehara@nirs.go.jp (H. Yonehara).

0531-5131/ © 2004 Elsevier B.V. All rights reserved.
doi:10.1016/j.ics.2004.10.014

elucidated yet. In Japan, some kinds of building materials rich in natural radionuclides of thorium series, such as clay and soil plaster, are commonly used for construction and/or surface finishing of wall in traditional Japanese style houses. Measurements with different types of monitor for thoron and its decay products were carried out to investigate general characteristics of thoron in Japanese living environments.

## 2. Methods of measurement

Different types of monitors were employed for measuring concentrations of thoron and its decay products. Description of the monitors employed in this study is shown in Table 1. Active method for thoron measurements and that for decay products were adopted for continuous measurements to investigate variation of concentration with time. Passive methods were employed for the measurement to investigate long-term averaged concentrations. Long-term measurements of thoron concentrations, averaged over a period of about 3 or 4 months in houses, were carried out by a radon and thoron discriminative measuring method using two types of RADOPOT® [1]. One of the detectors is original an RADOPOT, which is sensitive almost only to radon. The other is a modified monitor of RADOPOT with 4 holes (12 mm in diameter) on the wall of the pot, which is sensitive to both radon and thoron. Radon and thoron concentrations can be obtained from the etch pit densities recorded on the two types of detector during exposure. The long-term averaged concentration of thoron decay products was measured by the method developed by W. Zhuo and T. Iida [2]. In this method, deposition rate of Po-212 on a film was measured by etch pit density on CR-39 detector.

## 3. Results of the measurements

Thoron concentrations were measured with passive detector, RADOPOT, in houses located in western part of Japan, including two areas where a radioactive spa is located.

Table 1
Types of monitor employed for the measurements of thoron and its decay products

| Subject of measurements | Monitor employed | Principle of measurements | Period of measurements |
|---|---|---|---|
| Thoron ($^{220}$Rn) | RAD7 (Durridge, US) | Active method Measurement of radon and thoron by spectral analysis of decay products collected with electrostatic collection | Sampling period: 3–10 h Sampling flow rate: 650 ml/min |
| | RADOPOT® (Radosys, Hungary) | Passive method Discriminative measurement of radon and thoron with CR-39 etched-track detectors in two types of plastic pot, high and low ventilation rate | Exposure period: 3–4 months |
| Thoron decay products | WLx (Pylon Electronics, Canada) | Active method Measurement of decay products of radon and thoron collected on a filter by a solid state detector 25 mm in diameter | Sampling period: 3–10 h |
| | Deposition rate monitor (W. Zhuo and T. Iida) | Passive method Measurement of etched-track on CR-39 detector caused by Po-212 deposited on a film using aluminized plastic film as absorber for energy discrimination | Exposure period: 3–4 months |

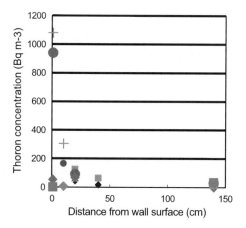

Fig. 1. Thoron concentration at different distances from the wall surface measured with passive monitor.

Fig. 1 shows thoron concentrations obtained by the measurements with the passive monitors at different distances from a wall surface in the rooms of Japanese traditional houses. From the result of the thoron measurements in these houses, it was found that thoron concentrations ranged from almost 0 to more than thousands Bq m$^{-3}$. Extremely high concentrations can be found in the places near wall surface, and it decreases drastically as distance from the wall surface increases. The equilibrium equivalent thoron

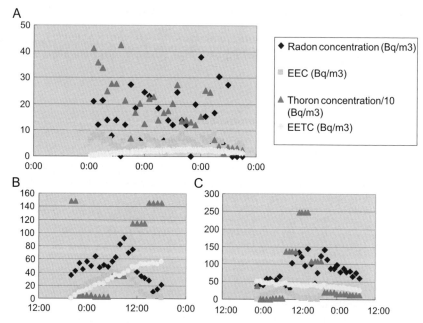

Fig. 2. Temporal variation in concentrations of radon/thoron and their decay products. (A) General house with clay wall in Kyoto prefecture. (B) Bedroom with a wall made of plasterboard in Gifu prefecture. (C) Resting room with a clay wall in Gifu prefecture.

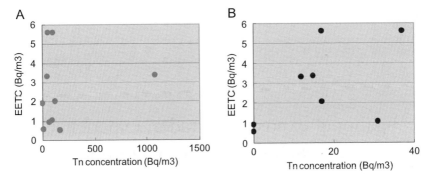

Fig. 3. Correlation between long-term averaged concentrations of thoron and its decay products measured by passive monitors. (A) Thoron concentration near the wall. (B) Thoron concentration in the center of the room.

concentrations (EETC) were also measured in those houses. The concentrations were found to be ranged from nearly 0 to about 6 Bq m$^{-3}$. No large change of the EETC with distance from the surface of the wall was found.

Fig. 2 shows variations of thoron concentrations with time, measured by active method, RAD7, near the wall surface in the general house in Kyoto prefecture and in two rooms in a house near a radioactive spa in Gifu prefecture, respectively. From the results, it is notable that thoron concentrations were found to be changed drastically with time. The concentration of thoron decay products was also found to be changed slowly with time.

The effective dose from radon and thoron exposure in these houses was estimated using values for dose conversion coefficients adopted by UNSCEAR 2000 report for the decay products and the report 1988 for radon and thoron gas. The annual effective doses from both radon and thoron exposures range from 0.6 to 2.2 mSv/y, and ratio of dose from exposure to thoron decay products to the total dose ranges from 0.12 to 0.78.

## 4. Conclusion and discussion

High thoron concentrations were found in Japanese traditional houses with clay wall. In some of the houses, the contribution of thoron decay products to effective dose was larger than that of radon decay products. Concentrations of thoron and its decay products were found to be changed with time. This reveals that long-term measurements would be necessary to estimate the annual dose. Fig. 3 shows the relationship between the long-term averaged concentrations of thoron and EET, both of which were measured by passive methods. There is no explicit correlation between the concentrations. In order to estimate annual effective dose from thoron exposure, it is important to measure the long-term averaged concentration of the decay products.

## References

[1] W. Zhuo, et al., A simple passive monitor for integrating measurements of indoor thoron concentrations, Rev. Sci. Instrum. 73 (8) (2002) 2877–2881.
[2] W. Zhuo, T. Iida, Estimation of thoron progeny concentrations in dwellings with their deposition rate measurements, J. Health Phys. 35 (3) (2000) 365–370.

# Occupational exposure to radon-experience and approach to regulatory control

Eckhard Ettenhuber*

*Federal Office for Radiation Protection, Germany*

**Abstract.** Based on comprehensive studies of workers exposure to radon work activities were identified within which the presence of radon can lead to significant radiation exposures of workers. The concept of radiation protection developed for these work activities was specified in the Radiation Protection Ordinance. Remediation has priority over the inclusion in the control system of radiation protection (monitoring). If the average radon concentration during the annual working time of 2000 h exceeds 1000 Bq m$^{-3}$ measures have to be carried out in order to decrease the exposure as low as achievable taking into account the details of the situation. If exposures cannot be decreased below the mentioned level, the exposure of the worker is included in the institutional control of radiation protection and has to be monitored in order to document that the limits of radiation protection are kept. © 2004 Elsevier B.V. All rights reserved.

*Keywords:* Radon; Radiation exposure; Work activities; History of surveillance; Remediation of workplace; Measuring method; Concept for institutional control; National regulation

## 1. Occupational exposure to radon in the system of radiation protection control

In the past workers employed in jobs involving incidental radiation exposure due to natural radiation sources were not included in the control system although the exposures in doing these jobs are in the same order of magnitude or, in many cases, higher than the exposures of workers included in the institutional control. It was generally considered not possible or not desirable to control a radiation exposure due to a natural radiation source. This understanding has completely been changed. The most important component of radiation exposure at work due to natural sources is the exposure to radon.

---

* Tel.: +49 1888 333 4200; fax: +49 1888 333 4205.
 *E-mail address:* eettenhuber@bfs.de.

0531-5131/ © 2004 Elsevier B.V. All rights reserved.
doi:10.1016/j.ics.2004.12.009

ICRP, IAEA, ILO, WHO and the Commission of European Communities acknowledged the importance of low radon levels in providing healthy conditions at work and published recommendations on radiation protection. If exposures incurred at work are the result of situations that can be regarded as being the responsibility of the operating management they should be included in the system of radiation protection [1].

## 2. Investigations of the occupational exposure to radon

Systematic investigations of the radon exposure in workplaces were started in the early 1970s, first in uranium mines and mills. Later the investigations were extended to other underground work activities. Step by step other workplaces were included in which increased concentrations of radon were expected, e.g. tourist mines, water works. The measurements were carried out at typical workplaces in order to provide information necessary for mitigation. Taking into account the individual occupancy times the individual exposure of workers were assessed. At the beginning the potential alpha energy concentration was mostly measured. Later on measurements of both the radon concentration and the time-integrated radon concentration using track-etch detectors were carried out in order to get experience of several techniques.

If necessary measure were taken in order to reduce the exposure. Optimisation of works and improvements in ventilation have proved to be the most effective protective measure.

## 3. National regulations of radiation protection

The ICRP recommendations on control of exposures to radon at home and at work [1] and the Basic Safety Standards [2,3] are the basis for national regulations of radiation protection that are specified in the in the Radiation Protection Ordinance-Strahlenschutzverordnung (StrlSchV) [4]. According to reference [3] the area was extended within which the Ordinance is operative. Human activities which involve the presence of natural radiation sources and lead to a significant increase in the exposure of workers or members of the public ('work activities') are now included.

According to reference [3] in the Ordinance regulations were specified to identify work activities of concern, to implement corrective measures to reduce exposures and to carry out radiation protection measures including monitoring.

Based on the studies described before working in underground mines, shafts, tunnels tourist mines and show caves were identified as work activities of concern as well as working in waterworks and in radon spas. These work activities are generally included in the system of radiation protection control outlined by the Ordinance. Although most of the workers exposed to radon during their work are employed in above ground workplaces such as small factories, shops, offices, etc., these workplaces are not included in the protection system by the Ordinance. According to reference [1] exposures at these workplaces are not regarded as being the responsibility of the operating management.

In the Ordinance the radon exposure (PRn) of $2 \cdot 10^6$ Bq h m$^{-3}$ is specified as the action level. It corresponds to an average concentration of 1000 Bq m$^{-3}$ assuming an annual working time of 2000 h. For the equilibrium factor of 0.4, this action level corresponds to an annual effective dose of 6 mSv. It should be noted that the competent authority can establish other action levels if the relevant equilibrium factor is significantly different from

0.4. The dose limits specified in the Ordinance for practices are in force for work activities, too. They can be adapted to the problem of exposure to radon. For practical purposes and, assuming an equilibrium factor of 0.4, the annual effective dose limit of 20 mSv can be equated with the radon exposure (PRn) of $6 \cdot 10^6$ Bq h m$^{-3}$. In addition, the working lifetime dose is limited to 400 mSv.

If a workplace of concern can be classified as a type of workplace specified in the Ordinance, it is, in principle, subject to control and the employer has to fulfil the legal obligation of radiation protection. However, the measures necessary for radiation protection depend on the level of exposure. A stepwise approach to control the radon exposure at work is stipulated (see Fig. 1).

At first the employer has to determine the possible radon exposure taking into account measurements and occupancy times. If the exposure is below the mentioned action level, in normal cases, specific measures to reduce radiation exposures are not necessary, since the measures being already in place for normal occupational hygiene reasons may usually provide sufficient radiological protection. If the action level is exceeded the employer has to take measures suitable to keep radiation exposure as low as possible taking into account all circumstances. Experience has shown that in many cases remedial measures are successful in reducing radon concentrations below the action level. If the measures are ineffective or, considering the relevant circumstances, not reasonably practicable, the employer has to notify the authority of the circumstances, the number of exposed workers, and the measures carried out. Should the situation arise the authority will demand additional measures. If, in spite of all measures, exposure cannot be decreased below the action level, the employer has to adopt an appropriate system of radiation protection for both the workplaces and the employees in accordance with the requirements specified for practices. A major component of the control system is the monitoring of the exposure of workers in order to document that the exposure is below the limit. Details of the approach to monitor exposure are not

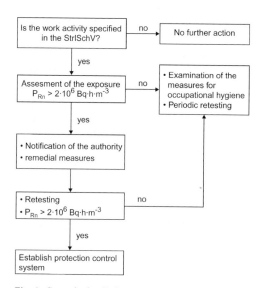

Fig. 1. Control of radiation exposure at work activities.

specified in the Ordinance. The employer can choose by himself the method appropriate for the present circumstances. Devices applied to monitoring have to be calibrated. Periodic investigations are carried out by the Federal Office for Radiation Protection (BfS) to estimate the accuracy and precision in comparison with traceable standards.

Taking into account the circumstances the authority can lay down the type of measurements to be applied and, if necessary, that the measurements have to be implemented by qualified laboratories. Further the employer is obliged to record the monitoring results, to calculate the annual effective dose and to hand over the results to the authority and to the governmental dose register. Mostly the employer is not well acquainted with the problems of radiation protection. Therefore technical instructions and information letters have been worked out in order to qualify the employers in doing properly the protection of workers.

## 4. Summary

On the basis of experience with the surveillance of occupational exposure to radon for many years and considering the international recommendations a pragmatic approach was developed to control the exposure of workers to radon at work. It is focussed on the decrease in high exposures to radon at work within which the exposure can be regarded as being the responsibility of the operating management. Taking into account the specified action level, the workplaces have to be identified in which interventions should be undertaken to reduce radon exposures. Only if exposures cannot be decreased below the action level, a control system has to be established for the workplaces and the exposed workers. The actual measures of control can be adopted to the specific conditions of the work activity taking into account the protective measures specified for practices in the Ordinance. In this way unnecessary or unrealistic measures can be avoided. If the employers are aware of the problem and their responsibility the radiation exposure at work activities can be reduced in an optimised way. However, this approach can be considered only as a milestone on the way to get acceptable conditions regarding the exposure to radon at all workplaces. At normal above ground workplaces such as workplaces in factories, shops, public buildings and offices elevated concentrations of radon can occur and systematic investigations should be carried out. At these workplaces the exposure conditions are similar to those in rooms people live in. Therefore these workplaces need another treatment than the workplaces within which exposure can be influenced by the operating management. In order to avoid imbalances between the protection against radon at home and at work the protection concept, action levels, etc., developed for the protection against radon at home should be adopted for the radiation protection at these work activities. The regulations planned in Germany for the ultimate solution of the radon problem will consider this principle.

## References

[1] ICRP, Protection Against Radon-222 at Home and at Work. ICRP Publication 65, Ann. ICRP 23 (2) (1993).
[2] IAEA, International basic safety standards for radiation protection, IAEA Saf. Ser. 115 (1996) (Vienna).
[3] European Communities. Council Directive 96729 EURATOM of 13. May 1996, Off. J. Eur. Communities No.I. 159 39 (29) (1996 June).
[4] Verordnung über den Schutz vor Schäden durch ionisierende Strahlung-Strahlenschutzverordnung (StrlSchV). BGBl. I S. 1714.

www.ics-elsevier.com

# Radon retrospective measurements

## Christer Samuelsson*

*Department of Radiation Physics, The Jubileum Institute, Lund University, Sweden*

**Abstract.** The techniques for estimating radon exposures to man in retrospect will be overviewed. The focus will be on radon ($^{222}$Rn) issues connected to exposures in dwellings. Potential retrospective radon metrologies are based on alpha damage in plastics, trapping of $^{210}$Pb within porous objects, and alpha recoil implantation into hard surfaces. The use of plastics is hampered by the short exposure time and problematic reconstruction of exposure geometry, handling habits, etc. The strength of the porous volume trap approach is that the growth rate of $^{210}$Pb is not disturbed by the complex indoor behaviour of airborne short-lived decay products. The most common radon retrospective approach today is based on the natural implantation of $^{210}$Pb atoms into glass or other hard indoor surfaces. Large-scale applications of the glass method have been many since the development of in-situ detectors for non-destructive determination of implanted activities. The implantation depth following a single alpha decay is about 50 nm in vitreous glass and this short range makes the surface trap approach sensitive to disturbances from contaminants on the surface. © 2004 Elsevier B.V. All rights reserved.

*Keywords:* Radon; Retrospective; Activity; Alpha; Recoil; Implantation; Domestic; $^{210}$Pb; $^{210}$Po; Glass; Surface; Epidemiology

## 1. Introduction

When assessing contemporary radon ($^{222}$Rn) risks to humans, the individual radon progeny exposure history is of major importance. In the following short survey of the field of retrospective radon (progeny) exposure assessments, I will assume that the endpoint detriment is lung cancer and limit the problem area to radon in homes.

## 2. Radon retrospective options

There is a clear indication from the miner studies of a latency period of roughly 5 years and that the lung cancer induction rate diminishes with time beyond that period [1].

---

\* Tel.: +46 46 173121; fax: +46 46 127249.
*E-mail address:* christer.samuelsson@radfys.lu.se.

0531-5131/ © 2004 Elsevier B.V. All rights reserved.
doi:10.1016/j.ics.2004.11.162

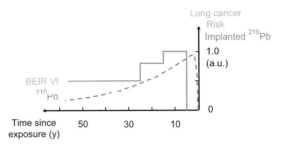

Fig. 1. Radon exposure memory. Relative lung cancer risk (full line) [1] and implanted $^{210}$Po activity (broken line) as a function of time since radon exposure.

Observing the contemporary lung cancer incidence, the ideal retrospective system should be able to mimic the risk induction curve of Fig. 1. The existence of a latency period implies that any retrospective signal that sees exposures only a few years back in time is not useful as such. Contemporary exposure measurements are valuable only if they correlate well with radon exposures decades back in time.

The perhaps most obvious way to try to reconstruct radon exposure histories is to place detectors in all houses occupied in the past by the individual. Due to personal habits, time spent indoor, and so on, one and the same radon gas concentration in a house may correspond to very different Potential Alpha Energy Concentration (PAEC)-values inhaled by the individual occupant. This fact, together with the high costs connected to an all-house approach, is the main reason for looking for alternative retrospective methods, methods that can hopefully get nearer to the inhaled activity by an individual and be more cost-effective.

A prospect along this line has been suggested by Fleicher [2], the etching of alpha tracks in glasses with plastic lenses. A pair of spectacles has the potential of revealing personal exposure, but unfortunately, the wearing time of spectacles is typically only a few years. This short longevity in combination with the problematic conversion from track densities to integrated PAEC has prevented the use of plastic glasses in radon epidemiological studies.

As illustrated in Fig. 1, long-lived radon progenies have a long memory and good prospects signalling radon exposures far back in time. Different retrospective methods based on long-lived radioactivity will be discussed in the following section.

## 3. Retro methods based on long-lived radon decay products

The airborne mixture of radon gas and short-lived decay progeny atoms in a given moment will, within a few weeks time, rest as $^{210}$Pb atoms, slowly decaying with a half-life of 22.3 years to $^{210}$Bi and then to the alpha emitting granddaughter $^{210}$Po. The build-up of $^{210}$Pb activity is a potential retrospective monitor for airborne radon only if we can find a type of trap, a substrate, that

1st Imperative: Persistently stores a certain and constant fraction of the created $^{210}$Pb atoms.
2nd Imperative: Is resistant over decades towards disturbances from the outside.
3rd Imperative: Contains only $^{210}$Pb atoms initially originating from indoor airborne radon.

These three Imperatives are crucial for a successful retrospective application and will be referred to frequently in the following discussion.

It could be added that the substrate must be common in all type of dwellings and easy to evaluate in order to suite large-scale investigations. Preferably, the readout process should be inexpensive and leave the substrate intact.

### 3.1. The human body

The human body is the ideal personal trap for $^{210}$Pb and for bone tissue the biological half-life is long enough to be useful. But regrettably the 3rd Imperative is not met. The variable intake of $^{210}$Pb and $^{210}$Po from diet, smoking, direct inhalation, etc., dominates the content of $^{210}$Pb in bone tissue and typically the part originating from domestic airborne radon is only 1–2% [3].

### 3.2. Porous objects

The radon concentration in the pore air of many objects in a dwelling follows very closely the level in the free airspace. At the subsequent decay of radon, the decay products will be trapped onto interstitial surfaces of objects like mattresses and other spongy materials. Wooden materials have also been tested, but the background activity levels are too variative [4]. The strength of the volume trap approach is that the signal is not disturbed by the complex indoor behaviour of airborne short-lived decay products. If a suitable object can be found and the radon exposure time is accurately given, the volume trap method can predict the radon exposure value very accurately. The conversion factor found by Oberstedt and Vanmarcke [5] was approximately 0.05 mBq of $^{210}$Pb per cm$^{-3}$ of pore air for a radon gas exposure of 1 kBq m$^{-3}$ year. In order to manage a practical sample size at moderate exposure levels, access to radiochemical techniques is necessary. The normal procedure is to chemically extract the granddaughter $^{210}$Po while monitoring the separation yield with an added $^{208}$Po tracer. The polonium isotopes are then deposited onto a silver disc and analysed with alpha spectrometry.

Two circumstances have hampered a widespread use of volume traps as retrospective radon monitors. Firstly, the costly demand on radiochemical laboratory resources, which make it practically impossible to conduct in-situ analysis, and secondly, the sparse availability of suitable objects in many dwellings. The porous object may be present but too precious or expensive to destroy or take sub samples from.

### 3.3. Indoor surfaces

The concept of utilising $^{210}$Pb or $^{210}$Po in surface dust as a radon retrospective monitor [6,7] failed to get a widespread attention, presumably due to the fact that surface dust as such do not meet the 2nd Imperative above. In 1988, this author explained the crucial role of the alpha recoil implantation process and presented experimental results from three dwellings showing a linear relationship between the estimated radon gas exposure and the implanted $^{210}$Po activity in six different glass objects [8]. The implanted activity could only be detected on the side of the glass sheet facing the room, and hence the $^{210}$Po signal was derived purely from airborne activity (3rd Imperative). The paper in 1988 caused a boom in radon retrospective research along the glass-polonium idea. Many research groups developed track-etch devices intended for measurement of implanted $^{210}$Po in-situ (see, for example, Refs. [9–12]). Some of these track-etch retro devices may also, besides implanted $^{210}$Po, register the local plate out rate of short-lived progenies onto the glass sheet analysed [13,14]. The advancement of these cost-effective in-situ retro devices has made it possible to practice the glass-polonium technique in large-scale radon epidemiological investigations in homes [13,15–17].

Some authors have found a stronger positive correlation between implanted $^{210}$Po and lung cancer, than for radon gas based measurements [17,18]. These findings support that radon gas exposure may not be the optimum quantity in the search for radon induced lung cancers in dwellings. Research and investigations in connection to the glass-polonium technique have clearly indicated that the method can yield very erratic results due to interferences from surface dirt and unpredictable local variations in short-lived daughter plate out. In practice, the glass-polonium method has evidently problems fulfilling the 1st Imperative. By defining criteria for which type of glass samples to accept, analysing several samples per person, correcting for local plate out variations, etc., the influence from non-dose related variations can be minimized. Lagarde [17], for instance, analysed two glass samples for each individual and used only samples with exposure periods in excess of 15 years. Steck [13] found a correlation of $r^2=0.5$ between total radon exposure estimated from contemporary radon gas measurements and the historical average reconstructed from implanted $^{210}$Pb surface activity. When corrections based on the result from the open track-etch film monitoring plate out rates were included, the correlation increased to $r^2=0.7$.

## 3.4. Alpha recoil implantation

The recoil ranges of $^{214}$Pb and $^{210}$Pb in soda-lime glass (density 2.5 g cm$^{-3}$) are 44 and 52 nm, respectively. Due to double alpha recoil implantation, the activity of $^{210}$Pb will be

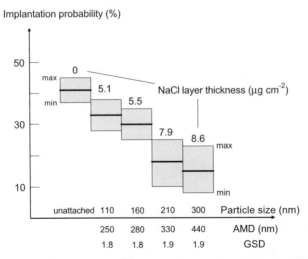

Fig. 2. Alpha recoil implantation probabilities of $^{214}$Pb into soda-lime glass as a function of size for NaCl aerosol particles. AMD and GSD are the activity median diameter and the geometrical standard deviation, respectively. The thickness of a corresponding homogeneous NaCl layer is also given (data from Refs. [19,21]).

distributed down to a depth of 95 nm from a surface contaminated by $^{218}$Po atoms [19,20]. This short range of alpha recoils may prohibit the use of the glass-polonium technique in aerosol-rich and dirty environments. If a variant part of the recoils ends up in dirt particles on a glass surface and not in the glass proper, the 1st Imperative above will be violated. An increased dust load on indoor glass surface with age, can explain the common observation that old glass objects have a tendency to show lower amounts of implanted activity than expected.

The experimental implantation probability for attached and unattached $^{218}$Po residing on a glass surface is given in Fig. 2. The decreasing implantation probability into the glass with particle size is a combined effect of $^{214}$Pb atoms recoiling into the particle to which $^{218}$Po is attached and the implantation into inactive particles nearby. Analytical and Monte Carlo calculations indicate that for clean surfaces the implantation probability for attached activity species will always exceed 25%, or 21% if the fraction implanted within 5 nm from the surface is excluded [19].

## 4. Conclusions

Radon retrospective measurements based on volume and surface trapping of $^{210}$Pb have been proven feasible in both small and large scale epidemiological studies. Restoring past radon exposures with contemporary risks in mind, only exposures older than about 5 years are of interest. Contemporary radon gas levels or signals from recently obtained retro-objects are risk relevant only if the correlation is with older exposure levels, but the information obtained can be used for correction measures. Retro-objects should preferably be at least 10 years old and must be chosen selectively in order to minimize erroneous results. In small scale applications with access to radiochemical laboratory resources, volume traps should be considered as they are less prone to disturbances from non-dose factors. In large scale applications in-situ measurements of implanted $^{210}$Po in flat glass surfaces with track-etch retro-detectors is an established technique, but the short recoil range of about 50 nm in glass is a problem. The implanted polonium level in a surface is therefore affected by historical surface contaminants and in addition the unpredictable local behaviour of airborne short-lived radon daughters. Due to this complexity, a more-than-one sample-per-individual approach is advantageous.

## Acknowledgement

This project is supported by the Swedish Radiation Protection Authority, Stockholm.

## References

[1] Committee on Health Risks of Exposure to Radon (BEIR VI), National Research Council, National Academy Press, Washington, DC, 1999.
[2] R.L. Fleicher, et al., Personal radon dosimetry from eyeglass lenses, Radiat. Prot. Dosim. 97 (3) (2001) 251–258.
[3] P.L. Salmon, V.I. Berkovsky, D.L. Henshaw, Relative importance of inhalation and ingestion as sources of uptake of $^{210}$Pb from the environment, Radiat. Prot. Dosim. 78 (4) (1998) 279–293.
[4] J. Paradaens, H. Vanmarcke, The usability of wood as a volume trap for the purpose of retrospective radon exposure assessment, Health Phys. 76 (6) (1999) 657–663.

[5]  S. Oberstedt, H. Vanmarcke, Volume traps—a new retrospective radon monitor, Health Phys. 70 (2) (1996) 222–226.
[6]  S. Schery, Dept. Physics, New Mexico Inst, Min. Tech., Socorro, NM, USA 1982. Personal communication. 2004.
[7]  R.S. Lively, E.P. Ney, Surface radioactivity from the deposition of Rn-222 daughter products, Health Phys. 52 (4) (1987) 411–415.
[8]  C. Samuelsson, Retrospective determination of radon in houses, Nature 334 (1988) 338–340.
[9]  D.J. Steck, R.S. Lively, E.P. Ney, Epidemiological implications of spatial and temporal radon variations, Proc. 29th Hanford Symp. on Health and the Environment, 1990, pp. 889–904.
[10] J.A. Mahaffey, et al., Estimating past exposure to indoor radon from household glass, Health Phys. 64 (4) (1993) 381–391.
[11] R. Falk, et al., Retrospective assessment of radon exposure by measurements of Po-210 implanted in surfaces using an alpha track detector technique, Environ. Int. 22 (1996) S857–S861.
[12] F. Trotti, et al., CR-39 track detectors applied to measurements of Po-210 embedded in household glass, Environ. Int. 22 (1996) S863–S869.
[13] D.J. Steck, R.W. Field, The use of track registration detectors to reconstruct contemporary and historical airborne radon ($^{222}$Rn) and radon progeny concentrations for a Radon-Lung Cancer Epidemiologic Study, Rad. Measur. 31 (1999) 401–406.
[14] J.P. Mc Laughlin, C. Walsh, The Simultaneous Measurement of Radon Progeny Surface Deposition and Associated Airborne Activities. NRE-VII 20-24 May 2002, Rhodes, Greece (To be publ in Radiation in the Environment, Elsevier).
[15] M.C.R. Alavanja, et al., Residential radon exposure and risk of lung cancer in Missouri, Am. J. Public Health 89 (7) (1999) 1042–1048.
[16] R. Falk, K. Almrén, I. Östergren, Experience from retrospective radon exposure estimations for individuals in a radon epidemiological study using solid-state nuclear detectors, Sci. Total Environ. 272 (2001) 61–66.
[17] F. Lagarde, et al., Glass-based radon-exposure assessment and lung cancer risk, J. Expo. Anal. Environ. Epidemiol. 12 (2002) 344–354.
[18] M.C.R. Alavanja, J.A. Mahaffey, R. Brownson, RE: "Residential Radon Gas Exposure and Lung Cancer: The Iowa Radon Lung Cancer Study", Am. J. Epidemiol. 152 (9) (2000) 895.
[19] B. Roos, Studies on the Alpha-Recoil Implantation of $^{214}$Pb and $^{210}$Pb in Glass Surfaces—Implications for Retrospective Radon Measurements. PhD thesis 2002, Lund University, Sweden.
[20] B. Roos, H.J. Whitlow, Computer simulation and experimental studies of implanted $^{210}$Po in glass resulting from radon exposure, Health Phys. 84 (1) (2003) 72–81.

# Thoron versus radon: measurement and dosimetry

Naomi H. Harley[a,*], Passaporn Chittaporn[a], Riasp Medora[b], Richard Merrill[c], Waraporn Wanitsooksumbut[d]

[a] Department of Environmental Medicine, New York University School of Medicine, 550 First Avenue, New York, NY 10016, United States
[b] Oak Ridge Associated Universities, United States
[c] Fluor Fernald Radiation Control Section, United States
[d] Thai Office of Atomic Energy for Peace, United States

**Abstract.** Two new instruments were developed for the U.S. Department of Energy Environmental Science Management Program (EMSP) to perform more detailed exposure assessment measurements at Fernald, OH. Fernald is a former uranium processing facility undergoing remediation. The instruments are a miniature radon and thoron detector, and a miniature particle size spectrum analyzer. Ongoing measurements of both thoron ($^{220}$Rn) and radon ($^{222}$Rn) gas are now being made at four locations: (1) at Fernald; (2) at the New York City National Weather Service site, used as a quality control (QC) site; (3) at a private home in Bangkok used as a QC site; and (4) at a research center and rare earth development facility processing monazite ore near Bangkok. Particle size distribution measurements are presented for two locations: at Fernald and at the Rare Earth Facility near Bangkok. Continuous radon and thoron measurements were made at all sites for at least 1 year and the bronchial dose is reported. © 2004 Published by Elsevier B.V.

*Keywords:* Radon and thoron measurement; Radon and thoron dose; Particle size measurement

## 1. Radon and thoron measurements

The New York University (NYU) miniature passive alpha track detector is used for all measurements. The detector (called 4Leaf) utilizes four separate chambers for duplicate measurements of radon and total gas (radon plus thoron). It can be used as a personal or area detector. Each chamber has a 9×9-mm square solid state nuclear alpha track film (CR-39)

---

* Corresponding author. Tel.: +1 212 263 5287.
  *E-mail address:* naomi.harley@med.nyu.edu (N.H. Harley).

0531-5131/ © 2004 Published by Elsevier B.V.
doi:10.1016/j.ics.2004.09.055

for detection. Thoron is calculated by signal difference between the two sets of measurements. A more detailed description of the 4Leaf radon/thoron detector is reported [1].

## 2. Particle size measurements

The particle size sampler developed at NYU is an integrating miniature particle (IMP) size sampler. There is a ZnS alpha phosphor annulus covering an impaction stage under the inlet jets, followed by up to six fine-mesh stainless steel screen filters (200–500 mesh) and a 0.8-μm Millipore membrane exit filter to collect all residual particles. A low flow pump (4–6 l/min) draws the atmosphere sampled through the sampler for periods of up to 2 months to yield an integrated particle size distribution. The operations at Fernald are described elsewhere with a more complete description of the particle size sampler [1]. Radon gas is present in all atmospheres, and most of its short-lived decay products attach immediately to the ambient aerosol particles and decay to long-lived $^{210}Pb/^{210}Po$. The $^{210}Po$ is alpha-counted directly on all filters and is an excellent aerosol particle tracer. Activity buildup measurements insure that only $^{210}Po$ is being counted. If there is any discrepancy in the 138-day buildup half time of $^{210}Po$, alpha spectrometry is also performed to identify any other radionuclides present.

## 3. Results

The results of the 4Leaf measurements at the National Weather Service site in Central Park, New York City; at one raffinate pit at Fernald; at a private home in Bangkok; and the average at a Rare Earth Research Facility in Bangkok are shown in Table 1. The IMP particle size spectrum colocated with the 4Leaf near one raffinate pit is shown in Fig. 1. The

Table 1
Effective dose and average radon and thoron concentrations at four locations

| Location | Measurement date | | $^{222}Rn$ (Bq/m$^3$) | S.D. | $^{220}Rn$ (Bq/m$^3$) | S.D. | $^{222}Rn$ (EEC) | $^{220}Rn$ (EEC) | Effective dose (mSv/year) | |
|---|---|---|---|---|---|---|---|---|---|---|
| | | | | | | | | | $^{222}Rn$ | $^{220}Rn$ |
| Central Park, New York City | 12/2003 to | 10 cm | 9.7 | 5 | 24 | 4 | 5.8 | 0.12 | 0.10 | 0.010 |
| National Weather Station | 7/2004 | 87 cm | 8.8 | 2 | 21 | 4 | 5.3 | 0.11 | 0.10 | 0.008 |
| Fernald former U processing plan Near pit | 10/2002 to 5/2004 | | 40 | 8 | 36 | 9 | 24.0 | 0.18 | 0.43 | 0.014 |
| Bangkok home | 12/2001 to 3/2004 | | | | | | | | | |
| Outdoor | | | 40 | 5 | 10 | 3 | 24.0 | 0.05 | 0.43 | 0.004 |
| Bedroom | | | 16 | 1 | 55 | 13 | 9.6 | 0.28 | 0.17 | 0.022 |
| Bath | | | 23 | 7 | 14 | 3 | 13.8 | 0.07 | 0.25 | 0.006 |
| Bangkok Rare Earth Facility | 1/2002 to | | 30 | 5 | 270 | 100 | 18.0 | 1.35 | 0.32 | 0.108 |
| Nearby home | 3/2004 | | 12 | 2 | 47 | 12 | 7.2 | 0.24 | 0.13 | 0.019 |

UNSCEAR (2000) dose factors.
$^{222}Rn$ (EEC) bronchial dose=9 nSv/Bq/m$^3$ h.
$^{220}Rn$ (EEC)=40 nSv/Bq/m$^3$ hr.
$F_{eq}$ ($^{220}Rn$) outdoors=0.005.
$F_{eq}$ ($^{220}Rn$) indoors=0.02.
$F_{eq}$ ($^{222}Rn$) outdoors=0.6.
$F_{eq}$ ($^{222}Rn$) indoors=0.4.

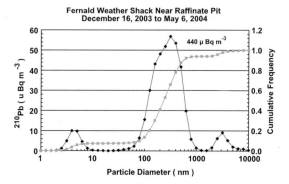

Fig. 1. Particle size distribution at one raffinate pit at Fernald, Ohio.

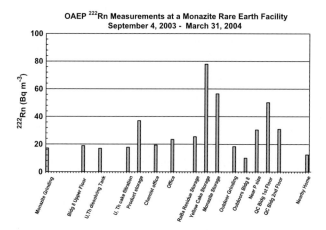

Fig. 2. Radon concentrations at a Rare Earth Facility, Bangkok.

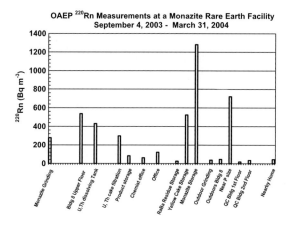

Fig. 3. Thoron concentrations at a Rare Earth Facility, Bangkok.

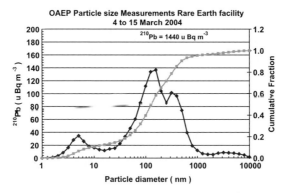

Fig. 4. Particle size distribution at the Rare Earth Facility, Bangkok.

cumulative frequency plot of the size spectrum on Fig. 1 is useful in the calculation of lung dose as the fraction contributing to each size mode.

The radon and thoron measurements at the Rare Earth Facility near Bangkok are shown in Figs. 2 and 3. One particle size spectrum at the Rare Earth Facility is shown in Fig. 4. The backup Millipore filter was analyzed for radionuclide composition by alpha spectrometry and alpha energy was consistent with 5.3 MeV $^{210}$Po. However, this is also the energy for $^{224}$Ra in the thorium 232 series. However, the buildup half time of alpha activity on this filter was 140 days—evidence that the activity is that of $^{210}$Po.

## 4. Dosimetry

The results of the dose calculations are shown in Table 1, along with the average $^{222}$Rn and $^{220}$Rn concentrations at the four locations. Dose conversion factors are taken from UNSCEAR (2000) and are in units of the equilibrium equivalent concentration (EEC).

We have measured thoron decay products and thoron gas both indoors and outdoors for over 2 years [2]. These data showed a stable thoron equilibrium factor $F_{eq}$ of 0.005 outdoors and 0.02 indoors, and these are used in the thoron dose calculations.

## Acknowledgment

Research support from USDOE EMSP contract DE FG02 03ER63661 is gratefully acknowledged. The authors would like to thank Dr. Isabel M. Fisenne and Ada Kong for performing the alpha spectrometry measurements on samples to determine their nuclide composition.

## References

[1] N.H. Harley, et al., Airborne particle size distribution measurements at USDOE Fernald, American Chemical Society Monograph (2004) (in press).
[2] N.H. Harley, et al., $^{220}$Rn (thoron) lung dosimetry using $^{220}$Rn gas measurements, Health Physics (2004) (submitted for publication).

# Rn–Tn discriminative measurements and their dose estimates in Chinese loess plateau

Y. Yamada[a,*], S. Tokonami[a], W. Zhuo[a], H. Yonehara[a], T. Ishikawa[a],
M. Furukawa[a], K. Fukutsu[a], Q. Sun[b], C. Hou[b], S. Zhang[b], S. Akiba[c]

[a] National Institute of Radiological Sciences, 4-9-1, Anagawa, Inage-ku, Chiba 263-8555, Japan
[b] National Institute for Radiological Protection, Beijing, China
[c] Kagoshima University, Kagoshima, Japan

**Abstract.** Numerous epidemiological studies have been carried out to clarify the health effects caused by indoor radon exposure. Some of passive monitors for radon are sensitive to thoron, and thus the results of measurements might be possibly overestimated as radon concentration. Recently, we developed a simple radon–thoron discriminative monitor by using two passive detectors and a special method for concentration measurements of thoron decay products, and applied them to field measurements in conducted in Chinese loess plateau where cave dwellings are widely distributed. Our new monitors revealed the presence of high thoron concentration there. The highest concentration of 1471 Bq/m$^3$ was observed and the mean was 240 Bq/m$^3$. The mean concentration of thoron decay products was 2.2 Bq/m$^3$. It was very low, compared with the expected concentration from the equilibrium factor of 0.1 adopted by UNSCEAR 2000. On the other hand, the mean radon concentration was 73 Bq/m$^3$. The effective dose estimated from our discriminative measurements was 2.4 mSv/year, which is below a half of 5.9 mSv/year estimated from the results of the nondiscriminative measurements. This means the significance of radon and thoron discriminative measurement and the direct measurement of thoron decay products. © 2004 Elsevier B.V. All rights reserved.

*Keywords:* Radon; Thoron; Decay product; Equilibrium equivalent concentration; Effective dose

## 1. Introduction

Radon is a ubiquitous indoor air pollutant that is found worldwide. Its sources are soil, building materials, groundwater, etc. Radon has many isotopes, but the nuclides of $^{222}$Rn and $^{220}$Rn (thoron) are our concern because of their presence in our human environment and the possibility of their health effects on the public. The contribution of each nuclide to

\* Corresponding author. Tel.: +81 43 206 3097; fax: +81 43 206 4097.
E-mail address: yj_yamad@nirs.go.jp (Y. Yamada).

0531-5131/ © 2004 Elsevier B.V. All rights reserved.
doi:10.1016/j.ics.2004.09.054

radiation exposure is quite different when half-life, radiation type and physical form (solid or gaseous) are considered. Radon's half-life of 3.8 days is long enough for it to enter indoor environment and cause an increase in the indoor concentration, but is relatively too long to enter the respiratory tracts and to irradiate the cells. In the uranium decay series, decay products called "radon progeny or daughters" contribute to radiation exposure. In the thorium decay series, the half-life of thoron is only 56 s, and thus, the transferred range is limited to the neighbourhood of the source. It is reported that the thoron concentration strongly depends on a distance from the source [1]. In general, the presence of thoron is not recognized. Thus, the exposure to thoron has not been considered to be a serious problem and not much attention has been paid to the exposure in most radon surveys. However, we consider that thoron is very important for its role in risk evaluation for the following two reasons. The first concerns the equipment used for taking radon measurements. Some radon monitors are sensitive to thoron and thus the values measured as radon concentration may possibly be affected by the presence of thoron [2]. The second is that area with high levels of thoron over 200 Bq/m$^3$ was found in recent surveys [2,3]. In this area, a large-scale case-control study was conducted by a joint LIH (China)–NCI (USA) team. Their conclusion is that high levels of residential radon increase the risk of lung cancer, and that increased lung cancer risks may equal or exceed the extrapolations based on the data for miners [4]. This paper reports results of radon and thoron survey conducted in Chinese loess plateau, and discusses problems in the dose evaluation.

## 2. Materials and methods

Recently, we developed a simple radon-thoron discriminative passive monitor [5]. The monitor has two diffusion chambers with different air exchange rates. By analyzing sensitivities to radon and thoron of the chambers, their concentrations are evaluated separately. The lowest detection limits are 3.5 Bq/m$^3$ for radon and 13 Bq/m$^3$ for thoron in case of an exposure period of 90 days [6]. In addition, we have the deposition rate monitor to estimate an equilibrium equivalent thoron concentration (EETC) for dose evaluation [7]. The monitor selectively detects high energy alpha particles emitted from $^{212}$Po, which is one of thoron decay products, by using a Mylar film with enough thickness for discrimination of high energy alpha particles. The lowest detection limits are 0.08 Bq/m$^3$ for 90-days exposure. A CR-39 was used as the detector in both monitors. Radon surveys

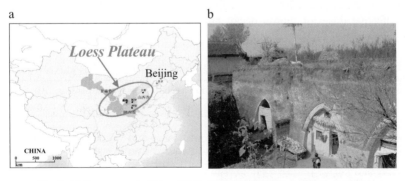

Fig. 1. (a) Survey area in China. (b) Typical cave dwelling in Gansu province.

using these monitors were conducted in the Chinese loess plateau as shown in Fig. 1a. The typical cave dwellings are shown in Fig. 1b. Number of inhabitants of the dwellings are over three million. In our radon survey, about 300 measurements were made in three provinces of Shaanxi, Shanxi, and Gansu for a long-term exposure period of 1 year. The monitors were suspended from ceiling at the center of cave. In Gansu province, size measurements of carrier aerosols attaching radon and/or thoron decay products were preliminarily made with a screen-type diffusion battery (SDB) [8].

## 3. Results and discussion

The results of the measurements of concentrations of indoor radon, thoron and EETC in three provinces are summarized in Table 1. The mean concentration of indoor radon and thoron over three provinces were 73 and 240 Bq/m$^3$, respectively. In all dwellings surveyed, thoron was detected out. The maximum thoron concentration was 1471 Bq/m$^3$ in Gansu province. In most of dwellings, the radon concentration was lower than the thoron one. The correlation between radon and thoron concentrations was positive, but it was very weak (correlation coefficient, $R=0.20$) as shown in Fig. 2a. The mean concentrations of each province were different, but the concentration ratios of thoron to radon were nearly the same among them. They ranged from three to four. The mean EETC was 2.2 Bq/m$^3$. It was extremely lower than expected from the thoron concentration of 240 Bq/m$^3$. If equilibrium factor for thoron were assumed to be 0.1 as adopted by UNSCEAR 2000 [9], much higher EETC should be expected to be 24 Bq/m$^3$. Ratio of EETC to thoron concentration, which is equilibrium factor for thoron, was calculated for all dwellings. The frequency distribution was shown in Fig. 2b. The mean ratio was $0.012\pm0.006$, and, in 91% of the measured dwellings, the ratios were found below 0.02. EETC should not be estimated by using the assumed equilibrium factor because the thoron concentration strongly depends on a distance from the source. The thoron concentration obtained by radon–thoron discriminative measurement should be used for evaluation as a potential of thoron emanation.

The effective dose caused by exposure to radon and thoron decay products estimated by using the following equation:

$$H_{Rn} = EERC \times t \times DCF_{Rn},$$

$$H_{Tn} = EETC \times t \times DCF_{Tn},$$

where $H_{Rn}$ is the annual effective dose for radon decay products (mSv/year), $H_{Tn}$ is the annual effective dose for thoron decay products (mSv/year), EERC is the equilibrium

Table 1
Radon, thoron and thoron decay product concentrations (Bq/m$^3$) in the Chinese loess plateau

| Survey area | Luliang in Shanxi | | Yan'an in Shaanxi | | Qingyan in Gansu | | All | |
|---|---|---|---|---|---|---|---|---|
| | Mean[a] | Range | Mean | Range | Mean | Range | Mean | Range |
| Number of houses | 97 | | 96 | | 102 | | 295 | |
| Radon concentration | 53 | 19~136 | 75 | 20~195 | 91 | 21~229 | 73 | 19~229 |
| Thoron concentration | 160 | 10~658 | 202 | 26~865 | 351 | 30~1471 | 240 | 10~1471 |
| EETC | 1.4 | 0.3~3.5 | 2.3 | 0.7~4.9 | 2.8 | 0.8~5.7 | 2.2 | 0.3~5.7 |

[a] Refers to arithmetic mean.

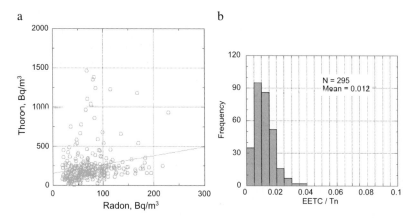

Fig. 2. (a) Correlation between Rn and Tn concentrations. (b) Frequency distribution of EETC/Tn ratio.

equivalent radon concentration (Bq/m$^3$), EETC is the equilibrium equivalent thoron concentration (Bq/m$^3$), $t$ is the time in hours of indoor exposure in a year (=7000 h), DCF$_{Rn}$ is the dose conversion factor for radon (=9 nSv/h/Bq m$^3$), and DCF$_{Tn}$ is the dose conversion factor for thoron (=40 nSv/h/Bq m$^3$). These dose conversion factors are provided in UNSCEAR [9]. In this study, EERC was calculated by using equilibrium factor of 0.4 for radon, which is also provided by UNSCEAR. Using the data shown in Table 1, the annual effective dose in the three provinces of Shaanxi, Shanxi, and Gansu was estimated to be 2.4 mSv/year. Compared with the thoron concentration of 240 Bq/m$^3$, the EETC was not so high that the effective dose due to thoron decay products was only 0.6 mSv/year. The other of 1.8 mSv/year was due to radon decay products. If radon–thoron discriminative measurements were not made in this survey area, the estimated effective dose should be quite different from our result of 2.4 mSv/year. In the LIH–NCI study, the 'Radtrak' monitor, which is non-radon–thoron discriminative, was used. The indicated radon concentration when 'Radtrak' monitor were used was calculated to be 234 Bq/m$^3$ (=73+0.67×240) by using the Tokonami's conversion factors [10]. The 234 Bq/m$^3$ expected for nondiscriminative measurement is close to the observed concentration of 223 Bq/m$^3$ in the LIH–NCI study. When the concentration of 234 Bq/m$^3$ was used as the radon concentration, the annual effective dose would be estimated to be 5.9 mSv/year. Even if radon–thoron discriminative measurements were made, no information on the EETC might lead misunderstanding of the dose evaluation. If the equilibrium factor of 0.1 were assumed for thoron, the effective dose of 6.7 mSv/year for thoron would be added to 1.8 mSv/year for radon. The total dose would be 8.5 mSv/year. As mentioned above, the presence of thoron would affect both radon measurements and dose evaluation. The significance of thoron measurements should be recognized.

The LIH–NCI study [4] found that high levels of residential radon increase the risk of lung cancer. If the lung cancer was caused by the low dose around 2.4 mSv/year, it means the increase of the risk factor for lung cancer. According to our preliminary size measurements of indoor aerosols, bimodal size distributions were often observed in Chinese loess plateau. The smaller peak was found to be around 10 nm. The dose conversion factor, DCF, from concentration to exposure dose depends on the size of aerosols attached to radon and thoron decay products [11]. The DCF in 10 nm is about 10 times larger than that in 100 nm [12]. If the size of radon and thoron decay product aerosols were considered in the dose estimation, the effective dose due to their decay products would increase, and, conversely, the risk factor of lung cancer would decrease.

## References

[1] M. Doi, et al., Health Phys. 66 (1994) 43–49.
[2] B. Shang, et al., in: A. Katase, M. Shimo (Eds.), Radon and Thoron in the Human Environment, World Scientific, Singapore, 1997, pp. 379–384.
[3] J. Wiegand, et al., Health Phys. 78 (2000) 438–444.
[4] Z. Wang, et al., Am. J. Epidemiol. 155 (2002) 554–564.
[5] W. Zhuo, et al., Rev. Sci. Instrum. 73 (2002) 2881–2887.
[6] S. Tokonami, et al., Radiat. Prot. Dosim. 103 (2003) 69–72.
[7] W. Zhuo, T. Iida, Jpn. J. Health Phys. 35 (2000) 365–370.
[8] Y. Yamada, et al., Radiat. Prot. Dosim. 88 (2000) 329–334.
[9] UNSCEAR, UNSCEAR 2000 Report, United Nations, New York, 2000.
[10] S. Tokonami, T. Sanada, M. Yang, Health Phys. 80 (2001) 612–615.
[11] A. Birchall, A.C. James, Radiat. Prot. Dosim. 53 (1994) 133–140.
[12] T. Ishikawa, et al., Jpn. J. Health Phys. 36 (2001) 329–338.

www.ics-elsevier.com

# Radon dosimetry and its implication for risk

## A. Birchall*, J.W. Marsh

*National Radiological Protection Board, Didcot, Oxon, UK*

**Abstract.** The major source of human exposure to radiation is from natural background, and the largest component of this arises from the inhalation of the short-lived daughters of radon gas ($^{222}$Rn). It is therefore important to be able to quantify the risk from this exposure. The risk from exposure to radon daughters can be determined in two different ways. Firstly, by using statistics on the excess lung cancer incidence in miners exposed to high levels of radon gas: the so-called *epidemiological* approach. Secondly, by calculating the effective dose (Sv) received per unit exposure, and multiplying this by the risk per Sv: the so-called *dosimetric* approach. When, in 1994, the ICRP Publication 66 Human Respiratory Tract Model (HRTM) was first used in the latter approach, the estimates of risk ($8.4 \times 10^{-4}$/WLM) exceeded those of the epidemiological approach ($2.8 \times 10^{-4}$/WLM) by a factor of 3. Since then, there have been many attempts to reconcile these two approaches, bearing in mind that if any of the ICRP weighting factors (e.g. tissue or radiation weighting factors) were changed by a factor of 3, to make these two approaches agree, this would have a significant effect on the dosimetry of other radionuclides, and may not be justified by other experimental evidence. This paper re-examines these two approaches, and the likely uncertainties associated with each, in the light of recent scientific knowledge. Recent risk estimates using the epidemiological approach ($\sim 5 \times 10^{-4}$/WLM) are nearly twice those made in 1994, while a recent detailed analysis using the dosimetric approach gives a risk about 15% lower than the 1994 study ($\sim 7 \times 10^{-4}$/WLM). Based on these current estimates, the two approaches are broadly consistent. It is observed that a small change in the weighting factor for the lung, from 0.12 (rounded by ICRP from 0.11) to 0.10 is all that is needed to make these two approaches agree almost exactly. © 2004 Elsevier B.V. All rights reserved.

*Keywords:* Radon; Exposure; Risk; Dosimetry; Epidemiology

## 1. Introduction

Radon ($^{222}$Rn) is a naturally occurring radioactive noble gas. It is formed within the decay chain of $^{238}$U, and can diffuse several meters before decaying into its shorter-lived

---

\* Corresponding author. Tel.: +44 1235 210840; fax: +44 1235 822891.
*E-mail address:* alan.birchall@vodafone.net (A. Birchall).

0531-5131/ © 2004 Elsevier B.V. All rights reserved.
doi:10.1016/j.ics.2004.11.158

decay products $^{218}$Po, $^{214}$Pb, and $^{214}$Bi (and $^{214}$Po). Radon can enter buildings through the floor, and in cases where ventilation is poor, the radon concentrations can build up to high levels, exposing the occupants via inhalation. From a radiological point of view, radon itself is not a major source of concern: since it is a noble gas, most of the radon inhaled is exhaled again. However, when radon decays, the decay product formed is usually positively charged, and rapidly attaches itself to ambient aerosols, which readily deposit in the lungs to deliver a dose.

Evidence for the risks from radon come from epidemiological studies of miners in high-radon mines and of people exposed in the home, and from experimental studies of animals exposed to radon. Based on extrapolation to lower exposures, radon is calculated to be the number one cause of lung cancers amongst non-smokers, and also accounts for the largest component (50%) of naturally occurring background dose, causing an estimated 2500 deaths per year in the UK and about double this in Japan. It is the biggest geological hazard, responsible for more deaths than earthquakes! It is therefore clearly important to be able to quantify the risk associated with exposure to radon.

## 2. Assessing the risks from exposure to radon

The risk from radon can be assessed in two ways. The first way, the *epidemiological* method is a direct approach which uses estimates of radon exposures of groups of miners, together with observed excess occurrences of lung cancers within the group, to estimate the risk per unit exposure. The second way, the *dosimetric* method involves first calculating the equivalent dose to the lungs from unit exposure, using a dosimetric model. From this, the effective dose (Sv) can be derived, and when multiplied by the ICRP detriment [1] per Sv (5.6%/Sv) gives the total detriment/WLM, which for lungs manifests itself as fatal lung cancers.

In 1994, the ICRP published [2] a Human Respiratory Tract Model (HRTM) which is currently used with systemic models to calculate dose coefficients (doses per unit intake). These dose coefficients have been incorporated into the legislation governing radiation protection [3] in most European countries. Although the HRTM was designed to be applicable to the short-lived decay products of radon, ICRP recommends that it is not used to calculate risks from radon exposures. Instead, ICRP [4] suggest that the more direct epidemiological approach should be used. However, in 1994, it was shown [5] that application of the HRTM to radon exposure resulted in a risk of $8.4 \times 10^{-4}$/WLM.

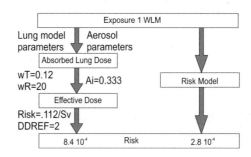

Fig. 1. Comparison of the risk per unit exposure based on both approaches in 1994.

Using the epidemiological approach, ICRP [4] in 1993 used a model in which the relative risk decreased with time since exposure together with a base line cancer risk that corresponded to a world population, to give a risk estimate of $2.8\times10^{-4}$/WLM. This estimate, which was also in line with that calculated in BEIR IV [6], is a factor of 3 times lower than that based on the dosimetric approach. Both approaches are summarised in Fig. 1.

## 3. Are the two approaches compatible?

From a scientific point of view, these estimates are remarkably similar, considering that the two approaches are completely different. However, the difference still presents a practical problem, and over the last 10 years, much effort has been made in trying to resolve this discrepancy. In order to determine the uncertainty in the dosimetric risk estimate, an uncertainty analysis was performed [5] by assigning probability distributions to parameter values performing Monte Carlo simulations. The analysis showed that although the uncertainty in the risk estimates were large, it was extremely unlikely that the dosimetric approach would lead to estimates of risk as low as 5 mSv per WLM (the value needed for consistency with the epidemiological approach) without altering at least one of the ICRP weighting factors ($w_T$(lung), $w_R$(alpha)), or the lung region apportionment factors ($A_{BB}$, $A_{bb}$, $A_{AI}$). These conclusions were supported by a similar analysis [7], based on a study of domestic exposure [8] (Fig. 2).

## 4. Implications of changing ICRP risk/weighting factors

Clearly, if either the radiation weighting factor for alpha particles is reduced from 20 to 7, or the tissue weighting factor for the lung is reduced from 0.12 to 0.04, then the effective dose would also reduce by a factor of 3 bringing it into line with the epidemiological estimate. However, this would have serious implications for the dosimetry of other radionuclides. ICRP has recently re-examined these weighting factors [9] and concluded that there is insufficient evidence for a change from the current defaults. Similarly, if the risk per Sv were reduced by a factor of 3, or if the apportionment factor for AI were increased and those for BB and bb were decreased, then the two approaches would agree, but this would also have important radiological implications.

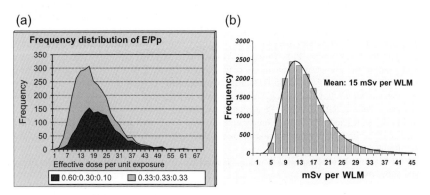

Fig. 2. Uncertainty analyses giving the effective dose per unit exposure (a) in mines (b) in homes.

## 5. A possible way forward

Over the years, much work has been done on trying to reconcile these two approaches. In particular, a major European study was undertaken to re-examine all of the parameter values used in the dosimetric approach [8]. As a result of this study, it can be shown [10] that the best estimate of dose per WLM is 12.5 mSv (compared to 15 mSv previously). ICRP derives [1] a detriment for the lungs of 0.111 (Table B-20, p136) which is rounded to 0.12. ICRP is currently reviewing its choice of weighting factors, and in this paper, we tentatively suggest that unless other scientific evidence becomes available, this figure should be reduced from 0.12 to 0.10. This would have only minor implications for the dosimetry of other radionuclides, but would reduce the dose per WLM to 10.4 mSv, and the corresponding risk to $5.8 \times 10^{-4}$/WLM.

Now looking at the epidemiological side: BEIR VI [11] suggests a higher central value of the risk per unit exposure than BEIR IV, but does not actually give a value. However, the US EPA [12] has applied the BEIR VI model to a typical US population to give a risk of $5.1 \times 10^{-4}$/WLM. A parallel analysis by Lubin et al. [13] arrived at a similar value for indoor exposure. Recently, the EPA [14] has revised its risk estimates based on more recent mortality data, and now suggests a value of $5.4 \times 10^{-4}$/WLM.

It can be concluded that if the tissue weighting factor for the lung is reduced from 0.12 to 0.10, then this, combined with the latest information of radon dosimetry parameters brings the dosimetric risk to $5.8 \times 10^{-4}$/WLM. This is now very similar to the latest epidemiological estimate currently recommended by the US EPA of $5.4 \times 10^{-4}$/WLM.

## References

[1] ICRP Publication 60. 1990 Recommendations of the international commission on radiological protection, Ann. ICRP, vol. 21, 1–3, Pergamon Press, Oxford, 1990.
[2] ICRP Publication 66. Human respiratory tract model for Radiological Protection, Ann. ICRP, vol. 24, 1–3, Pergamon Press, Oxford, 1994.
[3] European Commission, Council Directive 96/29/EURATOM, Official Journal of the European Communities, (1996) L159 29.6.1996.
[4] ICRP Publication 65. Protection against Rn-222 at home and at work, Ann. ICRP, vol. 23, 2, Pergamon Press, Oxford, 1993.
[5] A. Birchall, A.C. James, Uncertainty analysis of the effective dose per unit exposure from radon progeny and implications for ICRP risk-weighting factors, Radiat. Prot. Dosim. 53 (1–4) (1994) 133–140.
[6] BEIR IV, Health risks of radon and other internally deposited alpha emitters, National Research Council. National Academy Press, (1988) ISBN 0-309-03789-1.
[7] J.W. Marsh, et al., Uncertainty analysis of the weighted equivalent lung dose per unit exposure to radon progeny in the home, Radiat. Prot. Dosim. 102 (3) (2002) 229–248.
[8] Commission of the European Communities (CEC) Fourth Research development Framework Programme, European Union contract F14P-CT95-0025, Risk of assessment of exposure to radon decay products (1999).
[9] ICRP Publication 92. Relative biological effectiveness (RBE), quality factor (Q) and radiation weighting factor ($W_R$), Ann. ICRP, vol. 33, 4, Pergamon Press, Oxford, 2003.
[10] J.W. Marsh, A. Birchall, K. Davis, Comparative dosimetry in homes and mines. Estimation of K factors, Presented at the 7th International Symposium on the Natural Radiation Environment (NRE-VII), 20–24 May 2002, Rhodes, Greece.
[11] BEIR VI, Health effects of exposure to radon, National Research Council, National Academy Press, (1999) ISBN 0-309-05645-4.

# Reference fields and calibration techniques for Rn-220 measuring instruments

## E. Gargioni*, D. Arnold

*Physikalisch-Technische Bundesanstalt (PTB), Braunschweig, Germany*

**Abstract.** One of the main concerns of any developer of a thoron reference chamber is to maintain a uniform and homogeneous gas distribution. However, the use of fans and air circulation systems often produces turbulent flows in the chamber, which can instead generate undesired non-uniformities in the atmosphere. As an example, the homogeneity studies at the calibration facility of the Physikalisch-Technische Bundesanstalt (PTB) are presented. The PTB facility consists of a stainless-steel reference chamber with a volume of approximately 0.1 m$^3$ and several electro-deposited $^{228}$Th sources. The turbulent flow created by the fan inside the chamber allows homogeneous $^{220}$Rn fields to be created only at a distance from the source greater than 15 cm.
© 2004 Elsevier B.V. All rights reserved.

*Keywords:* Thoron; Activity concentration; Calibration

## 1. Introduction

Several techniques for measuring the activity concentration of thoron ($^{220}$Rn) and its progenies in working and living areas have been developed in the past [1], and in most of the cases the methodologies used for $^{222}$Rn were simply extended to the monitoring of $^{220}$Rn, although in some situations systematic errors could seriously affect the measurements. While the spatial distribution of $^{222}$Rn and its progenies is uniform in most conditions, that of $^{220}$Rn, due to its relatively short half-life ($T_{1/2}$=55.6 s), strongly depends on the distance from the emanating source [1–3]. Moreover, the longer-lived progenies, $^{212}$Bi ($T_{1/2}$=60.6 min) and $^{212}$Po ($T_{1/2}$=10.6 h), are never in equilibrium with the mother nuclide. This implies that the concentration of the $^{220}$Rn progenies cannot be derived from that of the gas itself using the equilibrium factor, as in the case of $^{222}$Rn. In order to develop a new methodology to assess the risk of thoron exposure, both the gas itself and

---
* Corresponding author. Tel.: +49 531 592 6602; fax: +49 531 592 6405.
 E-mail address: elisabetta.gargioni@ptb.de (E. Gargioni).

0531-5131/ © 2004 Elsevier B.V. All rights reserved.
doi:10.1016/j.ics.2004.09.020

its progenies should be measured indoors with two distinct procedures [4]. In this respect, the calibration of $^{220}$Rn measuring devices and the development of accepted measures for quality control play an important role.

In the past few years, a number of facilities for calibrating thoron measuring systems have been developed [5–9]. The basic set-up, common to all systems, consists of an airtight reference volume in which the thoron atmosphere is produced, a thoron source or thoron supply system, and several devices to monitor the thoron activity concentration and the environmental parameters. Active thoron measuring instruments can be externally connected to the chamber by means of sampling ports, while passive detectors must be put inside. The major concern of any developer of a thoron reference chamber is to create a uniform and homogeneous atmosphere. The use of air circulation systems, however, produces turbulent flows in the chamber, which can generate undesired non-uniformities in the gas distribution [10]. Moreover, a turbulent air circulation can affect the response of passive detectors, which work on the principle of gas diffusion and, due to the high air velocity, are often calibrated in conditions not found in practice. Up to now, extensive studies on the thoron distribution inside the chamber volume are missing and the effect of turbulent flows on the response of passive thoron monitors has not been investigated thoroughly. At the PTB reference chamber, the thoron space distribution was investigated under typical operating conditions. The results of this study show that in some regions of the chamber volume the gas is not uniformly distributed and care should be taken during the calibration of active and passive systems.

## 2. Homogeneity studies at the PTB thoron chamber

### 2.1. Materials and methods

The metrological facility for $^{220}$Rn developed at the PTB in the past few years consists of a cylindrical stainless-steel chamber with a nominal volume of 0.1 m$^3$, in which thoron gas fields with different activity concentrations are produced by means of $^{228}$Th exhalation sources. A source, 3 cm in diameter, is placed at the bottom, along the axis of the reference chamber, and a fan at the top mixes the air. The experimental set-up is schematically shown in Fig. 1.

The characterization of the $^{220}$Rn atmosphere was performed using a multi-wire ionization chamber (MWIC) operating in pulse mode and is described elsewhere [9].

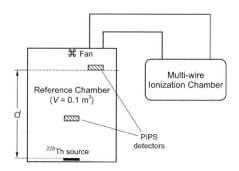

Fig. 1. Schematic diagram of the PTB thoron reference chamber. Chamber radius=22.5 cm, height=63.5 cm.

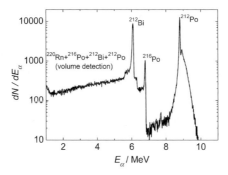

Fig. 2. A typical alpha spectrum measured with the PIPS detector in air inside the thoron reference chamber. Measuring time $t=3 \cdot 10^5$ s, source-to-detector distance $d=25$ cm.

The uniformity of the $^{220}$Rn reference fields was studied by measuring the $^{216}$Po counts directly deposited on the surface of two Passivated-Implanted Planar Silicon (PIPS) detectors placed at several distances from the source, ranging from 1.5 to 44 cm. With the PIPS detectors, alpha-particle spectra were measured in air. The count rate of the $^{216}$Po peak is related to the local $^{220}$Rn concentration. $^{216}$Po ($T_{1/2}=0.145$ s) is in fact the first $^{220}$Rn daughter and, due to its short half-life, is always in equilibrium with $^{220}$Rn.

### 2.2. Results

A typical alpha-particle spectrum measured in air inside the thoron chamber with a PIPS detector is shown in Fig. 2. The sharp peaks are produced by the progenies directly deposited on the detector surface, while the continuum part at low energies originates from alpha particles emitted in the volume around the detector, which deposit in air only a fraction of their energy. The distance $d$ from the $^{228}$Th exhalation source was varied between 1.5 and 44 cm. The temporal stability of the thoron atmosphere during the measurements was checked by recording the $^{216}$Po peak in the spectrum every 3600 s. The count rate in the peak resulted to be constant with time, within the statistical uncertainty. Therefore, the sum spectrum was analyzed to study the thoron distribution in each point. In Fig. 3, the measured $^{216}$Po counts at different distances $d$ from the source are presented for one PIPS detector positioned along the axis of the chamber and for another one placed on

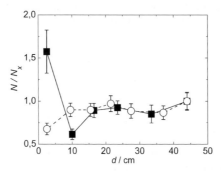

Fig. 3. Results of the thoron spatial distribution inside the PTB reference chamber. The $^{216}$Po counts $N$, relative to the value $N_x$ at $d=44$ cm, were measured along the axis (full squares) and along one side of the reference chamber (open circles). Measuring time $t=1.6 \cdot 10^5$ s. The error bars represent the standard uncertainty of the ratio.

the side of the chamber where the MWIC is sampling the air (see Fig. 1). The analysis of the measured spectra is performed by means of an analytical fitting procedure, using Gaussian functions and a semi-empirical low-energy function, which represents the continuous pulse distribution due to alpha decay events in the volume around the detector. The count rate $N$ was normalized to the value $N_x$, measured at 44 cm from the source. The measurements at 1.5 cm on the chamber axis were not possible due to the detection of the alpha-particles directly emitted by the $^{228}$Th source in the same energy range. The results show that the atmosphere can be considered to be uniform only at a distance from the source greater than approximately 15 cm. The concentration is decreasing between 2.5 and 10 cm along the axis and, at the same time, it is increasing between 2.5 and 10 cm along the side. This shows that a thoron cloud with higher concentration is present in that area of the chamber, probably due to a steady-state vortex created by the fan and by the presence of the detectors themselves. However, further investigations for a better understanding of the role of turbulence inside the chamber, also in the presence of different detectors, are desirable.

## 3. Conclusions

A simple method was used at the PTB to check the spatial distribution of $^{220}$Rn in the reference chamber. The recording of $^{216}$Po directly deposited on the surface of solid-state detectors show that care must be taken when positioning the instruments under study with respect to the thoron source or supply system, not only in walk-in chambers [10,11] but also in smaller volumes, particularly because the presence of different devices in the chamber can modify the air flow and thus the gas distribution. Finally, future work is required in order to improve the calibration techniques for thoron measuring systems and to study the effect of turbulent flows on the response of passive dosimeters.

## References

[1] C. Nuccetelli, F. Bochicchio, The thoron issue: monitoring activity, measuring techniques and dose conversion factors, Radiat. Prot. Dosim. 78 (1) (1998) 59–64.
[2] M. Doi, et al., Spatial distribution of thoron and radon concentrations in the indoor air of a traditional Japanese wooden house, Health Phys. 66 (1) (1994) 43–49.
[3] T. Yamasaki, Q. Guo, T. Iida, Distributions of thoron progeny concentrations in dwellings, Radiat. Prot. Dosim. 59 (2) (1995) 135–140.
[4] L. Tommasino, R. Falk, K. Fujimoto, Pinocchio, Tengu and Trolls are very able to protect themselves from thoron exposure, Radiat. Prot. Dosim. 104 (2) (2003) 99–101.
[5] I.M. Fisenne, A.J. Cavallo, Radon, thoron and progeny exposure facility, in: N.A. Chieco (Ed.), 28th edition, EML Procedures Manual. Report HASL-300, vol. I, US DOE, New York, 1999.
[6] K.N. Yu, V.S.Y. Koo, Z.J. Guan, A simple and versatile $^{222}$Rn/$^{220}$Rn exposure chamber, Nucl. Instrum. Methods, A 481 (2002) 749–755.
[7] S. Tokonami, et al., Sensitivity to thoron on passive radon detectors, Proc 10th IRPA Congress, Hiroshima, 2000.
[8] H. Möre, R. Falk, L. Nyblom, A bench-top calibration chamber for $^{220}$Rn activity in air, Environ. Int. 22 (1) (1996) S1147–S1153.
[9] E. Gargioni, A. Honig, A. Röttger, Development of a calibration facility for measurements of the thoron activity concentration, Nucl. Instrum. Methods, A 506 (2003) 166–172.
[10] W. Zhuo, et al., Simulation of the concentrations and distributions of indoor radon and thoron, Radiat. Prot. Dosim. 93 (4) (2001) 357–368.
[11] T. Ishikawa, Effects of thoron on a radon detector of pulse-ionization chamber type, Radiat. Prot. Dosim. 108 (4) (2004) 327–330.

ELSEVIER

www.ics-elsevier.com

# Dosimetric considerations for environmental radiation and NORM

Steven L. Simon*

*Division of Cancer Epidemiology and Genetics, National Cancer Institute, National Institutes of Health, 6120 Executive Blvd., Bethesda, MD 20892-7238, USA*

**Abstract.** The physical theory of dosimetry for environmental radiation, including radiation emitted from naturally occurring radioactive materials (NORM), is no different than that required in any other setting where doses are estimated. However, the application of such theory to environmental radiation and NORM may require considerations that differ from dose estimation elsewhere. This is especially true if the intent is to provide estimated doses for epidemiologic analyses. It should be realized that metrics of radiation dose for radiation protection purposes are generally not the same as for analytic epidemiologic studies which require estimates of absorbed dose to specific organs of identified persons. In addition, exposures to environmental radiation and NORM typically involve radiation fields that vary considerably over space, and the patterns of an individual's movements, as well as the types of buildings in which they reside and work, can significantly affect the dose received from external radiation. Realistically describing the spatial variation of environmental exposure rates is a difficult challenge for environmental dosimetry, rather than the physical principles that are relatively well understood. This publication will review these ideas in the context of improving estimated doses from high background radiation studies. © 2004 Elsevier B.V. All rights reserved.

*Keywords:* Dose; Dosimetry; Natural background radiation; NORM; Epidemiology

## 1. Introduction

Radiation doses from ionizing radiation in areas of the world with high natural background radiation (HNBR) have been reported for several decades. However, the publications over the years have reported "dose" in terms of various metrics including exposure, dose equivalent, equivalent dose, effective dose equivalent, and effective dose. Similarly, many publications have not carefully defined whether the reported "dose" was

---

\* Tel.: +1 301 594 1390; fax: +1 301 402 0207.
*E-mail address:* ssimon@mail.nih.gov.

0531-5131/ © 2004 Elsevier B.V. All rights reserved.
doi:10.1016/j.ics.2004.11.016

intended to represent dose to air, dose to a specific tissue/organ or the whole body, dose to a representative person, or dose to an identified person. Lack of clarity regarding the definition of reported "doses" makes comparison of the findings from different studies difficult. Moreover, in some cases, improper dose units have been used to infer epidemiologic conclusions.

It is clear that different types of investigations may require different metrics of dose and different levels of detail to assess dose. The range of detail included in dose estimation can extend from simple to relatively complex, depending on whether the goal is simple documentation of the air kerma at a specific location or an estimation of the organ absorbed dose received by identified persons. Dose estimation in studies of HNBR should include both external and internal dose where relevant, though due to space considerations, discussion in this paper is limited to external dose.

It is the purpose of this paper to discuss issues relating to choice of an appropriate metric of radiation dose and the means to obtain external doses to specific organs, particularly for studies of HNBR. These considerations can be used to advance the quality of estimated doses towards the level required for analytic epidemiologic studies.

## 2. Discussion

### 2.1. Choice of metric of radiation exposure

Though there is little discussion in the literature about requirements of dosimetry for epidemiologic purposes, the International Commission on Radiological Protection has long noted that "Both equivalent dose and effective dose are quantities intended for use in radiological protection, including the assessment of risks in general terms...For estimation of the likely consequences of an exposure of a known population, it will sometimes be better to use absorbed dose...relating to the exposed population" [1]. Moreover, the U.S. National Academy of Sciences reviewed criteria to be considered in dose reconstruction for epidemiologic uses [2] and concluded that dose estimates should be reported as annual organ absorbed doses from both low-LET and high-LET radiation. Because radiation- and tissue-weighting factors for equivalent dose and effective dose have been developed for radiation protection, rather than for research purposes, and because those factors are subject to change over time, research studies to quantify health risks are better served by estimates of absorbed dose to specific organs. Investigators with an interest in the relationship between high background radiation exposure and health risks should seriously consider estimating organ-specific absorbed doses (Gy) on the basis of identified individuals.

### 2.2. Instrument "dose" to organ dose

In this discussion, only two types of measurements will be discussed: (1) measurements of dose rate or time-integrated dose in air, and (2) measurement of time-integrated dose on the body of individuals. The first technique might be accomplished by various types of meters and detectors such that the instrument does not receive significant backscatter from a person's body and does not significantly perturb the radiation field. The second technique might be accomplished by an integrating dosimeter (e.g., TLD, film badge, etc.) attached to the body of a person with the intention that the detector receives backscatter from the body.

"In air" measurements may be reported by an instrument in a variety of metrics and units, depending on the calibration of the device: Roentgen $s^{-1}$, rad $s^{-1}$ or rem $s^{-1}$, or mGy $s^{-1}$ or mSv $s^{-1}$. In this case, the measurement pertains to a point in space even if the device is calibrated in absorbed or equivalent dose. The measurement represents, at best, a dose or dose rate that might be received at that specific location.

"On body" measurements are generally reported as a form of equivalent dose (*personal dose equivalent* refers to a calibration on an ICRU slab of a specified depth, generally 10 mm). Here, the measurement includes scattered radiation typical of the body and reflects an integration of the dose rates experienced as the individual moves throughout a spatially varying radiation field.

The value of "in air" or "on body" measurements will depend on their purpose, though neither are the preferred quantity for epidemiological studies, i.e., absorbed dose to specific organs [1]. "In air" or "on body" measurements can be used to derive organ doses, however, different calculations and assumptions are necessary. If exposure rate is measured (e.g., by a pressurized ionization chamber) and expressed in units of C $kg^{-1}$ $s^{-1}$, the measured value is directly related to the air kerma rate (ignoring the small correction for radiative losses) through $W/e$, the mean energy required to form an ion pair in air. In that case, the absorbed dose to a specific organ or tissue can be estimated as:

$$D_T = \dot{X} t (W/e)(D_T/K_a) \quad (1)$$

where, $D_T$=tissue absorbed dose (Gy), $\dot{X}$=exposure rate (C $kg^{-1}$ $s^{-1}$, where 1 R $s^{-1}$=2.58×10$^{-4}$ C $kg^{-1}$ $s^{-1}$), $t$=time (s) spent at location with exposure rate as described, $W/e$=mean energy expended in air to form an ion pair ≅ 34 J/C, $K_a$=air kerma (Gy).

If an "on body" measurement is reported, where the detector has been calibrated to the ICRU slab, the reported *personal dose equivalent* may be converted to air kerma and the organ absorbed dose estimated:

$$D_T = H_p(10)[K_a/H_p(10)](D_T/K_a) \quad (2)$$

where $H_p(10)$=personal dose equivalent (mSv).

Representative values of the coefficients for Eqs. (1) and (2) are provided in Table 1 [3]. The values presented are for a rotationally symmetric exposure geometry and for three energies: (i) 0.186 MeV (energy of the primary gamma ray emitted by $^{226}$Ra), (ii) 0.43 MeV (average energy of the $^{238}$U chain in equilibrium), and (iii) 0.64 MeV (average energy of the $^{232}$Th chain in equilibrium).

Table 1
Coefficients [3] for estimating absorbed organ doses in high background studies

| Source | Average gamma energy (MeV) | Air kerma per unit $H_p(10)$[a] (Gy $Sv^{-1}$) | Organ dose per unit air kerma[b] (Gy $Gy^{-1}$) | | | |
|---|---|---|---|---|---|---|
| | | | Red bone marrow | Breast | Lung | Thyroid |
| Ra-226 | 0.186 | 0.66 | 0.84 | 0.89 | 0.91 | 1.14 |
| U-238 chain (in equil.) | 0.43 | 0.78 | 0.79 | 0.85 | 0.86 | 1.03 |
| Th-232 chain (in equil.) | 0.64 | 0.82 | 0.79 | 0.86 | 0.86 | 1.02 |

[a] $[K_a/H_p(10)]$ in Eq. (2).
[b] $[D_T/K_a]$ in Eqs. (1) and (2).

## 2.3. Increasing realism of external dose estimations

Estimates of absorbed (external) dose to specific organs of individuals, as required by epidemiologic studies, must account for the dose obtained both in and outdoors by accounting for the air kerma rate at all locations where an individual spends significant amounts of time as well as the proportions of time spent at each location. It is well known that when indoors, the building can provide shielding against radiation emitted from the soil; however, the building can potentially contribute to the external dose if it is made from earthen materials derived from a HNBR area. To add necessary realism to the estimated external dose for an individual, Eq. (1) can be rewritten in the form of a summation of exposure rates at all locations where time is spent (assuming here that only a single choice of the energy dependent ratio, $[D_T/K_a]$, is needed).

$$D_T = \left[ \sum_{i=1}^{n} \dot{X}_{\text{indoors},i} t_{\text{indoors},i} + \sum_{j=1}^{m} \dot{X}_{\text{outdoors},j} t_{\text{outdoors},j} \right] (W/e)(D_T/K_a) \qquad (3)$$

where $i$ refers to indoor locations with significantly different exposure rates, $j$ refers to outdoor locations with significantly different exposure rates, $\dot{X}$=exposure rate (C kg$^{-1}$ s$^{-1}$) either indoors or outdoors, $t_{\text{inside},i}$ refers to the number of seconds per day spent indoors at location $i$, $t_{\text{outside},j}$ refers to the number of seconds per day spent outdoors at location $j$.

## 3. Concluding remarks

While knowledge about HNBR areas has increased substantially over the years [4], estimation of doses in many studies still needs to be improved for several reasons. First, imprecise definitions of the metric and improper units of reported "doses" can lead to confusion and error in making comparisons among different sites. The definitions of weighted dose metrics, e.g., *effective dose* [1], are subject to change over time and include modifying (weighting) factors to the absorbed dose that are not relevant when performing analytic epidemiologic studies designed to determine health risks. Estimates of absorbed dose to specific organs for individuals are necessary for epidemiologic analyses and can be made from measurements of exposure, air kerma, or personal dose equivalent. Increased realism in organ dose estimates can be accomplished by properly accounting for spatial variations in the radiation field including the differences in air kerma rate indoors and outdoors and the time spent at each location, or by obtaining measurements from personal radiation monitoring devices worn by individuals.

## References

[1] ICRP, Recommendations of the International Commission on Radiological Protection, ICRP Publication 60, Pergamon, Elsevier Science, NY, 1990.
[2] NAS/NRC, Radiation Dose Reconstruction for Epidemiologic Uses, National Academy of Sciences/National Research Council, National Academy Press, Washington, DC, 1995.
[3] ICRP, Conversion Coefficients for Use in Radiological Protection Against External Radiation, International Commission on Radiological Protection, Publication 74, Pergamon, Elsevier Science, NY, 1997.
[4] UNSCEAR, Sources and Effects of Ionizing Radiation, Report to the General Assembly, Vol. 1, United Nations Scientific Committee on the Effects of Atomic Radiation, New York, 2000.

ELSEVIER

www.ics-elsevier.com

# Measurements to determine the radiological impact of uranium and thorium in soils in the darling scarp

## L.F. Toussaint*

*Murdoch University, School of Mathematical and Physical Sciences, Western Australia 6004, Australia*

**Abstract.** In the early 1970s, it was noticed that the naturally occurring gamma radiation levels measured over the pisolitic laterite in the Darling scarp were some three times higher than gamma levels over sand in the Perth coastal plain in Western Australia. Analyses of soil samples taken in the scarp indicated that the main contributor was from the thorium decay chain (0.25 Bq g$^{-1}$) with a lesser contribution from the uranium decay chain (0.07 Bq g$^{-1}$). The thorium concentration exceeded the world range reported by the United Nations Scientific Committee on the Effects of Atomic Radiation Report to the General Assembly. It was considered prudent to carry out further radiological assessment from all likely radiation exposure pathways, including gamma, radon gas and thoron gas in the scarp. A survey of radon concentrations in homes showed the Darling scarp levels to be higher than the coastal plain. Later, passive monitors developed by NIRS in Japan showed that scarp thoron levels were higher than for the coastal plain. The total annual dose to individuals living in the scarp area was found to be 4.6 mSv while the dose to those living on the coastal plain was found to be 1.9 mSv. © 2004 Elsevier B.V. All rights reserved.

*Keywords:* Gamma; Radon; Thoron; Radium; Thorium, NORM

## 1. Introduction

The city of Perth is situated on the coastal plain of Western Australia and the Darling scarp lies to the east of the Perth. The Perth coastal plain is generally lime and quartz sand while the Darling scarp is comprised of pre-Cambrian rock, overlain by granite, lateritic gravel and laterite subsoils. In the early 1970s, measurements of natural gamma radiation in parts of the Darling scarp indicated levels that were some three times higher

---

* Tel./fax: +8 9457 1080.
  *E-mail address:* billt@iinet.net.au.

0531-5131/ © 2004 Elsevier B.V. All rights reserved.
doi:10.1016/j.ics.2004.09.026

than those over the coastal plain. More detailed studies were subsequently carried out. These studies included the measurement of radiation dose from various exposure pathways with the aim of assessing the likely total radiation dose to those living in the area.

## 2. Gamma radiation surveys

### 2.1. Relationship between gray and sievert for environmental conditions

For environmental situations, the relationship between absorbed dose rate in air ($\mu$Gy/h) to body dose rate ($\mu$Sv/h) as quoted by UNSCEAR [1] is:

$$0.7 \times \text{absorbed dose-rate in air} = \text{body dose-rate}$$

The absorbed dose rate in air $D$ at 1 m above a semi-infinite plane may be found from the application of the equation:

$$D = 0.427 \times \text{U238} + 0.662 \times \text{Th232} + 0.043 \times \text{K40} \tag{2.1}$$

where $D$ is the absorbed dose rate in air (nGy/h); U238 is the activity concentration of uranium-238 (Bq kg$^{-1}$); Th232 is the activity concentration of thorium-232 (Bq kg$^{-1}$); and K40 is the activity concentration of potassium-40 (Bq kg$^{-1}$).

### 2.2. First studies using portable high pressure ionization monitoring instruments

Yeates and King [2] used a portable high pressure ionization monitoring instrument containing 30% argon and 70% nitrogen at a pressure of 45 atmospheres for the gamma measurements over areas of Naturally Occurring Radioactive Material (NORM). The gamma radiation was found to vary from 0.002 $\mu$Gy/h (over limestone in the coastal plain) to 0.35 $\mu$Gy/h (for the Darling scarp).

### 2.3. Mobile study using a high pressure ionization monitoring instrument

A gamma radiation study using a Reuter Stokes RSS-111 high pressure ionization instrument mounted in a vehicle [3] recorded gamma data every 5 s onto magnetic tape. The recorded gamma radiation levels varied from about 0.06 $\mu$Gy/h (Perth coastal plain) to about 0.3 $\mu$Gy/h (scarp). The highest gamma radiation level (0.6 $\mu$Gy/h) was found over a small area in the scarp.

### 2.4. Large-scale gamma survey in homes using TLDs

A national survey by Langroo et al. [4] measured gamma radiation levels in homes throughout Australia with thermoluminescent (TLD) gamma monitors. The results of this survey indicated a mean gamma dose rate of 1000 $\mu$Sv/year and an approximate normal distribution. For the Darling scarp, the annual dose rate in homes was 1300 $\mu$Sv/year. This figure is less than that predicted by the hourly dose rate measured by other gamma measuring instruments outside homes. The reason for this difference is because of the construction of most modern Australian homes. A sand pad some 20 cm thick is used as a building base and then a concrete floor is poured over the top of this. This has a tendency to shield the higher terrestrial gamma radiation.

## 3. Analysis of soil in the darling scarp

For soil and gravel samples (pisolitic laterite) collected in the scarp, gamma spectroscopy indicated that the main contributor was from the thorium decay chain [5]. The activity concentration range of 0.04–0.50 Bq g$^{-1}$ (mean 0.25 Bq. g$^{-1}$) exceeded the UNSCEAR quoted range of 0.01–0.05 Bq g$^{-1}$ (mean 0.03 Bq. g$^{-1}$). For uranium, the measured scarp activity concentrations were in the 0.02–0.11 (0.07) Bq g$^{-1}$ range. The range quoted by UNSCEAR was 0.01–0.05 (0.03) Bq g$^{-1}$. The application of Eq. (2.1) showed that there was good agreement between calculated and measured gamma levels. Typical concentrations in the Darling scarp for uranium and thorium of 0.07 Bq g$^{-1}$ and 0.25 Bq g$^{-1}$, respectively, gave a predicted gamma level of 0.19 μGy/h.

## 4. Radon measurements

Langroo et al. [4] conducted an Australia-wide radon in homes survey over a 1-year monitoring period using the passive CR-39 monitors. The median radon concentration was 11 Bq m$^{-3}$ for Western Australia with the distribution being approximately log-normal [6]. In a few homes in the Darling scarp, the radon concentration exceeded 100 Bq m$^{-3}$. The results from a Western Australian survey [7] indicated a mean radon concentration of 11.6 Bq m$^{-3}$ (coastal plain) and 28.9 Bq m$^{-3}$ (Darling scarp—Postcode 6076) which relates to an approximate annual dose of 0.19 mSv (coastal plain) and 0.48 mSv (scarp).

## 5. Thoron measurements

Passive monitors developed by the National Institute for Radiological Sciences (NIRS) in Japan have been used by Doi et al. [8] to estimate thoron concentrations in traditional earth-walled or "tsuchi-kabe" Japanese homes. These monitors were used in the study of thoron concentrations in the Darling scarp and the coastal plain [9]. The mean coastal plain thoron concentration inside houses was found to be 3.4 Bq m$^{-3}$ and the Darling scarp in-house concentration was 14.0 Bq m$^{-3}$. Measurements outside scarp homes were found to be much more variable and higher (mean 76.3 Bq m$^{-3}$) due to the short half-life of thoron. The approximate annual thoron dose contribution indoors was thus 0.68 mSv for the coastal plain and 2.8 mSv for the scarp.

## 6. Discussion

For those living in the Darling scarp area, the estimated annual radiation dose was found to be 1.3 mSv from the gamma contribution, 0.48 mSv from the radon contribution and 2.8 mSv from thoron exposure, giving a total annual dose of 4.58 mSv. For those living on the coastal plain, the estimated annual radiation dose was 1.0 mSv from the gamma contribution, 0.19 mSv from the radon contribution and 0.68 mSv from thoron exposure, giving a total annual dose of 1.87 mSv. For Western Australian-type housing construction (concrete floor over a sand pad) in the scarp, the indoor thoron levels were considerable less than those outdoors. There were large variations in thoron concentrations which are primarily due to the relatively short half-life of thoron and the monitor-to-thoron source distances involved.

Although the inhabitants of the scarp area receive a higher radiation dose than those living in the coastal plain, the low dose levels and small population of the area suggest that there is insufficient statistical power for a meaningful epidemiological study to be undertaken. From the magnitude of the dose assessed, however, and studies (such as those by Mifune et al. [10]) in other parts of the world, the dose is unlikely to result in any serious health consequences. It would nevertheless be prudent to ensure that any gravels and soils used for rammed earth housing construction are assessed radiologically prior to any widespread use [11]. This may be done initially by calculating an "Activity Index" [12] based on the radionuclide content of the building material.

## Acknowledgements

Sincere thanks and appreciation is expressed to Professor Hiroshige Morishima, Dr. Suminori Akiba, the organizing committee and Kinki University National Institute of Radiological Sciences for making this conference possible. Thanks and appreciation is also expressed to fellow friends and colleagues in both Japan and Australia for their assistance and cooperation.

## References

[1] UNSCEAR Sources, Effects and Risks of Ionising Radiation. United Nations Scientific Committee on the Effects of Atomic Radiation (1983, 1993 and 1998). Report to the General Assembly, with annexes (New York, NY, United Nations).
[2] D.B. Yeates, B.E. King, Estimation of the gamma-ray natural background radiation dose to an urban population in Western Australia, Health Physics, vol. 25, Pergamon Press, 1973 (Oct.), pp. 373–379.
[3] L.F. Toussaint, Background radiation in Western Australia, Radiation Protection in Australia 3 (4) (1985) 151–155 (The Journal of the Australasian Radiation Protection Society).
[4] M.K. Langroo, et al., A nationwide survey of $^{222}$Rn and gamma radiation levels in Australian homes, Health Physics 6 (1991 (December)) 753–761.
[5] Z.J. Alach, et al., Radionuclide Concentrations in the Darling Scarp of Western Australia, Radiation Protection in Australia 14 (2) (1996 (April)) 35–38 (The Journal of the Australasian Radiation Protection Society).
[6] K. Fujimoto, Correlation between indoor radon concentration and dose-rate in air from terrestrial gamma radiation in Japan, Health Physics 75 (3) (1998) 291–296.
[7] L.F. Toussaint, Radon concentrations in western Australian homes, Radiation Protection in Australia 15 (1) (1998) 15–19 (The Journal of the Australasian Radiation Protection Society).
[8] M. Doi, S. Kobayashi, K. Fujimoto, A passive measurement technique for characterisation of high-risk houses in Japan due to enhanced levels of indoor radon and thoron concentrations, Radiation Protection Dosimetry, vol. 45, Nuclear Technology Publishing, 1992, pp. 425–430, No. 1/4.
[9] L.F. Toussaint, et al., The measurement of thoron concentrations in Australia using the Japanese passive R–T dosimeter, in: Akira Katase, Michikuni Shimo (Eds.), Proceedings of the Seventh Tohwa University International Symposium, Radon and Thoron in the Human Environment. Fukuoka, Japan, World Scientific Press, 1998, pp. 373–378.
[10] M. Mifune, et al., Cancer mortality survey in a spa area (Misasa, Japan) with a high radon background, Japanese Journal of Cancer Research, vol. 83, Japanese Cancer Association, 1992, No. 1.
[11] M. Walsh, P. Jennings, A study of environmental radon levels in rammed earth dwellings in the south west of Western Australia, Radiation Protection in Australia 19 (2) (2002) 67–73 (The Journal of the Australasian Radiation Protection Society).
[12] R. Mustonen, et al., Enhanced Radioactivity of Building Materials. Report prepared for the European Commission under contract No 96-ET-003, STUK, Radiation and Nuclear Safety Authority, Helsinki, Finland, 1997.

www.ics-elsevier.com

# Cancer and non-cancer epidemiological study in the high background radiation area of Yangjiang, China

Jianming Zou[a,*], Zufan Tao[b], Quanfu Sun[b], Suminori Akiba[c], Yongru Zha[a], Tsutomu Sugahara[d], Luxin Wei[b]

[a]*Guangdong Prevention and Treatment Center for Occupational Disease, Guangzhou 510300, China*
[b]*National Institute for Radiological Protectivy, Chinese Center for Disease Control and Prevention, Beijing 100088, China*
[c]*Department of Public Health, Kagoshima University Faculty of Medicine, Kagoshima 890-8520, Japan*
[d]*Health Research Foundation, Kyoto 606-8225, Japan*

**Abstract.** The major objective of this study is to examine cancer mortality risk associated with low-level radiation exposure occurring in the high background radiation area (HBRA) in Yangjiang of Guangdong Province, China. The average annual effective doses received by the inhabitants from natural sources of external and internal exposures in HBRA are estimated to be 2.10 and 4.27 mSv, respectively, and the corresponding doses in the control area (CA) to be 0.77 and 1.65 mSv. We analyzed the mortality of non-cancer diseases as well in order to shed light on the comparability of the HBRA and the CA. We examined mortality for cancer and non-cancer diseases during the period 1979–1998. The prospective mortality study followed 125,079 subjects during the period 1979–1998, accumulated 1,992,940 person-years (PYs) at risk, and ascertained 12,444 deaths, including 1202 cancer deaths. The mortality of all cancer showed no difference between the HBRA and the CA [relative risk (RR)=1.00; 95% confidence interval (CI), 0.89 to 1.14]. When cancer deaths were limited to persons with pathological diagnosis, the RR changed only slightly (RR=0.99; 95% CI, 0.78 to1.26). The RR was not evidently modified by sex or age or follow-up period (1979–1986, 1987–1998). In site-specific cancer mortality analysis, only cancer of the esophagus showed a statistically significant excess in the HBRA (RR=2.61; 95% CI, 1.11 to 7.66). However, the observed excess mortality of esophageal cancer did not show a monotonic increase with external radiation dose or cumulative lifetime dose. The RR comparing non-cancer mortality in the HBRA with that in the CA was 1.06 (95% CI, 1.01 to 1.10), which was a statistically significant increase. However, the excess was limited to those aged under 50 and the latter half of the observation period (the period 1987–1998), suggesting that the excess mortality may be due to recent changes in lifestyles of the

* Corresponding author. Tel./fax: +86 20 89024250.
E-mail addresses: zoujm-gz@163.net (J. Zou), qfusun@public3.bta.net.cn (Q. Sun).

0531-5131/ © 2004 Published by Elsevier B.V.
doi:10.1016/j.ics.2004.11.167

younger generations. In the cause-specific analysis of non-cancer deaths, disease of the digestive organs showed a statistically significant increase. This appears to be mainly due to liver diseases. In conclusion, the present study showed no increase of cancer mortality in the HBRA. However, it is difficult for the present study to support or deny the possibility that the radiation-related cancer risk associated with chronic exposure to low-dose radiation may be different from the risk associated with high dose ranges or high dose-rate exposure. The observed increase of non-cancer mortality in the present study is unlikely to be attributable to radiation exposure. © 2004 Published by Elsevier B.V.

Keywords: Cancer; Non-cancer; Epidemiology; Natural radiation; Low dose

## 1. Introduction

An epidemiological study in the high background radiation area (HBRA) of Yangjiang in Guangdong province, China, has been carried out since 1972. The primary objective of this study is to examine the cancer mortality risk associated with naturally occurring high background radiation exposure by comparing mortality rates here to those in the neighbouring control area (CA). In addition, the mortality of non-cancer diseases is compared between the HBRA and the CA in order to shed further light on the comparability of the two areas. The average annual effective doses received by the inhabitants from natural sources of external and internal exposures in the HBRA were estimated to be 2.10 and 4.27 mSv, respectively, and the corresponding doses in the CA to be 0.77 and 1.65 mSv.

In 2000, we reported the results of a cancer mortality study covering the period 1979–1995 [1]. In this paper, we present the results from the analysis of combined data for the periods from 1979 to 1986, and from 1987 to 1998, adding 3 years to our previous report.

## 2. Materials and methods

The mortality data for the period 1979–1986 were collected by a prospective follow-up survey of dynamic populations, consisting of around 80,000 inhabitants in the HBRA and as many in the CA. The mortality data for 1987–1998 were obtained from a prospective follow-up survey of a fixed cohort consisting of 106,517 individuals alive as of the January 1, 1987. The methods of mortality follow-up survey are described in detail elsewhere [2]. In brief, trained local census takers surveyed the hamlets of the study areas to collect information on deaths and migrations among the inhabitants in each hamlet. The collected information was recorded in the demographic survey sheet prepared for each household in the hamlet. The members of task group on cancer mortality then visited the studied areas and reviewed the survey sheets. In order to ascertain the cause of death, they visited all the major hospitals in the study area, and reviewed medical records of the deceased and extracted relevant information. If necessary, they revisited the local village doctors and the family members or next of kin to collect further information on cause of death. The underlying cause of death thus was ascertained and coded according to the 9th revision of the International Classification of Diseases (ICD-9).

The cumulative individual dose to each cohort member was estimated on the basis of measurements of external exposure to natural radiation source. The external doses were estimated by environmental dose rate measured with scintillation survey-meters and converted into the annual absorbed doses considering the occupancy factors specific for both sex and age. For the internal doses, 4.27mSv for the HBRA and 1.65 mSv for the CA, two fixed annual values, were used in estimating the individual lifetime cumulative dose. Based on the hamlet-specific average external dose rates, the cohort members were categorized into four dose groups for internal comparison: the high, intermediate, and low dose-rate groups in the HBRA, and the control group in the CA. The average annual effective (external) doses ($10^{-5}$ Sv/a) for each group were 246.07 (224.10–308.04), 210.19 (198.07–224.09), 183.31 (125.29–198.06, and 67.92 (50.43–95.67), respectively [3].

The estimates of relative risk (RR) and excess relative risk coefficients (ERRs per Sievert) and their 95% confidence interval (CI) were obtained from Poisson regression analysis using the AMFIT in Epicure.

## 3. Results

Through 1979 to 1998, 1,992,940 person-years (PYs) at risk were accumulated with the follow-up of 125,079 subjects, 12,444 deaths including 1202 cancer deaths, 1204 deaths from external causes (injury and poisoning), and 10,038 cases from non-cancer disease except external causes were identified.

### 3.1. Cancer mortality

Cancer deaths yielded a total of 1202 cases, making up 9.7% of all causes of death. The first-five leading cancer sites, accounting for 68.5% of all cancer deaths, were cancers of the liver, nasopharynx, lung, and stomach, and leukaemia.

The sex- and age-adjusted RRs (95% CI) for certain special types of cancers by dose group are presented in Table 1. The RR for overall cancers was 1.00 (95% CI, 0.89 to 1.14), indicating that there was no difference in overall cancer mortality between the HBRA and the CA. The RRs for site-specific cancers of the nasopharynx, stomach, colon, liver, lung, bone, and female breast were at a level of less than one; the RRs for cancers of the esophagus, rectum, pancreas, skin, cervix uterus, thyroid, brain, and central nervous system, as well as those for leukaemia and lymphoma, were larger than one. However, only cancer of the esophagus showed a statistically significant excess in the HBRA,

Table 1
Estimates of relative risks for major cancer sites by dose-rate group (1979–1998)

| Site of cancer | CA | RR (95% CI) for HBRA | | | |
|---|---|---|---|---|---|
| | RR | Low Group | Interm. Group | High Group | Subtotal |
| All cancers | 1.00 | 1.08 (0.93–1.26) | 1.00 (0.86–1.17) | 0.92 (0.78–1.08) | 1.00 (0.89–1.14) |
| Leukaemia | 1.00 | 0.82 (0.35–1.86) | 1.20 (0.57–2.56) | 1.07 (0.47–2.39) | 1.03 (0.56–2.02) |
| Solid cancers | 1.00 | 1.10 (0.94–1.28) | 0.99 (0.85–1.16) | 0.91 (0.77–1.08) | 1.00 (0.88–1.14) |
| Liver | 1.00 | 1.07 (0.80–1.42) | 0.84 (0.62–1.14) | 0.74 (0.53–1.02) | 0.89 (0.70–1.13) |
| Nasopharynx | 1.00 | 0.96 (0.66–1.37) | 0.98 (0.68–1.40) | 0.88 (0.60–1.28) | 0.94 (0.71–1.26) |
| Lung | 1.00 | 0.93 (0.57–1.50) | 0.67 (0.39–1.12) | 1.04 (0.64–1.68) | 0.87 (0.60–1.30) |
| Stomach | 1.00 | 0.91 (0.55–1.48) | 0.98 (0.60–1.57) | 0.79 (0.46–1.33) | 0.90 (0.61–1.34) |
| Esophagus | 1.00 | 2.55 (0.91–8.21) | 3.30 (1.26–10.24) | 1.88 (0.60–6.37) | 2.61(1.11–7.66) |

Table 2
Relative risks of non-cancer deaths excluding external causes

| Factors | | Control | | HBRA | |
|---|---|---|---|---|---|
| | | Cases | RR | Cases | RR (95% CI) |
| All | | 2847 | 1.00 | 7191 | 1.06 (1.01–1.10)* |
| Period | 1979–1986 | 1233 | 1.00 | 3094 | 1.02 (0.95–1.09) |
| | 1987–1998 | 1614 | 1.00 | 4097 | 1.09 (1.02–1.15)* |
| Sex | Female | 1347 | 1.00 | 3353 | 1.05 (0.98–1.11) |
| | Male | 1500 | 1.00 | 3838 | 1.06 (1.03–1.13)* |
| Age | 0–39 | 285 | 1.00 | 992 | 1.31 (1.15–1.50)* |
| | 40–49 | 90 | 1.00 | 305 | 1.28 (1.01–1.62)* |
| | 50–59 | 208 | 1.00 | 525 | 1.00 (0.85–1.17) |
| | 60–69 | 551 | 1.00 | 1288 | 1.06 (0.95–1.17) |
| | 70+ | 1713 | 1.00 | 4081 | 1.01 (0.95–1.07) |

\* $P<0.05$.

RR=2.61 (95% CI, 1.11 to 7.66), and the trend test for the dose-rate group was not statistically significant ($P=0.063$). The others showed no statistically significant difference between the HBRA and the CA. Among different dose-rate groups, the RRs did not show any monotonic trend, and slightly decreased cancer mortality was observed in the highest dose-rate group.

The ERR per Sv for all solid cancers associated with cumulative lifetime dose, which was the sum of external and internal radiation doses, was estimated to be −0.06 (95% CI, −0.60 to 0.67) for the entire HBRA and the CA.

*3.2. Non-cancer mortality*

The RR for non-cancer diseases excluding external cause mortality in the HBRA was 1.06 (95% CI, 1.01 to 1.10), and was significantly higher than that in the CA, but the excess was limited to those aged under 50 and in the latter half of the observation period (1987–1998) as shown in Table 2. In cause-specific analysis, a statistically significant increase in the HBRA was observed in viral hepatitis (RR=4.79; 95% CI, 1.73 to 19.83; $P=0.001$) and chronic liver diseases (RR=1.50; 95% CI, 1.17 to 1.95; $P=0.001$). On the other hand, the mortality of tuberculosis, the most common chronic infection in the study area, was lower in the HBRA than in the CA (RR=0.63; 95% CI, 0.54 to 0.74; $P\leq0.001$).

## 4. Discussion and conclusion

The current study provided a similar result to a previously reported one that the mortality for all cancers showed no significant difference between the HBRA and the CA; relative risk was estimated to be 1.00 (95% CI, 0.89 to 1.14). In site-specific analysis, an excess of esophageal cancer deaths was found in the HBRA. However, the RR for esophageal cancer did not show any increase in relation to the increase in radiation dose. This increase is difficult to explain by radiation exposure in view of recent risk estimates obtained from a study of A-bomb survivors [4], so further study is therefore necessary.

The accuracy of diagnosis is another concern in the present study. In this study, 26% of cancer deaths were from liver cancer, which is well known for its difficulty in diagnosis.

On the other hand, for cancer of the nasopharynx, which is the second most common cancer in the studied area (18%), the false-positive and false-negative rates can be assumed to be low because the evident clinical signs of nasopharyngeal cancer make its differential diagnosis relatively easy. On top of that, most types of cancer in the study area is fatal because cancer treatment is prohibitively expensive for most of the farmers in the area. This background information suggests that the numbers of false-positive and false-negative cases in ascertaining all-cancer deaths were small in the present study. It should also be noted that the sensitivity and specificity of ascertaining cancer deaths in the present study are unlikely to show any differences between the HBRA and the CA. The overall conclusion regarding the problem of the accuracy of diagnosis in present study is that our RR estimates may be biased toward the null value due to relatively low accuracy of diagnosis, but the magnitude of such biases is unlikely to be large. In fact, the RR of all cancers based on the diagnoses of pathology or hematology was not much different from that on the basis of all kinds of diagnoses: 0.99 (95% CI, 0.78 to 1.26) vs. 1.00 (95% CI, 0.89 to 1.14).

Another concern of significance here was the excess of non-cancer mortality in the HBRA, which is limited to those aged 50 years and younger, but it did not show a monotonic increase of mortality according to external radiation dose or cumulative lifetime dose. As we know, the follow-up study of atomic bomb survivors showed an excess of non-cancer deaths (ERR per Sv=0.14, 90% CI, 0.09 to 0.19), but the excess is much smaller than cancer risk [4]. On top of that, RR comparing the HBRA and the CA is expected to increase with age because lifetime radiation dose from natural sources is expected to increase with age. Taken together, these findings suggest that the statistically significant excess of non-cancer mortality does not seem to be attributable to natural radiation exposure but may be due to recent changes in lifestyle among those aged less than 50 years.

In conclusion, the present study showed no statistically significant increase of the mortality due to overall cancer in the HBRA; no radiation-related excess of site-specific cancer in the HBRA was found. The excess of esophageal cancer in the HBRA was not likely associated with exposure to high natural radiation. The observed increase of non-cancer mortality in the HBRA is unlikely to be attributable to radiation exposure; it may be due to recent changes in lifestyle of younger generations.

## References

[1] Z.F. Tao, et al., Cancer mortality in the high background radiation areas of Yangjiang, China, during the period between 1979 and 1995, J. Radiat. Res. 41 (2000) 31–41 (Supplement).
[2] Z.F. Tao, et al., Study on cancer mortality among the residents in high background radiation area of Yangjiang, China, in: L.X. Wei, T. Sugahara, Z.F. Tao (Eds.), High Levels of Natural Radiation: Radiation Dose and Health Effects, Elsevier, Amsterdam, 1997, pp. 249–254.
[3] Y.L. Yuan, et al., Recent advances in dosimetry investigation in the high background radiation area in Yangjiang, China, in: L.X. Wei, T. Sugahara, Z.F. Tao (Eds.), High Levels of Natural Radiation: Radiation Dose and Health Effects, Elsevier, Amsterdam, 1997, pp. 223–233.
[4] D.L. Preston, et al., Studies of mortality of atomic bomb survivors. Report 13: solid cancer and noncancer disease mortality: 1950–1997, Radiat. Res. 160 (2003) 381–407.

# What did we learn from epidemiological studies in high background radiation area in India

P. Jayalekshmi[a,*], P. Gangadharan[a], V.S. Binu[a], R.R.K. Nair[a], M.K. Nair[a], B. Rajan[a], S. Akiba[b]

[a]*Karunagappally Cancer Registry, Kerala, India*
[b]*Faculty of Medicine, Kagoshima University, Kagoshima, Japan*

**Abstract.** Cancer causing potential of chronic exposure to high natural radiation seen in Chavara–Neendakara coast of Karunagappally taluk in Kerala, India is studied by the Natural Background Radiation Cancer Registry, Karunagappally since 1990. Population survey of 359,619 people recorded socio-demographic (SD) information and radiation levels measured inside (71,674) and outside (76,942) houses. Radiation level between and within the panchayats showed large variations. About 23,000 population were exposed to >4 mSv. Cancer pattern and trends were examined in relation to radiation levels. During 1990–2000 lung cancer was the predominant cancer among males (age adjusted rates (AAR) −18.2), followed by esophagus (AAR −6.3). Among females, cancers of the breast (AAR −14.3), cervix (AAR −13.4) and thyroid (AAR −4.3) were the major cancers. Only marginal differences in cancer incidence and its pattern were seen in different areas when broadly classified according to radiation level zones. Increase of incidence of lung cancer with radiation was seen among males but not in females. This needs in-depth studies. Role of confounding and competing risk factors need to be assessed with larger data sets. © 2004 Published by Elsevier B.V.

*Keywords:* Natural radiation; Cancer

## 1. Introduction

A large population is exposed to high natural radiation emitted by the black sands in Karunagappally on the Kerala coast, 100 km north of Trivandrum, the capital city of Kerala, India. In 1990, Regional Cancer Centre (RCC), Trivandrum initiated studies in Karunagappally to investigate the relation between chronic natural radiation exposure and occurrence of cancer. This is the first ever epidemiological study of cancer in the area. The study area is the entire Karunagappally taluk, which has an area of 192 km$^2$ and a

* Corresponding author. Tel.: +91 476 2685203; fax: +91 471 2447454.
  *E-mail address:* qln_nbrrkply@sancharnet.in (P. Jayalekshmi).

0531-5131/ © 2004 Published by Elsevier B.V.
doi:10.1016/j.ics.2004.11.192

population of 4,10,056 (2001 Census). Population density is 2000+ per km$^2$ and annual growth rate 0.799% (1991–2001). Majority of people are agriculturists, fishermen, cashew workers, fish processing, factory workers, coir makers, etc. Literacy rate is high (>85%). There is no dedicated cancer treatment or diagnostic center in Karunagappally. Almost 100,000 people live in the high radiation zone. Radiation source is thorium in the sands. Migration is negligible in this taluk and people have lived here for centuries.

## 2. Methodology

To meet the study objectives, the activities undertaken were as follows: (1) enumeration of the total population of Karunagappally taluk—to obtain the socio-demographic (SD) profile and prevalence of known risk factors; (2) radiation level measurements inside and outside of houses—to evaluate their role in causing the cancer burden; (3) cancer registration, which started from January 1, 1990. Surveillance activities were also undertaken from the beginning. By house-to-house visits, SD information was obtained from 359,619 persons. With technical advice from Environmental Assessment Division of BARC, gamma radiation level of inside of 71,674 houses and outside of 76,942 houses were measured. Cancer case recording has been undertaken continuously without break.

## 3. Observations

The SD survey showed findings that 60% men and 15% women above the age of 15 were tobacco habitués. Smoking was more prevalent in men and chewing in women. Female smokers were <4% and majority of them were old age women. Age at marriage of girls has been increasing over the years. About 65% of houses were made of bricks, cement and had concrete terrace. Well water is the major source for drinking and bathing. Diet and dietary practices varied between communities. More than 90% were non-vegetarians. Radiation level measurements showed variations between panchayats and even between wards in the same panchayat. Exposure level changed with mobility of subjects. Outside house levels varied from 0.3 mGy/year to 74.2 mGy/year in different areas [1]. House occupancy time was assessed and was used to obtain the exposure dose.

Cancer registry followed an active cancer registration method. By a house visit of the deceased, information on cause of death was ascertained. During the period 1990–2000, 3623 cancer cases were identified. Crude incidence rates were 91.5 for males and 71.0 for females per 100,000 and age adjusted rates (AAR) for males and females were 104.2 and

Table 1
Cancer incidence rates in Indian registries (NCRP) [2,3]

| Place | AAR male | AAR female |
|---|---|---|
| Delhi 1997–1998 [2] | 120.9 | 134.8 |
| Barshi 1997–1998 [2] | 43.9 | 51.7 |
| Mumbai 1997–1998 [2] | 117.3 | 127.9 |
| Chennai 1997–1998 [2] | 111.3 | 125.2 |
| Bangalore 1997–1998 [2] | 91.9 | 114.8 |
| Trivandrum Urban 1998–1999 [3] | 93.8 | 90.8 |
| Trivandrum Rural 1998–1999 [3] | 85.5 | 70.2 |
| Karunagappally 1998–2000 | 104.5 | 71.2 |

Table 2
Radiation levels (mSv) in panchayats and cancer incidence

| Panchayats | Population | Minimum | Median | Maximum | AAR | |
| --- | --- | --- | --- | --- | --- | --- |
| | | | | | Male | Female |
| Chavara | 35,504 | 0.900 | 3.280 | 27.680 | 121.0 | 76.3 |
| Neendakara | 14,256 | 0.790 | 3.600 | 34.200 | 84.5 | 77.4 |
| Alappad | 22,778 | 0.560 | 1.970 | 17.970 | 99.0 | 65.9 |
| Panmana | 42,251 | 1.000 | 3.040 | 21.460 | 89.7 | 67.4 |
| Thekkumbhagom | 14,980 | 0.590 | 1.320 | 18.350 | 112.4 | 78.7 |
| Karunagappally | 41,552 | 0.500 | 1.490 | 5.150 | 101.6 | 68.0 |
| Clappana | 19,326 | 0.590 | 1.310 | 3.790 | 86.0 | 83.0 |
| K.S. Puram | 38,731 | 0.300 | 1.730 | 8.200 | 115.6 | 68.6 |
| Thevalakkara | 35,086 | 0.370 | 0.760 | 9.310 | 93.8 | 75.1 |
| Thodiyoor | 38,508 | 0.290 | 0.790 | 2.780 | 87.1 | 66.9 |
| Thazhava | 33,450 | 0.340 | 0.980 | 6.100 | 108.8 | 63.9 |
| Oachira | 23,267 | 0.320 | 0.590 | 6.230 | 121.8 | 88.2 |

74.8, respectively. Overall, cancer incidence in the area is not alarmingly different from other population groups in India. Comparable data from other registries in India are shown in Table 1.

Cancer incidence rate among men was higher in Karunagappally than in Trivandrum but females had a lower rate in Karunagappally than in Trivandrum. The leading cancers in men were lung (16.7%), oesophagus (5.8%) and stomach (5.3%). In females, cancer of the breast (19.4%), cervix (17.1%) and thyroid (6.6%) were the leading types. For a preliminary analysis, the radiation levels seen in panchayat areas were grouped arbitrarily

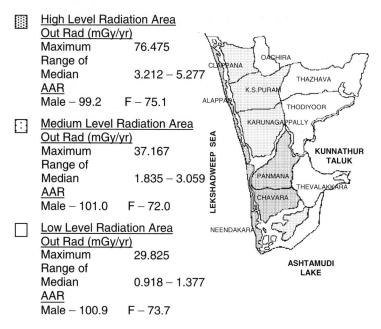

Map 1. Karunagappally taluk—panchayats: outside house radiation levels (out rad) and cancer incidence rates.

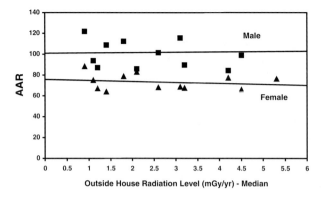

Fig. 1. Radiation levels in panchayats and cancer incidence: all cancer male and female, Karunagappally 1990–2000.

as the high, medium and low levels. The crude and AAR obtained for the three areas are shown in Table 2. With regard to radiation, no marked variation was seen in AAR of all cancers in males or females. The highest median radiation level was observed in Neendakara (3.6 mSv) and the lowest level in Oachira (0.59 mSv) (Map 1).

Cancer incidence in relation to median radiation levels in panchayats is shown in Fig. 1. No increase in overall cancer incidence was noted among males and females with regard to radiation levels.

Preliminary analysis did not indicate any increased risk of developing cancer in relation to radiation levels. Since 1999, RCC collaborates with Health Research Foundation (HRF) of Japan for continuation of investigations and estimation of individual radiation dose. For this study a cohort approach was taken for risk estimates and 4 panchayats with high background radiation and 2 with low radiation were selected as the radiation study cohort. These cohort groups are followed up regularly. Periodic migration surveys are planned. Mortality data is regularly collected. Estimation of cumulative dose is now underway.

## 4. Discussion

The observations so far demonstrated no apparent increase of all cancer in relation to radiation either in males or females. However, lung cancer in men showed some increase. In this context, the roles of risk factors like tobacco use and diet have to be probed further with better assessment of radiation exposure. For this the collaboration with HRF is contemplated along with more studies in varied fields. It is essential to continue incidence studies without a break to obtain sizable numbers of cancer cases and person-years for cancer risk estimation.

## References

[1] P. Gangadharan, et al., Technical Report of Natural Background Radiation Cancer Registry, Karunagappally 1990–1999. Regional Cancer Centre 2004.
[2] A. Nandakumar, K.T. Thimma Setty, Two-Year Report of the Population Based Cancer Registries 1997–1998, National Cancer Registry Programme ICMR2002.
[3] A. Mathew, B. Vijayaprasad, Cancer Incidence and Mortality in Trivandrum (1998–1999), Population Based Cancer Registry, Regional Cancer Centre, Trivandrum, 2002.

# Mortality and morbidity from cancer in the population exposed to high level of natural radiation area in Ramsar, Iran

Alireza Mosavi-Jarrahi[a,b,*], Mohammadali Mohagheghi[b], Suminori Akiba[c], Bahareh Yazdizadeh[b], Nilofar Motamedi[d], Ali Shabestani Monfared[e]

[a]*Deptartment of Social Medicine, Shaheed Beheshti University of Medical Sciences, Tehran, Iran*
[b]*The Cancer Institute Research Center, Tehran, Iran*
[c]*Kagoshima University, Kagoshima, Japan*
[d]*Mazanderan University of Medical Sciences, Sari, Iran*
[e]*Babol University of Medical Sciences, Babol, Iran*

**Abstract.** The aim of this study was to investigate if a higher incidence or mortality from cancer is seen among the population living in the high-background radiation area of Ramsar. The Ramsar city was divided into areas of high level of natural radiation area (HLNRA) and normal level of natural radiation area (NLNRA). Data regarding morbidity and mortality from cancer were obtained from different sources. Standard incidence ratios (SIR) were used to compare the morbidity experience of HLNRA to those of NLNRA. Standard mortality ratios (SMR) were calculated for two regions of the HLNRA (low exposure and high exposure) using national mortality rates. The incidence ratio for females was 1.5, indicating a slight increase in incidence in the HLNRA compared with that in the NLNRA. The mortality was slightly higher for females living in HLNRA (SMR=1.3 in high exposure area and 1.2 in low exposure area). Neither SMRs nor SIRs showed a statistically significant difference from the unity. No increase in incidence or mortality was seen among the male population. We have concluded that more epidemiologic studies are needed before a solid conclusion can be made. © 2004 Elsevier B.V. All rights reserved.

*Keywords:* Background radiation; Mortality; SMR; Cancer; Ramsar; Iran

---

* Corresponding author. Postal address: P.O. Box 15875-4194 Tehran, Iran. Tel.: +98 21 693 3399; fax: +98 21 642 8655.
*E-mail address:* rmosavi@yahoo.com (A. Mosavi-Jarrahi).

0531-5131/ © 2004 Elsevier B.V. All rights reserved.
doi:10.1016/j.ics.2004.11.109

## 1. Introduction

Ramsar, an Iranian city located in Caspian littoral, is one of the few areas in the world where very high levels of natural radiation, as high as 100 [Gy h-1, were recorded in some spots [1]. Studies of the effect of natural radiation on the health of people living in this area have been a special interest to scientists for a long time. Cytogenetics studies of the inhabitants in the area of Ramsar have established a higher frequency of chromosomal aberrations [2,3]. The aim of this study was to investigate if a higher incidence or mortality from cancer is seen among the population living in the high background radiation area of Ramsar compared to normal radiation area.

## 2. Materials and methods

The total area of Ramsar was divided into two regions: high level of natural radiation area (HLNRA) and normal level of natural radiation area (NLNRA). The HLNRA included the following four villages: Chaparcar-Sar, Talesh-Mahaleh, Sadat-Mahaleh (Sadat Rostai and Sadat-Shahr) and Ramak. The NLNRA included all other areas of the Ramsar city. Since the level of radiation differed in villages in HLNRA, the villages were further divided into two categories; Talesh-Mahalleh, Chaparsar, Sadat Rostai and Ramak were considered as HLNR high exposure area, and Sadat-Shahr as HLNR low exposure area. Table 1 shows the population distribution stratified for this study.

Mortality data in the HLNRA for the period 2001–2003 was obtained from the local health centers. Information about morbidity from cancer was obtained from pathology centers and local hospitals as well as Tehran Cancer Data System of the Cancer Institute Research Centre. The morbidity data included just incidence cases of cancer for the calendar year of 2003. No cancer morbidity data prior to 2003 were available. The population of the NLNRA was estimated using census data of 1996 accounting for the population growth rate used to estimate national mortality rates. The population for the HLNRA was obtained from the office of Primary Health Care Networks, where a track of population in the villages is kept and updated. Incidences of cancer for both areas were adjusted to the world population structure. Standard incidence ratios (SIR) were used to compare the cancer incidence in the HLNRA with that of the NLNRA. Standard mortality ratios (SMR) were used to compare the mortality experience of the HLNRA with that of the general population using age and sex specific mortality rates published by the Ministry of Health [4].

Table 1
The population distribution of the Ramsar (urban and rural area) living in NLNRA and HLNRA (high exposure and low exposure)

| Age group | HLNRA | | NLNRA |
|---|---|---|---|
| | High exposure | Low exposure | |
| 0–14 | 637 | 1689 | 44,317 |
| 14–59 | 1776 | 4189 | 11,342 |
| >60 | 609 | 1403 | 3554 |
| Total | 3022 | 7281 | 59,213 |

## 3. Results

During the calendar year of 2003, a total of 43 incident cases of cancer occurred in the Ramsar area, of which, 36 cases were in the NLNRA and seven cases in the HLNRA. While the distribution of cancer was comparable between sexes in the NLNRA (52% for female and 48% for male), a marked difference was seen in the HLNRA (85% in female and 15% in male). The age adjusted incidence rate of cancer was slightly higher among female population of the HLNRA compared to the NLNRA (incidence ratio, IR=1.5). Table 2 shows the age adjusted incidence rates of cancer for the two areas of HLNRA and the NLNRA.

The six cases of cancer that occurred in the female population of HLNRA included three cases of breast, one cervix, one multiple myeloma and one case of unknown primary site. In the NLNRA, neoplasm of hemopoietic and lymphoproliferative systems, with a relative frequency of 32%, ranked first, and GI cancers, with a frequency of 18%, ranked second.

During the period 2001–2003, a total of 19 deaths due to cancer occurred in the HLNRA, corresponding to crude cancer mortality rate of 64 per 100,000 populations. The risk of mortality due to cancer was slightly higher among the female living in the both low and high exposure areas of HLNRA compared with general population (SMR=1.3 and 1.2 for high and low exposure area, respectively). The risk was even higher in the age group 15–50 among female population of HLNRA (Table 3). The SMRs did not show any increase of mortality among the male population of HLNRA (Table 3). None of the SMRs showed a statistically significant difference from the unity.

## 4. Discussion

Correlation between natural background radiation and cancer morbidity and mortality has been studied in various countries in the world. Early studies proved a positive association (Italy [5] and Poland [6]), but some studies failed to show any correlation (India [7] and Ireland [8]). Our study showed a slight increase in cancer morbidity among females living in the area with higher levels of natural background radiation and no increase of cancer among males. The discrepancy between male and female may support an association between high radiation and cancer due to the fact that females are more exposed to radiation compared to males because of more local and indoor stays. A lower incidence of cancer in the male population of HLNRA compared with NLNRA has no explanation except that we may have missed cases in the incidence part of our study or because the incidences were estimated based on just a 1 year follow-up, and such estimates are subject to precision errors (lack of power).

Table 2
The incidence of cancer (per 100,000 populations) for the different areas of Ramsar

| Area | Incidence in male | Incidence in female | Incidence in both sexes |
|---|---|---|---|
| High level of natural radiation area | 21.1 | 124.7 | 71.6 |
| Normal level of natural radiation area | 71.2 | 80.7 | 81.5 |
| Age adjusted incidence ratios | 0.30 | 1.5 | 1.13 |

Table 3
Standard mortality ratios due to cancer for the two areas of high level of natural radiation

| Age group | HLNR high exposed area | | HLNR low exposed area | |
|---|---|---|---|---|
| | Male | Female | Male | Female |
| <15 | 0 | 0 | 0 | 0 |
| 15–50 | 1.8 | 2.0 | 2.4 | 2.8 |
| >50 | 0.9 | 1.4 | 0.6 | 1.0 |
| All ages | 0.9 | 1.3 | 0.8 | 1.2 |

The mortality part of our study showed again a slight increase of mortality in the HLNRA compared to national mortality rates. A larger increase in mortality seen in the female compared with the male, consistent with the incidence data, could be explained with the fact that females are more exposed to radiation due to their indoor and local stays. The magnitude of the SMR is very sensitive to the mortality rates used. The rates for calculation of SMRs were derived from pooled mortality for 18 provinces in Iran. Cancer incidence and mortality have relatively wide variations in Iran, and such variations need to be considered in interpreting the reported SMRs.

In summary, while a slight increase was detected in cancer mortality and morbidity among the female population living in the HLNRA, no solid conclusion can be made due to the fact that none of our finding was statistically significant, addressing the need for further epidemiologic studies with more person–years of follow-up.

## References

[1] M. Sohrabi, A.R. Esmaili, New public dose assessment of elevated natural radiation areas of Ramsar (Iran) for epidemiological studies, Int. Congr. Ser. 1225 (2002) 15–24.
[2] M. Ghiassi-Nejad, et al., Long-term immune and cytogenetic effects of high level natural radiation on Ramsar inhabitants in Iran, J. Environ. Radioact. 74 (1–3) (2004) 107–116.
[3] T.Z. Fazeli, et al., Cytogenetic studies of inhabitants of a high level natural radiation area of Ramsar, Iran, Proc. Int. Conf. on high levels of Natural Radiation, Ramsar, Iran, 3–7 Nov. 1990, IAEA Publ. Series, 1993, pp. 459–464 (Vienna).
[4] Mortality in the 18 provinces in Iran, Naghavi, Office of Development.
[5] L. Gianferrari, et al., Mortality from cancer in a area of high background radiation, Bull. WHO (26) (1962) 696.
[6] S. Plewa, J. Alecsanndrwecz, K. Janiki, Environment and leukaemia mortality. III: distribution of leukaemia morbidity and background ionizing radiation of the environment, Pol. Arch. Med. Wew. (32) (1962) 844–849.
[7] A.R. Gopal-Ayengar, et al., Evaluation of long term effects of high background radiation on selected populations of the Kerala coast, in: 4th International Conference on Peaceful uses of Atomic Energy, vol. 2, 1972, pp. 31–51.
[8] S.P.A. Allwright, et al., Natural background radiation in cancer mortality in the Republic of Ireland, Int. J. Epidemiol (12) (1983) 414–418.

ELSEVIER

International Congress Series 1276 (2005) 110–113

www.ics-elsevier.com

# Pattern of cancer mortality in some Brazilian HBRAs

## Lene H.S. Veiga[a,]*, Sérgio Koifman[b]

[a]*Institute of Radioprotection and Dosimetry, Brazilian Nuclear Energy Commission, Brazil*
[b]*National School of Public Health, Oswaldo Cruz Foundation, Brazil*

**Abstract.** Among residents of Brazilian High Background radiation Areas, there is great concern about radiation-related health effects and there is also a common certitude that cancer incidence is higher in those areas than in other Brazilian areas with normal background radiation. This paper aims to present an overview of Brazilian High Background Radiation Areas and evaluate whether cancer mortality among residents from Poços de Caldas, Araxá, and Guarapari is higher than would be expected when applying mortality rate of their respective States. Results show that cancer mortality from the Brazilian HBRAs, Poços de Caldas, and Guarapari is higher than would be expected for their respective reference population. On the other hand, cancer mortality for the Araxá population is lower than would be expected. © 2004 Elsevier B.V. All rights reserved.

*Keywords:* Natural radiation; Cancer; Mortality; Poços de caldas; Araxá; Guarapari

## 1. Introduction

Some Brazilian locations have been globally recognized as High Background Radiation Areas (HBRAs); included among these are Poços de Caldas, Araxá, and Tapira, all located in Minas Gerais State and Guarapari, located in Espirito Santo State [1,2]. As a consequence of this recognition, residents of those areas have great concern about radiation-related health effects. There is a common sense among these populations that cancer incidence is higher than in other Brazilian areas with normal background radiation. Nevertheless, no previous cancer statistics have been shown to support this hypothesis and, until the present time, no health effect study had been conducted in Brazilian HBRAs. Most of the data concerning natural radiation exposure in those areas were obtained during

---

* Corresponding author. Mailing address: Av. Salvador Allende, s/n, Recreio, Caixa Postal 37750, Rio de Janeiro, RJ, CEP 22642-970, Brazil. Tel.: +55 21 3411 8089; fax: +55 21 2442 2699.
*E-mail address:* lene@ird.gov.br (L.H.S. Veiga).

0531-5131/ © 2004 Elsevier B.V. All rights reserved.
doi:10.1016/j.ics.2004.11.046

the late 1970s. Recently new assessments were performed at Poços de Caldas [3] and Guarapari [4]. Those results indicated that the great urbanization process in Guarapari changed the radiation exposure pattern; the external radiation exposure is at present lower than that in the past. The radiation level in Guarapari can be considered normal, except in the hot spots on the beaches and in the fishing village of Meaipe [4]. At Poços de Caldas, it was shown that only rural areas could be considered as high natural background radiation areas. The radiation dose in urban areas can be considered normal.

Therefore, this paper aims to assess whether there is an excess of cancer mortality among residents from Poços de Caldas, Araxá, and Guarapari in comparison with a reference population. The State of Minas Gerais was used as the standard population for Poços de Caldas and Araxá, and the Espirito Santo State for Guarapari.

## 2. Methodology

Mortality data on cancer and all other causes of death were examined for Poços de Caldas, Araxá, and Guarapari as well as for the States of Minas Gerais and Espirito Santo, which were used as reference areas. Mortality data from 1991 to 2000, obtained from the Brazilian National Mortality System, were evaluated for both sexes and for the following age intervals: <1, 1–4, 5–9, 10–14, 15–19, 20–29, 30–39, 40–49, 50–59, 60–69, 70–79, and 80 and over. Standardized mortality ratios (SMRs) for every city were estimated as the ratio of deaths observed to those expected. Expected numbers of death for the cities of Poços de Caldas and Araxá were obtained by multiplying the sex and age group stratum's population by different cancer sex–age-specific death rates for the Minas Gerais State. The expected numbers of deaths for Guarapari City were obtained by applying cancer sex-age-specific death rates for Espirito Santo State.

## 3. Results

Table 1 presents the observed and expected numbers of deaths for selected cancer and all causes of death for cities of Araxá and Poços de Caldas, respectively, using the specific general mortality for Minas Gerais State as the standard population. For the Araxá population, mortality for all causes of death was significantly higher than expected from the reference population (SMR=118, CI=115–121). Nevertheless, total cancer did not

Table 1
Standardized mortality ratios for Araxá and Poços de Caldas City, both genders, 1991–2000

| International classification of diseases | | Cause of death | ARAXÀ | | POÇOS DE CALDAS | |
|---|---|---|---|---|---|---|
| ICD-9 | ICD-10 | | Obs | SMR (95% CI) | Obs | SMR (95% CI) |
| 001–999 | A00–Z99 | All causes | 5059 | 118 (115–121) | 8355 | 115 (113–118) |
| 140–239 | C00–D48 | All cancer sites | 479 | 100 (60–156) | 1122 | 141 (133–149) |
| 150 | C15 | Esophagus | 17 | 60 (35–97) | 32 | 65 (44–92) |
| 151 | C16 | Stomach | 42 | 79 (56–106) | 133 | 143 (120–170) |
| 161 | C32 | Larynx | 9 | 96 (44–183) | 21 | 126 (78–193) |
| 162 | C33–C34 | Lung | 37 | 79 (56–109) | 113 | 138 (114–166) |
| 174 | C50 | Female Breast | 23 | 77 (49–116) | 78 | 166(131–207) |
| 185 | C61 | Prostate | 35 | 132 (92–184) | 57 | 120(91–156) |
| 204–208 | C91–C95 | All leukemias | 19 | 100 (60–156) | 49 | 154 (114–204) |

Table 2
Standardized mortality ratios for Guarapari City, both genders, 1991–2000

| International classification of diseases | | Cause of death | Obs | SMR 95% CI |
|---|---|---|---|---|
| ICD-9 | ICD-10 | | | |
| 001–999 | A00–Z99 | All causes | 3942 | 95 (92–98) |
| 140–239 | C00–D48 | All cancer sites | 468 | 109 (99–119) |
| 150 | C15 | Esophagus | 42 | 160 (115–216) |
| 151 | C16 | Stomach | 90 | 169 (136–208) |
| 161 | C32 | Larynx | 9 | 105 (48–199) |
| 162 | C33–C34 | Lung | 77 | 156 (123–195) |
| 174 | C50 | Female breast | 28 | 121 (80–175) |
| 185 | C61 | Prostrate | 48 | 197 (145–261) |
| 204–208 | C91–C95 | All leukemias | 19 | 99 (59–155) |

exceed the expectation and no statistically significant excess was observed for any selected cancer sites.

For Poços de Caldas, mortality for all causes and all cancers were significantly higher than expected. Among single cancer sites, stomach, lung, breast, and leukemia showed statistically significant excess. Larynx and prostate showed non-significant excesses ranging from 20% to 26%, whereas esophagus cancer was below the expectation. Among leukemia subtypes, lymphocytic and myelocytic leukemia presented a nonstatistically significant excess (SMR=120, CI=60–215 and SMR=123, CI=74–193, respectively), whereas for cellular type not specified, statistically significant excess was observed (SMR=242, CI=146–378).

Table 2 presents the standardized mortality ratio for the Guarapari population, using the specific general mortality for Espirito Santo State. Mortality for all causes was significantly lower than expected (SMR=95, CI=92–98), whereas a nonsignificant excess of 9% was observed for all cancer mortality. Among single cancer sites, high statistically significant SMRs were observed for esophagus, stomach, lung, and prostate cancer. Mortality from leukemia was close to the rate expected, whereas larynx and breast cancer were slightly higher than expected, although not statistically significant.

## 4. Discussion

Studies of geographical variation need to be interpreted with caution because many factors other than environmental exposure can contribute to such variation in the recorded frequency of disease or death. Mortality data can be affected by varying qualities of cause of death certification and differences in survival among regions. The imprecision in the information produces a dilution effect and biases the results. Genetic and ethnic factors may confound geographical variations, and migration patterns may also affect geographical comparisons if there are substantial inward or outward movements.

Despite the higher cancer mortality observed for some cancer sites in the Poços de Caldas population, the estimated doses from natural radiation at Poços de Caldas indicated that only the rural population would be highly exposed. For the Guarapari population, the higher cancer mortality cannot be related to the radiation exposure, based on the fact that radiation levels in Guarapari city can be considered normal, with high spots only at the beaches, and levels are not significant concerning chronic public exposure.

The observed excess cancer mortality in Poços de Caldas and Guarapari HBRAs must be seen as the result of a very preliminary study, and further analysis should include other important variables, such as socioeconomic status, smoking, and dietary habits, as well as other aspects of environmental exposure such as pesticide use in agricultural activities (mainly for Poços de Caldas). Quality of cause of death certification in those regions must be also assessed.

## 5. Conclusion

This report represents the first time that cancer mortality has been assessed for Brazilian HBRAs. It was observed that cancer mortality for Poços de Caldas and Guarapari is higher than would be expected for their respective reference populations, Minas Gerais State and Espirito Santo State, respectively. On the other hand, cancer mortality for the Araxá population is lower than would be expected.

Nevertheless, natural radiation levels at those regions cannot be associated with this excess cancer mortality. Other aspects of environmental exposure must be investigated in a more consistent study.

### References

[1] T.L. Cullen, Review of the Brazilian investigations in areas of high natural radioactivity. Part I: Radiometric and dosimetric studies, in: T.L. Cullen, E. penna Franca (Eds.), Proc. of the International Symposium on High Natural Radioactivity, Poços de Caldas, Brazil, 20 Junho 1975, Academia Brasileira de Ciências, RJ, 1975, p. 49.
[2] E. Penna Franca, et al., Status of investigations in the Brazilian areas of high natural radioactivity, Health Physics 11 (1965) 671–699.
[3] L.H.S. Veiga, et al., Preliminary indoor radon risk assessment at the Poços de Caldas Plateau, MG-Brazil, Journal of Environmental Radioactivity 70 (2004) 161–176.
[4] I. Sachet, Caracterização da radiação gama ambiental em áreas urbanas utilizando uma unidade móvel de rastreamento. PhD thesis. Instituto de Biociências Nucleares, Universidade Estadual do Rio de Janeiro (in Portuguese).

# WHO network and international collaborative project on residential radon risk

## Zhanat Carr*, Michael Repacholi

*Radiation and Environmental Health Programme, World Health Organization Headquarters, Switzerland*

**Abstract.** This paper describes WHO's project on assessment of health risks from residential radon exposure and development of an evidence base for guidelines and policies for Member States. The project will assess global burden of disease associated with in-door radon exposure and provide tools for risk communication on radon risks and mitigation procedures. The purpose of the paper is to draw attention and encourage participation of the international scientific community and national authorities to the new initiative of WHO. © 2004 Elsevier B.V. All rights reserved.

*Keywords:* Ionizing radiation; Residential radon; Global burden of disease; WHO guidelines and recommendations

## 1. Introduction

Exposure to radon in the home and workplace constitutes one of the greatest risks to natural background ionizing radiation. From the 1998 US National Academy of Sciences report, the US EPA estimates that between 15,000 and 25,000 lung cancer deaths are caused in the USA each year from radon [1].

In order to reduce radon risk, it is important that national authorities have methods and tools based on solid scientific evidence and sound public health policy. To this end, WHO is working towards developing guidelines to limit exposure to radon, provide effective mitigating measures where the risks are greatest and risk communication tools to inform the public on this issue.

---

\* Corresponding author. WHO, SDE/PHE/RAD, 20 Avenue Appia, Geneva-27, 1211, Switzerland. Tel.: +41 22 791 3483; fax: +41 22 791 4123.

*E-mail address:* carrz@who.int (Z. Carr).

0531-5131/ © 2004 Elsevier B.V. All rights reserved.
doi:10.1016/j.ics.2004.10.009

The purpose of this paper is to draw attention of the international scientific community and national authorities to this new WHO initiative and to encourage their participation.

## 2. Project description

The WHO Radon program will include:

- Creation of a worldwide database on national residential radon levels, action levels, and regulations related to radon mitigation, as well as a list of authorities/institutions dealing with radon issues and research.
- Estimation of the global burden of disease (GBD) associated with exposure to residential radon.
- Preparation of public health guidance to limit exposure to radon.
- Communication of results of the project to government authorities and the general public to raise awareness of radon and its health consequences.

GBD is a tool for national public health authorities to quantify the burden of disability and mortality for major diseases or disease groups associated with specific risk factor [2]. WHO has been assessing GBD for some 20 major environmental risk factors, but this has never been conducted for ionizing radiation risk factors [3].

Current risk assessments for exposure to residential radon are based on two major methods—projection of risk estimates derived from occupational studies or from studies based on radon measurements in homes. Attributable risks from the BEIR VI report were applied to 1990 lung cancer mortality for the U.S. population from the National Canter for Health Statistics to estimate the Years of Lost Life (YLL) from lung cancer in the US. As the attributable risks are based on U.S. levels of residential radon exposure and smoking prevalence (due to the strong interaction between radon exposure and smoking), they cannot be directly projected to other regions of the world. However, these results suggest a potentially large global burden of disease.

It remains unclear to what extent radon from drinking water contributes to the risk from radon level in air. Deaths attributed to inhaled radon originating from drinking water comprises much lower number. Nevertheless, the estimate for radon in drinking water is subjected to uncertainties, as there were no studies on measurements of radon level in drinking water of private water supplies, on cost-effectiveness of methods for decreasing radon level in drinking water and what impact they may have.

This project will provide the information enabling Member States to create national radon programs to fulfill the following objectives:

- identifying advisory levels for radon in homes
- developing guidelines on mitigation in radon-risk areas
- creating public awareness and reduction of radon risks.

To achieve these aims, WHO is forming a network of key partner agencies around the world working on radon to participate in the project.

## 3. Project activities and time-frame

This project it scheduled for completion in a 3-year time frame with the following components:

(I) WHO to hold an international workshop on residential radon on order to
- Review the latest information on the health effects of radon
- Discuss and agree upon the scope, extend, methodology, major components, responsible experts, and time frame of the project
- Identify gaps in information for better risk assessment of radon exposure, and adopt most suitable risk model to be applied for GBD assessment
- Compile information on radon exposures worldwide and identify potentially "at risk" countries for which exposure information is not available
- Review and summarise the most effective mitigation measures and current public health policies used by national authorities in MS
- Provide a summary of information, conclusions and recommendations that can be used as a basis for the development of policy and advocacy
- Form a network of partner agencies to provide oversight and expertise to the activities within the project.

(II) Consultant to prepare reports, based on the international workshop results, and arrange a small working group meeting to
- Prepare draft meeting report
- Draft a summary paper for peer review publication
- Develop a public health policy and advocacy documents
- Develop WHO information sheet and press release

(III) Personnel to prepare a GBD protocol using the draft developed by the Canadian partners. Send for review and incorporate comments. Hold a working group meeting to finalise the radon GBD protocol.

(IV) Expert group to gather information from countries where good information on exposures to radon is available. Establish process of data gathering under an agreed protocol in four selected, "at risk" countries, that are not currently collecting radon information, and compile a global database for GBD analysis.

(V) WHO experts to conduct analysis of the database and a GBD assessment of radon in homes worldwide. Host a final meeting to discuss and finalise the results. Discuss and agree on any updates of proposed public health policy and advocacy. Prepare the following publications:
- Peer review scientific article
- WHO Environmental Burden of Disease series monograph
- Update WHO information sheet and press release
- Prepare section in the WHO World Health Report
- Policy and advocacy documents for distribution to WHO Member States

WHO encourages participation of scientists and national authorities to deal with this important public health issue. For further information, contact the corresponding author.

## References

[1] US National Research Council, Board on Effects of Ionising Radiation VI Report on Health Effects of Exposure to Radon 1999, http://www.nap.edu/books/0309056454/html/index.html.
[2] C.J.L. Murray, J.A. Salomon, C.D. Mathers, A.D. Lopez (Eds.), Summary Measures Of Population Health: Concepts, Ethics, Measurement And Applications, WHO, Geneva, 2002.
[3] World Health Report 2002, http://www.who.int/whr/2002/en/.

# Commentary on information that can be drawn from studies of areas with high levels of natural radiation

Cardis Elisabeth*

*International Agency for Research on Cancer (IARC), 150 cours Albert Thomas, 69 372 Lyon Cedex 08, France*

**Abstract.** Studies of the health of populations living in areas of high levels (HL) of natural radiation are a potentially important source of information on the effects of chronic low-dose rate exposures to ionising radiation. The current paper presents a brief overview of the status of studies in Brazil, China, India, and Iran. At present, only limited information on health risks can be drawn. Issues of concern in the interpretation of the studies are discussed, and recommendations are made to improve the informativeness of studies in the future. Where feasible, the conduct of analytical epidemiological studies—in particular, case control studies—with individual estimates of doses to specific organs from internal and external exposures and with individual information on known and possible risk factors for the diseases of interest should provide important information on health risks from low-level chronic radiation exposures. This information can be maximised by the adoption of common study protocols and common dosimetric approaches and by coordination of efforts across countries. © 2004 Elsevier B.V. All rights reserved.

*Keywords:* Low-dose radiation; Epidemiology; Dosimetry; Health effects

## 1. Introduction

Ionising radiation is one of the most studied and ubiquitous carcinogens in our general environment. Guidelines for ionising radiation protection have existed at the multinational level since the 1940s. The main basis for radiation protection, however, is the study of atomic bomb survivors—a population exposed primarily at high-dose rates—while the primary public health concern is the protection of persons with relatively low dose protracted exposures such as are received by the public in the general environment in high background radiation areas (HBRA) and by workers occupationally.

---

* Corresponding author. Tel: +33 4 72 73 85 08; fax: +33 4 72 73 80 54.
*E-mail address:* cardis@iarc.fr.

0531-5131/ © 2004 Elsevier B.V. All rights reserved.
doi:10.1016/j.ics.2004.11.111

The use of data from populations with high-dose rate acute exposures, like the atomic bomb survivors, to estimate the effects of the relatively low-dose rate chronic exposures of environmental and occupational concern necessitates the use of models to extrapolate risks from high doses to low doses, from high exposure rates to low exposure rates, and from acute exposures to protracted or chronic exposures. These models are, inevitably, subject to uncertainty and have been the object of controversy for several decades.

Studies of the health of populations living in areas of high levels (HL) of natural radiation are a potentially important direct source of information on the magnitude of health effects of chronic low-dose rate exposures to ionising radiation.

To provide a sound evaluation of these effects, however, epidemiological studies of populations living in HBRA must fulfil several important criteria: they must cover very large numbers of subjects with a wide range of exposures, the follow-up for mortality and/or morbidity must be complete and nonselective, and precise and accurate individual dose estimates (or markers of exposure) must be available. The feasibility and the quality of epidemiological studies will also largely depend on the existence and the quality of basic population-based registration and the ability to trace study subjects and to obtain information about other potential important risk factors for the diseases of interest on an individual basis.

These requirements are particularly important given the nature of the questions being addressed, namely, the effects of relatively low doses and hence the quantification of a priori relatively small risks.

The information presented at the Conference in the last days, and, in particular, in the current session, concerned areas of HBRA in four countries. The status of the studies in these different areas varies a great deal.

- China and India have very comprehensive and long-standing programmes of studies in these areas. A tremendous amount of work have been carried out, particularly, in recent years, by the local investigators and by their Japanese collaborators (with important support from the Health Research Foundation Japan) to establish well-defined cohorts of subjects living in these areas, to set up mechanisms for their follow-up for cancer mortality and other endpoints, to improve individual estimates of doses, and, most recently, to set up nested case control studies which will allow the analysis of the effect of other potential risk factors and confounding factors.
- In Iran and Brazil, however, work is only just starting. Although environmental and in-house exposure rate measurements exist showing, in some instances, elevated levels of radiation, more work is needed to evaluate the feasibility of informative studies, and, if appropriate, to set up cohorts and evaluate individual doses and plan epidemiological studies.

## 2. Summary of characteristics and results of studies of HBRAs presented

Table 1 summarises the characteristics of the main populations of interest in the HBRAs presented at the meeting. The largest population studied is that living in Karunagappally (Kerala area, India). The total population of the area is large, 385,103. Of these, 359,619 have been personally interviewed in the framework of a sociodemographic survey. Information is available from this survey on demographic and lifestyle factors. External

Table 1
Summary of characteristics of populations living in areas of HBRA

|  | Brazil [2,3] | China | India | Iran |
|---|---|---|---|---|
| Size of population in HBRA | Poços de Caldas 6000 Araxà 1300 | Cohort with external dose estimates 125,079 | Interviewed 359,619 Home measured 76,942 | Ramsar total 60–70,000 Talesh Mahalleh 1000 |
| Source of exposures | Radon, Th-232, U-238 | Th-232, U-238 | Monazite sands: Th-232, ... | Hot springs: $Ra^{226}$ and decay products |
| Reported dose distribution mSv/year—mean (range) |  |  |  |  |
| Total effective dose (internal+external) | Poços de Caldas 7.19 Araxà (1.2–6.1) | 6.4 |  |  |
| External | Poços de Caldas 1.32 Araxà (1.2–6.1) | 2.1 (1–3) | Out. 2.1 (0.5–76) Ins. 1.8 (0.5–57) | 6 (0.6–135) |
| Internal | Poços de Caldas 5.87 Araxà NA | 4.3 | NA | 2.4–71[a] |

Doses are expressed as effective dose in mSv—India: median whole body doses are presented in mGy not mean effective doses.
[a] Range of mean values from 12 regions.

radiation measurements were made inside 71,674 houses and outside 76,942 houses, and the area is covered by the Regional Cancer Centre, Trivandrum. The median external whole body doses are of the order of 2 mGy per year, similar to those estimated in China and Brazil but lower than in Iran. No information was provided on internal dose levels.

In China, a large cohort, including 125,079 subjects living in HBRAs and a neighbouring control area (CA) of Guangdong Province has also been assembled since 1987. External dose have been derived for each hamlet in the region based on external radiation measurements inside 5990 houses (one-third of the houses in each hamlet) and on outdoor measurements in 526 hamlets (in 53 villages). Occupancy factors were obtained from a questionnaire survey conducted from 1991 to 1993 on 5291 subjects living in over 88 hamlets. Individual external effective dose estimates were derived for each subject in the cohort based on the hamlet-specific indoor and outdoor air kerma rates and on sex- and age-specific occupancy factors [1]. Migration data were taken into account in the dose estimation. Individual internal dose estimates are not, however, currently available.

The populations living in HBRAs in Brazil and Iran are small. Although the population of Ramsar area is of the order of 60–70,000, only a small proportion of these live in the area with high natural radiation levels (HL) [4–6]: 3022 in four villages (including 1000 in Tallesh Mahalleh, where very high levels with estimated annual external effective doses ranging up to 135 mSv have been reported).

Table 2 summarises the results on all cause and all cancer mortality presented at the meeting. Mortality from cancer did not differ between areas of HBRA and control areas (CA) in China in analyses of mortality over 20 years. There was also no association

Table 2
Summary of study characteristics and results

|  | Brazil | China | India | Iran | |
|---|---|---|---|---|---|
| Size of studied cohort | Araxà 90,000 | HBRA 89,694 | 359,619 | HL | 3022 |
|  | Poços de Caldas 120,000 | CA 35,385 |  | LL | 7 281 |
| Follow-up period | 1991–2000 | 1979–98 | 1990–2001 | 1998–2001 | |
| Number of deaths | Araxà 5059 | 8,905 |  |  | |
|  | Poços de Caldas 8,355 |  |  |  | |
| Number of cancer deaths | Araxà 479 | 855 | Incident cases 3623 |  | 43 |
|  | Poços de Caldas 1122 |  |  | HL | 7 |
| RR/SMR for cancer (95% CI) | Araxà 1.00 (0.6–1.5) | 1.00 (0.89–1.14) No relation with external radiation dose | No relation with external radiation dose | Women | 1.3 HL |
|  | Poços de Caldas 1.41 (1.3–1.5) |  |  |  | 1.2 LL |
| RR/SMR for noncancer | Araxà 1.18 (1.1–1.2) | 1.06 (1.01–1.10) |  | Men | 0.8 HL |
|  | Poços de Caldas 1.15 (1.1–1.2) |  |  |  | 0.9 LL |

between all cancer mortality and external radiation dose [7]—it is noted, however, that dose–response analyses are based on three relatively low-dose groups (<1.98, 1.98–2.24, and >2.24–3.1 mSv/year) that do not differ much and do not discriminate between subjects with very different doses. A significant difference was seen, however, between HBRA and CA with respect to noncancer mortality—including cerebrovascular diseases, tuberculosis, viral infections, and diseases of the digestive system (in particular, chronic liver disease) [7]. It is difficult to interpret the results as shown as information on risk factors for these diseases (including smoking, HBV infection, alcohol consumption, and consumption of some foodstuffs) have not been taken into account. Previous studies have not shown much difference, however, in the distribution of a number of risk factors between the HBRA and CA areas [8].

In India, no association was seen between cancer incidence and mean external effective radiation dose level in an ecologic study [9]. In Iran, the standardised mortality ratio (SMR) for cancer was higher than the national mortality both in HL and LL (low level) areas in women only. It is noted that, during the very short study period (3 years), only one death from cancer was seen in men in the HL areas compared with six in women, suggesting a possible underascertainment of deaths in this study [6]. In Brazil, a significantly elevated standardised mortality ratio was seen for cancers (all types combined) in Poços de Caldas and for noncancer mortality in both study areas compared to mortality rates in Minas Gerais State [10]. As the study population is substantially larger than that which is thought to be living in HBRA and including cities, it is difficult to interpret this result.

To better assess the effects of radiation exposure, lung cancer case control studies have been carried out both in China and in India, nested within the cohorts mentioned above. The Chinese study included 63 cases and 126 controls, and indoor radon concentrations were determined with passive detectors (Tao—presentation in Conference). The Indian study included 205 cases and 615 controls [11]. A significant effect of smoking was seen in both studies. No association with external radiation dose was seen in either study, although an odds ratio of 2.3 (95% CI 0.9–5.7 based on 14 exposed cases) was seen for subjects living in areas of India where external dose was greater than 10 mGy/year compared to subjects living in less-exposed areas. Information about internal dose to the lung is unfortunately not available in either of these studies; as this is likely to be much higher than the dose from external exposures, it is difficult to interpret the results of these studies. No significant association was seen, however, between indoor radon concentration and risk of lung cancer in the Chinese study (Tao—comment at Conference).

## 3. Issues for consideration in the interpretation of HBRA studies

There are a number of important issues that must be considered in interpreting the results presented here. These include study design, study power, and dosimetry.

The studies discussed aim to evaluate the health effects of low-level chronic radiation exposure in high background radiation areas. The effects of low-level exposures (which include those received in HBRAs) are, by definition, expected to be small. It is therefore important to consider the attributable risk that might be expected if current risk estimates derived from atomic bomb survivors and other high-dose populations are applicable here. Based on the dose distributions presented in Table 1 (which are shown as effective dose and hence may greatly underestimate dose to the lung), one would expect about a 3% increase in all cancer risk in the HBRA area in China (i.e., about 25 extra cancer deaths out of 855) and a 1% to 1.5% increase in genetic risk in India (where studies of such endpoints are foreseen).

Given the small size of the expected risk, it is therefore critical to identify health effects that can reasonably be studied in such populations and to use appropriate and sensitive epidemiological study designs. The conduct of nested case control studies—with individual dose assessment and collection of individual information on known and possible risk factors for the diseases of interest—will be a useful tool for the evaluation of health risks from low-level chronic radiation exposures.

Furthermore, as dose from internal exposures varies dramatically with the target organ, it will be very important to adequately characterise risks to estimate dose to specific organs from all sources (internal and external), taking into consideration exposures received inside and outside dwellings, occupancy factors, and movements across study areas.

In addition, as dose is time-dependent and as doses in these areas tend to be low, analyses would be more powerful if they used continuous lagged cumulative doses to specific organs for each subject rather than groupings.

## 4. Conclusions and recommendations

In conclusion, it is difficult currently to evaluate the existence and possible magnitude of a health risk associated with residence in HBRAs. The steps that have been taken in China and India, including the establishment of cohort studies, provide a model framework for studies of low-dose risks from high background radiation and could be used as a model in other areas of the world such as Brazil and Iran.

Before starting such work, however, it is essential to evaluate formally the feasibility of such studies and the potential information that can be obtained for radiation risk estimation.

If such studies were then judged to be both feasible and informative, there would be great benefit from more concerted or coordinated efforts across countries and the development and use of common study protocols. This includes protocols for dosimetry (as the comparability and adequacy of available dose estimates are unclear at present [12]) and for study design and analysis.

The use of common core protocols across countries would allow direct comparison of results across studies and may also allow combined analyses, which would maximise the information from these studies.

Different mechanisms for such coordination include, depending on resources, the creation and periodic meetings of specific informal multinational working groups in epidemiology, dosimetry, and related disciplines or the establishment of a formal multinational collaborative study, coordinated either by one of the concerned countries, by Japan (whose involvement in these studies in recent years has contributed to the establishment of similar protocols in different countries), and/or by an international organisation specialising in health research.

## References

[1] Q.F. Sun, et al., Excess relative risk of solid cancer mortality after prolonged exposure to naturally occurring high background radiation in Yangjiang, China, J. Radiat. Res. 41 (2000) 43–52 (Suppl.).
[2] L.H.S Veiga, et al., Estimating Indoor radon risk at a Brazilian area of high natural radiation, J. Environ. Radioact. 70 (3) (2002) 161–176.
[3] IRD—Institute of Radiation Protection and Dosimetry, 2002. Technical Report, Rio de Janeiro, Brazil (in Portuguese).
[4] M. Sohrabi, A.R. Esmaeli, New Public Dose Assessment of Elevated Natural Radiation Areas of Ramsar (Iran) for Epidemiological Studies, in: W. Burkart, M. Sohrabi, A. Bayer (Eds.), Procds. of 5th IC on High Levels of Natural Radiation and Radon Areas 4–7 September 2000, Elsevier, 2002, pp. 15–24.
[5] M. Sohrabi, M. Babapouran , New Public Dose Assessment from Internal and External Exposures in Low and Elevated Level Natural Radiation Areas of Ramsar, Iran, These proceedings paper ICS 1276.
[6] A. Mosavi-Jarrahi, et al., Mortality and morbidity from cancer in the population living in high-level natural radiation area of Ramsar, Iran. These proceedings.
[7] J. Zou, et al., Cancer and non-cancer mortality among inhabitants in the high background radiation areas of Yangjiang, China (1979–1998). These proceedings.
[8] Y.R. Zha, et al., Confounding factors in radiation epidemiology and their comparability between the high background radiation areas and control areas in Guangdong, China, in: L.X. Wei, T. Sugahara, Z.F. Tao (Eds.), High Levels of Natural Radiation 1996: Radiation Dose and Health Effects, Elsevier, Amsterdam, 1997, pp. 263–269.
[9] B. Rajan, et al., Natural background radiation cancer registry. These proceedings./or Jayalekshmi et al—these proceedings (manuscript No. 1276146v0).
[10] L.H.S. Veiga, S. Koifman, Pattern of cancer mortality in some Brazilian HBRAs. These proceedings—manuscript 1276036.
[11] V.S. Binu, et al., The risk of lung cancer in HBR area in India—a case-control study These proceedings manuscript 12761451.
[12] S.L. Simon, Dosimetric considerations for environmental radiation and NORM. These proceedings.

# High-level doses brought by cosmic rays

## Kazunobu Fujitaka*

*National Institute of Radiological Sciences, 4-9-1, Anagawa, Inage, Chiba 263-8555, Japan*

**Abstract.** Among many factors which bring high-level radiation in our environment, cosmic rays hold a unique status because we cannot catch sources. Nevertheless, we are exposed tremendously in case we go into space. Even on the ground surface, we are exposed to high-level radiation if we climb high mountains. One of the most important things which people are interested in will be airplane dose. If one flies to anywhere in the world, it will bring large dose. Of course, if we are astronauts, for example, we are target of cosmic ray exposure. After all, we will be unable to go beyond Pluto, the end of solar system, because we will receive 70 Sv which equals to cancer therapy dose, and no cell will be alive. © 2004 Published by Elsevier B.V.

*Keywords:* Cosmic rays; Airplane dose; International space station; Supernova

## 1. Introduction

There are various radiations around the earth. They are galactic cosmic rays, solar protons, and Van Allen belt particles. As a good instruction, I wish to emphasize the former two. Hematogenous tissue is the center of consideration. In our environment, cosmic rays are treated as entities with practically no shielding and no absorption effects, while the remaining components of radiation will suffer from both; such gamma rays are of low energy in general, below 3MeV. However, cosmic rays are different. It ranges widely, but the peak lies at around 0.6 GeV. Moreover, as the silhouette of Mt. Fuji indicates, the cosmic ray dose rate doubles for every 1500 m elevation.

Cosmic rays are a kind of radiation, of charged particles, and of high energy, but their source seems to be above our head. Dr. Victor Hess, a balloon explorer, received the Nobel prize later on that finding. To make it easy to remember, the dose rate in an airplane is 100

---

\* Tel.: +81 43 206 3230; fax: +81 43 251 4836.
*E-mail address:* fujitaka@nirs.go.jp.

0531-5131/ © 2004 Published by Elsevier B.V.
doi:10.1016/j.ics.2004.11.045

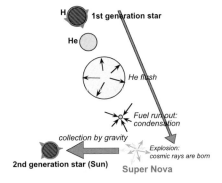

Fig. 1. Birth of cosmic rays.

times higher than that on the ground. Furthermore, the dose rate in a spacecraft is more than 10 times larger than that in an airplane.

## 2. Origin of cosmic rays

We cannot be free from the big bang theory. When the creation occurred about $1.37 \times 10^{10}$ years ago, cosmos was enlarged tremendously within a very short time. Present is here, and there are many Nebulas, like the Milky Way. They have a very old story. This picture is originally made by Prof. Katsuhiko Sato, and I have rewritten some. Fig. 1 is a cartoon of nuclear reactions in a star. When hydrogen is burnt out, Helium got fired, and the star size becomes very large because of high efficiency of burning, but explosion cannot be maintained due to shortage of fuels. As star size is determined by balance between the outward pressure and inward gravity, the star suddenly shrank. As this shrinking occurred so rapidly, materials also burst. That is a supernova. Then, collecting materials by gravity, second generation star is born.

What we see in the environment are not primary cosmic rays but are secondary cosmic rays (Fig. 2). When a primary cosmic rays attack the atmospheric nuclei components, nitrogen, oxygen, or argon, it generates charged pions and neutral pions, as well as

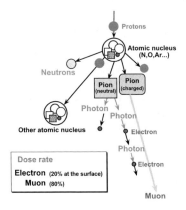

Fig. 2. Observed are secondary cosmic rays.

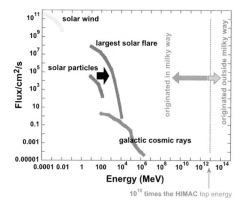

Fig. 3. Energy spectrum in space.

neutrons [1]. Charged pions promptly decay into two photons within $10^{-7}$ s, and the photon produces electron pair, and the electron pair produce photons, and vice versa. Charged pions can live slightly longer, say $10^{-6}$ s, and come down to our environment, and they are changed into muons. Then, about 80% of cosmic ray dose is of muon origin, and the remaining 20% is of electron origin. Airplanes are exposed to these electrons, and people on the ground are exposed to muons. It is very important to know that there is no single particle which is called as "cosmic ray." However, there are many particles worthy to be called as "cosmic ray family."

Fig. 3 shows energy spectra of cosmic rays. In the lowest energy region, there is a solar wind, and the dotted line is the upper bound which Milky Way can accelerate. This peak is about $10^{10}$ times the peak energy of what our accelerator HIMAC can generate. However, much higher energy is possible only if it is accelerated by a different mechanism. Of course, hydrogen is dominant followed by helium.

When a strong solar wind blows, geomagnetic lines of force is strengthen by plasma pressure, and cosmic rays of low energy will be refracted. In polar region, where

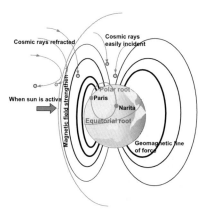

Fig. 4. Geomagnetic field.

geomagnetic lines of force are vertical, cosmic rays can move almost freely along the lines, and airplanes fly in such an environment (Fig. 4). However, in equatorial region, the lines are horizontal, and especially low-energy cosmic rays are stopped from entering. Therefore, flights in polar area will receive larger doses, while flights in equatorial region give smaller doses.

## 3. Radiation in airplanes

This map is based on geomagnetic coordinates. You see that Narita is located at almost equal latitude of Casa Blanca, Africa. This picture shows that air flights to New York from Narita will receive the largest dose, but the value will be decreased if magnetic pressure is added. However, low latitude cities, like Honolulu, changed just a little. I want to add one more, which is "solar flare." In case solar flare occurs, cosmic ray dose will be decreased due to increment of magnetic pressure. Such dose decrement is well known as Forbush Decrease in USA or European countries but is seldom seen in Japan. I only have one measured data indicating that. This map shows how much we are exposed to cosmic rays if we fly from Narita. As you see, the largest dose will be obtained if you fly to New York. Data were obtained by the same silicon pocket dosimeter and by myself and were calibrated practically. Fig. 5 shows how many times we can fly observing the current laws. The dose limit stipulated by Japanese law will be 1 mSv in 1 year [2]. I said "will be" because there is no written form of law as far as cosmic rays are concerned. In this case, you may fly to New York 11 times in a year; but, inside Japan, you can fly to anywhere freely.

## 4. Radiation in space

In the surrounding of international space station, the dose rate will be 1 mSv/day if the sun does not burst into flare. Fig. 6 indicates how much we are to be exposed in space missions. This small circle is the dose of return flight to New York, and this is the dose limit of the whole annual natural radiation, 2.4 mSv. Furthermore, this large circle represents the dose which astronauts receive in return flight to Mars. NASA declared that they would send human to Mars 30 years later, while the similar flight is planned by ESA

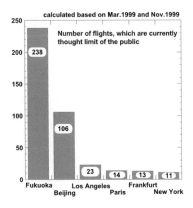

Fig. 5. Numbers of permissible return flights.

Fig. 6. Estimated dose in various missions (mSv).

with the name of Aurora project. They need half a year to go to Mars and stay there for 1.5 years and require another half a year to come back to the Earth. It will bring 1 Sv in total. JAXA prepared the table indicating how much the dose will be allowed in each age interval [3]. Their results imply that men or women of more than 40 years old will be assigned as Mars astronauts. Everything is based on normal solar status. Exposure to go back to Mars will be totaled to 1 Sv. Pluto, a boundary star of solar system, will require 70 Sv. Then, what does the 70 Sv mean? It is the cancer treatment dose. Cancer cell will be killed in this dose. Moreover, normal cell will be also killed. In other words, we will be unable to go beyond Pluto.

## 5. Conclusion

Cosmic rays vary in geomagnetic latitude, altitude, and season. Thus, I wish to suggest that mankind will be unable to go beyond Pluto. It simply means that mankind will have to remain in the solar system forever. Furthermore, airplane has a potential danger. As you know, space activity in USA is being stopped. Then, we asked Russia to launch our detectors into the space. To our team, Russia, Austria, and USA joined together under auspices of NIRS-ICCHIBAN Project. Very recently, China also offered us an opportunity, and we accepted it because it is a kind of bargain.

## References

[1] M.A. Pomerantz, Cosmis Rays, Van Nostrand Reinhold, New York, 1971.
[2] International Committee on Radiological Protection, Pub. 60, Pergamon Press, Oxford, 1990.
[3] Report for manned space mission exposure treatment (in Japanese), Dec., 2001, National Space Advance Agency.

# Factors affecting cosmic ray exposures in civil aviation

## G.M. Kendall*

*National Radiological Protection Board, Chilton, Didcot, Oxon OX11 0RQ, United Kingdom*

**Abstract.** Cosmic rays come from the Sun and other sources in our galaxy and beyond. Primary cosmic rays undergo nuclear reactions when they strike atoms in the atmosphere, generating a complex variety of secondary particles. These can give radiation doses to people on the ground but also in aeroplanes where dose rates are higher. Doses from cosmic rays depend on duration of exposure, altitude, latitude, and calendar period. The first of these is self-evident: at a given dose rate, the dose increases with duration of exposure. At ground level, the atmosphere provides considerable shielding, but during flight we leave part of this shielding below us. Latitude (distance from the equator) is also important. The Earth's magnetic field helps to deflect cosmic rays away. This protection is greatest at the equator and least at the poles so that the intensity of cosmic rays in the atmosphere increases as one moves to higher latitude. The magnetic field generated by the solar wind also affects the number of cosmic ray particles reaching the earth. The higher the flux of particles from the Sun, the lower the cosmic ray doses at aircraft altitudes. The intensity of this solar wind varies with calendar period. © 2004 Published by Elsevier B.V.

*Keywords:* Cosmic radiation; Occupational exposure; Radiation; Dose

## 1. Introduction

Doses from cosmic rays are dependent on latitude. UNSCEAR [1] estimates that the mean effective dose to populations at about 50° North from cosmic rays at ground level is approaching 0.35 mSv a year, of which perhaps a fifth is from neutrons. Doses are higher at the poles and lower at the equator. Cosmic ray doses also depend strongly on height above sea level and populations living at altitude contribute disproportionately to the mean

---

\* Tel.: +44 1235 822729; fax: +44 1235 833891.
  *E-mail address:* gerry.kendall@nrpb.org.

0531-5131/ © 2004 Published by Elsevier B.V.
doi:10.1016/j.ics.2004.08.102

effective dose to the world population, which UNSCEAR estimates to be approaching 0.5 mSv, with about 25% from neutrons.

Because doses from cosmic rays are very dependent on altitude and latitude, doses to passengers and aircrew are very variable. I will discuss the factors involved in what follows.

## 2. Sources of cosmic rays

Cosmic rays can be divided into two broad types, those which come from the sun and those which come from our galaxy or beyond. The solar cosmic rays generally have energies of up to 100 MeV, with a few having energies up to 10 GeV. However, galactic cosmic rays, which are probably accelerated by the shock waves driven by exploding supernova can be much more energetic, with a very small proportion having energies above $10^{20}$ eV. Cosmic rays are largely protons with some alpha particles and smaller proportions of heavier nucleii.

The very energetic primary cosmic rays undergo a variety of nuclear reactions when they strike atoms in the atmosphere. These generate a complex variety of secondary particles including neutrons and pions which in turn interact and lose energy as they penetrate to lower altitudes. By the time they reach ground level, most of the surviving particles are muons, but at aircraft altitudes, more complex mixtures are encountered. Some exotic particles may be formed in initial high-energy interactions of cosmic rays, but, for practical purposes the doses originate from more mundane particles or rather their secondary (or later) collision products. These products fall into three categories: nucleons (protons and neutrons), electromagnetic (electrons and photons), and muons. The relative importance of the various components varies with depth in the atmosphere with the nucleons of relatively greatest importance at high altitudes and the muons most important close to sea level. However, it is important to remember that dose rates overall rise sharply at higher altitudes. It is, incidentally, a manifestation of the theory of relativity that muons survive down to sea level; they can do this only because of the time dilation resulting from their high kinetic energies.

One important but specialised point should be mentioned. From time to time the sun emits bursts of protons. Under certain circumstances the number incident on the atmosphere can increase sharply. The particles are of relatively modest energy and do not penetrate deeply but they could cause a significant increase in the dose above about 50,000 ft. Concorde carried a device to detect and assess such rare events.

## 3. Factors affecting cosmic rays doses

Doses from cosmic rays depend on duration of exposure, altitude, latitude and calendar period.

It is self-evident that the longer the time spent in a radiation field the greater the dose. Conversely, spending a shorter time in a field may compensate for a higher dose rate. Thus passengers flying by Concord across the Atlantic did not necessarily receive a higher radiation dose because, although the dose rate was higher, the journey time was shorter.

The atmosphere plays a vital role in protecting us from cosmic rays. At ground level, the atmosphere typically provides as much shielding as 3 ft of lead but during flight we leave part of this shielding below us and cosmic ray doses increase. Modern passenger planes are generally designed to fly higher than did the previous generation of aircraft and so they generally have less atmospheric shielding.

Another very important point is that the earth's magnetic field will deflect the charged primary cosmic rays. It is more effective at deflecting particles of lower energy and also those which arrive at the equator; there is little deflection at the poles. This means that the intensity of cosmic rays in the atmosphere increases as one moves to higher latitude. The strength of the earth's magnetic field is affected by its interactions with the "solar wind" of protons from the sun. These are, of course, themselves a component of the cosmic rays. The intensity of the solar wind varies with calendar period. Perhaps counterintuitively, the higher the flux of particles from the Sun, the lower the cosmic ray doses at aircraft altitudes.

The latitude-dependent protection offered by Earth's magnetic field means that the mixture of particles striking the upper atmosphere varies from the poles to the equator. The fact that lower energy particles are both deflected more readily by Earth's magnetic field and more readily attenuated in the atmosphere means that the difference in doses between the equator and the poles decreases with height. At sea level, there is only modest variation in dose with latitude. At typical aircraft heights, it varies significantly.

People undertaking space flight could be exposed to very high doses from cosmic rays unless special precautions were taken to shield them.

## 4. Control of doses from cosmic rays

In Publication 60 [2], ICRP moved towards including natural radiation sources within the general system of radiation protection. ICRP did not recommend that the existing systems of control for artificial radiation sources should simply be transferred to natural exposures. Indeed, they noted that since radiation is ubiquitous, one might reach a situation in which all workers were subject to a regime of radiological protection. In the context of occupational exposure, ICRP first recommended that attention should be restricted to exposures which might reasonably be regarded as the responsibility of the operating management. This provision would exclude cosmic rays at ground level and radiation from nuclides in the undisturbed earth's crust, for example. ICRP also recognised that it was desirable to allow national authorities discretion in the scope of the controls which should be applied.

ICRP specifically identified certain circumstances as ones where natural sources should be considered as part of occupational exposure. These included the operation of jet aircraft and spaceflight.

ICRP Publication 60 provided the impetus for a revision of the European Basic Safety Standards Directive [3]. This, in Title VII, covered natural radiation exposures. Title VII contains three short articles which build on the advice of ICRP. A document "Radiation Protection 88" [4] provides technical guidance on the practical implementation of Title VII. This guidance suggests that aircrew (both those on the flight deck and cabin crew) constitute the group most at risk. National authorities should ensure that assessments are made of the likely exposures of those aircrew for whom they are responsible. The guidance suggests that for some aircrew doses are unlikely to exceed 1 mSv in a year; no controls are necessary under these circumstances. For aircrew receiving annual doses in the range 1–6 mSv, there should be individual estimates of dose.

## 5. Estimating doses to passengers and crew

At aircraft altitudes of about 18 km and below, the primary cosmic rays have given way to a great variety of secondary (or later) particles, some of which will be of very high energies. This is not a situation that arises elsewhere in radiation protection, except perhaps around high-energy particle accelerators. Both measurements and theoretical calculations are used to try to characterize the radiation fields involved. It is beyond the scope of this paper to go into these questions, but they have been discussed, for example, by McAulay et al. [5].

The radiation fields to which aircrew are exposed are complex; however, they are predictable and consistent (with a proviso about solar flare events at very high altitudes). The European guidance suggests that for conventional jet aircraft estimates of dose can normally be made using computer-generated estimates of doses for the routes flown. These computer-derived dose estimates will normally be moderately cautious overestimates of long-term mean doses. That this is the case can be confirmed by occasional measurements.

A significant fraction of the dose from cosmic rays to aircrew can be delivered by high-energy protons. An important parameter is the radiation weighting factor assigned to these particles. At the time of ICRP Publication 60, high-energy protons were not thought to be important for practical radiological protection and a very conservative factor of five was recommended. In ICRP Publication 92 [6], a reduction to a radiation weighting factor of 2 is flagged.

## References

[1] United Nations Scientific Committee on the Effects of Atomic Radiation, Sources and Effects of Ionizing Radiation; UNSCEAR 2000 Report to the General Assembly with Annexes, United Nations, New York, 2000.
[2] International Commission on Radiological Protection, 1990 Recommendations of the International Commission on Radiological Protection, Annuals of the ICRP 21, ICRP Publication, vol. 60, Elsevier Science, New York, 1991.
[3] Commission of the European Communities, 1996. Council Directive 96/29/EURATOM of 13 May 1996 Laying Down the Basic Safety Standards for the Protection of the Health of Workers and the General Public Against the Dangers Arising from Ionising Radiation. Official Journal of EC, Series L, No. 159 of 1996.
[4] European Commission, Recommendations for the implementation of Title VII of the European Basic Safety Standards Directive (BSS) concerning significant increases in exposure to natural radiation sources, Radiation Protection, vol. 88, Office for Official Publications of the European Communities, Luxembourg, 1997.
[5] I.R. McAulay, et al. Exposure of air crew to cosmic radiation, EURADOS Report 1996-01 Office for Official Publications of the European Communities; European Radiation Dosimetry Group, Luxembourg, 1996.
[6] International Commission on Radiological Protection, Relative biological effectiveness (RBE), quality factor (Q), and radiation weighting factors (wR), ICRP Publication 92, Annuals of the ICRP, vol. 33, Pergamon, Oxford, 2003.

# Properties, use and health effects of depleted uranium

W. Burkart*, P.R. Danesi, J.H. Hendry

*IAEA—International Atomic Energy Agency, Department of Nuclear Sciences and Applications, P.O. Box 100, Vienna, Austria*

**Abstract.** Depleted uranium (DU) has been claimed to contribute to health problems both in military personnel directly involved in war actions as well in military and civilian individuals who resided in areas where DU ammunition was expended. Due to the low specific radioactivity and the dominance of alpha-radiation, no acute health risk can be attributed to external exposure to DU. Internalised DU is both chemo- and radio-toxic. The major risk is from inhalation of DU dust or particles with less than 10 μm aerodynamic-equivalent diameter, formed when DU ammunitions hit hard targets (aerosol formation) or during weathering of DU penetrators. One major conclusion is that for all post-conflict situations, the inhaled DU quantities (central estimates) produced radiation doses that would be only a fraction of those normally received by the lung from natural radiation. Hence no long term lung effects due to these DU amounts can be expected. These conclusions also hold for whole-body exposure from ingestion of DU in local food and water. © 2004 Elsevier B.V. All rights reserved.

*Keywords:* Depleted uranium; Internal radiation exposure; Health impact; DU properties; DU field studies

## 1. Introduction

The most relevant properties, use and health effects of depleted uranium available up to 2001 were reviewed by IAEA [1]. Relevant and extensive reviews on the topic were also published by WHO [2] and The (UK) Royal Society [3]. Since then, some additional information on the subject has been published mainly as a result of field studies conducted in the Balkans and Kuwait, and there has been renewed interest in DU generated by reports claiming that DU ammunition was used again during the conflicts in Afghanistan (2002) and Iraq (2003). In this paper, the most relevant information reported in Ref. [1] is summarized, and the health effects are updated in the light of new studies reported in the period 2002–2004. References already quoted in Ref. [1] are not listed.

---

* Corresponding author. Tel.: +43 1260021600; fax: +43 1260029227.
E-mail address: w.burkart@iaea.org (W. Burkart).

0531-5131/ © 2004 Elsevier B.V. All rights reserved.
doi:10.1016/j.ics.2004.09.047

## 2. Properties of uranium

Uranium is a heavy, silvery-white, ductile and slightly paramagnetic metal, which is pyrophoric when finely divided. It is slightly softer than steel and reacts with water when present in a finely divided state. In air, it easily oxidizes and becomes coated with a layer of oxide. All isotopes of uranium are radioactive. Naturally occurring uranium contains three isotopes, namely $^{238}$U, $^{235}$U, and $^{234}$U. $^{235}$U and $^{238}$U are the origin of two different decay chains. The most abundant naturally occurring uranium isotope, $^{238}$U, has the longest half-life and, consequently, the lowest specific activity. Because of its higher specific activity, $^{234}$U contributes as much as $^{238}$U to the radioactivity of natural uranium although the weight percentage of this isotope is extremely small (0.006%). Natural uranium is considered a weakly radioactive element. In addition, uranium is categorized as a heavy metal with chemotoxic potential. All natural uranium isotopes emit alpha particles having little penetrating power. As a result, uranium principally represents an internal radiation hazard. Depleted uranium (DU) is distinguished from natural uranium by lower relative concentrations of $^{235}$U (<0.7%) and $^{234}$U. In typical DU, the content of $^{235}$U is about one-third of its original value (0.2–0.3%). Consequently, the activity of DU is about 60% of the activity of natural uranium. Depleted uranium may contain traces of $^{236}$U and $^{239+240}$Pu from cross-contamination. Metallic uranium is 65% more dense than lead, has a high melting point (1132 °C) and is highly pyrophoric. These properties and the relative high availability and low cost have led to various civilian and military applications of DU.

## 3. Military use of DU

DU penetrators ignite on impact (especially with steel), due to the high temperature generated. In addition, the projectile sharpens as it melts and pierces heavy armor. The DU dust which may be formed during impact can be dispersed and contaminate the environment. Most of the dust particles are smaller than 5 μm in size and spread according to wind direction. After an attack with DU ammunition, this will be deposited on the ground and other surfaces as partially oxidized DU fragments of different size and as uranium oxide dust. The majority of the penetrators that impact on soft targets (e.g. sand or clay) can penetrate the ground and do not generate significant dust contamination. Inhalable particles (<10 μm) of DU oxides are formed on the ground during the corrosion process of the DU penetrators. Ammunition containing DU was used in recent conflicts: 1991—Iraq and Kuwait (Gulf War I), 1995—Bosnia–Herzegovina; 1999—Kosovo, 2002—Afghanistan, 2003—Iraq (Gulf War II).

## 4. Exposure pathways and health effects

### 4.1. External exposure

Only the beta and gamma components of DU contribute to external dose. Potential health effects from external exposure are limited to skin contact. The dose rate to the skin in direct contact with a piece of pure DU is caused by beta particles and is about 2 mSv/h, and much less in the case of dust. Nevertheless, direct skin contact with DU should be prevented.

## 4.2. Internal exposure

Health effects related to internal exposure may result from either chemical or radiological toxicity. Measurement of uranium excreted in urine is a sensitive method for determining the amount of DU inhaled. For retrospective assessments of exposure, where only a small fraction of the amount initially inhaled remains in the lungs, it is more practical to measure DU in the environment and to use individual monitoring only when environmental measurements indicate significant levels of contamination.

### 4.2.1. Ingestion

DU can enter the body in the form of uranium metal from fragments and as uranium oxides. In the body fluids, uranium is dissolved as uranyl ions. Uranium is absorbed into the blood, transported, and retained in body tissues and organs. Ingestion of DU is not considered the major exposure pathway. However, direct ingestion of contaminated soil must be taken into consideration, in particular for children. Only about 0.2% of ingested, insoluble DU is absorbed into the blood. The fraction absorbed into the blood is rapidly cleared, with approximately 90% leaving the body in urine within the first week after intake. The rest will be distributed to tissues and organs. In particular, about 10% is deposited in the kidneys, and most of this is eliminated in a few weeks. About another 15% is deposited in bone. Uranium remains much longer in the bone compartment, to the extent that after 25 years, 1% is still present.

### 4.2.2. Inhalation

Inhalation of dust is considered the major pathway for DU exposure both in combat and non-combat situations. Respirable particles of DU oxides, formed during the corrosion of penetrators that did not hit hard targets, could also be inhaled [4]. For persons entering an armored vehicle shortly afterwards, the aerosols generated at impact may lead to considerable inhalation exposures. About 95% of inhaled particles greater than 10 μm aerodynamic-equivalent diameter (AED) are deposited in the upper respiratory tract. Most of these are cleared to the pharynx, and to the gastrointestinal tract. Particles below 10 μm AED can reach deeper pulmonary regions (bronchioles and alveoli) and be retained for a considerable time. The amount of DU that is absorbed into the blood and deposited in tissue and organs depends mainly on particle size and the solubility of the uranium-containing particle. 'Soluble' chemical forms are absorbed within days while 'insoluble' forms generally take months to years. Toxic chemical effects are more likely to be associated with the more soluble forms of uranium while radiation effects are more likely to be associated with the insoluble forms, such as particles that are deposited in the lung and local lymph nodes and retained for extended periods of time. The kidneys are the critical organ for uranium chemotoxicity.

### 4.2.3. Embedded fragments

Wound contamination can occur during combat activities or in case of accidental bruising of skin on contaminated surfaces. In the latter case, after wound cleaning, the resulting exposure to DU can be expected to be negligible. However, embedded fragments not removable by surgical means result in chronic, internal exposure.

## 5. Radiation doses, risk and cancer

Calculations, using the measured DU concentrations in top soil in one of the most severely DU-hit locations in Kuwait, assuming an overall residence time of 1 year, suggest that the annual dose received by adults and children from the inhalation of DU particles would amount to around 10 µSv. This level of dose is vastly lower than the annual dose from natural irradiation (2.4 mSv) and the additional annual dose limit for the public. For individuals spending 12 h inside a DU contaminated tank, the corresponding dose is about 0.28 mSv, again much less than natural (or recommended maximum additional) annual doses for the public. According to the literature, the annual committed effective dose from embedded fragments can be estimated to 0.1 mSv.

There are reports of biomarkers as indicators of DU exposure. There was an association of HPRT mutations with high uranium levels in US Gulf War Veterans who were victims of 'friendly fire' and an increase in unstable chromosome aberrations in a group of UK Gulf War and Balkan's War Veterans [5]. It is difficult to detect an increased cancer risk due to radiation at doses lower than 10–50 mSv acute dose and 50–100 mSv protracted dose because the excess risk is small in comparison to spontaneous rates of cancers of the same type [6]. Therefore, no direct experimental or epidemiological evidence can be obtained.

## 6. Field studies in which IAEA participated

Field and environmental assessments, involving in-situ measurements and analysis of many environmental samples in the laboratory, have been conducted jointly by IAEA, UNEP, WHO, national institutions and international experts in Kosovo (November 2000) [7], Serbia and Montenegro (October 2001) [7], Kuwait (February 2002) [8] and Bosnia and Herzegovina (October 2003) [7]. The general conclusion of these studies was that no widespread contamination of the ground surface by DU was detectable. Detectable ground surface contamination by DU was only measured in small areas about a meter away from where DU penetrators were found. DU fragments and DU oxides particles (S type) of respirable size were found dispersed in the ground around and beneath penetrators lying on the surface. The general conclusion of these studies was that no significant risk of contamination is expected in the short- and medium-term. Analyses of urine samples provided by people working western Kosovo and peacekeeping troops were consistent with this conclusion. All results indicated no substantial DU exposure among these groups.

## References

[1] A. Bleise, et al., J. Environ. Radioact. 64 (2003) 93–112.
[2] Depleted Uranium: Sources, Exposure and Health Effects, WHO Report, WHO/SDE/PHE/01.1, April 2001.
[3] The Health Hazards of Depleted Uranium Munitions, Royal Society (UK) Reports, Part I (2001) and II (2002).
[4] P.R. Danesi, et al., J. Environ. Radioact. 64 (2003) 143–154.
[5] H. Schroder, et al., Radiat. Prot. Dosim. 103 (2003) 211–219.
[6] D.J. Brenner, et al., Proc. Natl. Acad. Sci. U. S. A. 100 (2003) 13761–13766.
[7] UNEP Post-Conflict Environmental Assessments: Depleted Uranium in Kosovo, Geneva 2001; Depleted Uranium in Serbia and Montenegro, Geneva, 2002; Depleted Uranium in Bosnia and Herzegovina, Geneva, 2003.
[8] Radiological Conditions in Areas of Kuwait with residues of Depleted Uranium; Radiological Assessment Report Series, Vienna, 2003.

# Public health and environmental aspects of DU

## J.P. McLaughlin*

*Department of Experimental Physics, University College Dublin, Belfield, Dublin 4, Ireland*

**Abstract.** Depleted uranium (DU) ammunition has been used in a number of conflicts most notably during the Gulf War and the Balkan conflicts of the 1990s and, more recently, in Afghanistan and in Iraq. In the case of the Balkans, a number of very comprehensive post-conflict environmental assessments of DU have been carried out by United Nations Environment Programme (UNEP). It was concluded that in most exposure scenarios there DU is unlikely to cause any significant physical health effects in the general population. A summary is given here of exposure pathways and their possible health effects. On the other hand, the health consequences of the psychosocial impact on communities arising from DU bombardment is not known and needs to be addressed. Public radiophobia regarding DU may be largely irrational but for the scientific and radiation protection communities to consider it as being outside their remit raises questions of an ethical nature and may be shortsighted in the long term. The establishment of a coherent and effective communication strategy to inform and educate the public on the known effects and the management of DU contamination in present and future conflict areas would appear to be a sensible and ethically proper approach to dealing with this issue. © 2004 Elsevier B.V. All rights reserved.

*Keywords:* Depleted uranium; Health risks; Risk communication

## 1. Introduction

Depleted uranium (DU) is a by-product of the processing of natural uranium ore to produce enriched uranium for the nuclear power industry or for nuclear weapons. Some DU also originates from the reprocessing of spent nuclear fuel. Apart from its many nonmilitary commercial applications, it is the consequences of its military use as a weapon that has brought it to public attention and is the focus of this account. In its military application, it is used as protective armour on tanks but it is its self-sharpening and pyrophoric properties that have made it a preferred material to be used as armour piercing incendiary ballistic

---

* Tel.: +353 1 716 2229; fax: +353 1 283 7275.
  *E-mail address:* james.mclaughlin@ucd.ie.

0531-5131/ © 2004 Elsevier B.V. All rights reserved.
doi:10.1016/j.ics.2004.10.029

penetrators. In this latter capacity, DU ammunition has been extensively used in a number of conflicts in recent years commencing with the Gulf War in 1991. It was subsequently used in the Balkan conflicts, in Afghanistan and, most recently, in the Iraq War of 2003. Comprehensive reports on the possible health and environmental effects of DU have been produced, in recent years, by reputable and respected scientific bodies [1–3]. All of these reports, while they may differ in emphasis and in detail, are in agreement that for the majority of likely scenarios the long-term post-conflict health and environmental impacts of DU are minimal and manageable. Notwithstanding the reassurances that can be derived from these studies, the general public in conflict areas, where DU has been used for military purposes, perceive its presence as a significant risk factor in their lives.

## 2. Public health and environmental behaviour

As there is already in existence a large number of reports dealing with the environmental and possible health effects of DU, here only an overview is given of the main features of these effects as they pertain to the public and only in a post-conflict context. DU has both chemical and radiological toxicity properties for which the lung and the kidneys are the main target organs. The main pathways for exposure are by external exposure, inhalation and ingestion [4]. If DU penetrators are handled external contact exposure of the skin due to radiation at an equivalent dose rate of about 2.5 mSv/h will take place. Deterministic effects should, however, not occur but to avoid unacceptable risks of skin cancer it is recommended that prolonged (>250 h) skin contact with DU should be avoided. Inhalation exposure can only occur if a penetrator has already been aerosolised by its impact with armour or other hard materials. Inhalation can be the major pathway during and in the immediate aftermath of conflict. The effects of any intakes of DU should be viewed in the context of the natural uranium content of the human body (circa 60 μg) arising from normal intakes of air, food and water [1]. While about 98% of ingested DU should be excreted in the faeces, in the case of inhalation depending on its chemical form and solubility in excess of 20% may be absorbed in the blood of which the majority will be excreted in the urine within a few days. Maximum intakes of DU recommended by bodies such as the WHO are arrived at on the basis of chemical toxicity in the case of the kidney and on the basis of radiation dose to the lung in the case of inhalation. In general terms, it can be stated that the evidence available strongly suggests that long-term significant health effects in members of the public due to intakes of DU are unlikely. It is also worth noting that there is no evidence that uranium, in any form, has caused cancer in humans. It has, however, been suggested that DU exposure might be associated with a statistically significant increase in Hodgkin's lymphoma observed in a large cohort of Italian soldiers who had served in the Balkans [5]. Recent in vitro studies into the effect of low levels of DU on human osteoblast cells have, however, demonstrated that genomic instability can result [6]. The observed genomic instability was found to be expressed in delayed reproductive cell death and micronuclei formation. It is not unreasonable to assume that similar effects may take place in human cells due to the normal body burden of about 60 μg of natural uranium. While these in vitro results are not health effects and it is very premature to make extrapolations to humans, the possibility that such uranium induced instabilities in genetic behaviour may be linked to cancer cannot be excluded.

In the Balkans, most DU penetrators did not impact on heavy armour or other hard targets and survived almost intact either on the surface or buried in the ground. Both the UNEP teams and other observers have found that near surface recovered DU penetrators had within 7 years suffered a decrease in mass of about 25% by corrosion in the soil [3]. This suggests that, in the climatic and soil conditions of the Balkans, all such buried penetrators can reasonably be expected to be totally corroded within 25–35 years. At specific sites, such dissolution coupled to migration of the corrosion products could lead to contamination of drinking water. In contrast to its behaviour in the Balkans, the environmental behaviour of DU used in the Iraq war in 2003 can be expected to be very different. In the Iraq war, unlike in the Balkans, it appears that many DU antitank projectiles were aerosolised and intact projectiles lie clearly visible on the desert surface in many places. Due to the dry desert conditions in the Iraq battle zones, such DU will be subject to very slow corrosion rates thereby reducing the possibility of water or food contamination. On the other hand, unconfirmed reports suggest that DU together with other metal debris of battle is being collected as scrap and may therefore end up in manufactured metal objects, some of which might even be used by the public in their homes. Such possibilities highlight the need for studies into the ultimate fate of DU in the aftermath of conflicts, which may occur in many different climatic and geochemical environments, and for proper communication channels to be established with local health and other public authorities. In addition, there is a need to decontaminate sites and buildings targeted by DU. This latter aspect may be of particular importance in Iraq where it is reported, but not as yet independently verified, that buildings which may have been contaminated by airborne DU particles have been re-occupied without decontamination.

## 3. Public perception and risk communication

It is a truism that uranium, in any form, nearly always gets a "bad press". When mention of uranium is made in the media, it is usually in some negative context such as that of nuclear weapons or the possibility of their proliferation, etc. Because of its role in the development and use of nuclear weapons, it was almost unavoidable that uranium, in any form, would be incorrectly perceived by the public as being a highly radioactive, dangerous and carcinogenic material. Therefore it is not surprising that DU is also viewed in these terms by the public. The public are unaware of the large body of scientific work, which shows that in most likely scenarios DU remaining in the environment in the aftermath of a conflict does not give rise to significant risks. This, in no small measure, is due to a failure by the scientific community to establish effective risk communication channels with the media and the public. Public radiophobia will in future years be strengthened by the dissemination of the knowledge that small amounts of plutonium isotopes ($^{239+240}$Pu up to 6 kBq/kg) ,even though they are not of radiological significance, are present in DU penetrators arising from reprocessed nuclear fuel contamination in the DU production plants [7,8]. The radiophobia and fears regarding DU may be largely irrational but for the scientific community to ignore their existence and to consider them as being outside their remit raises questions of an ethical nature and is ultimately shortsighted. It is not, however, only in communities in conflict areas, where DU weapons have been used, that such misperceptions are present. Objections to the use of DU have become part of antinuclear rhetoric in many developed countries. In addition to their depiction of DU as a material of radiological health significance, antinuclear groups argue that the existence of DU weapons, manufactured as they are from a waste product in the nuclear power fuel cycle, is at variance with the contention that nuclear power production is a peaceful activity. It is also argued in antinuclear circles that reports by the scientific community, which emphasise that DU in the environment presents very low risk to the public, gives support to its continuing military use. The existence and political potency of such arguments, however spurious or simplistic they may be, should not be ignored and require a measured and coherent objective response at an international level. An essential part of such a response should be the establishment of a DU risk communication strategy and programme. As a risk communication programme is not simply a matter of disseminating information, the development of such a programme would require input from not only physical scientists but also from those with expertise in specialities such as psychology, sociology, the media, etc.

## 4. Conclusion

DU, both as armour and munitions, has been and will continue to be increasingly used in warfare for the foreseeable future. On the basis of current scientific knowledge in most feasible scenarios, its long-term effects on local populations and the environment, both from radiological and chemical toxicity perspectives, are not likely to be significant. DU usage may, however, have a significant psychosocial impact derived from misperceptions of risk, radiophobia and the absence of proper post-conflict risk communication programmes in affected areas. The not unreasonable hypothesis that psychosocial effects occurring in the aftermath of DU bombardment may cause increased stress and associated physical health effects has not yet been tested or verified. Nevertheless, it seems appropriate and long overdue that a programme of DU risk communication and contamination management should be initiated and developed by the international scientific community.

## References

[1] N.H. Harley, et al., Review of the Scientific Literature as it Pertains to Gulf War Illnesses, vol. 7, Depleted Uranium, National Defence Research Institute (U.S.), RAND, 1999.
[2] The Royal Society, The Health Hazards of Depleted Uranium Munitions: Part I, The Royal Society, London (2001 May).
[3] UNEP, Depleted Uranium in Bosnia and Herzegovina, United Nations Environment Programme, Geneva (2003 May).
[4] A. Bleise, P.R. Danesi, W. Burkart, Properties, use and health effects of depleted uranium (DU): a general overview, Journal of Environmental Radioactivity 64 (2003) 93–112.
[5] M. Grandolfo, et al., Depleted Uranium: some remarks on radiation protection, Radiation in the Environment Series, The Natural Radiation Environment VII, vol. 7, Elsevier, 2004 (in press).
[6] A.C. Millar, et al., Genomic instability in human osteoblast cells after exposure to depleted uranium: delayed lethality and micronuclei formation, Journal of Environmental Radioactivity 64 (2003) 247–259.
[7] J.P. Mc Laughlin, et al., Actinide analysis of a depleted uranium penetrator from a 1999 target site in southern Serbia, Journal of Environmental Radioactivity 64 (2003) 155–165.
[8] E.R. Trueman, S. Black, D. Read, Characterisation of depleted uranium (DU) from an unfired CHARM-3 penetrator, Science of the Total Environment 327 (2004) 337–340.

# A comparison of human exposure to natural radiation and DU in parts of the Balkan region

Z.S. Zunic[a,*], K. Fujimoto[b], I.V. Yarmoshenko[c]

[a]*Institute of Nuclear Sciences "Vinca", P.O. Box 522, 11000 Belgrade, Serbia and Montenegro*
[b]*National Institute of Radiological Sciences, Chiba, Inage-ku, Anagawa, Japan*
[c]*Institute of Industrial Ecology of Ural Brunch of Academy of Science of Russia, Ekaterinburg, Russia*

**Abstract.** The paper presents the field results from six selected rural communities in the Southeast (Kalna), South (Niska Banja, Gornja Stubla, Borovac) and West (Uzice, Han Pijesak) parts of the Balkan region, where general population exposures to unmodified and technologically enhanced natural radioactive material (TENORM), including ammunition containing depleted uranium (DU) or reprocessed DU as well as indoor high radon, have been investigated. Since the only definitive evidence of DU contamination in soil samples could tell the disruption of the natural isotopic abundance of U-238 and U-235, it is important to distinguish between different sources of radiation burden to the population involved. The results of this work confirm that there is a constant need for developing comprehensive information about the spatial distribution of exposures in an adequately fine geographical grid, and for the accumulative dose estimation up to the present that clearly presented the uncertainty ranges and probability distribution of the estimated individual and population doses.
© 2004 Elsevier B.V. All rights reserved.

*Keywords:* Enhanced radioactive area; NORM; TENORM; Depleted uranium; Balkan

## 1. Introduction

Due to the existence of many radioactive mineralization zones, the Balkans contains a large number of areas having high concentrations of primordial naturally radioactive isotopes in soils and rocks. The Natural Radiation Environment (NRE) in the area of the former Yugoslavia has been regionally surveyed in detail since 1951 by geological and aero radiometric methods and a total of 279 natural radiation anomalous zones were identified

---

\* Corresponding author. Tel.: +381 11 2458 222x367; fax: +381 11 344 24 20.
*E-mail address:* zzunic@verat.net (Z.S. Zunic).

0531-5131/ © 2004 Elsevier B.V. All rights reserved.
doi:10.1016/j.ics.2004.11.194

[1]. Out of these zones, 146 may be characterized as mainly uranium radioactive, 103 as thorium radioactive and 30 as mixed. In addition, whatever radiological burden that may be due to any naturally occurring radioactive material (NORM) and the technologically enhanced naturally occurring radioactive materials (TENORM) should also be considered. Attempts were made in Yugoslavia to identify rural populations receiving elevated natural radiation exposures that might serve as potential cohorts for a planned future health study. This and other related considerations formed the basis of the natural radiation monitoring programme proposal presented at the first Yugoslav Nuclear Society Conference (YUNSC) in 1996 [2]. The programmes presented have the following goals: (a) identification, mapping and estimation of the geogenic origin of the elevated indoor radon and thoron areas; (b) an integrated application of standardized experimental measurement procedures combined with the theoretical knowledge originating from the research experience of other countries where similar projects have already been implemented. In order to achieve the second goal, an international collaboration in the area of high natural radiation exposures assessment has been realized by Yugoslav scientists and research groups from Japan (NIRS, Chiba), and 10 scientific institutions of the European countries. The military operations in Western Balkan Countries (Bosnia and Herzegovina and Serbia and Montenegro) triggered the third goal: (c) the need to launch comparative studies of the environmental radiation baseline levels in non-affected and affected areas in the Balkan region in the years 1995 to 1999. The measurements took place mostly in Serbia, Kosovo, Adriatic Coast of Montenegro and Bosnia and Herzegovina in the region of Han-Pijesak, in order to obtain the experimental results of different representative samples on human exposures to NORM and TENORM. The aim of this presentation is to give an overview of the experimental results and conclusions obtained so far for the investigated rural areas of the Balkans.

## 2. Field description

The first attempt at an integrated survey of population exposure due to natural radiation sources in Yugoslavia was carried out over the period from March 1997 to February 1999 at four locations having diverse geochemical characteristics [3–5]. The additional results of the field measurements were performed in eight new communities over the period 2000 to 2004 years. Out of these eight field sites, seven have been investigated in Serbia (Borovac, Bratoselce, Reljan, Pljackovica, Niska Banja, Uzice and Usce) and one in Bosnia and Herzegovina (Republika Srpska-Han Pijesak). In this work, out of all 12 field sites investigated so far, the unpublished data of 2 previous field locations (Kalna and Gornja Stubla) and new results regarding four newly investigated field locations (Niska Banja, Borovac, Uzice and Han Pijesak) were presented.

The Han Pijesak community (population 5500) is located 1.5 km south from an artillery storage and military barracks which were heavily attacked in 1995, resulting in approximately 730 kg of reprocessed DU deposited in that area. The bedrock of this area consists of limestone [6]. The bedrock of the Uzice region in Western Serbia consists mostly of bauxite, which can contain elevated quantities of thorium ore [7]. The third rural community, Niska Banja (population 6500) in South Serbia, is a radon spa. Besides Borovac village the villages of Bratoselce, Reljan and a mountain Pljackovica, all situated in South Serbia, were targeted in 1999 [6]. The village Borovac has 200 inhabitants. The area has a sandy soil, of depth from 0.5 up to 2 m thick. The bedrock consists of a coarse-grained

Table 1
Parameters of log-normal distribution of indoor radon/thoron concentration for sites surveyed

| Site | Arithmetic mean (Bq/m$^3$) | Geometric mean (Bq/m$^3$) | GSD | Maximum (Bq/m$^3$) |
|---|---|---|---|---|
| Kalna | 188/n.a. | 156/n.a. | 1.8/n.a. | 810/n.a. |
| Niska Banja | 1093/83 | 292/69 | 4.7/1.8 | 13,354/269 |
| Gornja Stubla | 447/404 | 336/311 | 2.2/2.5 | 6010/945 |
| Borovac | 40/73 | 37/65 | 1.5/1.7 | 64/118 |
| Uzice | 52/79 | 43/61 | 1.9/2.4 | 158/143 |
| Han Pijesak | 35/54 | 29/40 | 1.8/2.4 | 128/251 |

granite rock mainly composed of quartz and feldspar. Two wells are situated in the hillside, about 100 m east of the targeted area, and a third well is fed by a natural spring at the end of a small creek. At least one of these wells was situated directly in line with the direction of drainage from the targeted area. The water from these wells is transported through pipes to the village of Borovac for drinking.

## 3. Methods

The concentrations of radon were measured using etched nuclear track detectors techniques and charcoal canisters. In addition to contemporary radon measurements, retrospective assessment of radon exposure in most of the dwellings were made on the basis of the determination of its long-lived decay product Po-210 concentrations in both glass surfaces traps and in porous volume traps (mostly sponge materials), although these results are not presented in this work. Air kerma rates and thoron measurements were carried out in the majority of the cases.

Some soil and water samples were collected at the targeted sites. And samples of well water and hair were also obtained from the villages located near the targeted sites to find out the effect of DU in the rural communities. Those samples were treated radiochemically and analysed by alpha spectrometry or mass spectrometry.

## 4. Results

Results of indoor radon and thoron measurements were used to assess annual indoor radon concentrations using seasonal normalization [8]. Table 1 presents average and maximal values of annual indoor radon and thoron concentrations for six sites with parameters of the log-normal distribution of indoor concentrations.

Table 2
Distribution of individual annual effective dose due to indoor radon and thoron progenies exposure (in brackets population average annual effective dose)

| Dose interval (mSv) | Kalna (4.7 mSv) | Niska Banja (29 mSv) | Gornja Stubla (17 mSv) | Borovac (2 mSv) | Uzice (2.4 mSv) | Han Pjesak (1.6 mSv) |
|---|---|---|---|---|---|---|
| 0–1 | 1% | 2% | 0% | 2% | 8% | 24% |
| 1–5 | 67% | 32% | 4% | 97% | 85% | 75% |
| 5–10 | 27% | 21% | 23% | 0.4% | 7% | 1% |
| 10–15 | 5% | 11% | 24% | 0.01% | 0.4% | 0.1% |
| 15–20 | 1% | 7.1% | 17% | 0% | 0.1% | 0% |
| >20 | 0.2% | 27% | 32% | 0% | 0% | 0% |

In order to relate exposures to NORM sources of radiation to those from TENORM sources, the annual effective dose due to indoor radon and thoron exposures were evaluated. Recent UNSCEAR values of 9 nSv/(Bq h m$^{-3}$)$^{-1}$ for radon EEC and 40 nSv/(Bq h m$^{-3}$)$^{-1}$ for thoron EEC have been applied for effective dose assessment. The following average annual effective doses and their distributions are derived (see Table 2) using the concentrations and distribution of indoor radon and thoron assuming the equilibrium factors of 0.4 and 0.05 for indoor radon and thoron, respectively.

A few soil samples collected from the targeted sites demonstrated the effect of DU in terms of abundance abnormality observed in the ratio between U-235 and 238. Some other samples are now under further investigation.

## 5. Conclusion

1. A number of Balkan regions (such as Niska Banja and Gornja Stubla) have to be considered as extremely high natural radiation areas.
2. Areas affected by contamination by ammunition containing DU (Han Pijesak, Borovac et al.) are situated within non-elevated natural radiation areas.
3. Inhomogenic distribution of contamination in the surface soil within radius of 50 cm.
4. Presence of plutonium in the penetrators found in the Borovac.
5. Physical and mental health examination should be also carried out in the communities near the targeted sites.

## Acknowledgements

The assistance of the Ministry for Science, Technologies and Development through Research Contract P-1965 is gratefully acknowledged.

## References

[1] S. Jankovic, The deposits of Serbia (Yugoslavia): regional metallogenic settings, environments of deposition and types, Belgr.: Min. Geol. Fac. (1990), p. 760 (in Serbian).
[2] Z.S. Zunic, J.P. Mc Laughlin, M. Kovacevic, Research proposal for national radon programme in Yugoslavia, Vinca Bull. 2/1 (1997) 521–530.
[3] Z.S. Zunic, et al., Integrated natural radiation exposure studies in stable Yugoslav rural communities, Sci. Total Environ. 272 (1–3) (2001 May 14) 253–259.
[4] Z.S. Zunic, et al., Field experience of indoor thoron gas measurements in a stable rural community in Yugoslavia, 10th International Congress of the International Radiation Protection Association; 2000 May 14–19; Hiroshima, Japan, 2000.
[5] Z.S. Zunic, et al., An overview of field investigations in indoor population exposure by natural sources of radiation in Yugoslavia before environmental contamination of depleted uranium in Kosovo, IRPA Regional Congress on Radiation Protection in Central Europe; 2001 May 20–20; Dubrovnik, Croatia, 2001.
[6] UNEP, Depleted uranium in Serbia and Montenegro post-conflict environmental assessment in the Federal Republic of Yugoslavia, United Nations Environment Program, Geneva, 2002.
[7] J. Paridaens, et al., Correlation between Rn exposure and Po-210 activity in Yugoslavian rural communities, in: W. Burkhart, M. Sohrabi, A. Bayer (Eds.), High levels of natural radiation and radon areas: radiation dose and health effects, Excerpta Medica International Congress Series, vol. 1225, 2002, pp. 87–93.
[8] Z.S. Zunic, et al., Statistical analysis of radon survey results, in: J. Peter, G. Schneider, A. Bayer (Eds.), Proceedings of the 5-th International Conference on High Levels of Natural Radiation and Indoor Radon Areas; 2000 Sep 4–7; Munich, Germany, BFS, Salzgitter, 2002, pp. 197–200.

# Summary

# Summary of HBRA epidemiological studies

## Quanfu Sun[a,*], Zhanat Carr[b]

[a]National Institute for Radiological Protection, Chinese Centre for Disease Control and Prevention, China
[b]Radiation and Environmental Health, World Health Organization, Geneva, Switzerland

**Abstract.** Presentations at a session called "What did we learn from epidemiological studies in HBRAs?" of the 6th International Conference on High Levels of Natural Radiation and Radon Areas are summarized, and several suggestions and comments are made on the methodology used in the HBRA studies. © 2004 Elsevier B.V. All rights reserved.

*Keywords:* Natural radiation; Low dose; Cancer risk; Non-cancer risk

## 1. Introduction

There were four papers presented in the session called "What did we learn from epidemiological studies in HBRAs?" that was chaired by Dr. Tao from the National Institute for Radiological Protection, China, and Dr. Rajan from the Regional Cancer Center in Trivandrum, India. The purpose of this paper is to summarize the session and to put it in perspective in terms of the larger picture.

## 2. Review of findings of Chinese, Indian, Iranian, and Brazilian research groups

The session was started with a paper entitled "Cancer and non-cancer mortality among inhabitants in the high background radiation area of Yangjiang, China", given by Dr. Zou. He summarized the methodology of the Yangjiang HBRA cohort study, and reported the main results of cancer and non-cancer analysis. The strength of this cohort study is that there were 2 million person-years accumulated and 1202 cancer deaths ascertained. Individual doses were estimated based on hamlet-specific environmental exposure and

---

* Corresponding author. 2 Xinkang Street, Deshengmenwai, Beijing 100088, China. Tel.: +86 10 6238 9931; fax: +86 10 6238 8008.
*E-mail address:* qfusun@public3.bta.net.cn (Q. Sun).

0531-5131/ © 2004 Elsevier B.V. All rights reserved.
doi:10.1016/j.ics.2004.11.133

age–sex-specific occupancy factors. The annual effective doses appeared to be 2.1 mSv from external and 4.3 mSv from internal exposure for the exposed cohort. The cohort consequently was categorized into four groups for which doses nearly overlapped. The overall cancer mortality risk was not increased in the HBRA when compared with that in the control area. The only cancer that showed a statistical increase in the HBRA was oesophageal cancer. This finding may not be associated with the radiation exposure received in the HBRA, because of the influence of other risk factors (e.g., lifestyle and diet, including tobacco smoking, pickled vegetables, and salted foods). China, together with some other Asian countries, is a part of the world's "oesophageal cancer belt", showing top cancer incidence rates for this specific site [1]. The study also compared the risk of non-cancer diseases in the HBRA with that in the control area, and reported an increase among the subjects younger than 50. This increase was attributed to lifestyle changes in younger generations. It is interesting to note that decreased mortality of all types of tuberculosis in the HBRA was reported and apparently linked with the threefold exposure to natural background radiation. In a relevant poster paper by Sun, it was suggested that the decreased mortality from tuberculosis may be related to the enhanced response of the immune system that may have been induced by low-dose radiation exposure. A similar effect was reported in radiotherapy patients but not among the Japanese A-bomb survivors [2]. Perhaps differences in the level of local health care, lifestyle, and socio-economic status should be explored as potential confounding factors for this unusual phenomenon.

The following study, "Epidemiological study in HBR area in Kerala, India", was presented by Dr. Jayalekshmi from the Karunagapally Cancer Registry. The cancer registry contains the data for 1993–1997 obtained for 360,000 persons. The rate adjusted to the world population was used to express the results. It was reported that lung cancer incidence rate (1990–2000) was linearly correlated with outdoor dose of radiation per year. However, no adjustment for smoking was mentioned. The analysis did not employ internal exposure estimates and cumulative doses, despite substantial dose assessment efforts being carried out in parallel and reported in other sessions of the conference (see WS2 materials). Further, the doses of external exposure (7–10 mGy/year) are nearly overlapping with the limits for occupational exposure to ionizing radiation [3]. Nevertheless, tremendous work was accomplished to collect data on confounding factors for all the subjects covered with the cancer registry, and a wide spectrum of outcomes, including genetic effects such as Down's syndrome, was investigated. On the other hand, a sub-cohort consisting of six *panchayats* (villages) was established in 1999, and detailed data on occupancy factors and migration were obtained. The results of analysis using more advanced statistical methods and adjusted for confounding factors are being anticipated with great interest.

A nested case-control study on male lung cancer was conducted in Kerala and presented by Dr. Binu. The number of lung cancer cases was 205, and three controls were selected for each case. Outdoor external doses were measured with pocket dosimeters. The study showed no evidence to support a relationship between external radiation and lung cancer risk.

The third paper, entitled "Mortality and morbidity in the population living in high level natural radiation area of Ramsar, Iran", was presented by Dr. Mosavi-Jarrahil. The authors examined cancer registry data for 2003, and the cancer mortality data for the year of 2001–

2003, and calculated the Standardized Incidence Ratio and the Standardized Mortality Ratio, taking the general population of Iran as a reference population. Cancer incidence and mortality rates showed slight increases in the HLNRA when compared with NLNRA. On the other hand, no such increase was observed among men. The authors pointed out that the system of cancer registration in Ramsar remains undeveloped, representing a lot of room for future work. Given the weakness of the methodology and case ascertainment, small sample size and low statistical power, these findings rather do not allow for convincing conclusions at the moment. However, these findings may be considered as initial attempts toward large investigative efforts that may be made in this area of Iran.

The last presentation in this session, entitled "Cancer mortality pattern in some Brazilian HBRAs", was presented by Dr. Veiga from Brazil. This was the first attempt at evaluation of cancer mortality in the HLBRAs of Brazil to date. The analysed data were for the period 1991–2000. SMR was used to compare the risk in the exposed area with that in the Brazilian general population. The preliminary analysis showed that residents in HBRA experienced a significantly elevated mortality from stomach cancer and somewhat increased mortality from esophageal, lung, and pancreatic cancer and leukaemia. The authors suggested that the increased risk of some cancers could be associated with confounding factors including cigarette smoking, alcoholic drinking, dietary intakes, and pesticide use. Future studies focusing on better exposure assessment and collection of information on confounding factors are anticipated.

Thanks to the Health Research Foundation, two special posters were displayed in the second floor of November Hall. More detailed information about the Chinese Yangjiang study and the Indian Kerala study was given there; in particular, the methodology of the individual dose estimation by the Chinese group, which was not presented in detail at this conference.

In addition to the papers given in this session, there were two additional papers on epidemiological studies of residential radon. Dr. Tao from China presented at the satellite meeting of the conference the results of a case-control study on lung cancer in Yangjiang HBRA (63 deceased cases). Dr. Yoshinaga from the National Institute for Radiological Sciences (Japan) reported at the oral Session "Epidemiology/Radiation Protection" the results of a case-control study on childhood leukaemia (248 cases). Neither study found a statistically significant association with exposure to residential radon. Dr. Yoshinaga used indoor radon concentration of a 6-month passive measurement in his analysis, and measured the indoor EMF field. Small sample size was one of the major limitations of the Yangjiang study.

## 3. Some considerations on the ongoing HBRA studies

To date, the Chinese and Indian areas having naturally elevated background radiation levels have been studied most vigorously. Joint ongoing binational efforts between Japanese and Chinese research institutions remain the best example of international collaboration in the area of HBRAs. It is worth mentioning that the Chinese group analyzed the risk of non-cancer diseases, and that their Indian colleagues studied risks of cancer and genetic/congenital diseases with their Health Audit Program. The in-depth investigations of the Indian researchers are impressive. Brazil and Iran research groups are making an effort to improve their respective situations with dosimetry and health statistics that should allow for setting up in the future better and larger epidemiological studies to estimate the health risks of the HBRAs.

To compare risk estimations across studies and to pool the data in the future, a unified approach to study design and methods of quantitative estimation of cancer risk based on individual dose estimation is needed. It should be emphasized that it is important to have adequate grounds for comparison of confounding factors between the exposed and control areas, or between exposed and non-exposed population groups. A categorization of exposure status based solely on the external radiation may be inaccurate when there is a big overlap of internal exposure between two areas or among the dose groups. Each of the HBRA research groups has conducted a lot of work on dosimetry in recent years, mainly in terms of the external radiation. It is crucial to pay more attention to the internal exposure when talking about the risk of certain types of cancer, lung cancer in particular. Also, it is import that some attention should be paid to thoron, as well as radon, in the areas with elevated thoron exposure.

The results presented in the epidemiology session of the 6th HLNRRA Conference provided new insight into the various possible effects of natural radiation on the health of HBRAs residents. Great contributions and some achievements have been made, though more progress remains to be made.

## References

[1] B.W. Stewart, P. Kleihues (Eds.), World Health Organization, (WHO), International for Research on Cancer (IARC). World Cancer Report, IARC Press, Lyon, 2003, pp. 223–227.
[2] A. Safwart, The immunobiology of low-dose total-body irradiation: more questions than answers, Radiat. Res. 153 (5) (2000 May) 599–604.
[3] International Commission on Radiological Protection (ICRP), Recommendations of the International Commission on Radiological Protection, Pergamon Press, New York, 1990, p. 1991.

www.ics-elsevier.com

# Summary of dosimetry (radon and thoron) studies

## S. Tokonami*

*National Institute of Radiological Sciences, Chiba, Japan*

**Abstract.** The 6th International Conference on High Levels of Natural Radiation and Radon Areas (6HLNRRA) conference was summarized from the viewpoint of dosimetry studies. In this part, radon and thoron studies were focused on. Here are three points to be mentioned as follows: (1) importance of radon exposure; (2) inconsistency in dose conversion factors (DCFs) between dosimetric and epidemiological approaches; (3) underestimation of thoron in the past studies. © 2004 Elsevier B.V. All rights reserved.

*Keywords:* Dosimetry; Radon; Thoron; Dose conversion factor

## 1. Introduction

The 6th International Conference on High Levels of Natural Radiation and Radon Areas (6HLNRRA) was held in Kinki University from September 6th to 10th, 2004. Many participants from three major research fields, i.e., dosimetry, biology, and epidemiology were comprehensively discussed on the health effect of radiation to human beings. In this paper, the 6HLNRRA conference was summarized from the viewpoint of dosimetry studies. In particular, radon and thoron studies were focused on because they were the main contributors of the dose from natural radiation sources. This paper describes three points on their studies as follows: (1) importance of radon exposure; (2) inconsistency in dose conversion factors (DCFs) between dosimetric and epidemiological approaches; (3) underestimation of thoron in the past studies.

## 2. Importance of radon exposure

It is well known that there was a high lung cancer incidence in miners. Many evidences can be found out among cohort studies. In addition, it can be said that there may be a suggestive lung cancer incidence due to residential radon. Several evidences have been

---

* Tel.: +81 43 206 3105; fax: +81 43 206 4098.
  *E-mail address:* tokonami@nirs.go.jp.

0531-5131/ © 2004 Elsevier B.V. All rights reserved.
doi:10.1016/j.ics.2004.09.056

given by case-control studies. As one of the most recent studies, the study of Lubin et al. [1] can be introduced. This study describes two pooled results analysed with China (Shenyang and Gansu) case control studies. The excess relative risk per 100 Bq m$^{-3}$ was estimated to be 0.13 (95% confidence interval: 0.01, 0.36).

Evaluation of radon exposure for lung cancer study has currently been improved. In many epidemiological studies, contemporary radon concentrations have been measured so far. Since the radon concentration in the past is more important from the viewpoint of risk assessment, however, retrospective measurements are needed and now available [2]. This technique is based on the $^{210}$Po surface activity from glass objects. The technique is expected to provide more precise information on the past radon concentration. However, it seems there will be difficulties in calibration and avoiding affects of smoking.

## 3. Inconsistency in dose conversion factors

There are two approaches in the dose assessment due to radon progeny inhalation: (1) dosimetric approach and (2) epidemiological approach. When calculating the dose conversion factor (DCF) from the two approaches, however, two approaches give us two different results. The DCF by the dosimetric approach was estimated to be 15 mSv WLM$^{-1}$, when the ICRP Publication 66 Respiratory Tract Model was developed. On the other hand, the DCF by the epidemiological approach was estimated to be 5 mSv WLM$^{-1}$ based on the conversion convention in ICRP Publication 65. There was a three-time difference between the two estimates. However, two approaches are now getting closer than 10 years ago [3].

Fig. 1A illustrates the radon progeny dose conversion factor as a function of particle size, derived from the ICRP 66 model. The data were calculated with the breathing rate of 0.78 m$^3$ h$^{-1}$ for environmental exposure. The DCF for unattached radon progeny with an activity median diameter (AMD) of 1 nm is much higher than that for attached progeny with an ordinary AMD of 100–300 nm. Recently, people are gradually paying attention to air quality in their living environment. In fact, the air quality makes different DCF. Because clean air (less aerosols) results in high dose conversion factor, characteristics of radon progeny aerosols should be categorized in various environments for more dose assessment.

## 4. Underestimate of thoron in the past studies

Why had thoron been underestimated in the past? The following reasons can be considered: (1) there is difficulty in calibration and measurement and that (2) there are no

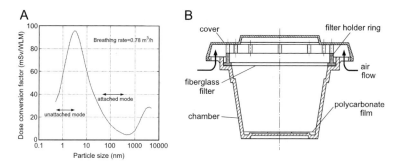

Fig. 1. (A) Radon progeny dose conversion factor. (B) Sectional view of the KfK monitor.

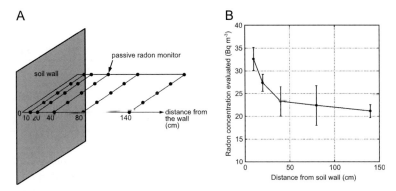

Fig. 2. (A) Geometric arrangement of the KfK monitor. (B) Detection response of the KfK monitor.

epidemiological data on thoron. As the current topic, however, significance of thoron has been recognized. People realized the importance of discriminative measurement of radon and thoron. Several efforts on the calibration are being made in Germany, India, Japan, Spain, USA, and so on.

Most of radon measurements have been made without discrimination so far. A case study can be shown here [4]. A passive radon detector was examined from the viewpoint of confounding influence of thoron on radon measurements. The detector was made in Germany (KfK monitor, Fig. 1B), which was used in the first survey in Japan. Five detectors each were placed at distance of 10, 20, 40, 80, and 140 cm from the soil wall in a Japanese house as shown in Fig. 2A. Fig. 2B illustrates the detection response of the passive radon detector. Consequently, the radon concentration increased as the detector approached the wall. Table 1 summarizes the sensitivity of alpha track-etch detectors examined by the author [5–7].

National Cancer Institute (NCI) had conducted a case-control study in Gansu province, China from 1994 to 1998 [8]. This is an evidence that residential radon might result in lung cancer. Many dwellings are located in the Chinese loess plateau. Indoor radon concentrations were measured by an alpha track-etch detector (Radtrak). There might be problems related to thoron contamination on radon measurements because the detector was sensitive to thoron as shown in Table 1. Table 2 shows the comparison of the survey result with other studies [9]. High radon concentrations were observed in the NCI study, whereas the radon concentration in the brand-new NIRS study was obviously lower than that from NCI. Although thoron and its progeny concentrations were not measured in the NCI study,

Table 1
Sensitivity of alpha track-etch detectors

| Detector | Radon (tracks cm$^{-2}$ kBq$^{-1}$ m$^{-1}$) | Thoron (tracks cm$^{-2}$ kBq$^{-1}$ m$^{-1}$) | References |
|---|---|---|---|
| KfK monitor | 0.85 | 0.70 | Tokonami et al. [4] |
| Radtrak | 2.81 | 1.88 | Tokonami et al. [4] |
| SSI/NRPB | 2.2 | 0.06 | Unpublished data |
| Regular Radpot | 2.62 | 0.10 | Tokonami et al. [7] |
| Modified Radpot | 2.64 | 1.32 | Zhuo et al. [6] |

Table 2
Comparison of the survey result with other studies

| Items | NCI, 2002 | NIRS, Brand-new | Wiegand et al. [10] | NIRS, 2004 |
|---|---|---|---|---|
| Study area province | Pingliang, Qingyang Gansu | Qingyang Gansu | Yan'an Shaanxi | Yan'an Shaanxi |
| Radon (Bq m$^{-3}$) | 223 | 92 | 92$^a$ | 76 |
| Thoron (Bq m$^{-3}$) | No data | 354 | 215$^a$ | 255 |
| EETC (Bq m$^{-3}$) | No data | 2.6 | 21.5$^b$($F$ of 0.1 used) | 2.2 |
| ERR per 100 Bq m$^{-3}$ | 0.19 (CI: 0.05, 0.47) | To be studied | No data | To be studied |

high thoron concentrations were found out, but its progeny concentrations were fairly low compared to its parent radionuclide. Similar results were given by Wiegand et al. [10]. They also found high-thoron areas in the Chinese loess plateau. The equilibrium factor of 0.1 was assigned to obtain the thoron progeny concentration based on the UNSCEAR manner. However, the NIRS study showed that the equilibrium factor by a factor of 1/10.

Thoron issues are summarized as follows: (1) There is a unique spatial distribution of thoron. High thoron concentrations are observed near the source but low far from the source. (2) Is there any significance of the equilibrium factor of thoron? The factor depends on the detector location. It seems to be meaningless to estimate thoron progeny concentrations with the equilibrium factor. Therefore, thoron progeny concentrations should be directly determined. (3) Intercalibration exercise is necessary using the worldwide network.

## 5. Conclusion

(1) Importance of radon exposure has to be realized again. Because there is suggestive lung cancer incidence due to residential radon, data should be accumulated and analyses be promoted.
(2) There is inconsistency in dose conversion factors between dosimetric and epidemiological approaches. Two estimates are getting closer in mine studies. These estimates should also be considered in the case of residential radon studies.
(3) Thoron issues should be considered more. Discriminative measurements are necessary. Behavior of thoron and its progeny should be understood. Thoron dosimetry should be studied more. Reliability of thoron measurements should be improved using intercalibration by worldwide network.

## References

[1] J.H. Lubin, et al., Int. J. Cancer 109 (2004) 132–137.
[2] F. Bochicchio, et al., Radiat. Meas. 36 (2003) 211–215.
[3] A. Birchall, J.W. Marsh, Abstracts on 6HLNRRA, vol. 51, 2004.
[4] S. Tokonami, et al., Proc. 9th International Conference on Indoor Air Quality and Climate, 2002, pp. 665–669.
[5] S. Tokonami, et al., Health Phys. 80 (2001) 612–615.
[6] W.H. Zhuo, et al., Rev. Sci. Instrum. 73 (2002) 2877–2881.
[7] S. Tokonami, et al., Radiat. Prot. Dosim. 103 (2003) 69–72.
[8] Z. Wang, et al., Am. J. Epidemiol. 155 (2002) 554–564.
[9] S. Tokonami, et al. Radiat. Res. in press.
[10] J. Wiegand, et al., Health Phys. 78 (2000) 438–444.

# Summary of biological studies

## Seiji Kodama*

*Division of Radiation Biology and Health Science, Research Center for Radiation and Radioisotopes, Research Institute for Advanced Science and Technology, Osaka Prefecture University, 1-2 Gakuen-cho, Sakai, Osaka 599-8570, Japan*

**Abstract.** The biological effects of low dose and low dose rate radiation have been extensively studied on the inhabitants in the high background area of natural radiation and experimental animals exposed to low dose (rate) radiation. Although the evidence to show the unique biological effect of low dose (rate) radiation has been accumulated, we still stand far from the comprehensive interpretation of the phenomena. More convincing data are expected to fill out a lack of understanding of the biological effect of low dose (rate) radiation. © 2004 Published by Elsevier B.V.

*Keywords:* Biological effect; High background radiation area; Chromosome aberration; Cancer; Radio-adaptive response; Bystander effect

## 1. Biological studies in the high background radiation area (HBRA)

The difficulty to discuss the biological effect of low dose radiation is due to a lack of convincing data to be referred. The health effect study on the inhabitants in the high background radiation area (HBRA) is one of informative approaches to resolve this issue [1,2]. Wang et al. (China) demonstrated the results of their chromosome study in the HBRA in South China. The level of natural radiation ($^{238}$U, $^{232}$Th, and $^{226}$Ra) at the HBRA was three to five times higher than that in the control area (CA). The chromosome study revealed that the frequency of unstable aberrations, i.e., dicentrics and rings, increased in a dose dependent manner, and that there was no threshold for the induction of those aberrations. This result indicates that the chromosome aberration is a sensitive and reliable biomarker to detect damage to humans induced by low dose and low dose rate of radiation. In contrast to the data obtained from unstable aberrations, the result concerning stable aberrations, i.e., translocations, detected by whole chromosome painting with

---

* Tel./fax: +81 72 254 9855.
  *E-mail address:* kodama@riast.osakafu-u.ac.jp.

probes for chromosomes 1, 2, and 4, indicated that the frequency was much higher than that of unstable aberrations, and that there was no dose dependency and no difference between the CA and the HBRA. However, a significant difference was found in the frequencies between the CA smokers and the CA nonsmokers, suggesting that smoking was responsible for a factor increasing stable chromosome aberrations. On the other hand, the epidemiological study demonstrated that there was no significant difference in cancer mortality in the inhabitants between the CA and the HBRA. It is a well-accepted concept that cancer cells develop in multistep processes with accumulating gene mutations, implying that more genetic damage is correlated with more induction for cancers. However, the health effect study in the HBRA in South China demonstrated that the increased cytogenetic damage, represented as unstable chromosome aberrations, did not result in increasing susceptibility to cancer induction. This discrepancy may be explained from the point of view on the possible difference in the biological features between stable and unstable aberrations (Table 1).

In general, dicentrics and translocations are produced in the peripheral lymphocytes in an equal frequency by radiation. The lymphocytes with unstable aberrations are eliminated during cell proliferation due to the impairment of cell division, while those with stable aberrations are accumulated through cell division (Table 1). This implies that unstable aberrations which can be detected in the peripheral lymphocytes are induced in the cells at the terminally differentiated stage in cell lineage, and thus, that those cells with unstable chromosome aberrations do not contribute to carcinogenesis (Table 1). In contrast, stable aberrations, of which frequency is not different between the CA and the HBRA, and which are produced in stem cells or stem-like cells, can be transmissible to the progenies, contributing to the induction for cancers (Table 1). The reason that the dose–effect relationship was not found in stable aberrations may be explained by the induction of the radio-adaptive response in cycling hematopoietic cells but not in noncycling lymphocytes in the $G_0$ phase [3].

In contrast to the interpretation mentioned above, recent studies on cancer cytogenetics demonstrated that unstable aberrations such as dicentrics played a role in carcinogenesis of solid tumors. For example, the frequency of anaphase bridge increased during the development of solid tumors in mice and humans because of the formation of dicentrics triggered by telomere dysfunction [4,5]. This evidence indicates that the telomeric instability and the subsequent formation of dicentrics may contribute to developing cancers. Therefore, the issue about the relationship between cancer incidence and chromosome aberrations raised by the study in the HBRA still contains unsolved questions.

Table 1
Comparison of the biological features between stable and unstable chromosome aberrations

| Feature | Aberration | |
| --- | --- | --- |
| | Stable | Unstable |
| Clastogen | Radiation/Chemicals | Radiation |
| Transmissible trait | Yes | No |
| Contribution to carcinogenesis | Yes | No (?) |
| Adaptive response | Yes | No |

Thampi et al. (India) reported the cytogenetic study on newborns in the HBRA in Kerala, India. The average dose received by the population is about four times the normal background radiation level. The radioactivity is primarily due to Thorium ($^{232}$Th) content and its decay products ranging from 8–10% of the monazite sand. The investigation of newborn malformations from 1995 to 2003 indicated that no radiation effect was found in the frequencies of malformation, stillbirths, and Down syndrome. The karyotype analysis revealed that the overall incidence of constitutional karyotype abnormalities was 4.85±0.45 per 1000 newborns, and that no difference was observed between the CA and the HBRA, indicating that the high-level natural radiation in Kerala has no heritable effect on humans.

The cytogenetic study on the inhabitant in the HBRA in Ramsar, Iran is on going. In this unique area, the hot springs are the main sources for the distribution of natural radionuclides, i.e., $^{226}$Ra and its decay products. A preliminary report by Zakeri et al. (Iran) indicated that increased chromosome aberrations and hormonal imbalance were observed in some inhabitants in the HBRA. Further studies are needed to confirm these results.

## 2. Biological effects of low dose radiation-animal studies

Tanaka et al. (Japan) reported the cytogenetic study of mice exposed to gamma irradiations with low dose rates, i.e., 0.05, 1, and 20 mGy/day. The prolonged radiation exposure with a low dose rate (20 mGy/day for 8 Gy) induced numerical chromosome aberrations such as monosomy, trisomy, and micronuclei. Because the numerical chromosome changes are not induced directly by acute exposure to radiation, this result suggests that the chronic radiation exposure with a low dose rate may induce genomic instability, and thus, it is suggestive for the interpretation of the human cytogenetic study in the HBRA.

It is well known that a low dose preirradiation induces cellular resistance to the subsequent high dose irradiation as the adaptive response [6]. There were two reports demonstrating that the radio-adaptive response was observed in the survival of mice received with whole body irradiation. One was that the preirradiation dose given in fractions was more effective than a single dose irradiation reported by Kale et al. (India). The other was that the low dose rate (1 mGy/min) preirradiation to mice was also effective to induce the radio-adaptive response of bone marrow stem cells measured by the number of spleen colonies reported by Shiraishi et al. (Japan). These results suggest that the similar radio-adaptive response is expected in humans in the HBRA. Indeed, a preliminary report by Mortazavi et al. (Iran) indicated the radio-adaptive response in lymphocytes of the inhabitants in the HBRA, Iran.

Nomura and Sakai (Japan) demonstrated a unique possibility that a low dose rate irradiation improved the effectiveness of insulin on the glucose metabolism in the diabetes model mice. The underlying mechanism, however, remains unknown.

## 3. Implication to the risk assessment of radiation

Cancer risk at low doses (for example, less than 100 mSv) is a matter of debate because we still do not have any convincing data to solve the uncertainty of that issue. The radio-

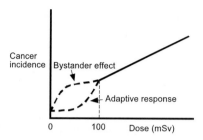

Fig. 1. Cancer risk estimation in low doses. The adaptive response may decrease the risk, while the bystander effect may increase the risk in low doses of radiation.

adaptive response may reduce the risk in low doses (Fig. 1). On the other hand, Hei (USA) and Suzuki et al. (Japan) demonstrated the existence of the bystander effect in the cell exposed to microbeam irradiation. The evidence of the bystander effect implies that the targets for the biological effects are larger than expected and suggests the increase for the risk in low doses (Fig. 1).

Further progress to understand the biological effect of low dose and low dose rate radiation is expected in future study.

## References

[1] Z. Tao, et al., Cancer mortality in the high background radiation areas of Yangjiang, China during the period between 1979 and 1995, J. Radiat. Res. 41 (2000) 31–41 (suppl.).
[2] T. Jiang, et al., Dose–effect relationship of dicentric and ring chromosomes in lymphocytes of individuals living in the high background radiation area in China, J. Radiat. Res. 41 (2000) 63–68 (suppl.).
[3] Z.Q. Wang, S. Saigusa, M.S. Sasaki, Adaptive response to chromosome damage in cultured human lymphocytes primed with low doses of X-rays, Mutat. Res. 246 (1991) 179–186.
[4] K.L. Rudolph, et al., Telomere dysfunction and evolution of intestinal carcinoma in mice and humans, Nat. Genet. 28 (2001) 155–159.
[5] D. Gissenlsson, et al., Telomere dysfunction triggers extensive DNA fragmentation and evolution of complex chromosome abnormalities in human malignant tumors, Proc. Natl. Acad. Sci. U. S. A. 98 (2001) 12683–12688.
[6] G. Olivieri, J. Bodycote, S. Wolff, Adaptive response of human lymphocytes to low concentrations of radioactive thymidine, Science 223 (1984) 594–597.

# Overall summary and comments

## Michiaki Kai*

*Department of Health Sciences, Oita University of Nursing and Health Sciences, Japan*

**Abstract.** The conference focused on three areas, Yangjiang in China, Kerala in India, and Ramsar in Iran, where the residents are exposed to high levels of natural radiation. The overall summary and comments will be made from the viewpoints of cancer risk at low dose rate and radiological protection. © 2004 Elsevier B.V. All rights reserved.

*Keywords:* Natural radiation; Epidemiology; Cancer risk; Low dose rate; Radon; Thoron

## 1. Why do we study on areas of high-level natural radiation and radon?

There are many areas showing high levels of natural radiation and radon in the world. This conference focused on three areas, Yangjiang in China, Kerala in India, and Ramsar in Iran. In these areas, epidemiological survey has already started or will start. The main aim of these studies is to investigate cancer risk from chronic exposure at low dose rate. Current risk estimate at low dose and low dose rate has given rise to much controversy because it is based both on epidemiological data available from acute exposure in atomic bomb survivors and several assumptions such as dose and dose rate effectiveness factor (DDREF). Furthermore, the lung cancer risk due to indoor radon is of much concern to the public. The investigations of areas with high levels of natural radiation and radon are expected to provide information for risk estimation having a more scientific basis.

---

\* Tel.: +81 97 586 4435; fax: +81 97 586 4387.
*E-mail address:* kai@oita-nhs.ac.jp.

0531-5131/ © 2004 Elsevier B.V. All rights reserved.
doi:10.1016/j.ics.2004.11.157

## 2. Epidemiological cohorts

The cohorts in China and India, who were exposed to radiation levels about three to five times higher than the usual natural level, consist of more than 100,000 inhabitants and show no excess cancer risk, although in Kerala, no individual dose measurement was estimated in cohort analysis. In Ramsar, a large difference of external doses among rooms was observed.

The cohort studies with a large number of residents will use the indirect method for individual dose assessment, using an ambient dose by environmental survey and occupancy factor, by which good correlation was obtained in China.

## 3. Chromosome study

The results of chromosome study in China are of particular concern because unstable types of chromosome aberration increase with total dose even at low dose rate, and because dicentric aberrations of more than 90% are observed without fragments. These findings suggest induction of the delayed type of chromosome aberration and its biological meaning of radiation carcinogenesis even if unstable types are unrelated to cancer. The reason why no excess translocations as a stable type have been observed may be the less than background level of translocation without excess exposure, which is calculated to be equivalent to 0.4 Sv at age 65. Consequently, a statistical increase of the translocation is unable to be detected at 0.2 Sv of total dose in China. In Germany, exposure to high indoor radon causes the increase of chromosome aberration with time-integrated concentration. A comparison of the same time-integration concentration shows the increase with time and suggests inverse dose-rate effects. This is consistent with findings in the cohort study of uranium miners.

## 4. Radon dosimetry

The internal exposure due to radon and its decay products is characterized by highly localized and chronic irradiation to various tissues of the respiratory tract. The currently predominant approach to radon dosimetry is to take a radon gas measurement and estimate the dose assuming an equilibrium state of the gas and its decay products. Recent studies show thoron (Rn-220) contributes to radon effective dose. The equilibrium factor in Rn-220 is lower than that in Rn-222. Therefore, the dose is estimated to be higher if no discrimination is made between Rn-222 and Rn-220. This indicates that Rn-220 gas measurement will be needed both in epidemiological study and even in radiation protection. Interestingly, it is indicated that the passive methods for measurement of Rn-220 possibly cause inaccurate dose estimates.

The dose estimation from radon takes an epidemiological approach; a dosimetric approach is taken in radiation protection. The difference of the previous estimates between these approaches was a factor of three. However, the latest estimates show the same effective dose of 10.4 mSv per WLM. One of the key issues of radon dosimetry is reliability of the radiation weighting factor of alpha particles and the tissue weighting factors in the ICRP respiratory tract model. The accuracy of both approaches in estimating effective doses remains uncertain.

If no effective dose is considered, a direct approach to risk estimate without RBE assumption is desirable even in radiation protection because it offers less uncertainty.

## 5. Lung cancer risk of indoor radon

Some pooled analyses in case-control studies of indoor radon continue to be done in Europe and North America. The pooled analysis in Germany shows significant increase of lung cancer at more than 140 Bq/m$^3$. The estimate of excess relative risk in this study, 0.12 per 100 Bq/m$^3$, is consistent with the BEIR-IV report. The quality of exposure assessment may be still low although the smoking effects can be modified in the analysis. The pooled analyses on indoor radon probably give a better perspective than extrapolation from uranium miner cohorts.

## 6. Radiation protection issues in radon

High occupational exposure to radon is being reduced in various workplaces from the viewpoint of radiation protection. The effective doses are higher than 20 mSv/year for 0.7% of the workers in the water supply facilities of Germany. Workplace monitoring is shifting toward individual monitoring, in which time-integrated radon concentration is a preferable metrics. We should continue to look for particular occupations to be considered and effective ways to reduce radon concentration.

## 7. Further studies needed for better understanding of low dose risk

The biological issue of radon risk is a bystander effect that suggests underestimation of low dose risk in the current estimate using a linear-nonthreshold model. On the other hand, the adaptive response observed in various endpoints possibly indicates a reduction of low dose risk. Theoretical understanding of radiation carcinogenesis is a critical issue in quantitative risk estimation. In addition, good radon dosimetry and cohort study contribute to a better understanding of lung cancer risk from indoor radon exposure. Investigation in such areas should continue in areas with high levels of natural radiation and radon.

# Contribution of high natural background radiation area studies to an evolved system of radiological protection

Masahito Kaneko*

*Radiation Effects Association, Tokyo, 1-9-16 Kajicho, Chiyoda-ku, Tokyo 101-0044, Japan*

**Abstract.** At the start of the 21st century, the most important task of the radiation protection community is to demonstrate that workers and members of the public have been adequately protected by the current dose limits. There is no evidence that natural radiations are causing adverse health effects in those high background radiation areas exceeding 10 mSv/y in India, Brazil and Iran. Evidence increasingly shows that there are threshold effects in risk of radiation. An evolved system of radiological protection with a "practical" threshold concept will be an alternative to the current system based on the linear-no-threshold (LNT) hypothesis. "Practical" thresholds may be defined as dose levels below which induction of detectable radiogenic cancers or hereditary effects are not expected. If the current dose limits are assumed to be below "practical" thresholds, there may be no need of "justification" and "optimization" (ALARA) principles for occupational and public exposures in normal situations. © 2004 Elsevier B.V. All rights reserved.

*Keywords:* High natural background radiation; Epidemiology; Threshold; Linear-no-threshold hypothesis; Radiological protection

## 1. Introduction

Based on the linear-no-threshold (LNT) hypothesis, the ICRP classifies harmful radiation health effects into two types, "deterministic" effects with thresholds and "stochastic" effects without thresholds, and recommends the general principles; justification of a practice, optimization of protection (ALARA principle), and individual dose and risk limits [1]. However, the assumption that radiation is harmful at all levels of dose has been regarded as a proven scientific fact by public opinion, mass media,

---

* Tel.: +81 3 5295 1781; fax: +81 3 5295 1486.
  *E-mail address:* mkaneko@rea.or.jp.

0531-5131/ © 2004 Elsevier B.V. All rights reserved.
doi:10.1016/j.ics.2004.10.008

regulatory agencies and many scientists, and has brought about "radiophobia". Dose limits based on subjective "acceptable risks" are not accepted by all people. The LNT hypothesis seems to result in endless over-regulation and waste of resources to take over-protective actions against radiation for no health benefits. High natural background radiation area studies will contribute to the establishment of an evolved system of radiological protection based on a "practical" threshold concept.

## 2. Reappraisal of the biological basis of the ICRP's policy

### 2.1. Some findings about low dose radiation health effects

(a) The 40-year follow-up studies on the genetic effects of atomic bomb radiation of Hiroshima–Nagasaki have demonstrated that there is no statistically significant effect of parental exposure to radiation of 0.4 to 0.6 Gy on any of the genetic indicators studied [2].

(b) Although there are high radiation areas exceeding 10 mSv/y in India, Brazil and Iran, there is no evidence that natural radiations are causing adverse health effects among the inhabitants [3].

(c) The final report of the large study of United States nuclear shipyard workers funded by the US Department of Energy showed that there was significantly lower total mortality in the exposed groups (both the <5 and ≥5 mSv groups) than in the non-radiation workers who engaged in similar work [4].

(d) The large combined study of nearly 96,000 United States, UK and Canadian nuclear workers showed no excess (negative) risk of total cancer mortality. The study purports to show a statistically significant dose response trend in leukemia mortality, but it was based largely on a few cases with cumulative doses above 400 mSv who worked at a reprocessing plant at which there could have been exposures to chemicals [5].

(e) The 100 years of observation on British radiologists revealed no statistically significant increase in cancer mortality among radiologists who were first registered after 1920 compared to other male physicians in England. Their mortality was significantly lower than that of all male medical practitioners. Moreover, there was no evidence of an effect of radiation on diseases other than cancer even in the earliest radiologists whose average lifetime dose was estimated to be 20 Sv [6].

### 2.2. Evidence contradicts LNT hypothesis

The existence of bio-defensive mechanisms such as DNA repair, apoptosis, adaptive response and immune system contradicts the LNT hypothesis. Evidence increasingly shows that there are threshold effects in risk of radiation [2,7–9].

### 2.3. Threshold or no threshold

For deterministic effects, the ICRP adopts clinically observable thresholds. However, the ICRP discusses only 'a real threshold' for stochastic effects and says: "there can be a real threshold in the dose–response relationship for those types of cancer only if the defence mechanisms are totally successful at small doses" (ICRP 60, paragraph 68) [1].

A malignant transformation of a single cell is not synonymous with a tumor or cancer. Cancer induction is such a complex matter that it almost certainly cannot be adequately described by a simple linear model [10]. There seems to be no rationale of distinguishing cancer risk assessments from those applied to other hazards, so-called deterministic effects. In fact, adverse health effects are only observable above certain dose levels.

## 3. An evolved system of radiological protection

### 3.1. Requirements for an evolved system of radiological protection

(1) The system shall be based on sound science.
(2) The system shall be coordinated with that of other "hazardous" substances.
(3) The system shall be simple and easily understandable for ordinary people, jury and judges.

### 3.2. Introduction of a "practical" threshold concept

(a) Instead of the current classification of stochastic and deterministic radiation effects, "practical" thresholds are introduced, below which induction of detectable radiogenic cancers or hereditary effects is not expected. According to Zbigniew Jaworowski, the Polish representative to the UNSCEAR, the practical threshold to be proposed could be based on epidemiological data from exposures in medicine, the nuclear industry, and regions with high natural radiation [11].
(b) What we need in practical radiation protection is "virtually safe dose". Regardless of the debate over whether there are thresholds or not for stochastic effects, the public should be aware that there are practical thresholds for radiation induced adverse health effects.
(c) As any workers and members of the public do not gain benefits from radiation exposures (excepting intentional irradiation for medical purposes), their radiation doses should be kept below "practical" thresholds. In such a case there will be no need of "justification" and "optimization" (ALARA) principles for protection against normal exposures. Then the ethical issue of "justification" to allow benefit to society to offset radiation detriments to individuals and also the ethical issue of "optimization" to exchange health or safety for economical gain can be resolved. Only this approach—introduction of the concept of "practical" thresholds—can be said to be based on an individual-oriented philosophy and satisfies the egalitarian principle of ethics.

### 3.3. Proposed dose limits

There seems to be no clear evidence of deleterious health effects from radiation exposures at the dose limits (50 mSv/y for workers and 5 mSv/y for members of the public), which have been adopted worldwide in the second half of the 20th century. The following dose limits are assumed to have been set below certain "practical" thresholds.

- Worker: 50 mGy/y (500 mGy/y for partial body or single organ exposure)
- Public: 5 mGy/y (50 mGy/y for partial body or single organ exposure).

Table 1
Bands of concern about individual whole body doses in a year

| Band of concern | Description | Level of dose[a] (mGy/y) | Protective actions |
|---|---|---|---|
| Band 3 | High (Serious) | >100 | Justify the exposure |
| Band 2 | Low (Normal) (Natural background radiation) | 1–100 | (Administrative dose limits) Worker: 50 mGy/y Public: 5 mGy/y |
| Band 1 | Very low (Negligible) | <1 | No protective action |

[a] Whole body low-LET radiations.

In case of high LET-radiations such as alpha particles and neutrons, the above dose limits should be multiplied by 1/10.

The proposed dose limits may be most appropriate to be positioned in the 2nd Band of concern in the Table 1.

## 4. Conclusion

At the start of the 21st century, the most important task of the radiation protection community is to demonstrate that workers and members of the public have been adequately protected by the current dose limits. High natural background radiation area studies are expected to contribute greatly to the people's understanding of the truth of low-level radiation health effects, as well as other epidemiological studies. The author believes that the introduction of a "practical" threshold concept will be essential for an evolved system of radiological protection, simple and easily understandable [12].

## References

[1] ICRP, 1990 Recommendations of the International Commission on Radiological Protection. Publication 60, Annals of the ICRP, vol. 21, No. 1–3, Pergamon, Oxford, 1991.
[2] S. Kondo, Evidence that there are threshold effects in risk of radiation, J. Nucl. Sci. Technol. 36 (1999) 1–9.
[3] M. Krishnan Nair, et al., Population study in the high natural background radiation area in Kerala, India, Radiat. Res. 152 (1999) S145–S148.
[4] G.M. Matanoski, Health Effects of Low-Level Radiation in Shipyard Workers. Final Report, DOE/EV/10095-T2, National Technical Information Service, Springfield, VA, 1991, http://cedr.lbl.gov/shipyard.pdf.
[5] E. Cardis, et al., Effects of low doses and low dose rates of external ionizing radiation: cancer mortality among nuclear industry workers in three countries, Radiat. Res. 142 (1995) 117–132.
[6] A. Berrington, et al., 100 years of observation on British radiologists: mortality from cancer and other causes 1897–1997, British J. Radiol. 74 (2001) 507–519.
[7] T.D. Luckey, Physiological benefits from low levels of ionizing radiation, Health Phys. 41 (1982) 771–789.
[8] E.J. Calabrese, L.A. Baldwin, Toxicology rethinks its central belief, Nature 421 (2003 Feb 13) 691–692.
[9] M. Pollycove, L.E. Feinendegen, Radiation-induced versus endogenous DNA damage: possible effect of inducible protective responses in mitigating endogenous damage, Human Exp. Toxicol. 22 (2003) 290–306.
[10] G. Walinder, Has radiation protection become a Health Hazard. Nuclear Training and Safety Center, Nykoping, Sweden, Medical Physics Publishing, Madison, WI, USA, 2000.
[11] Z. Jaworowski, Radiation risk and ethics, Phys. Today 52 (9) (1999) 24–29.
[12] M. Kaneko, An evolved system of radiological protection, Proceedings of the 11th International Congress of the International Radiation Protection Association, Madrid, Spain, 23rd–28th, May, 2004.

# Problems with HBRA studies in health risk assessment

## G.M. Kendall*

*National Radiological Protection Board, Chilton, Didcot, Oxon OX11 0RQ, United Kingdom*

**Abstract.** Epidemiological studies in High Background Radiation Areas (HBRAs) offer the prospect of large study populations and therefore appear attractive as a way of obtaining information about the effects of low doses of ionising radiation. However, bias and confounding are problems in all epidemiological studies and their residual effects may make it very difficult to detect relative risks lower than about 1.4. The easiest studies to propose will often be of ecological design in which data are aggregated over a population, for example, those living in a particular area, rather than cohort or case/control studies, which use data for individuals. Ecological studies are subject to problems of correlations between aggregated disease rates and aggregated measures of exposure which do not arise with studies which use individual data on disease, exposures and other risk factors. UNSCEAR suggests that only cohort and case/control studies should be used in quantitative assessments of radiation risks. © 2004 Published by Elsevier B.V.

*Keywords:* Epidemiology; Natural background radiation; High Background Radiation Areas; Cancer

## 1. Introduction

The real or imagined dangers from radiation are of enormous public interest. Epidemiology gives good information, better than for most carcinogens, on the effect of reasonably high doses of radiation—for protracted exposures, from a few hundred mSv down to 100 mSv and possibly to 50 mSv [1]. But for radiation, in contrast to most chemical carcinogens, we assume that there is no threshold below which exposures are harmless. So we need to extend as far as possible the region of dose

---

\* Tel.: +44 1235 822729; fax: +44 1235 833891.
*E-mail address:* gerry.kendall@nrpb.org.

where we have direct information and to make judgements about the dose response relationship below this. Studies of the effects of variations in natural background, particularly in High Background Radiation Areas (HBRAs), are an attractive prospect for extending the range of doses over which epidemiology can provide information on radiation risks. Perhaps the main advantage is that, very often, large populations will be available for study. But such studies also have significant problems.

## 2. Epidemiology and low dose population studies

The essence of the epidemiological approach is to compare rates of death or disease in the population under study and in a control group. However, epidemiology is not an experimental science, in which all variables apart from those under study can be held constant. Despite the best efforts of epidemiologists, factors such as bias and confounding are a constant danger [2,3]. It has been suggested that epidemiological methods may reach a limit in detecting relative risks below 1.3 or 1.4, far above the risks normally expected from a few mSv of radiation [2].

Confounding arises when there is an exposure, which is associated with both the factor that is being investigated and the disease under study. This would give rise to an apparent relationship between the factor being investigated and the disease, even if the factor did not cause the disease. For example, in occupational studies of radiation risk, higher doses are likely to have been incurred by those who have been in employment longest. The Healthy Worker Effect is likely to be weaker in those who have been employed for a long time than those employed for shorter periods, so disease rates may be higher in those with higher doses for reasons unconnected with radiation.

Epidemiological studies can be of cohort, case/control and ecological (or correlation) design. The last of these is attractive, since it does not require the collection of data about each individual in the study. However, such studies are particularly prone to bias owing to the lack of data on individual exposures and confounders [3].

Studies of natural background are subject to difficulties in selecting appropriate comparison groups and selection factors for place of residence as well as bias and confounding [2]. The easiest studies to propose will often be of ecological design.

## 3. A cautionary tale

Radon might appear to be the component of background radiation where population studies would be most fruitful. Radon typically gives the largest fraction of natural background and its effects are clearly seen in cohort and case/control studies. But ecological studies of radon exposures proved puzzling. A study by Cohen [4] based on about 1600 US counties gave a strong negative correlation between mean radon concentrations and mean lung cancer rates. For many years, this was an unexplained anomaly. However, Puskin [5] recently found similar negative correlations for other smoking related cancers, regardless of whether the organs concerned receive significant doses from radon. This strongly suggests that Cohen's results were a consequence of negative correlation between radon levels and smoking between the US counties.

## 4. Conclusions

Epidemiological studies designed to examine the effects of radiation exposure by looking at populations living in areas of high natural background are attractive, mainly because large populations can be studied. But a number of factors complicate the interpretation of such studies, most particularly if they are of ecological design. Muirhead [6] has a fuller discussion. UNSCEAR [3] suggests that only cohort and case/control studies should be used in quantitative assessments of radiation risks.

## References

[1] D.J. Brenner, et al., Cancer risks attributable to low doses of ionising radiation: assessing what we really know, PNAS 100 (2003) 13761–13766.
[2] J.D. Boice Jr., C.E. Land, D. Preston, Ionizing radiation, in: D. Schottenfeld, J.F. Fraumeni Jr. (Eds.), Cancer Epidemiology and Prevention, Oxford University Press, New York, 1996, pp. 319–354.
[3] United Nations Scientific Committee on the Effects of Atomic Radiation. Sources and Effects of Ionizing Radiation; UNSCEAR 2000 Report to the General Assembly with annexes; New York, United Nations 2000.
[4] B.L. Cohen, Test of the linear no-threshold theory of radiation carcinogenesis for inhaled radon decay products, Health Physics 68 (1995) 157–174.
[5] J.S. Puskin, Smoking as a confounder in ecologic correlations of cancer mortality rates with average county radon levels, Health Physics 84 (2003) 526–532.
[6] C.R. Muirhead, Uncertainties in assessing health risks from natural radiation, including radon, Proceedings of the 5th International Symposium on HLNRRA, Munich September 2000, In Excerpta Medica International Congress Series, vol. 1225, Elsevier, 2002, pp. 231–237.

www.ics-elsevier.com

# New public dose assessment from internal and external exposures in low- and elevated-level natural radiation areas of Ramsar, Iran

## Mehdi Sohrabi*, Mozhgan Babapouran

*National Radiation Protection Department, Atomic Energy Organization of Iran, Tehran, Islamic Republic of Iran*

**Abstract.** A detailed dosimetry study was carried out in 12 regions of Ramsar having some elevated-level natural radiation areas (ELNRA) for the determination of effective doses of public from internal and external exposures indoors and outdoors. The $^{222}$Rn levels were measured by using the AEOI passive radon diffusion chambers indoors of about 500 houses in two separate studies in autumn and in winter, and also in some locations outdoors. The $^{222}$Rn and its progeny and equilibrium factors were also determined indoors and outdoors of some buildings by the active Pylon AB-5 system based on which the effective equivalent dose ($E$) of the public was determined. The annual mean effective equivalent dose ($\bar{E}$) in different regions due to $^{222}$Rn ranges from 2.48 to 71.74 mSv with maximum levels up to 640 mSv determined in one house in Talesh Mahalleh. The annual $\bar{E}$ value from external exposures, as determined in 1000 locations outdoors and in 800 locations indoors, were reported in another paper before which range from 0.6 (in normal background areas) to 6 mSv (in the ELNRA) with maximum values up to 135 mSv in the above-stated house with a high radon level. This paper mainly reports the $^{222}$Rn levels indoors and outdoors and the effective equivalent doses due to $^{222}$Rn and its progeny and concludes some results of total annual effective equivalent doses of public due to external and internal exposures. © 2004 Elsevier B.V. All rights reserved.

*Keywords:* Ramsar; Iran; Public; Annual effective dose; Natural radiation; Elevated level; $^{222}$Rn and its progeny; Gamma radiation; External dose; Internal dose; Assessment; Terrestrial; Cosmic ray; Measurement; Calculation; Indoor; Outdoor

---

* Corresponding author. Presently: at the International Atomic Energy Agency, P.O. Box 200, Vienna, Austria. Tel.: +43 1 2600 21484; fax: +43 1 26007 21484.
  *E-mail address:* M.Sohrabi@iaea.org (M. Sohrabi).

0531-5131/ © 2004 Elsevier B.V. All rights reserved.
doi:10.1016/j.ics.2004.11.102

## 1. Introduction

Elevated-level natural radiation areas (ELNRA) of Ramsar are due to high $^{226}$Ra content water from 50 dynamic hot springs flowing into the areas and precipitating $^{226}$Ra and its daughter products into the soil and some travertine-type stones. In fact, the area is considered as being dynamic due to continuous flow of radioactivity causing increasing radioactivity levels in the ELNRA which is recently classified as a very high level natural radiation area (VHLNR) [1]. This area has the highest natural radiation exposure level compared to the other existing areas in the world [2] and has been the subject of some previous dosimetry and radiobiological studies [3,4]. However, a more detailed radiological study for assessing the effective equivalent doses of public from external and internal exposures seemed necessary for radiobiological and epidemiological studies. In this context, studies were carried out in a large number of houses and locations indoors and outdoors in 12 regions of Ramsar. The results on determination of external exposure levels indoors and outdoors in Ramsar and its ELNRAs were reported elsewhere [5]. This paper, however, reports the results of $^{222}$Rn measurements indoors of a large number of houses as well as some in outdoors in the regions, and assesses the effective equivalent doses of public due to internal exposures for assessing the total public annual effective doses from internal and external exposures.

## 2. Measurements and methods

The $^{222}$Rn levels were measured indoors and outdoors in 12 regions of Ramsar to determine the public effective equivalent doses from internal exposures. The AEOI passive radon diffusion chambers were used for $^{222}$Rn measurements indoor and also for some outdoor locations [6]. The active Pylon AB-5 system and Lucas chambers were also used extensively to determine level of $^{222}$Rn and its progeny as well as the equilibrium factor ($F$) indoors and outdoors of some buildings.

The radon chambers were calibrated in a known $^{222}$Rn field and a track to $^{222}$Rn level conversion factor of $15.1 \pm 0.6$ tracks cm$^{-2}$/kBq m$^{-3}$ day was determined and was inter-compared in the same field with that of the Karlsruhe passive radon diffusion chamber with a conversion factor of 16.2 tracks cm$^{-2}$/kBq m$^{-3}$ day.

The $^{222}$Rn levels indoors were determined once in autumn and once in winter in 525 rooms of 476 houses in 12 regions of Ramsar. These regions (no. of rooms) include Chaboksar (55), Katalom (22), Javaherdeh (20), Tonekabon (37), Lamtar (27), Lapsar (34), Sefid Tameshk (19), Sadat Mahale (107), Ghaemieh (15), Chaparsar (60), Ramak (37), Talesh Mahalleh (92). In Talesh Mahalleh, which has the highest exposure levels indoors and outdoors, more than 90% of houses as well as rooms in the 1st and 2nd floors of five buildings were surveyed.

## 3. Results of $^{222}$Rn level measurements

The $^{222}$Rn levels in 12 regions of Ramsar are shown in an increasing order in Table 1. The mean levels in different regions range from 64 Bq m$^{-3}$ in Chaboksar to 2255 Bq m$^{-3}$ in Talesh Mahalle in autumn and from 107 to 3235 Bq m$^{-3}$ in winter. The maximum levels range from 193 to 18 097 Bq m$^{-3}$ in autumn and from 256 to 31 080 Bq m$^{-3}$ in winter.

Table 1
The maximum and mean $^{222}$Rn level indoors in 12 regions of Ramsar and its ELNRA

| Regions (no. of rooms) | $^{222}$Rn Level (Bq m$^{-3}$) in autumn | | $^{222}$Rn Level (Bq m$^{-3}$) in winter | |
|---|---|---|---|---|
| | Max (Bq m$^{-3}$) | Mean (Bq m$^{-3}$) | Max (Bq m$^{-3}$) | Mean (Bq m$^{-3}$) |
| Chaboksar (55) | 276 | 64±55 | 316 | 116±56 |
| Katalom (22) | 193 | 66+46 | 256 | 107±57 |
| Javaherdeh (20) | 164 | 70±43 | 382 | 137±93 |
| Tonekabon (37) | 179 | 72±42 | 438 | 134±85 |
| Lamtar (27) | 278 | 78±67 | 319 | 124±71 |
| Lapasar (34) | 717 | 86± – | 1064 | 176± – |
| Sefid Tameshk (19) | 319 | 86±83 | 598 | 173±151 |
| Sadat Mahalleh (107) | 1293 | 247±219 | 1950 | 297±269 |
| Ghaemieh (15) | 934 | 299±259 | 649 | 308±189 |
| Chaparsar (60) | 3396 | 397± – | 1428 | 410± – |
| Ramak (37) | 3219 | 541± – | 2551 | 497± – |
| Talesh Magalleh (92) | 18,097 | 2255± – | 31,080 | 3235± – |

This maximum level determined in Talesh Mahalle is almost 10 times higher than that reported in another survey [7]. This high level indoor seems justified in the light of having high $^{226}$Ra content in the construction material of the wall of one room which shows about 0.10 mGy h$^{-1}$ gamma exposure on its surface. The reason for this discrepancy needs to be further verified.

As it has been concluded also in previous studies [5], Talesh Mahalleh has the highest $^{222}$Rn level and gamma exposure levels indoors and outdoors; for example about 30 times higher $^{222}$Rn level than those observed in Chaboksar and Katalom. Chaboksar, Katalom, Javaherdeh, Tonekabon, Lamtar, Lapasar, and Sefid Tameshk have mean $^{222}$Rn levels ranging from 64 to 86 Bq m$^{-3}$ in autumn and two times higher level in winter. Sadat Mahalleh, Ghaemieh, Chaparasr, and Ramak have mean $^{222}$Rn levels almost the same in both seasons. The mean level in Talesh Mahalleh in winter is about 1.4 times higher than that in autumn.

By the analysis of the $^{222}$Rn levels indoors in autumn and in winter, it can be concluded that while the $^{222}$Rn levels are expected to be higher in winter, they show some unexpected variations in the contrary. Table 2 shows the percentage of rooms which have either higher, lower, or equal $^{222}$Rn levels in autumn and in winter in the 12 regions. It is interesting to note that:

(i) about 80% of the rooms in Chaboksar, Katalom, Javaherdeh, Tonekabon, Lamtar, Lapasar, and Sefid Tameshk (with lowest $^{222}$Rn levels) have levels about two times higher in winter than in autumn;
(ii) about 50% of rooms in Sadat Mahalleh, Ghaemieh, Chaparasr, and Ramak with relatively higher $^{222}$Rn levels have either lower or higher levels in winter and in autumn, and
(iii) about 56% of the rooms in Talesh Mahalleh have higher values in winter.

It has been experienced that buildings with rooms having higher $^{222}$Rn levels are usually built directly on grounds with high $^{226}$Ra level soil. In winter, due to increased

Table 2
Percentage of rooms with higher, lower, or equal $^{222}$Rn levels in autumn and winter

| Regions (no. of rooms) | Percentage (%) of rooms with higher $^{222}$Rn level in winter | Percentage (%) of rooms with higher $^{222}$Rn level in autumn | Percentage (%) of rooms with equal $^{222}$Rn level in two seasons |
|---|---|---|---|
| Chaboksar (50) | 80 | 18 | 2 |
| Katalom (18) | 67 | 22 | 11 |
| Javaherdeh (20) | 70 | 20 | 10 |
| Tonekabon (33) | 82 | 15 | 3 |
| Lamtar (25) | 72 | 20 | 8 |
| Lapasar (22) | 82 | 5 | 13 |
| Sefid Tameshk (19) | 74 | 16 | 10 |
| Sadat Mahalleh (91) | 45 | 40 | 6 |
| Ghaemieh (12) | 50 | 50 | – |
| Chaparsar (53) | 51 | 45 | 4 |
| Ramak (26) | 50 | 50 | – |
| Talesh Magalleh (79) | 40 | 56 | 4 |

soil moisture, the radon emanation is suppressed which might equally compensate for other variations such as lower room ventilation rate, etc. This explanation is recommended, however, to be further verified. Considering almost equal $^{222}$Rn levels in autumn and in winter, in particular, in the ELNRA, it seems that the upper annual effective dose of public can be simply estimated by doubling the sum mean values of two seasons.

The $^{222}$Rn levels in the 1st and 2nd floors of five buildings in Talesh Mahalle showed that the level in the 1st floor is in general higher than that of the 2nd floor by a factor ranging from about 1 to 10 times. The higher ratio is in buildings directly built on the ground with high $^{226}$Ra-content soils. The equilibrium factor ($F$) was determined in few buildings which range from 0.39 to 0.73 with a mean value of 0.5. However, since the number of measurements was limited, the equilibrium factors of 0.4 and 0.8 were used for indoors and outdoors, respectively [8]. Moreover, the residence time indoor and outdoor in the regions were investigated through a questionnaire leading to respective values of 0.84±0.4 and 0.16±0.08. However, for the purpose of this study, the residence times of 0.2 and 0.8 were used as is also recommended by UNSCEAR [8].

The $^{222}$Rn levels in some locations outdoors showed mean $^{222}$Rn level of 10 Bq m$^{-3}$ in the normal background areas, 65 Bq m$^{-3}$ in the VHLNRA with the highest level of 580 Bq m$^{-3}$ in the yard of one house in Talesh Mahalle which has also the highest $^{222}$Rn level and gamma exposure indoor. The $^{222}$Rn levels in air surrounding the water streams from hot springs with high $^{226}$Ra level in the VHLNRA was 240 Bq m$^{-3}$ with a decrease to 140 Bq m$^{-3}$ and further to 84 Bq m$^{-3}$ as one moved away from the stream.

## 4. Effective equivalent dose ($E$) of public

The effective equivalent dose ($E$) of public from external exposure has been reported elsewhere [5]. The $E$ due to internal exposure from $^{222}$Rn is calculated here for outdoors ($E_{out}$) and indoors ($E_{in}$) using $^{222}$Rn level to dose conversion factor of 9 nSv h$^{-1}$ per Bq m$^{-3}$ (EEC) for inhalation and 1.5 μSv year$^{-1}$ per Bq m$^{-3}$ for radon dissolved in tissue recommended by UNSCEAR [8].

$$E_{out} = C_{Rno} \times 0.8(EEC) \times 9 \text{ nSv h}^{-1} \text{ per Bq m}^{-3}(EEC) \times 0.2(\text{occupancy factor})$$
$$\times 8760 \text{ h year}^{-1}$$

$$E_{in} = C_{Rni} \times 0.4(EEC) \times 9 \text{ nSv h}^{-1} \text{ per Bq m}^{-3}(EEC) \times 0.8(\text{occupancy factor})$$
$$\times 8760 \text{ h year}^{-1}$$

$$E_{i \& o} = [C_{Rno} \times 0.2] + [C_{Rni} \times 0.8] \times 1.5 \text{ μSv year}^{-1} \text{ per Bq m}^{-3}$$

$$E = E_{out} + E_{in} + E_{i \& o}$$

The quarterly, semiannual, and annual mean effective equivalent dose of the public, due to internal exposure from $^{222}$Rn indoors and outdoors in 12 regions are shown in Table 3. As discussed above, the effective equivalent dose due to internal exposure in Chaboksar, Katalom, Javaherdeh, Tonekabon, Lamtar, Lapasar, and Sefid Tameshk is relatively low with mean values $\bar{E}$ ranging from 0.45 to 0.61 mSv in autumn. The $\bar{E}$ values in the above stated regions are almost two times higher in winter. The maximum E in these regions varies from 1.83 to 8.4 mSv in autumn with almost two times higher values in winter. Sadat Mahalleh, Ghaemieh, Chaparasr, and Ramak have $\bar{E}$ values ranging from 1.6 to 3.6 mSv in autumn with almost the same value for winter. The $\bar{E}$ in Talesh Mahalleh in winter is about 1.4 times higher than $\bar{E}$ in autumn. It seems a good estimation to determine the upper annual E by doubling the sum of the internal doses of the two seasons. By considering the sum of mean values $\bar{E}$ of the two seasons, the upper annual $\bar{E}$ from internal exposure ranges from 2.4 in Chaboksar to 71.74 mSv in Talesh Mahalle and the maximum annual E ranges from 7.88 mSv in Chaboksar to 640 mSv in Talesh Mahalle.

Table 3
The minimum, maximum, and mean effective equivalent dose ($\bar{E}$) in autumn and winter, the semiannual $\bar{E}$, and the estimated mean annual $\bar{E}$ due to $^{222}$Rn indoors and outdoors of 12 regions of Ramsar

| Regions (no. of rooms) | Autumn | | Winter | | E (mSv) | |
|---|---|---|---|---|---|---|
| | Min–Max E (mSv) | $\bar{E}$ (mSv) | Min–Max E (mSv) | $\bar{E}$ (mSv) | Semiannual $\bar{E}$ (mSv) | Annual $\bar{E}$ (mSv) |
| Chaboksar (55) | 0.097–1.83 | 0.45±0.39 | 0.24–2.11 | 0.79±0.4 | 1.24 | 2.48 |
| Katalom (22) | 0.097–1.29 | 0.46±0.34 | 0.23–1.7 | 0.73±0.4 | 1.19 | 2.38 |
| Javaherdeh (20) | 0.097–1.1 | 0.49±0.31 | 0.19–2.54 | 0.92±0.65 | 1.41 | 2.82 |
| Tonekabon (37) | 0.097–1.18 | 0.49±0.30 | 0.190–2.86 | 0.90±0.59 | 1.39 | 2.78 |
| Lamtar (27) | 0.097–1.81 | 0.54±0.47 | 0.200–2.11 | 0.84±0.49 | 1.38 | 2.76 |
| Lapasar (34) | 0.097–0.47 | 0.59±– | 0.300–6.95 | 1.18±0.49 | 1.77 | 3.54 |
| Sefid Tameshk (19) | 0.097–8.18 | 0.61±– | 0.260–3.90 | 1.15±1.02 | 1.76 | 3.52 |
| Sadat Mahalleh (107) | 0.130–8.40 | 1.60±1.46 | 0.180–12.60 | 1.97±1.78 | 3.57 | 7.14 |
| Ghaemieh (15) | 0.360–6.12 | 1.98±1.72 | 0.490–4.22 | 2.00±1.27 | 3.98 | 7.96 |
| Chaparsar (60) | 0.13–22.10 | 2.70±– | 0.400–9.40 | 2.85±2.20 | 5.55 | 11.10 |
| Ramak (37) | 0.210–21.10 | 3.66±– | 0.410–16.8 | 3.37±– | 7.03 | 14.06 |
| Talesh Magalleh (92) | 0.48–118.00 | 14.87±– | 0.54–202 | 21.00±– | 35.87 | 71.74 |

## 5. Conclusions

The results of external exposure studies indoors and outdoors in Ramsar and its ELNRA [5] have shown that the annual effective doses of people living in normal background areas are below 1 mSv. In the ELNRA, out 45% of the people have potential annual doses below 1 mSv, about 30% from 1 to 5 mSv, about 20% from 5 to 20 mSv, and less than 5% from 20 to 135 mSv. For internal exposure, as seen in Table 3, the annual $\bar{E}$ from internal exposure ranges from 2.4 in Chaboksar to 71.74 mSv in Talesh Mahalle. The public annual $\bar{E}$ from internal dose in the first seven regions of Table 3, which are normal background areas, is relatively low ranging from 2.34 to 3.52 mSv year$^{-1}$ which are usually areas with normal gamma background. The annual $\bar{E}$ in Sadat Mahalleh, Ghaemieh, Chaparsar, and Ramak is relatively higher and ranges from 7.14 to 14.06 mSv. The total $\bar{E}$ from external and internal exposure have been calculated. The annual $\bar{E}$ from external exposure in normal and elevated radiation level areas ranges respectively from 0.7 to 6 mSv, and for internal exposure ranges from 2.4 to 71.74. This leads to the total annual $\bar{E}$ from external and internal exposures from 3.1 to 77.74 mSv. However, much higher total $\bar{E}$ has been detected as reported above. In conclusion, while only 5% of the populations in these areas are expected to receive relatively high exposures, the total $E$ is high enough to worth radiobiological and epidemiological studies in the area.

## References

[1] M. Sohrabi, Environments with elevated radiation levels from natural substances, Procds. of Int. Conf. on Restoration of Environment with Radioactive Residues, Arlington, Virginia, USA, 29 Nov.–03 Dec. (1999), printed by IAEA, STI/PUB/1092, paper IAEA-SM-359/1.7, IAEA, Vienna (2000) 113–134.
[2] M. Sohrabi, The state-of-the-art on worldwide studies in some environments with elevated naturally occurring radioactive materials (NORM), J. Appl. Radiat. Isot. 49 (1998) 169–188.
[3] M. Sohrabi, Recent radiological studies in high level natural radiation areas of Ramsar, in: M. Sohrabi, J.U. Ahmed, S.A. Durrani (Eds.), Procs. Int. Conf. on High Level Natural Radiation Areas, Ramsar, Iran, 3–7 Nov. (1990), IAEA Publ. Series, IAEA, Vienna, 1993, pp. 39–47.
[4] T.Z. Fazeli, et al., Cytogenetic studies of inhabitants of some high natural radiation areas of Ramsar, Iran, in: M. Sohrabi, J.U. Ahmed, S.A. Durrani (Eds.), Procs. Int. Conf. on High Level Natural Radiation Areas, Ramsar, Iran, 3–7 Nov. (1990), IAEA Publ. Series, IAEA, Vienna, 1993, pp. 459–464.
[5] M. Sohrabi, A.R. Esmaeli, New public dose assessment of elevated level natural radiation areas of Ramsar (Iran) for eidemiological studies, in: W. Burkart, M. Sohrabi, A. Bayer (Eds.), Proc. 5th Int. Conf. on High Levels of Natural Radiation and Radon Areas; Radiation Dose and Health Effects. 04–07 Sept. 2000, Elsevier Publ., 2002.
[6] M. Sohrabi, A.R. Solaymanian, Some characteristics of the AEOI passive radon diffusion dosimeter, Nucl. Tracks Radiat. Meas. 15 (1-4) (1988) 605–608.
[7] M. Sohrabi, et al., Determination of $^{222}$Rn levels in houses, schools and hotels of Ramsar by the AEOI passive radon diffusion dosimeters, in: M. Sohrabi, J. Ahmed, S.A. Durrani (Eds.), Procs. Int. Conf. on High Levels of Natural Radiation, Ramsar, IR Iran, 3–7 Nov. (1990), IAEA Proc. Series, IAEA, Vienna, 1993, 365–375.
[8] United Nations Scientific Committee on the Effects of Atomic Radiation, Sources and Effects of Ionizing Radiation, United Nations, New York, 1993.

# Oral presentations

# Changes of chromosome aberration rate and micronucleus frequency along with accumulated doses in continuously irradiated mice with a low dose rate of γ-rays

Kimio Tanaka*, Atsushi Kohda, Kazuaki Ichinohe, Tsuneya Matsumoto

*Department of Radiobiology, Institute for Environmental Sciences (IES), Rokkasho, Japan*

**Abstract.** Chronological changes of chromosome aberration rates (CAs) along with accumulated doses in exposed mice with a low dose rate of γ-ray were studied. CA and micronuclei in spleen cells were observed serially in mice continuously irradiated at 20 mGy/day up to 400 days for 8 Gy. Unstable types of CA rates were rapidly increased at 1 Gy, while micronucleus incidence increased from 5 Gy. Over these doses, their increase rates were saturated. FISH analysis in spleen cells of mice exposed to 8 Gy had higher incidence of monosomy and trisomy than in non-exposed mice. The number of cells with 2–3 micronuclei out of 10,000 spleen cells is also higher in 5 to 8Gy-exposed mice. These results indicate that prolonged γ-ray-irradiation at low dose rates of 20 mGy/day induces a delayed chromosome instability in mice. © 2004 Published by Elsevier B.V.

*Keywords:* Low dose rate; Gamma rays; Chromosome aberrations; Micronucleus; Genetic instability

## 1. Introduction

Chronically exposed people such as nuclear workers, medical radiologists, and residents in high back radiation areas have higher incidence of (CAs) than non-exposed peoples. However chronological changes of CA rates related to accumulated doses with a low dose rate of radiation have not been well studied. Precise analysis in human population is always limited because of confounding factors, such as smoking, medical radiation exposures, and so on, influence the results obtained. For this reason, animal experiment will be very important for risk assessment. Institute for Environmental Sciences (IES) has a unique facility that is designed to

---

\* Corresponding author. Tel.: +82 175 71 1754; fax: +82 175 71 1982.
*E-mail address:* kmtanaka@ies.or.jp (K. Tanaka).

0531-5131/ © 2004 Published by Elsevier B.V.
doi:10.1016/j.ics.2004.12.005

allow for keeping of mice under specific pathogen-free (SPF) conditions and continuous low dose rate $^{137}$Cs γ-ray irradiation with 0.05 mGy/day, 1 mGy/day, and 20 mGy/day, which are 20, 400, and 8000 times higher level than the natural background external radiation level, respectively. Using this facility, we have analysed the cancer incidence, life span shortening [1], cellular response [2], and oncogene alterations and so on. It is well known that delayed chromosomal instability is developed by radiation after several cells in vivo. Present studies are performed using irradiation of $^{137}$Cs-γ rays at 20 mGy/day (1 mGy/h) to obtain chronological change of CA rates related with accumulated dose and to observe whether or not chromosomal instability is induced in mice by continuous irradiation with a low dose rate.

## 2. Materials and methods

Continuous irradiation of SPF female C3H/HeNJcl mice started from 8 weeks of age by $^{137}$Cs-γ rays at a dose rate of 20 mGy/22 h/day up to 400 days for 1 to 8 Gy. Five mice were sacrificed at each total dose. Spleen cells were cultured for 48 and 72 h for chromosome analysis and micronucleus assay, respectively, with a stimulation by LPS (10 μg/ml), ConA (3 μg/ml), and 2-mercapto-ethanol. For micronucleus assay, cytocharasin B (0.5 μg/ml) was added at 24 h before harvesting to obtain binucleated cells. The 10,000 binucleated lymphocytes were enumerated to observe micronucleus-positive cells under the microscope. For chromosome analysis, unstable type aberrations and chromatid type aberrations were observed in 500–1000 metaphases. Interphase FISH analysis using centromere probes of mouse chromosomes of 5, 15, and 18 were also used for detecting cells with monosomy and trisomy in cultured spleen cells.

## 3. Results and discussion

The number of aberrant chromosomes in chromosome type aberrations per cell was increased rapidly to 11.1% at 1 Gy and s showed of almost a plateau or a slight increase over 1 Gy. Frequencies of CA and micronucleus in non-exposed mice were increased slightly with age. Incidence of ring chromosomes out of chromosome type of aberrations in exposed mice was about 5 times higher than non-exposed mice. Ring chromosome seems to be more accurate cytogenetic markers for detecting biological effects induced by low dose rate radiation. These chronological changes of (CAs) seem to be induced through a balance between development of (CAs) and life span of spleen cells. On the other hand, the pattern of chronological changes of micronucleus incidence was quite different from that of (CAs). Micronucleus incidence in spleen cells was increased from the dose more than 5 Gy. The numbers of cells with 2–3 micronuclei out of 10,000 observed cells were higher in mice irradiated with 5, 6, and 8 Gy than age-matched non-exposed mice. FISH analysis of spleen cells showed that several mice exposed to 8 Gy had higher number of cells with monosomy and trisomy. These numerical (CAs) seem to be induced indirectly after radiation exposure and thus the results indicate that prolonged γ-ray-irradiation at low dose rates of 20 mGy/day induces delayed chromosomal instability in mice. These results will be available for risk assessment in low dose rate radiation exposure. The work was carried out with financial support given by Aomori Prefecture.

## References

[1] S. Tanaka, et al., Radiat. Res. 160 (3) (2003) 376–379.
[2] T. Sugihara, et al., Radiat. Res. 162 (3) (2004) 296–307.

# Dose and dose rate effects of low dose ionizing radiation on activation of p53 in immortalized murine cells

Takashi Sugihara[a,*], Junji Magae[b], Renu Wadhwa[c], Sunil C. Kaul[c], Yasushi Kawakami[b], Tsuneya Matsumoto[a], Kimio Tanaka[a], Yoichi Oghiso[a]

[a]*Department of Radiobiology, Institute for Environmental Sciences, 1-7 Ienomae, Obuchi, Rokkasho, Kamikita, Aomori 039-3212, Japan*
[b]*Institute of Research and Innovation, 1201 Takada, Kashiwa, Chiba 277-0861, Japan*
[c]*Gene Function Research Center, National Institute of Advanced Industrial Science and Technology (AIST), Central 4, 1-1-1 Higashi Tsukuba, Ibaraki 305-8566, Japan*

**Abstract.** We examined the biological effects of low dose rate (LDR) of $^{60}$Co γ-ray irradiation (from 0.1 to 10 cGy/h) on the transcriptional activity of p53, using NIH/PG13Luc cells stably transfected with p53-dependent luciferase reporter plasmid. Microarray analysis revealed up-regulations of six P53-mediated genes (CDKN1A/p21, MDM2, SIP27, CCNG1/cyclin G1, EI24/PIG8 and Dinb/POLK) by exposure of cells for 72 h to LDR $^{60}$Co γ-ray irradiation. Using a real-time PCR, a significant elevation in the expression of CCNG1/cyclin G1, MDM2 and CDKN1A/p21 was observed at dose rates over 5 cGy/h. The dose rate dependency of these three p53-mediated genes was also observed. The expression of CCNG1/cyclin G1 at high dose rates (HDRs) γ-rays was higher than that for LDR. However, the expression of MDM2 for LDR γ-rays was higher than for HDR. Cells irradiated with γ-rays at 0.1 and 1 cGy/h appeared to show G1 phase arrest. Furthermore, G2 growth arrest was observed in cells irradiated with γ-rays at 5 and 10 cGy/h, being correlated with p53-mediated CCNG1/cyclin G1 up-regulation. These results imply that cellular response to radiation differs between the two LDRs used, 0.1–1 cGy/h and over 5 cGy/h. © 2004 Published by Elsevier B.V.

*Keywords:* p53; Low dose radiation; Low dose rate radiation; Microarray; Cell cycle

---

\* Corresponding author. Tel.: +81 175 71 1253; fax: +81 175 71 1270.
*E-mail address:* sugihara@ies.or.jp (T. Sugihara).

0531-5131/ © 2004 Published by Elsevier B.V.
doi:10.1016/j.ics.2004.11.183

## 1. Introduction

We investigated the transcriptional activation of p53 protein in murine cells continuously irradiated with γ-ray at LDRs using cells with stably integrated p53-responsive reporter plasmid. Differential effects on cell growth, cell cycle and p53-mediated gene expression in cells exposed to γ-ray at different LDRs between 0.1 to 10 cGy/h were studied.

## 2. Materials and methods

Murine immortal NIH3T3/PG13Luc cells were stably transfected with the p53-responsive luciferase reporter plasmid [1]. Cells for the analysis of gene expressions by microarray were irradiated at HDR of soft X-ray at a dose rate of 30 Gy/h for 16 mins. LDR and HDR irradiation exposures were carried out using $^{60}$Co (50,000 Ci) γ-rays. Cells for the analysis of gene expressions by microarray were irradiated at dose rates of 0.1, 1 and 10 cGy/h for 72 h. To determine the cellular effects (by cell cycle analysis, luciferase assay and real time PCR) of dose rate, the cells were irradiated at 0.1, 1, 5 and 10 cGy/h for 72 h. Flow cytometry (Epics XL, Beckman Coulter, Inc., Maiami, FL, USA) was performed to measure the DNA content of the cells as described previously. Murine cDNA chips with 8800 cDNA clones (Agilent technologies, Palo Alto, CA, USA) were used to detect the differential expression of these genes. Differential ratios of fluorescence intensities were obtained by competitive hybridization using Cy5- and Cy3-labeled cDNA.

## 3. Results and discussion

NIH3T3/PG13Luc was irradiated with different low dose rates (LDRs) of γ-ray for different durations. Cell cycle analysis of nonirradiated and irradiated cells by flow cytometry was done 72 h after irradiation. Compared to nonirradiated cells, irradiation at dose rates of 0.1 and 1 cGy/h significantly ($P<0.05$) increased the ratio of G1 phase cells, whereas the ratio G2/M phase cells decreased ($P<0.05$) at 1 cGy/h. Significant ($P<0.05$) increases in the ratio of G2/M phase cells were also observed in cells irradiated at 5 and 10 cGy/h. These findings indicate that radiation effects on the cell cycle during the G1 and G2/M phase in cells exposed to LDR irradiation differ between the lower dose rates of 0.1 and 1 cGy/h and at higher dose rates of 5 and 10 cGy/h.

We compared the p53-mediated gene expression profiles of cells irradiated at high dose rates (HDRs) of soft X-ray (30 Gy/h for 16 min; 8 Gy) or LDR of γ-ray irradiation (0.1, 1, 10 and 100 cGy/h for 72 h) with those of the nonirradiated cells using microarray analysis. The six of the eight up-regulated genes detected at HDR irradiation were well-known p53-mediated genes such as CDKN1A/p21, MDM2, SIP27, CCNG1/cyclin G1, EI24/PIG8 and Dinb/POLK. DCXR and EPHX1 have not been reported as p53-mediated genes.

## Acknowledgements

The work was supported by financial grants from Aomori Prefecture, Japan.

## Reference

[1] R. Wadhwa, et al., Cancer Res. 60 (2000) 6818–6821.

# Histologic assessment of regenerating rat liver under low-dose rate radiation exposure

## D.V. Guryev*

*Radioecology Department, Institute of Biology, Komi Scientific Centre, Ural division of RAS, 28 Kommunisticheskaya St., Syktyvkar, 167982, Russia*

**Abstract.** In this study of the Wistar rats model, the partial hepatectomy (PH 1/3) as a mitogen was used to display radiation-induced damages in active dividing hepatocytes. Thus, 30-days total exposure in dose 0.78 cGy caused the significant decrease of the liver index (liver and total body weights ratio) by 1 and 30 days after PH. Histologic data show the high rate of binucleated hepatocytes 1, 2, 14, and 30 days after irradiation and PH and this could result in adaptive response to low-dose rate radiation. The effect of liver index decrease obviously occurs as a result of both high percentages of degenerating liver cells all days after exposure and registered structural changes in parenchyma (piecemeal necroses, pathologic mitoses, etc.). In comparison, we demonstrated high rate of degenerating hepatocytes in only irradiated animals 14 and 30 days after the radiation exposure as a result of long-term response of the slowly proliferating cells system. Consequently, the cells functional activity determines the development pace of the radiation-induced long-term effects.
© 2004 Elsevier B.V. All rights reserved.

*Keywords:* Low-dose ionizing radiation; Liver regeneration; Radiation-induced damages; Adaptive response

## 1. Introduction

The biological effects of ionizing radiation exposure are well known to be closely associated with functional state of cells. Certain slowly proliferating tissues, such as liver, have ability to accumulate radiation-induced damages and to finally predetermine some long-term effects of whole organism. Thus, active proliferating liver cells could be used as a model system to register both a radiation-induced damages and mechanisms of the cells response to exposure hazard.

* Tel.: +7 8212 436301; fax: +7 8212 240163.
  *E-mail address:* guryev@ib.komisc.ru.

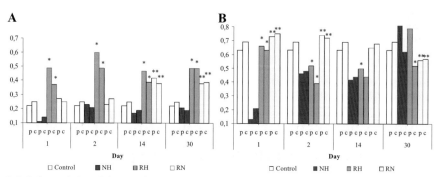

Fig. 1. Relative variation of degenerating (A) and binucleated (B) hepatocytes for each group throughout study. p, periportal; c, pericentral zones of liver tissue; *$P<0.05$ vs. NH group; **$P<0.05$ vs. control.

## 2. Materials and methods

110 adult male Wistar rats (248.5±26.1 g) were used. The radiation ($^{226}$Ra source) consisted of a 30-day total exposure (dose was 0.78 cGy). The rats were realized 30% hepatectomy right away after radiation treatment and were randomized into following groups: NH had a hepatectomy only; NL had laparotomy only; RH had irradiation and hepatectomy; RL had radiation and laparotomy. Animals were sacrificed 1, 2, 14, and 30 days after treatment. We analysed a liver index (liver and total body weights ratio) and histologic data as both quantitative variability of uni- and binucleated hepatocytes, degenerated liver cells, mitotic index, and qualitative changes in liver structure in periportal and pericentral zones [1,2].

## 3. Results and discussion

The rats liver index in NH and NL groups has been rehabilitated by 30th day after surgery, whereas animals of RH and RL groups had the significantly decreased index. Moreover, the means liver index comparison for RH and NH groups for the first days after surgery reveals the regeneration delay in irradiated animals. Histologic data of RH group for this point show the high rate of pathological mitoses and degenerating hepatocytes that allows proliferative processes to be unsuccessful (see Fig. 1A). We also registered the high rate of binucleated hepatocytes 1, 2, 14, and 30 days after irradiation and partial hepatectomy (see Fig. 1B). Normally, the proliferation was realizing due to these cells dividing and this checkpoint of the cell cycle does not inhibit cell functions [3]. Thus, we suggest that a high number of binucleated hepatocytes could be produced as a result of the cell dividing damage (acytokinesis) that could lead to adaptive response to low-dose rate radiation. Evidently, the increase of the DNA quantity reduces some DNA damages realization possibility. We demonstrated high rate of degenerating hepatocytes in only irradiated animals by 14 and 30 days after the radiation exposure as a result of long-term response of the slowly proliferating cells system (see Fig. 1A). Consequently, the cells functional activity determines the development pace of the radiation-induced long-term effects.

## References

[1] S.S. Shwarz, et al., Proc. of Inst. of Plant and Animals Ecology, Sverdlovsk, vol. 58, 1968, pp. 387.
[2] G.G. Avtandilov, Medical morphometry, Moscow, Medicine (1990) 384.
[3] I.V. Uryvaeva, Theor. Boil. 89 (1981) 4.

# Increased radiosensitivity of splenic T lymphocytes in pregnant mice

Yuka Igari*, Kazuyuki Igari, Hiroyo Kakihara, Fumio Kato, Akira Ootsuyama, Toshiyuki Norimura

*Department of Radiation Biology and Health, School of Medicine, University of Occupational and Environmental Health, Japan, 1-1 Iseigaoka, Yahatanishi-ku, Kitakyushu 807-8555, Japan*

**Abstract.** We used the T cell receptor (TCR) assay to evaluate the mutagenic risk of radiation in pregnant mice. Pregnant and delivered mice and litters were irradiated at various days during gestation and postdelivery, and the TCR assay was performed. The TCR mutation frequency (TCR-MF) of the pregnant and delivered mice increased in comparison with that of irradiated nonpregnant mice. On the other hand, an increased TCR-MF was not seen in the fetuses and litters. These results suggest that the radiosensitivity may be increased in mother mice during periods of pregnancy and postdelivery. © 2004 Elsevier B.V. All rights reserved.

*Keywords:* Pregnancy; T cell receptor; Mutation; Radiation

## 1. Introduction

In general, effects on the fetus are of concern when a pregnant mother is exposed to radiation. In this study, we focused on the effects on the mother after exposure. We investigated the sensitivity to mutagenic effects of radiation in pregnant mice using the T cell receptor (TCR) assay. The TCR assay is a sensitive method for detecting somatic mutation and an indicator for the genotoxic effects of radiation [1].

## 2. Materials and methods

Pregnant mice were irradiated at 2.5, 6.5, 10.5, 14.5, 16.5, and 18.5 days gestation. Groups of postdelivery mice and litters were irradiated at 1, 7, 14, 21, and 28 days after

---

\* Corresponding author. Tel.: +81 93 603 1611; fax: +81 93 692 0559.
*E-mail address:* yuka13@ninus.ocn.ne.jp (Y. Igari).

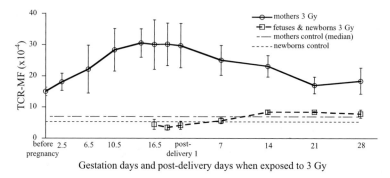

Fig. 1. TCR-MF 28 days after 3 Gy irradiation (mother mice and litters).

delivery. Each group was irradiated with 3 Gy of gamma radiation (dose rate, 93 cGy/min). All mice, including newborns that were irradiated at 16.5 and 18.5 days gestation, were sacrificed at 28 days after irradiation, and the TCR assay was performed. In nonirradiated pregnant mice groups, the assay was performed according to the day when the irradiated group was irradiated. Spleen cells were used as the sample and TCR mutation frequency (TCR-MF) was calculated as the number of $CD3^-4^+$ T cells divided by the total number of $CD3^+CD4^+$ T cells.

## 3. Results and discussion

### 3.1. TCR-MF of pregnant mice (Fig. 1)

There was no significant difference in spontaneous TCR-MF between the pregnant and nonpregnant mice. However, when compared on the basis of radiation-induced mutagenicity, the splenic T lymphocytes of pregnant and postdelivery mice were two times more sensitive than those of nonpregnant mice. This result suggests that pregnancy may increase the radiosensitivity of pregnant and postdelivery mice [2]. To see the relationship between pregnancy-related hormones and radiosensitivity, serum estrogen and progesterone levels were analysed. It seems that progesterone was effective in increasing of TCR-MF [3]. (Fig. 1).

### 3.2. TCR-MF of the fetuses and newborns (Fig. 1)

The TCR-MF of the irradiated litters, including the newborns that were irradiated at the fetal stage, was not increased. It seems that the fetuses and newborns are resistant to the mutagenic effects of radiation.

## References

[1] S. Kyoizumi, et al., Frequency of mutant T lymphocytes defective in the expression of the T-cell antigen receptor gene among radiation-exposed people, Mutat. Res. 265 (2) (1992) 173–180.
[2] M. Ricoul, B. Dutrillaux, Variation of chromosome radiation sensitivity in fetal and adult mice during gestation, Mutat. Res. 250 (1991) 331–335.
[3] M. Ricoul, L. Sabatier, B. Dutrillaux, Increased chromosome radiosensitivity during pregnancy, Mutat. Res. 374 (1997) 73–78.

# Effects of low dose-rate irradiation on the glucose metabolism in type II diabetes model mice

Takaharu Nomura*, Kazuo Sakai

*Low Dose Radiation Research Center, Nuclear Technology Research laboratory, Central Research Institute of Electric Power Industry, 2-11-1 Iwado-kita, Komae, Tokyo 201-8511, Japan*

**Abstract.** We examined the effects of low dose-rate irradiation on the levels of glucose and insulin in the mice. Ten-week-old female type II diabetes model (db) mice were irradiated with γ-rays at 0.30, 0.70 or 1.2 mGy/h. The level of plasma glucose and the level of plasma insulin was measured. After the irradiation for 24 days, in the control mice, the level of insulin was decreased to 50% of the initial level, while the decrease was significantly less in the irradiated mice. The mice were irradiated for 2 weeks, and then, after one night fast, they were subjected to a glucose tolerance test. In the irradiated mice, the glucose level was gradually increased. In the non-irradiated mice, the glucose level increased rapidly and the level was kept. These results suggest that the low dose-rate irradiation improved the effectiveness of insulin on the insulin metabolisms and the glucose metabolism in the db mice. © 2004 Elsevier B.V. All rights reserved.

*Keywords:* Type II diabetes model mouse; Low dose-rate irradiation; Blood glucose level; Blood insulin level; Inhibitory damaged in pancreatic β-cells

## 1. Introduction

We previously found that prolonged γ-irradiation at 0.70 mGy/h decreased the level of urine glucose in 3 out of 12 type II diabetes model mice (C57BL/KsJ-*db/db*) (db mice). This phenomenon suggests that insulin resistance may be improved by low-level irradiation. We hypothesised increase in the ability of antioxidants in pancreas after the irradiation and decrease in damage of glucose. To understand the underlying mechanism, we examined the effects of low dose-rate irradiation on the levels of glucose and insulin in the mice.

---

\* Corresponding author. Tel.: +81 3 3480 2111; fax: +81 3 3480 3113.
*E-mail address:* nomura@criepi.denken.or.jp (T. Nomura).

0531-5131/ © 2004 Elsevier B.V. All rights reserved.
doi:10.1016/j.ics.2004.12.004

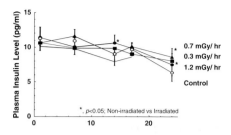

Fig. 1. Time course change in insulin level.

## 2. Materials and methods

Ten-week-old female db mice were irradiated with γ-rays at 0.30, 0.70 or 1.2 mGy/h. The level of plasma glucose was measured by an enzymatic assay [1] and the level of plasma insulin was measured by enzyme immunoassay method [2]. The mice were irradiated for 2 weeks, and then, after one night fast, they were subjected to glucose tolerance test (2 mg/kg, given orally).

## 3. Results

After the irradiation for 24 days, in the non-irradiated control mice, the level of insulin was decreased to 50% of the initial level, while the decrease was significantly less in the irradiated mice (Fig. 1).

After the glucose injection, in the irradiated mice, the glucose level was increased gradually in the course of 120 min, while in the non-irradiated control mice the glucose level increased rapidly and the level was kept from 10 to 120 min (data not shown).

## 4. Discussion

The suppression of the decrease in the insulin level (Fig. 1) suggested that the low dose-rate irradiation protected insulin-producing activity. We previously reported that endogenous antioxidant levels in mice were increased after low-dose irradiation [3]. The elevated antioxidant level might, in turn, protect insulin producing pancreatic β-cells against glucose radicals. Then, the level of insulin might be maintained and the glucose level was controlled. Under the experimental conditions we employed, no dose-rate dependence was observed among the irradiated mice; we need to analyze the glucose level after longer periods of irradiation.

## References

[1] I. Miwa, et al., Mutarotase effect on colorimetric determination of blood glucose with—D-glucose oxidase, Clin. Chim. Acta 37 (1972) 538–540.
[2] J.-L. Zhang, et al., Serum resistin level in essential hypertension patients with different glucose tolerance, Diabet. Med. 20 (10) (2003) 828–831.
[3] T. Nomura, et al., Suppression of ROS-related disease in mouse models by low-dose γ-irradiation. Biol. Effects Low Dose Radiat, Excerpta Medica 1211 (2000) 101–106.

# Adaptive response, split dose and survival of mice

## Ashu Bhan Tiku, R.K. Kale*

*School of Life Sciences, Jawaharlal Nehru University, New Delhi, India*

**Abstract.** The significance of radiation-induced adaptive response (AR) has been well recognized. Different systems have been tested for AR by small doses of radiation using various biological end points. The mechanisms underlying AR are still unclear. These aspects, including its relevance, have been discussed briefly in this presentation. © 2004 Published by Elsevier B.V.

*Keywords:* Adaptive response; Radiation; Split dose; Survival

## 1. Introduction

Ever since Olivieri et al. [1] have shown reduction in the number of chromosome aberrations in human lymphocytes, given a small conditioning dose prior to exposure to higher challenging dose of ionizing radiation, there is growing interest in the radiation-induced adaptive response (AR). Although different systems have been tested for radiation-induced AR using various biological end points, there are still uncertainties related to its possible biological implications and potential medical benefits. With this backdrop, an attempt has been made to discuss the phenomenological features and mechanisms of AR, as well as its relevance to human health.

## 2. Discussion

Various studies have characterized AR in terms of different biological endpoints, such as micronuclei, chromosome aberration, mutation frequency and cell survival. AR has also been demonstrated at biochemical level. However, its mechanism is not very well understood. The conditioning dose is suggested to induce various cellular and biochemical responses, leading to protection against subsequent higher doses of radiation. At the molecular level, very little information is available.

Since the data from epidemiological studies are still insufficient to define the implication of AR for human health, risk assessment and therapeutic measures, the

* Corresponding author. Tel.: +91 11 26704519; fax: +91 11 26717586.
*E-mail address:* rkkale@hotmail.com (R.K. Kale).

results from animal experiments, particularly survival studies, are extremely important to find out its relevance to human beings. Recently, preirradiation of mice with the conditioning dose has been shown to provide a significant protection in terms of survival against subsequent challenging dose [2]. Furthermore, AR was found to decrease as the time between the conditioning and challenging doses increased from 6 to 24 h. Decreased AR with time was also seen at biochemical level [3]. However, enhanced survival was reported in mice preirradiated with a conditioning dose 2 months before a second exposure to lethal dose [4]. A priming dose on gestation day 11 significantly increased the number of foetuses and reduced the incidence of congenital malformation caused by exposure to higher dose [5]. The postnatal physiological and neurological development of prenataly irradiated animal studies showed high postnatal mortality in prenataly adapted mice, and survivors suffered from various detrimental effects [6]. Thus, the time interval between the priming and challenging doses seemed to be important factor, as it influenced the AR.

Various conditioning doses (1 cGy to 1 Gy) have been tested for their ability to induce AR. Our study showed that the lower conditioning dose (0.25 Gy) was more effective than the higher conditioning dose (0.5 Gy). Decrease in the efficiency of conditioning dose to induce AR with increased magnitude is still not completely understood. Interestingly, the conditioning dose given in small fractions (e.g., 0.25+0.25 Gy) was more effective, but two higher fractions (0.5+0.5 Gy) almost abolished AR and increased the lethality. These findings support the idea that the cell requires a certain amount of signal within a given interval of time for AR to be induced.

The present findings and mechanisms are not sufficient to address the uncertainties related to AR, as well as its relevance to human health, risk assessment and clinical applications. For details, see BELLE Newsletter, 7 (3), 1999 and Ref. [2].

Although AR has been characterized quite well, further work, particularly on animal survival, is required to find its relevance to human health and to address the uncertainties associated with it.

## Acknowledgement

ABT is grateful to the Department of Science and Technology, New Delhi, for financial support under the Young Scientist Scheme.

## References

[1] G. Olivieri, J. Bodycote, S. Wolff, Adaptive response of human lymphocytes to low concentrations of radioactive thymidine, Science 223 (1984) 594–597.
[2] A.B. Tiku, R.K. Kale, Adaptive response and split dose effect of gamma radiation on the survival of mice, J. Biosci. 29 (2004) 111–117.
[3] A.B. Tiku, R.K. Kale, Radiomodification of glyoxalase I in liver and spleen of mice: adaptive response and split-dose effect, Mol. Cell. Biochem. 216 (2004) 79–83.
[4] M. Nose, et al., Rescue of lethally irradiated mice from hematopoietic death by pre exposure to 0.5 Gy X rays without recovery from peripheral blood cell depletion and its modification by OK432, Radiat. Res. 156 (2001) 195–204.
[5] B. Wang, et al., Adaptive response in embryogenesis: I. Dose and timing of radiation for reduction of prenatal death and congenital malformation during the late period of organogenesis, Radiat. Res. 150 (1998) 120–122.
[6] B. Wang, et al., Adaptive response in embryogenesis: II: retardation of postnatal development of prenataly irradiated mice, Radiat. Res. 152 (1999) 119–123.

# Heavy ion induced transformation of telomerase immortalized human small airway epithelial cells

Chang Q. Piao[a,*], Yong L. Zhao[a], M. Suzuki[b], A. Balajee[a], Tom K. Hei[a]

[a]Center for Radiological Research, College of Physicians and Surgeons, Columbia University, 630 West 168th Street, New York, NY 10032, USA
[b]International Space Radiation Laboratory, National Institute of Radiological Sciences. 4-9-1 Anagawa, Inage-ku, Chiba 263-8555, Japan

**Abstract.** The tumorigenic effects of galactic cosmic rays (GCR) have become a major concern to the members of the international space station program. Normal human small airway epithelial cells were immortalized by ectopic expression of hTERT. The hTERT-expressing cells were exposed to $^{56}$Fe ions which simulate GCR irradiation. The irradiated cells underwent a stepwise transformation process. They showed phenotypic alterations and loss of some chromosomes compared with control hTERT-expressing cells. © 2004 Elsevier B.V. All rights reserved.

*Keywords:* Transformation; Heavy ions; Telomerase; hTERT

## 1. Introduction

Understanding carcinogenesis and its mechanisms induced by high LET radiation such as radon and cosmic rays is essential for human risk estimation and radiation protection. However, the limited life span of normal human epithelial cells in culture represents a substantial barrier for studying the multi-step process of neoplastic transformation induced by carcinogens. It has been suggested that most of the immortalized human epithelial cells are infected by viruses such as SV40 or HPV which induce genomic instability. It is necessary to immortalize human epithelial cells without virus-incorporation to study transformation induced by irradiation.

---

* Corresponding author. Tel.: +1 212 305 0846; fax: +1 212 305 3229.
 *E-mail address:* cp16@columbia.edu (C.Q. Piao).

0531-5131/ © 2004 Elsevier B.V. All rights reserved.
doi:10.1016/j.ics.2004.12.003

## 2. Materials and methods

Primary cultured human small airway epithelial cells (SAEC) from Clonetics were transfected with the construct pBabest2 containing hTERT by retrovirus-mediated gene transfer. Exponentially growing hTERT-expressing SAEC cells were irradiated with a single 60 cGy dose of 1 GeV/nucleon $^{56}$Fe ions accelerated with the Alternating Gradient Synchrotron at the Brookhaven National Laboratory, Upton, NY.

## 3. Result and discussion

The parental SAEC and the SAEC transfected with the vector without hTERT senesced at about the tenth passages. In contrast, the SAEC transfected with vector containing hTERT (SAEC-hTERT) continually proliferated. Telomerase activity was over-expressed in the SAEC-hTERT cells but not detectable in the SAEC. A total of $5 \times 10^6$ of The SAEC-hTERT cells at passage 5 were irradiated with 60 cGy of $^{56}$Fe ions (SAEC-hTERT-Fe). The SAEC-hTERT cells as a control were cultured in parallel with SAEC-hTERT-Fe cells. The SAEC-hTERT and the SAEC-hTERT-Fe cells were continuously cultured for more than 150 PDs, confirming that both of them were immortal. After 100 PDs, the SAEC-hTERT-Fe cells exhibited morphological alterations. Their cell size became smaller and they lost contact-inhibition growth (Fig. 1A). The SAEC-hTERT-Fe cells had a higher proliferation rate and saturation density than SAEC-hTERT cells (SAEC-hTERT-Fe=$4.2 \times 10^5$/cm$^2$, SAEC-hTERT=$2.3 \times 10^5$/cm$^2$). In addition, SAEC-hTERT-Fe cells were anchorage-independent growth with a plating efficiency about 0.8% in soft agar, whereas no colonies were formed in the control SAEC-hTERT cells (Fig. 1B). Karyotype of the SAEC-hTERT-Fe cells was hypoploid, however diploid karyotypes were maintained in the control SAEC-hTERT cells (Fig. 1C). This suggests that some chromosomes were lost in the process of transformation.

It has been proposed that the life-span of some human somatic cells can be extended or immortalized by ectopic expressing hTERT [1,2]. Previously, studies from our laboratory have shown that human papillomavirus immortalized-human bronchial epithelial (BEP2D) cells can be malignantly transformed by high LET α-particles

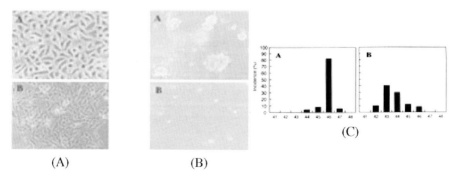

Fig. 1. (A) Loss of contact inhibition growth. Panel A: SAEC-hTERT; panel B: SAEC-hTERT-Fe. (B) Anchorage-independent growth. Panel A: SAEC-hTERT-Fe; panel B: SAEC-hTERT. (C) karyotype analysis. Panel A: SAEC-hTERT; panel B: SAEC-hTERT-Fe.

through a multi-stage process including genomic instability and clonal selection [3]. In this study, transformation was induced by $^{56}$Fe irradiation in the hTERT-expressing SAEC cells. The hypoploid karyotype is consistent with the potential loss of tumor suppressor genes in the carcinogenic process.

## References

[1] A.G. Bodnar, et al., Extension of life-span by introduction of telomerase into normal human cells, Science 279 (1998) 349–352.
[2] J. Yang, et al., Human endothelial cell life extension by telomerase expression, J. Biol. Chem. 274 (1999) 26141–26148.
[3] T.K. Hei, et al., Malignant transformation of human bronchial.

www.ics-elsevier.com

# Mechanisms of liver carcinogenesis by chronic exposure to alpha-particles form internally deposited Thorotrast

Lu Wang, Duo Liu, Takashi Shimizu, Manabu Fukumoto*

*Department of Pathology, Institute of Development, Aging and Cancer, Tohoku University, 4-1 Seiryo-machi, Aoba-ku, Sendai, Miyagi 980-8575, Japan*

**Abstract.** We analyzed mutations of the *p53* and the K-*ras* genes, microsatellite instability (MSI), in Thorotrast intrahepatic cholangiocarcinoma (ICC) and loss of heterozygosity (LOH) in Thorotrast ICC and angiosarcoma (AS). The major *p53* mutation suggested that reactive oxygen species are not likely involved in gene mutations of Thorotrast ICC. MSI in Thorotrast ICC was significantly more frequent than that in non-Thorotrast ICC and was partly attributed to the inactivation of the *hMLH1* mismatch repair gene via methylation. LOH pattern of Thorotrast ICC was partly shared with non-Thorotrast ICC but was different from that of AS. A BAS image analyzer revealed that the distribution of thorium deposits was always in living macrophages. We conclude that Thorotrast-induced cancers are developed through complex carcinogenic steps by the biological reaction during remodeling of the liver architecture. © 2004 Elsevier B.V. All rights reserved.

*Keywords:* Thorotrast; Human; Liver; Cholangiocarcinoma; Angiosarcoma

## 1. Introduction

Thorotrast is a colloidal solution of natural α-emitter, thorium dioxide, previously used as a radiological contrast medium. It caused hepatic malignancies by the local exposure to α-particles decades after administration. Thorotrast-induced tumors consist of hepatocellular carcinoma (HCC), intrahepatic cholangiocarcinoma (ICC) and angiosarcoma (AS) at nearly the same instance. Identification of genetic changes specific to Thorotrast tumors could contribute to the risk assessment of internally deposited α-emitters. The development of cancer is thought to be a consequence of multiple carcinogenic steps including

---

\* Corresponding author. Tel.: +81 22 717 8507; fax: +81 22 717 8512.
*E-mail address:* fukumoto@idac.tohoku.ac.jp (M. Fukumoto).

0531-5131/ © 2004 Elsevier B.V. All rights reserved.
doi:10.1016/j.ics.2004.09.045

proto-oncogene activation and tumor suppressor gene (TSG) inactivation. Genomic instability is observed in cells after radiation exposure. Allelotyping using microsatellites can be used to assess loss of heterozygosity (LOH) which is indicative of the presence of putative TSG. Anomalies of the DNA mismatch repair (MMR) system causes genomic instability termed microsatellite instability (MSI).

## 2. Experiments and results

The major *p53* mutation observed in Thorotrast ICC was the transition type, suggesting that reactive oxygen species are not likely involved in gene mutations. LOH analysis suggested that Thorotrast ICC develops from a stem cell that can differentiate into either bile duct epithelial cells or hepatocytes. LOH pattern suggested that minimally required genetic changes for the development of hepatobiliary cancers are common among different carcinogenic causes including radiation exposure [1]. LOH frequency of Thorotrast AS was higher than that of non-Thorotrast and tended to be higher in larger chromosomes and in the long arm compared with smaller chromosomes and the short arm. MSI induced in Thorotrast ICC was partly attributed to hypermethylation of the *hMLH1* promoter and the underlying mechanism inducing the MSI phenotype was different between Thorotrast and non-Thorotrast ICC [2]. MSI frequency was not different between Thorotrast and non-Thorotrast AS. Conglomerates of Thorotrast were mainly observed in the cytoplasm of macrophages in the fibrous connective tissues and in blood vessels, irrespective of tumor and non-tumor parts [3].

## 3. Conclusions

We conclude that inflammatory process during remodeling of the liver and cell to cell interaction contribute to radiation-induced carcinogenesis more than direct insults to DNA. However, direct DNA injury by radiation may more contribute to induction of AS than that of ICC. It is interesting to study methylation and transcription status of specific genes, especially those involved in genomic stability after irradiation, and to clarify whether MSI occurs at early stages and is maintained further, or fluctuates, during radiation-induced carcinogenesis. Although high incidence of ICC is characteristic to Thorotrast liver cancers, relative risk of AS by internally deposited α-particle emitters including plutonium is far more than that of ICC. Further pursuit of this line of research on radiation-induced AS is underway in our laboratory (Fig. 1).

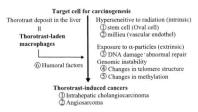

Fig. 1. Schematic diagram of Thorotrast-induced liver cancers. Cell to cell interactions between Thorotrast laden macrophages and target cells toward carcinogenesis during remodelling of the liver architecture caused by radiation is important.

## References

[1] D. Liu, et al., Allelotypic characteristics of Thorotrast-induced intrahepatic cholangiocarcinoma, Radiat. Res. 161 (2004) 235–243.
[2] D. Liu, et al., Microsatellite instability in Thorotrast-induced human intrahepatic cholangiocarcinoma, Int. J. Cancer 102 (2002) 366–371.
[3] A. Goto, et al., Microdistribution of alpha particles in pathological sections of tissues from Thorotrast patients detected by imaging plate autoradiography, Radiat. Res. 158 (2002) 54–60.

# Mangiferin, a glucosylxanthone, protects against the radiation-induced micronuclei formation in the cultured human peripheral blood lymphocytes

Ganesh Chandra Jagetia*, V.A. Venkatesha

*Department of Radiobiology, Kasturba Medical College, Manipal-576104, India*

**Abstract.** The effect of mangiferin (MGN), a glucosylxanthone, derived from *Mangifera indica* was studied on the radiation-induced DNA damage in the cultured human peripheral blood lymphocytes (HPBLs) by micronucleus assay, where HPBLs were treated with 0, 5, 10, 20, 50, 100 μg/ml of mangiferin 30 min before exposure to 3 Gy of $^{60}$Co γ-radiation. Treatment of HPBLs with 50 μg/ml reduced the radiation-induced micronuclei to the maximum extent. Irradiation of HPBLs to 0, 1, 2, 3, or 4 Gy resulted in a dose-dependent increase in the frequency of micronuclei, while treatment of HPBLs with 50 μg/ml MGN before exposure to 0, 1, 2, 3, or 4 Gy of $^{60}$Co γ-radiation resulted in a significant decline in the frequency of micronuclei when compared with the untreated irradiation group. © 2004 Elsevier B.V. All rights reserved.

*Keywords:* Mangiferin; Lymphocytes; Irradiation; Micronuclei

## 1. Introduction

High toxicity of known radioprotectors stimulated us to screen the radioprotective effect of mangiferin (MGN), a C-glucosylxanthone (1,3,6,7-tetrahydroxyxanthone-C2-beta-D-glucoside) purified from different parts of *Mangifera indica*, and *Anemarrhena asphodeloides*, which has been found to inhibit the proliferation of neoplastic cells in vitro and in vivo and antagonize the cytopathic effect of HIV [1] in the cultured human peripheral blood lymphocytes (HPBLs) exposed to radiation by micronucleus assay.

## 2. Materials and methods

The details of the human peripheral blood lymphocytes (HPBLs) culture are given elsewhere [2]. The HPBLs culture treated with 0, 5, 10, 20, 50, or 100 μg/ml of MCN

---
\* Corresponding author. Tel.: +91 820 2571201; fax: +91 820 2571919.
 *E-mail addresses:* gc.jagetia@kmc.manipal.edu, gcjagetia@rediffmail.com (G.C. Jagetia).

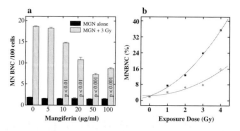

Fig. 1. Effect of mangiferin on the radiation-induced micronuclei formation in HPBLs exposed to γ-radiation. (a) Selection of optimum dose; (b) effect of 50 µg/ml of mangiferin on the radiation-induced micronuclei formation in HPBLs; squares, PBS+irradiation; circles, MGN+irradiation.

before exposure to 0 or 3 Gy γ-irradiation and 50 µg/ml MGN was found best, and it was used for further study, where HPBLs of PBS+irradiation group received 10 µl/ml of sterile PBS, while that of MGN+irradiation received 50 µg/ml of mangiferin before exposure to 0, 1, 2, 3, or 4 Gy γ-radiation 30 min after mangiferin treatment from a Tele Cobalt therapy source at a dose rate of 1 Gy/min. The HPBLs were transferred into a $CO_2$ incubator and allowed to grow for 72 h at 37 °C. The micronuclei were prepared as described earlier [2].

## 3. Result and discussion

MGN did not elevate the spontaneous frequency of MNBNC significantly. Treatment of HPBLs with different concentrations of MGN before exposure to 3 Gy resulted in a significant decline in the frequencies of micronuclei, and a greatest decline in the micronuclei (MN) was observed at 50 µg/ml MGN (Fig. 1a). The exposure of HPBLs to different doses of γ-radiation resulted in a dose-dependent elevation in the frequencies of micronuclei in the PBS+irradiation group (Fig. 1b). Treatment of HPBLs with 50 µg/ml MGN before exposure to different doses of γ-radiation reduced the frequency of micronuclei significantly compared to the PBS+irradiation group. The dose–response was linear quadratic for both groups.

The irradiation of HPBLs with different doses of γ-radiation not only increased MNBNC with one MN but also MNBNC bearing two and multiple MN. A similar observation has been made earlier [2,3]. Treatment of HPBLs with 50 µg/ml MGN before irradiation not only reduced the frequency of MNBNC with one MN significantly but also BNC bearing two and multiple MN, indicating that MGN has been able to inhibit the multiple sites of damage to DNA and complex chromosome aberrations. An identical effect has been reported earlier in HPBLs treated with Aegle *marmelos* or *Syzygium Cumini* extract before irradiation [2,3]. The radioprotective effect of MGN may be due to scavenging of radiation-induced free radicals, as it has been reported to scavenge hydroxyl radicals [4], or by inhibition of radiation-induced depletion of nonprotein sulfhydryl groups.

## References

[1] M.S. Zheng, Z.Y. Lu, Zhongguo Yao Li Xue Bao 10 (1989) 85–90.
[2] G.C. Jagetia, P. Venkatesh, M.S. Baliga, Mutagenesis 18 (2003) 387–393.
[3] G.C. Jagetia, M.S. Baliga, Toxicol. Lett. 132 (2002) 19–25.
[4] G.M. Sanchez, et al., Phytother. Res. 14 (2000) 424–427.

# Association of glutathion S-transferase and chromosomal aberrations as a means to determine occupational exposure

A. Movafagh*, F. Maleki, S.G. Mohammadzadeh, S. Fadaei

*Department of Genetic and Biochemistry, Shahid Beheshti Medical University, Tehran, Iran*

**Abstract.** It has been reported that exposure of radiotherapy workers to ionizing radiation causes chromosomal damages. Some of the damaged cells show a large number of chromosome aberrations such as dicentrics, rings and numerous acentric fragments. The influence of glutathione S-transferase (GST) on the rate and frequency of chromosomal aberrations in peripheral blood lymphocytes of 50 medical radiotherapy workers handling X-ray machines for more than 5 years (mean period 12 years), and 43 control individuals were subjected to our study. GST activity in serum was estimated by the improved Habig method, and the frequency of chromosomal aberrations in blood lymphocytes was evaluated by a conventional Trypsin G-banding technique. The present study showed that dicentrics, acentrics, followed by ring chromosomes and of total chromosomal aberrations, and GST activity were significantly higher among the radiotherapy workers when compared to controls ($P=0.04$). Increased GST concentration and chromosomal damages were correlated in the present study. Despite the limited number of blood samples, the results seem to indicate an association between chromosomal aberration and GST activity although it needs more data in this part of Iran to confirm our conclusion. © 2004 Elsevier B.V. All rights reserved.

*Keywords:* Chromosome aberration; Glutathion S-transferase; Occupational radiation; Iran

## 1. Introduction

The GST is located on chromosomes 1p13.3 and 22q11.2, and lack of GST activity in individuals with these deletions may result in increased risk of somatic mutation. GST may play an important role in cellular detoxification of ionizing radiation in human. Several lines of evidence indicate that ionizing radiation can induce persistent genetic instability in

---

* Corresponding author. Tel.: +98 281 3322003; fax: +98 2414138.
 *E-mail address:* movafagh_a@yahoo.com (A. Movafagh).

0531-5131/ © 2004 Elsevier B.V. All rights reserved.
doi:10.1016/j.ics.2004.12.011

exposed cells. Relatively high frequencies of chromosomal aberrations like rings and double minute chromosome were recorded among persons exposed to radiation [1].

## 2. Material and methods

GST activity was determined by the method of Habig [2]. From each individuals, approximately 1 ml heparinized blood was collected and mixed with 4 ml RPMI-1640 culture medium supplemented with 20% fetal bovine serum and 120 μg PHA. The sample was incubated at 37 °C under the aeration of 5% $CO_2$ for 72 °C h. The cultured cells were harvested and subjected to treatment with 75 mM KC for 20 min and fixed in 1:3 mixture of acetic acid and methanol. Metaphases were banded using giemsa-trypsin (GTG-) banding.

## 3. Results

GST activity was significantly higher in the occupationally exposed (radiotherapy workers) when compared to control individuals ($P=0.04$). Despite the limited number of subjects, the results seem to indicate an association between the occupational exposure due to radiotherapy, increased GST concentrations, and chromosomal aberrations. The mean frequencies of various chromosomal aberrations were higher among radiotherapy workers with over 10 years of occupational exposure ($P=0.05$) when compared to those workers with less than 10 years of exposure. GST activity was more frequent in males than in females.

## 4. Discussion

Chromosomal aberration frequency provides the most reliable and accurate biological marker of radiation dose to detect accidental exposure. Hence, IAEA in 1986 at Vienna suggested that chromosomal aberration must be used for biological dosiometry in human exposure to radiation. In our investigation, the majority of aberrations were acentric, dicentric and ring chromosome, as has also been reported by other studies. The yields of chromosome fragments of total aberrations were significantly higher among radiation workers with long-term exposure [3]. The data presented here and in other study [4] together indicate that occupational radiation exposure 2 or more years prior to blood testing has no significant influence on chromosomal damages. In addition to structural chromosome aberration, all types of occupational radiation exposure increase the variability in the number of chromosomes [5]. In conclusion, chromosome aberration frequency provides the most reliable and accurate biological marker of radiation dose inflicted by accidental exposure. Also the increase of the chromosome aberration observed in our radiotherapy workers underscored the need of adopting measures to avoid or minimize overexposure.

## References

[1] B. Karthikeya, et al., Comparison of intra- and intra chromosomal aberrations in blood, Radiat. Prot. Dosim. 103 (2003) 103–109.
[2] W.H. Habig, et al., Glutathion S transferase, the first enzymatic step in mercapturic acid formation, J. Biol. Chem. 249 (1974) 7130–7139.
[3] I. Hayata, et al., Cytogenetical dose estimation for 3 severely exposed patients in the JCO critically accident, J. Radiat. Res. 42 (2001) 149–155.
[4] A. Norman, et al., Effects of age, sex and diagnostic X-ray on chromosome damage, Int. J. Radiat. Biol. 46 (1984) 317–321.
[5] J. Dahle, et al., Induction of delayed mutations and chromosomal instability in after UVA and X-radiation, Cancer Res. 1 (2003) 1464–1469.

# Hormone levels associated with immune responses among inhabitants in HLNRAs of Ramsar-Iran

F. Zakeri[a,*], A. Kariminia[b]

[a]Radiobiology and Biological Dosimetry Division, National Radiation Protection Department, Iranian Nuclear Regulatory Authority, Tehran, Iran
[b]Immunology Department, Pasteur Institute of Iran, Tehran, Iran

**Abstract.** The purpose of this study was to determine whether the exposure to HLNR affects the hormone levels associated with immune responses in young and elderly subjects. The studies demonstrated a decreased level of DHEA and increased level of cortisol in the elderly HLNRA group which might be possibly associated with higher cumulative doses. There was a higher incidence of hypothyroidism in the HLNRA group, mostly in elder group. A significant increase of CD69 expression on CD4+ T lymphocytes was found in the young HLNRA group. Low levels of DHEA in the elder group were correlated to their increased IgE levels. © 2004 Elsevier B.V. All rights reserved.

*Keywords:* Adrenal stress hormones; High-level natural radiation; Immune responses; Thyroid hormones

## 1. Introduction

Radiation can act as stress for the triangle of immune, endocrine and nervous systems. Stress affects the output of adrenal stress hormones and hormonal balance is essential to the maintenance of good health and a strong immune system. When there is a prolonged stress of any type, the body prefers to make increasingly greater amounts of cortisol and less DHEA, and this imbalance results in reduced metabolism by suppressing production of active thyroid hormones and also, influence the maturation pathways and proliferation of T lymphocytes [1].

## 2. Materials and methods

Serum DHEA, cortisol, T3, T4, TSH, immunoglobulins and lymphocyte activation factor were measured among 72 individuals in Ramsar divided to two subgroups of less than 50

---

* Corresponding author. Postal address: Atomic Energy Organization of Iran, P.O. Box: 14155-4494, Tehran, Iran. Tel: +98 21 61384154; fax: +98 21 8009502.
*E-mail address:* fzakeri40@yahoo.com (F. Zakeri).

0531-5131/ © 2004 Elsevier B.V. All rights reserved.
doi:10.1016/j.ics.2004.11.043

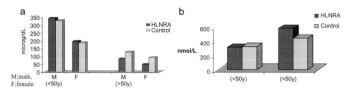

Fig. 1. (a) DHEA concentrations in studied groups. (b) Cortisol concentrations in studied groups.

years old and elder (50–90 years) group. Their annual effective dose equivalent ranged from 5 to 75 mSv/year. The results compared with those of 60 controls in normal background area.

## 3. Results

There was a significant decreased level of DHEA and increased level of cortisol in the elderly HLNRA group (Fig. 1a and b).

There was a higher incidence of hypothyroidism—mostly in elder HLNRA group—compared with the control group (8.3% vs. 4%). They were under the thyroid hormone-replacement therapy, so no major differences in serum T3, T4 or TSH were found in the studied groups. Total serum IgE was significantly increased in HLNRA group. Low levels of DHEA in the elder HLNRA group were correlated to increased IgE levels. Concentration of other immunoglobulins did not show significant differences between groups. A significant increase of CD69 expression on CD4+ stimulated T cells in the young HLNRA group and higher expression of CD69 on CD8+ cells in the elder HLNRA group were found.

## 4. Discussion

In the present study, DHEA was lower and cortisol was higher in elder HLNRA group than in their age-matched controls, which might be possibly associated with higher cumulative doses. Considering the metabolism, an elevated cortisol to DHEA ratio can cause suppression via alteration of thyroid hormone function [1], so it may be the cause of more evidences of hypothyroidism in elder HLNRA group. Higher concentrations of IgE were found in the HLNRA groups, and in the elder HLNRA group, it was correlated to increased IgE levels. DHEA may be one of the regulators of IgE synthesis [2] and increases the production of Th1-associated lymphokines [1]. These results indicated that decreased level of DHEA might promote a shift in Th1/Th2 ratio balance toward Th2-dominant immunity, leading to increased IgE level. Higher concentrations of IgE may probably be partly due to the environmental allergens or sulphur in hot springs. A significant increase of CD69 expression on the CD4+ stimulated T cells was found in the young HLNRA group. Although CD69 expression is a consequence of T-cell activation, it also acts as a negative regulator of antitumor responses [3]. Supplementary radio-epidemiological studies are needed to determine the health consequences of these effects.

## References

[1] C.N. Shealy, A review of dehydroepiandrosteron (DHEA), Integr. Physiol. Behav. Sci. 30 (1995) 308–313.
[2] N. Sudo, X.N. Yu, C. Kubo, Dehydroepiandrosteron attenuates the spontaneous elevation of serum IgE level in NC/Nga mice, Immunol. Lett. 79 (3) (2001 Dec 3) 177–179.
[3] E. Esplugues, et al., Enhanced antitumor immunity in mice deficient in CD69, J. Exp. Med. 197 (9) (2003 May 5) 1093–1106.

# Radioadaptive responses induced in lymphocytes of the inhabitants in Ramsar, Iran

S.M.J. Mortazavi[a,*], A. Shabestani-Monfared[b], M. Ghiassi-Nejad[c], H. Mozdarani[d]

[a]*Medical Physics Department, School of Medicine, Rafsanjan University of Medical Sciences (RUMS), Rafsanjan, Iran*
[b]*National Radiation Protection Department (NRPD), Iranian Nuclear Regulatory Authority (INRA), P.O. Box 14155-4494, Tehran, Iran*
[c]*Medical Physics Department, School of Medicine, Babol University of Medical Sciences, Babol, Iran*
[d]*Medical Genetics Department, School of Medical Sciences, Tarbiat Modarres University, Tehran, Iran*

**Abstract.** Ramsar, a city in northern Iran, has among the highest levels of natural radiation known to exist in an inhabited area. Twenty-two residents of high-level natural radiation areas and 33 residents from an adjacent normal-level natural radiation area participated in this study. In the first phase of the experiment, 15 healthy residents from high-level natural radiation areas and 30 healthy inhabitants of a nearby normal-level natural radiation area were studied. In the second phase, seven healthy residents with cumulative lifetime doses of up to 10 Sv were studied for assessing the induction of adaptive response in each study participant and obtaining complementary data. Cultured cells were given a challenge dose of either 2 Gy (first phase of the study) or 1.5 Gy (second phase of the study) of Co-60 gamma radiation. Overall data showed a significant radioadaptive response in the residents of high-level natural radiation areas. Results obtained in the second phase of the study showed that five out of seven inhabitants exhibited a reduction in induced chromosomal aberrations following exposure to a 1.5 Gy challenge dose of gamma radiation. As the cumulative dose increased from a few hundred mGy to 1 Gy, the magnitude of the induced adaptive response increased linearly.
© 2004 Published by Elsevier B.V.

*Keywords:* Adaptive response; Natural radiation; Ramsar

---

* Corresponding author. Academic Affairs' Office, Central Building of the Rafsanjan University of Medical Sciences, Rafsanjan, Iran. Tel.: +98 391 822 0097; fax: +98 391 822 0092.
 *E-mail address:* jamo23@lycos.com (S.M.J. Mortazavi).

0531-5131/ © 2004 Published by Elsevier B.V.
doi:10.1016/j.ics.2004.12.002

## 1. Introduction

The annual effective dose (excluding radon progeny contributions) in high-level natural radiation areas (HLNRAs) of Ramsar is a few times higher than the ICRP recommended annual effective dose limit for radiation workers [1]. Our preliminary results reported previously suggested that exposure to HLNR can induce radioadaptive response in lymphocytes of Ramsar residents [2–4]. In this paper, we report the effect of cumulative dose on the radioadaptive response.

## 2. Materials and methods

In the first phase of the study, venous blood samples were taken from 15 and 30 healthy volunteers of both sexes who lived in HLNRAs and NLNRAs, respectively. The maximum measured dose rate of natural radiation was 155 $\mu Sv\ h^{-1}$. In the second phase, seven healthy volunteers from HLNRAs and five healthy volunteers from a nearby NLNRAs with dose rates from 0.07 to 0.11 $\mu Sv\ h^{-1}$ served as controls. It should be noted that in the second phase, due to our selection criteria for study participants (inhabitants who received the annual doses higher than 300 mSv), only limited number of volunteers from HLNRAs were available for this study. Standard conditions for cell cultivation, irradiation and analysis of chromosome aberrations were used.

Cultured cells were given a challenge dose of either 2 Gy or 1.5 Gy of Co-60 gamma radiation at a dose rate of 114 mGy/s. Some of the culture flasks were sham irradiated to assess either the frequency of chromosomal aberrations induced by natural radiation alone in HLNRAs' residents and the spontaneous frequency of aberrations in NLNRA residents. After the challenge dose, all the culture flasks were incubated a further either 2 (first phase) or 6 (second phase) h.

For each data point, about 200 well-spread metaphases were blind scored for chromosomal aberrations. The number of chromatid-type aberrations was determined. Gaps (achromatic lesions smaller than the width of a chromatid) were included in the statistical analysis of the first phase, but in order to enhance the reliability of the results, these lesions were excluded in the statistical analysis of the second phase.

## 3. Results and discussion

The overall results of the first phase of our study showed a significant adaptive response when the cultured lymphocytes of the 15 inhabitants of HLNRAs were exposed to a 2 Gy gamma challenge dose (Table 1). These results confirm the previous results obtained in other in vivo human studies such as radiation worker studies.

Table 1
Mean frequency of chromatid aberrations in nonirradiated and irradiated cells of residents in HLNRAs and NLNRAs (control)

| Area | No. of participants | Maximum annual dose (mGy) | MCA[a] in nonirradiated cells (gaps included) | MCA[a] in cells exposed to 2 Gy Gamma rays (gaps included) | Induction of radioadaptive response |
|---|---|---|---|---|---|
| HLNRAs | 15 | 260 | $0.099 \pm 0.003$[b] | $0.111 \pm 0.003$[b] | Positive ($P<0.001$) |
| NLNRAs | 30 | 1.05 | $0.049 \pm 0.003$ | $0.167 \pm 0.004$ | ND[c] |

[a] Mean chromosome aberrations per cell.
[b] Mean ± S.E.
[c] Not determined.

Table 2
Frequency of chromosomal aberrations in nonirradiated and irradiated cells of the residents of HLNRAs and NLNRAs

| Study group | Sample size | Cumulative dose (mSv) | MCA* in nonirradiated cells[a] (gaps excluded) | MCA* in cells exposed to 1.5 Gy[@] (gaps excluded) |
|---|---|---|---|---|
| HLNRA | 7 | 2534 | $0.014 \pm 0.004$ | $0.106 \pm 0.015$ |
| NLNRA | 5 | 14.6 | $0.016 \pm 0.002$ | $0.188 \pm 0.020$ |
| $P$-value | | | Not significant | <0.001 |

[a] Mean±S.E.
* Mean chromosome aberrations per cell.

The results obtained in the second phase indicate that residents of areas with extraordinary levels of natural radiation (annual doses up to 260 mGy) show a significant radioadaptive response (Table 2). It was observed that the five persons who received cumulative doses of 360–950 mGy showed a significant radioadaptive response, while the two individuals with the highest cumulative doses (6800 and 8400 mGy) failed to show a significant radioadaptive response. That is, 70% showed an adaptive response from living in the HLNRAs. No participant from the HLNRA had an increase in radiation damage compared to the controls. We consider of great potential importance that high levels of natural radiation can serve as the priming or conditioning dose.

A relationship was found between the cumulative dose of each study participants and the magnitude of the induced radioadaptive response. Results of our experiments showed that high levels of natural radiation in inhabitants whose cumulative doses were up to 1 Gy significantly decreased radiation damage as measured by reduced chromosomal aberrations in irradiated lymphocytes. This can be considered a beneficial effect of high natural radiation.

Our findings on the biological effects of prolonged exposure to high levels of natural radiation in the inhabitants of HLNRAs of Ramsar showed no apparent harmful health effects. We have been reported previously that the health effects of prolonged exposure to high levels of natural radiation may contradict current ultra-conservative radiation protection regulations [5]. Governments should adopt public health measures and policies that are cost-effective in risk reduction by considering the financial, social and psychological impact on their citizens. Based on our results, we suggest that worldwide research on the residents of high-level natural radiation areas help scientists better justify if LNT model of radiation risk is appropriate as the basis for public health measures.

## References

[1] M. Sohrabi, World high level natural radiation and/or radon prone areas with special regards to dwellings, in: L. Wei, T. Suahara, Z. Tao (Eds.), Proceeding of the 4th International Conference on High Levels of Natural Radiation (ICHLNR), Beijing, China, 1996, 1997, pp. 3–7.
[2] S.M.J. Mortazavi, et al., How should governments address high levels of natural radiation and radon? Lessons from the Chernobyl nuclear accident, Risk: Health, Safety and Environment 13/1.2 (2002) 31–45.
[3] M. Ghiassi-nejad, et al., Very high background radiation areas of Ramsar, Iran: preliminary biological studies, Health Physics 82 (2002) 87–93.
[4] S.M.J. Mortazavi, P.A. Karam, High levels of natural radiation in Ramsar, Iran: should regulatory authorities protect the inhabitants? Iranian Journal of Science (Germany) 2 (2) (2002) 1–9.
[5] S.M.J. Mortazavi, M. Ghiassi-nejad, T. Ikushima, Do the findings on the health effects of prolonged exposure to very high levels of natural radiation contradict current ultra-conservative radiation protection regulations, in: T. Sugahara, O. Nikaido, O. Niwa (Eds.), Radiation and Homeostasis, Elsevier, Amsterdam, 2002, pp. 19–21.

# Elevated radon concentrations in a Pleistocenic cave operating as a show cave

C. Papastefanou*, M. Manolopoulou, S. Stoulos, A. Ioannidou, E. Gerasopoulos

*Atomic and Nuclear Physics Laboratory, Aristotle University of Thessaloniki, Panepistimiou, 54124 Thessaloniki, Greece*

**Abstract.** Radon measurements were carried out, using solid-state nuclear track-etch detectors (SSNTDs) with type CA 80-15 cellulose nitrate films, in a Pleistocenic cave at Petralona, in Halkidiki, northern Greece, 55 km distant from the city of Thessaloniki. Radon levels as high as 88 kBq m$^{-3}$ (2.38 nCi l$^{-1}$) recorded inside the cave is equivalent to 11.90 WL in terms of occupational exposure to radon and its decay products. Absorbed dose rates were measured using TL dosimeters, type TLD-200 (CaF$_2$-Dy), in a continuous monitoring program (integrated measurements). Dose rate levels as high as 110 nGy h$^{-1}$ were recorded inside the cave. © 2004 Elsevier B.V. All rights reserved.

*Keywords:* Radon concentration; Absorbed dose rate; Pleistocenic cave

## 1. Radon concentration measurements

Radon concentration measurements were performed by solid-state nuclear track-etch detectors, SSNTDs using Kodak CA 80-15 cellulose nitrate films. The detectors, plastic films of diameter 1.6 cm, were placed in a special holder of Plexiglass, in which they were supported by wire clips. The holder was placed in the center of upper side on the top of a cylindrical ceramic pot, 11 cm in diameter and 12 cm in height. The lower side was open for air entry. To avoid humidity effects, a heater with thermostat was set inside the ceramic pot. The method and the device used were reported by Savvides et al. [1]. The temperature inside the ceramic pot was about 10 °C above ambient. The exposure time of the detectors was about 30 days.

---

* Corresponding author. Tel.: +302310998005; fax: +302310206138.
 *E-mail address:* papestefanou@physics.auth.gr (C. Papastefanou).

0531-5131/ © 2004 Elsevier B.V. All rights reserved.
doi:10.1016/j.ics.2004.11.165

Table 1
Radon concentrations and radon levels in the cave

| Location | Radon concentration | | Radon level |
|---|---|---|---|
| | pCi l$^{-1}$ | kBq m$^{-3}$ | WL[a] |
| Cave entrance | 5 | 0.19 | 0.025 |
| Interior Mausoleum (Archanthropous site) | 345 | 12.77 | 1.725 |
| Interior Mausoleum (Excavations site) | 575 | 21.28 | 2.875 |
| Section B | 2380 | 88.06 | 11.90 |
| Precipice (Varathron) | 322 | 11.91 | 1.61 |
| Slaughter-House (Sphagion) | 456 | 16.87 | 2.28 |

[a] 1 pCi l$^{-1}$=0.005 working level, WL with equilibrium factor, $F$=0.5.

Table 2
Absorbed dose rate measurements inside the cave

| Location | Dose rate (nGy h$^{-1}$) | Annual dose rate (mGy year$^{-1}$) |
|---|---|---|
| Interior Mausoleum | 78 | 0.68 |
| Section A (1) | 45 | 0.40 |
| Section B (2) | 109 | 0.95 |

(1) Surface at stalagmites, (2) layers 24–26.

The results are summarizes in Table 1.

Thermoluminescence dosimeters, TLD-200 (CaF$_2$-Dy) of Harshaw, were set in several locations inside the cave for the measurement of the absorbed dose. The exposure time of dosimeters was about 30 days. The results are summarized in Table 2.

The mean absorbed dose rate in outdoor air 1 m above the ground surface is considered equal to 52 (20–126) nGy h$^{-1}$ and is due analytically to $^{238}$U series (mean=15, range=7–47) nGy h$^{-1}$, $^{232}$Th series (mean=20, range=7–42) nGy h$^{-1}$ and $^{40}$K (mean=17, range=6–37) nGy h$^{-1}$[2]. This means that the absorbed dose rates measured inside the cave are almost at background level.

## References

[1] E. Savvides, et al., A simple device for measuring radon exhalation from the ground, Int. J. Appl. Radiat. Isot. 36 (1985) 79–81.
[2] UNSCEAR, United Nations Scientific Committee on the Effects of Atomic Radiation, Sources and Effects of Ionizing Radiation, United Nations, New York, 2000.

www.ics-elsevier.com

# Radon in groundwater: analysis of causes using GIS and multivariate statistics
# A case study in the Stockholm county

K. Skeppström*, B. Olofsson

*Department of Land and Water Resources Engineering, Royal Institute of Technology (KTH), Stockholm, Sweden*

**Abstract.** This study investigated radon ($^{222}$Rn) in 1460 drilled wells in Stockholm County, taking into consideration a number of factors that were originally considered to be independent sources of information. A combined approach of GIS and multivariate statistical analyses were used. The results show that the following factors clearly affect the radon content: bedrock, distance to fracture, topography and the use of the well. © 2004 Elsevier B.V. All rights reserved.

*Keywords:* Radon; Groundwater; GIS; Multivariate statistics

## 1. Introduction

The occurrence of $^{222}$Rn in groundwater is principally linked to the type of rocks from which the water has been extracted [1]. Rocks with a high concentration of uranium, the parent element of $^{222}$Rn, are expected to have high radon concentrations [2,3]. The hypothesis of this research states that an integrated study including factors of geology, hydrology, topography and geochemistry and technical factors is necessary to characterize the major variables influencing the concentration of radon in drinking water. This study investigates how the concentration of radon in drinking water is correlated to altitude, landuse, geology, fracture zone and occurrence of uranium.

## 2. Materials and methods

Investigated factors were rasterised with a spatial resolution of 50 m in a GIS environment. The spatial analyst function of ArcGIS was used to derive the spatial pattern of each

---

* Corresponding author. Tel.: +46 8 790 86 84; fax: +46 8 208 946.
*E-mail address:* kirlna@kth.se (K. Skeppström).

0531-5131/ © 2004 Elsevier B.V. All rights reserved.
doi:10.1016/j.ics.2004.11.173

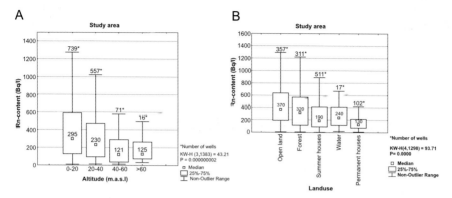

Fig. 1. (A) Box plots showing correlation of radon with altitude. (B) Box plots showing correlation of radon with soil types and landuse.

well from each factor map. The extracted data were analysed, using methods of Principal Component Analysis (PCA) and Kruskal–Wallis ANOVA by ranks in STATISTICA.

## 3. Results

Fig. 1 shows the results for two factors. It can be observed that radon concentration is higher at low altitudes (Fig. 1A) while wells used on a permanent basis have lower radon content than wells used in summerhouses (Fig. 1B). Strong positive correlations between the concentration of radon in groundwater and the following factors were observed ($p<0.001$): granite rocks, high uranium concentration and a shorter distance to a fracture zone. With respect to soil types, it was found that sand/gravels had the lowest radon content while clay the highest.

## 4. Discussions

At high altitude, water has a short residence time since they tend to flow to lower grounds. This means that there exists constant circulation of water at high altitude, leading to lower concentrations of radon. Water is almost stagnant at low altitude, and, depending on the geology type (e.g. clay), a higher radon concentration can be encountered there. Leached radioactive elements in bedrocks (e.g. uranium and radium) can also get carried by water in bedrock fractures to low altitudes, and eventually decay to radon at low altitude [2]. Regular use of a well in a permanent house gives rise to a continuous circulation of water, and, as a consequence, a lower radon concentration. Similar results have been reported in previous work [3]. This preliminary study clearly shows the influence of other factors on the radon concentration. Detailed investigations at a local scale including additional factors such as technical details of wells and geochemistry would be conducted in a next stage.

## References

[1] R.C. Cothern, P.A. Rebers (Eds.), Radon, Radium and Uranium in Drinking Water, Lewis Publishers, 1990.
[2] G. Åkerblom, J. Lindgren, Mapping of groundwater radon potential, Eur. Geol. 5 (1997) 13–22.
[3] G. Knutsson, B. Olofsson, Radon concentrations in drilled wells in the Stockholm region of Sweden, Bull. - Nor. Geol. Unders. (NGU) 439 (2002) 79–85.

# Current studies on radon gas in Thailand

P. Wanabongse[a,*], S. Tokonami[b], S. Bovornkitti[c]

[a]*Radiation Monitoring Group, Office of Atoms for Peace, Vibhavadi Rangsit Road, Bangkok 10900, Thailand*
[b]*Radon Research Group, National Institute of Radiological Sciences, Chiba, Japan*
[c]*The Academy of Science, the Royal Institute, Bangkok, Thailand*

**Abstract.** In Thailand, measurements on indoor radon at Phu Wiang district of Khon Kaen province, Saraphi district of Chiang Mai province and Na Mhom district of Songkhla province revealed the levels of $21\pm7$, $21\pm6$ and $52\pm17$ Bq m$^{-3}$, respectively. Measurements on dissolved radon in 75 natural hot spring samples collected from 22 provinces revealed the levels ranged from 0.8 to 7219.7 Bq l$^{-1}$. © 2004 Elsevier B.V. All rights reserved.

*Keywords:* Indoor radon; Dissolved radon; Natural hot spring; Thailand

## 1. Introduction

Three important locations were selected for indoor radon study, owing to the known uranium deposit in Khon Kaen province [1], high prevalence of lung cancer in Saraphi district of Chiang Mai province [2] and reported high indoor radon level (200 Bq m$^{-3}$) in Songkhla province [3]. A nationwide survey on dissolved radon in natural hot springs was carried out specifying the locations reported by the Royal Institute [4].

## 2. Materials and methods

Indoor radon concentration was measured with CR-39 detector (Fig. 1A). The detectors were installed in 188 houses for 186 days at Phu Wiang district of Khon Kaen province, in 50 houses for 99 days at Saraphi district of Chiang Mai province and in 25 houses for 36 days at Na Mhom district of Songkhla province. Evaluation of radon concentrations was conducted at the National Institute of Radiological Sciences, Chiba, Japan.

Dissolved radon in 200-ml natural hot spring sample was collected into a 270-ml Lucas scintillation cell, using a circular air-tight system (Fig. 1B). Radon concentration was

---

\* Corresponding author. Tel.: +662 562 0095; fax: +662 561 3013.
*E-mail address:* paiwan@hotmail.com (P. Wanabongse).

0531-5131/ © 2004 Elsevier B.V. All rights reserved.
doi:10.1016/j.ics.2004.11.009

Fig. 1. (A) CR-39 radon detector and (B) a circular air-tight system for measurement of dissolved radon.

determined from the counting efficiency of measuring device, taking into account the dilution of radon gas inside the system, and that the release of radon from water is temperature-dependent.

## 3. Results and discussion

### 3.1. Indoor radon

Indoor radon levels measured at Phu Wiang district of Khon Kaen province, Saraphi district of Chiang Mai province and Na Mhom district of Songkhla province were $21\pm7$, $21\pm6$ and $52\pm17$ Bq m$^{-3}$, respectively. They did not exceed the safety limit. Of note, the level at Na Mhom district was more than twice higher than the others.

### 3.2. Radon in hot spring water

Results of measurement for dissolved radon in natural hot springs are shown in Table 1. Water samples collected from the southern Thailand yielded higher radon levels, i.e., samples from Surat Thani province yielded the levels of 7219.7, 973.1 and 935.0 Bq l$^{-1}$. A sample from Chumpon, a nearby province of Surat Thani, yielded the radon level of 4515.4 Bq l$^{-1}$. Samples from other regions in the country yielded much lower dissolved radon levels.

## References

[1] F. Chotikanatis, P. Arunyakananda, Uranium sources in Thailand, J. R. Inst. Thail. 27 (2002) 100–104.
[2] V. Vatanasapt, et al., Cancer incidence in Thailand, 1988–1991, Cancer Epidemiol. Biomark. Prev. 4 (1995) 475–483.
[3] T. Bhongsuwan, et al., Radon risk assessment indoor/outdoor to public communities in Songkhla lake basin, Proceedings of the 8th Nuclear Science and Technology Conference, Bangkok, Thailand. Bangkok: Kurusapa Lat Praow, 2001 (Jun 20–21), pp. 757–768.
[4] The Royal Institute. Thai geography glossary: the Royal Institute Number, 4th printing (Revised). Thailand; 2002. pp. 246–249.

Table 1
Findings of dissolved radon in natural hot springs in Thailand

| Regions | Number of provinces surveyed | Number of samples | Radon levels (Bq l$^{-1}$) |
| --- | --- | --- | --- |
| Northern | 4 | 18 | 0.8–76.5 |
| Central and Eastern | 5 | 11 | 1.8–130.8 |
| Western | 4 | 9 | 2.0–171.1 |
| Southern | 9 | 37 | 0.9–7219.7 |

www.ics-elsevier.com

# Natural radioactivity in the high background radiation area at Erasama beach placer deposit of Orissa, India

D. Sengupta[a],*, A.K. Mohanty[a], S.K. Das[b], S.K. Saha[b]

[a]*Department of Geology and Geophysics, Indian Institute of Technology, Kharagpur, West Bengal, 721 302, India*
[b]*Radiochemistry Division, Variable Energy Cyclotron Centre, BARC, Bidhan Nagar, Kolkata, West Bengal, 700 064, India*

**Abstract.** The newly discovered Erasama beach placer deposits in southeastern coast of Orissa state has a high natural radiation environment. The concentrations of radioactive elements such as $^{232}$Th, $^{238}$U and $^{40}$K were measured by gamma ray spectrometry, using a High Purity Germanium detector. The studies show that these placer deposits have high natural radioactivity and external gamma dose rate levels. © 2004 Elsevier B.V. All rights reserved.

*Keywords:* Natural radiation; High background radiation area; Monazite; Beach sand; Orissa

## 1. Introduction

There are few regions in the world that are known for high background radiation areas (HBRAs), where the local geological controls and geochemical effects cause enhanced levels of terrestrial radiation [1]. In India, there are quite a few monazite sand bearing placer deposits causing high background radiation along its long coastline [1]. Radiological investigations were carried out in the southern coast of Orissa to measure the radiation dose rates and trace the minerals causing enhanced level of natural radiation [2]. The present study investigates the activity of radioactive elements, such as $^{232}$Th, $^{238}$U and $^{40}$K, in beach sand samples and the associated radiation dose levels in air, in this region.

---

\* Corresponding author. Tel.: +91 3222 283380; fax: +91 3222 255303.
*E-mail address:* dsgg@gg.iitkgp.ernet.in (D. Sengupta).

0531-5131/ © 2004 Elsevier B.V. All rights reserved.
doi:10.1016/j.ics.2004.12.010

## 2. Materials and methods

Erasama beach (Lat. 86°25′ –86°33′ N, Long. 20°1′ –20°11′ 38″ E) is a part of the eastern coast of Orissa state, India. The area extended over a length of 24 km and average width of more than 1 km, trending almost NE–SW, bounded by the Bay of Bengal in the southeast and coastal alluviums of Pleistocene age in the northwest side. Sand samples were collected from the beach by the grab sampling method at an interval of up to 1 km. About 1 kg of sand samples were collected from each location. The gamma ray spectrometric analysis of radionuclides was carried out using a coaxial HPGe detector. The radionuclides considered were as follows: $^{228}$Ac (209, 338, 911 keV), $^{212}$Pb (239 keV), $^{212}$Bi (727 keV) and $^{208}$Tl (583 keV) for the $^{232}$Th; $^{214}$Pb (295, 352 keV) and $^{214}$Bi for the $^{238}$U; and $^{40}$K (1460.8 keV).

## 3. Results and discussion

The absorbed gamma dose rates in air at 1 m above the ground surface for the uniform distribution of radionuclides ($^{232}$Th, $^{238}$U and $^{40}$K) were computed on the basis of guidelines provided by UNSCEAR [1]. The average concentrations of $^{232}$Th, $^{238}$U and $^{40}$K were found to be $2000\pm200$ Bq kg$^{-1}$, $350\pm50$ Bq kg$^{-1}$ and $200\pm40$ Bq kg$^{-1}$, respectively, for the total sands. The average activity concentrations of monazite and zircons were $220{,}000\pm500$ Bq kg$^{-1}$ and $2000\pm100$ for the $^{232}$Th and $30{,}000\pm250$ Bq kg$^{-1}$ and $3500\pm150$ Bq kg$^{-1}$ for the $^{238}$U, respectively. The absorbed dose rate varied from 650 to 3150 nGy h$^{-1}$ with a mean ($\pm$S.D.) value of $1925\pm718$ nGyh$^{-1}$. The annual external effective dose rates varied from 0.78 to 3.86 with a mean ($\pm$S.D.) value of $2.36\pm0.88$ mSv y$^{-1}$. However, the presence of $^{232}$Th in beach sand contributed a maximum of 91% (1750 nGy h$^{-1}$) to the total absorbed dose rate in air, followed by $^{238}$U of 8.5% (165 nGyh$^{-1}$), and the minimum contribution was by $^{40}$K of 0.5% (8 nGy h$^{-1}$). The annual external effective dose rates varied from 0.78 to 3.86 mSv y$^{-1}$ with a mean ($\pm$S.D.) value of $2.36\pm0.88$ mSv y$^{-1}$, which is quite similar to the HBRA the Chhatrapur beach placer deposit [2].

## 4. Conclusion

On the basis of higher levels of natural radioactivity and gamma absorbed dose rates in air, the newly discovered Erasama beach placer region can be considered as a high natural background radiation area. It can be compared with other high background radiation areas, such as in different coastal regions of India like Chhatrapur in southern parts of coastal Orissa, Ullal in Karnataka, Manavalakkurichy and Kalpakkam in Tamailnadu; coastal tracks of Tamilnadu, coastal areas of Kerala and south west coast of India; Guarapari in Brazil and Yangjiang of China.

## References

[1] UNSCEAR, Sources and effects of ionizing radiation. United Nations Scientific Committee on the Effect of Atomic Radiation, United Nations, New York, 1993, 2000.
[2] A.K. Mohanty, et al., Natural radioactivity and radiation exposure in the high background area at Chhatrapur beach placer deposit of Orissa, India, J. Environ. Radioact. 75 (1) (2004) 15–33.

ELSEVIER

www.ics-elsevier.com

# High radon areas in Norway

Terje Strand*, Camilla Lunder Jensen, Katrine Ånestad,
Line Ruden, Gro Beate Ramberg

*Norwegian Radiation Protection Authority, NRPA, P.O. Box 55, N-1332 Østerås, Norway*

**Abstract.** Norway is considered to be one of the most radon-affected areas in Europe. Based on results of nationwide surveys, the mean radon concentration in Norwegian dwellings has been estimated to be 89 Bq/m$^3$, and 9% of houses have an annual mean radon concentration exceeding the recommended action level of 200 Bq/m$^3$. Very high radon concentrations (50,000 Bq/m$^3$) have been recorded in houses located on highly permeable sediments. In some of these areas, more than 2/3 of the housing stock exceeds the action level. In most of these areas, the external background level and the concentration of radium in geological samples near the surface are moderate or low (less than 100 Bq/kg) and the high levels are due to transport of radon from large volumes of permeable material surrounding the construction. High indoor radon concentrations (above 10,000 Bq/m$^3$) have also been recorded in areas of exposed bedrock with elevated levels of radium and in areas of moderately permeable sediments containing radium rich rock fragments such as alum shale and uranium-rich granites. The identification of high radon areas in Norway is based on extensive municipal surveys with measurements most sparsely populated areas. In a sample, where 2–20% of houses in municipalities were randomly selected, depending on their population density (2% in the most densely populated areas and 20 in the most sparsely populated areas). So far, nearly 200 out of 435 municipalities have carried out radon surveys in accordance with the mapping strategy recommended by NRPA. In this paper, results and characteristics of some of the high radon areas will be briefly discussed. © 2004 Elsevier B.V. All rights reserved.

*Keywords:* Radon; Radon concentrations; Radon surveys

## 1. Introduction

In Norway, indoor radon concentrations are among the highest in Europe. This is partly explained by the geology due to the large occurrences of radium rich soil and bedrock (e.g., alum shale and uranium-rich granites), large occurrences of highly

---

* Corresponding author. Tel.: +47 67162500; fax: +47 67147407.
*E-mail address:* terje.strand@nrpa.no (T. Strand).

0531-5131/ © 2004 Elsevier B.V. All rights reserved.
doi:10.1016/j.ics.2004.10.027

permeable unconsolidated sediments (e.g., moraines and eskers), and the construction of buildings due to the cold climate. An additional factor is the extensive use of highly permeable light expanded clay aggregates in the foundation construction. Entry of radon from the building ground is the dominant source of indoor radon in Norway. However, in households with their water supply from deep drilled wells in uranium rich granites, the household water is sometimes an important contributor to the indoor radon level [1].

Several large-scale surveys have been carried out during the last 20 years, and up to date, approximately 75,000 dwellings have been measured by using etched track detectors over a period of at least 2 months in the heating season. The annual mean concentration in Norwegian dwellings is calculated to be 89 Bq/m$^3$, and 9% and 3% of the dwellings exceed 200 and 400 Bq/m$^3$, respectively [1]. There are significant geographical variations and in some municipalities more than 50% of the dwellings exceeds the action level (200 Bq/m$^3$) compared to <1% in other areas. The highest concentrations have been recorded in dwellings on highly permeable grounds [2].

Areas where more than 20% of the results exceed 200 Bq/m$^3$ are classified as high-risk areas, and in these areas, it is recommended that all dwellings are measured. Between 2000 and 2003, extensive surveys were undertaken in 158 of 435 municipalities (measurements were made in 37,200 dwellings). In 30 of them, more than 20% of theresults exceed 200 Bq/m$^3$ and high-risk areas were identified in 6 out of 10 municipalities.

## 2. High radon areas

High radon areas in Norway are characterised by (1) exposed bedrock with elevated levels of radium, (2) highly permeable unconsolidated sediments derived from all rock types and (3) moderately permeable sediments containing radium-rich rock fragments. The primary source rocks in Norway are the alum shale and uranium-rich granites with levels of radium up to 4000 and 500 Bq/kg, respectively. In typical alum shale areas, more than 50% of the results exceed 200 Bq/m$^3$ and the annual mean concentration of indoor levels as high as 10,000 Bq/m$^3$ has been recorded.

The highest concentrations have been measured in a residential area located on a highly permeable ice-marginal deposit in South Western Norway. In this area, the annual mean concentration of indoor levels as high as 50,000 Bq/m$^3$ has been recorded. In this area, 90% and 30% of 136 homes are estimated to have an annual radon concentrations exceeding 200 Bq/m$^3$ and 3000 Bq/m$^3$, respectively. Measurements by CR-39 etched track detectors were undertaken in three different periods of the year, and in some of the houses, the radon concentration was considerably higher in summer than in winter. Geochemical analysis of bedrock, groundwater and sediments, and comparisons between indoor radon values and soil radon values indicate that the indoor radon concentrations in this area are strongly affected by subterranean airflows caused by temperature differences between soil air and atmospheric air [3]. The airflows concentrate the radon-laden soil air towards the topographic highest part of the deposit in winter and towards the topographic lowest part in summer. Similar areas have been identified in other municipalities in South Eastern Norway.

## References

[1] C.L. Jensen, et al., The Norwegian radon mapping and remediation program, Proceedings of IRPA 11, Madrid, Spain, May 23–28, 2004 2004, paper 6a61.
[2] A.V. Sundal, et al., The influence of geological factors on indoor radon concentrations in Norway, Sci. Total. Environ. 323 (2004) 41–53.
[3] A.V. Sundal, et al., Geological and geochemical factors affecting radon concentrations in dwellings located on permeable glacial sediments—a case study from Kinsarvik, Norway, Environ. Geol. 45 (2004) 843–858.

# Levels of indoor radon, thoron, and their progeny in Himalaya

## R.C. Ramola*

*Department of Physics, HNB Garhwal University, Badshahi Thaul Campus, Tehri Garhwal 249 199, India*

**Abstract.** Levels of radon, thoron, and their progeny were measured in the houses of Garhwal and Kumaun Himalaya using LR-115 plastic track detectors. Radon and thoron concentrations were found to vary from 7 to 191 Bq/m$^3$, and from 1 to 145 Bq/m$^3$, respectively. The equilibrium factor between radon and progeny varies from 0.02 to 0.90, with an average of 0.26 for the region. The resulting dose rate due to radon, thoron, and their decay products varied from 0.04 to 1.89 μSv/h with an average of 0.96 μSv/h. In comparison to the international recommendations, the observed dose rates due to radon, thoron and progeny was found somewhat higher but well below the international recommendations. © 2004 Published by Elsevier B.V.

*Keywords:* Radon; Thoron; Himalaya; Radiation dose

## 1. Introduction

Radon, an intermediate decay product of uranium-238 series, presents all over the Earth in trace amounts. It contributes about half of the total radiation dose to the human environment. Thoron, an isotope of radon, does not contribute much to radiation level due to its short half-life, but is quite important while measuring the total radiation dose. A number of radon and thoron measurements were carried out all over the world in order to find out its health hazard effects to the general population [1,2]. In Himalaya, traditional houses are made of local stone and mud. Some houses are also built over the gneisses formation, which is a rich source of uranium. This paper presents the results of the measurements of radon, thoron, and their progeny in the houses of Garhwal and Kumaun Himalayas under a national radon survey program of India.

---

* Tel.: +91 1376 232856; fax: +91 1376 256056.
 *E-mail address:* rcramola@sancharnet.in.

0531-5131/ © 2004 Published by Elsevier B.V.
doi:10.1016/j.ics.2004.10.021

Table 1
Seasonal variation of radon, thoron, and progeny inside the houses of Garhwal and Kumaun Himalayas

|  | Winter | | Summer | | Rainy | | Autumn | |
| --- | --- | --- | --- | --- | --- | --- | --- | --- |
|  | Min | Max | Min | Max | Min | Max | Min | Max |
| Rn concentration (Bq/m$^3$) | 12 | 191 | 7 | 130 | 13 | 190 | 18 | 131 |
| Tn concentration (Bq/m$^3$) | 0.12 | 145 | 0.69 | 89 | 0.31 | 55 | 0.56 | 108 |
| mWL$_{Rn}$ | 0.04 | 30.9 | 0.69 | 11 | 0.67 | 14 | 0.66 | 24 |
| mWL$_{Tn}$ | 0.03 | 144 | 0.01 | 74 | 0.04 | 66 | 0.03 | 65 |
| Equilibrium factor ($F_{Rn}$) | 0.02 | 0.9 | 0.08 | 0.9 | 0.09 | 0.9 | 0.09 | 0.9 |
| Equilibrium factor ($F_{Tn}$) | 0.01 | 0.9 | 0.01 | 0.9 | 0.01 | 0.9 | 0.01 | 0.9 |
| Dose (µSv/h) | 0.05 | 1.47 | 0.03 | 0.87 | 0.03 | 0.80 | 0.03 | 0.85 |

## 2. Experimental method

Radon, thoron, and their progeny were measured in about 200 houses of Garhwal and Kumaun Himalayas by using LR-115 Type II plastic track detectors. Three small pieces of detector (2.5×2.5) cm were fixed in a twin-chamber radon dosimeter having three different modes, namely, bare, filter, and membrane modes. The bare mode gives the values of radon/thoron concentrations and their progeny, while the filter and membrane modes record the values due to radon/thoron and pure radon gas, respectively. The dosimeters were suspended inside the house at a height of 2.5 m from the floor. The detectors retrieved after 3 months were etched and scanned in the laboratory for track density measurements. The measured track densities for radon and progeny were then converted in radioactive unit by using calibration factors 125 tracks/cm$^2$/day=1 WL and $3.12 \times 10^{-2}$ tracks/cm$^2$/day=1 Bq/m$^3$, respectively [3]. Thoron and progeny were converted in appropriate unit by using the calibration factor obtained by the Environmental Assessment Division, Bhabha Atomic Research Centre, Mumbai, for the use of various Indian radon research groups under the collaborated research program of the Department of Atomic Energy [4].

## 3. Results and discussion

Levels of radon, thoron, and their progeny for different seasons of a year in the houses of Garhwal and Kumaun Himalayas are given in Table 1. Concentrations of radon and its progeny were found to be maximum during winter and minimum during summer. However, no systematic variations were observed for thoron and progeny.

The equilibrium factors for radon and progeny, and thoron and progeny vary from 0.02 to 0.9 (with an average of 0.26) and from 0.01 to 0.9 (with an average of 0.09), respectively. The resulting dose varies from 0.03 to 1.47 µSv/h, with the highest values in winter. Most of the values are found to be higher in winter, which is mainly because of poor ventilation condition in the houses during the winter season. A majority of houses are also mud houses, which also influence the concentrations of radon and thoron in the houses. In conclusion, the levels of radon, thoron, and resulting doses are somewhat higher in the region, but well below international recommendations.

## References

[1] Protection against radon-222 at home and work, ICRP Publication, vol. 65, Pergamon, Oxford, 1993.
[2] UNSCEAR, Sources, Effects and Risk of Ionizing Radiation, United Nations, New York, 1993.
[3] R.C. Ramola, et al., Calibration of LR-115 plastic track detectors for environmental radon measurements, Ind. Built Environ. 5 (1996) 364.
[4] T.V. Ramachandran, Proceedings of the 11th National Symposium on SSNTDs, Amritsar, India, 1998, pp. 50–68.

# Long-term measurements of radon progeny concentrations with LR 115 SSNTDs

K.N. Yu[a,*], D. Nikezic[a], F.M.F. Ng[a], B.M.F. Lau[a], J.K.C. Leung[b]

[a]*Department of Physics and Materials Science, City University of Hong Kong, Tat Chee Avenue, Kowloon Tong, Hong Kong*
[b]*Department of Physics, University of Hong Kong*

**Abstract.** We proposed a method to determine the equilibrium factor using LR 115. © 2004 Elsevier B.V. All rights reserved.

*Keywords:* Radon progeny concentration; Equilibrium factor; LR 115 detector

## 1. Introduction

Methods for long-term monitoring of concentrations of radon progeny or the equilibrium factor are still being explored. The equilibrium factor is defined as $F=0.015F_1+0.515F_2+0.380F_3$ where $F_1$, $F_2$ and $F_3$ are the ratios of the activity concentrations of $^{218}$Po, $^{214}$Pb and $^{214}$Bi(Po) to that of $^{222}$Rn.

## 2. Proxy equilibrium factor measurements with LR 115

The partial sensitivities $\rho_i$ of the bare LR 115 detector to $^{222}$Rn, $^{218}$Po and $^{214}$Po (i.e., the number of tracks per unit area per unit exposure) were studied through Monte Carlo simulations and were found to be the same for $^{222}$Rn, $^{218}$Po and $^{214}$Po. The total track density $\rho$ (track/m$^2$) on the detector is then $\rho=\rho_i(C_0+C_1+C_3)t$, where $C_0$, $C_1$ and $C_3$ are concentrations of $^{222}$Rn, $^{218}$Po and $^{214}$Po (Bq/m$^3$), and $t$ is the exposure time. For known $C_0$, the proxy equilibrium factor $F_p$ ($F_1+F_3=\rho/(\rho_i tC_0)-1$) [1] is found. Equilibrium factors were calculated from the Jacobi room model [2] and plotted with $F_p$ in Fig. 1.

---

* Corresponding author. Tel.: +852 27887812; fax: +852 27887830.
*E-mail address:* peter.yu@cityu.edu.hk (K.N. Yu).

0531-5131/ © 2004 Elsevier B.V. All rights reserved.
doi:10.1016/j.ics.2004.10.003

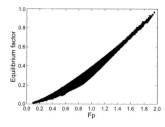

Fig. 1. Dependence of the equilibrium factor $F$ on the proxy equilibrium factor $F_p$ ($F_1+F_3$).

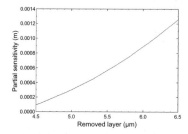

Fig. 2. The relationship between the partial sensitivity of the LR 115 detector from DOSIRAD and the removed active layer thickness.

Table 1
The partial sensitivities and track densities determined for our detectors and the derived values of $F_p$ and $F$ (from LR 115 detectors)

| Detector | Partial sensitivity ($10^{-4}$ m) | Track density ($10^6$ m$^{-2}$) | $F_p$ | Derived $F$ (from LR 115) | Experimental $F$ |
|---|---|---|---|---|---|
| 1 | 1.960 | 0.124 | 1.80 | 0.83–0.88 | 0.76±0.14 |
| 2 | 3.710 | 0.221 | 1.69 | 0.75–0.80 | 0.76±0.14 |
| 3 | 3.108 | 0.171 | 0.81 | 0.18–0.31 | 0.21±0.06 |
| 4 | 3.532 | 0.168 | 0.57 | 0.08–0.18 | 0.21±0.06 |
| 5 | 4.625 | 0.245 | 0.54 | 0.07–0.17 | 0.13±0.04 |
| 6 | 3.572 | 0.182 | 0.09 | 0.00–0.03 | 0.13±0.04 |

The experimental values of $F$ are also given for a comparison.

## 3. Experimental validation

LR 115 detectors were exposed in a radon exposure chamber and then etched in 2.5 N aqueous solution of NaOH at 60 °C under stirring. The detector thickness was measured by surface profilometry [3]. The relationship between $\rho_i$ and the removed layer was derived for LR 115 detectors from DOSIRAD being used and shown in Fig. 2. The partial sensitivities were determined from Fig. 2. The results are shown in Table 1. The derived $F_p$ values (from LR 115) are in agreement with the experimental values.

## Acknowledgment

The present research is supported by the CERG grant CityU1081/01P from the Research Grant Council of Hong Kong.

## References

[1] D. Nikezic, F.M.F. Ng, K.N. Yu, Theoretical basis for long-term measurements of equilibrium factors using LR 115 detectors, Appl. Radiat. and Isotopes 6 (2004) 1431–1435.
[2] W. Jacobi, Activity and potential energy of $^{222}$Rn and $^{220}$Rn daughters in different air atmosphere, Health Phys. 22 (1972) 441–450.
[3] C.W.Y. Yip, et al., Effects of stirring on the bulk etch rate of LR 115 detector, Radiat. Meas. 37 (2003) 197–200.

# Convenient methods for evaluation of indoor thoron progeny concentrations

Weihai Zhuo*, Shinji Tokonami

*Radon Research Group, National Institute of Radiological Sciences, Chiba 263-8555, Japan*

**Abstract.** Four simple methods for estimating the equilibrium-equivalent of thoron concentration indoors are introduced. Although these methods are still difficult in evaluating indoor thoron progeny concentrations accurately, their advantages of simplicity and low cost are preferable to large-scale and long-term surveys of indoor thoron progeny concentrations. © 2004 Elsevier B.V. All rights reserved.

*Keywords:* Thoron; Thoron exhalation rate; Thoron progeny; Equilibrium-equivalent thoron concentration

## 1. Introduction

High concentrations of indoor thoron ($^{220}$Rn) have been reported in some regions in Europe and Asia. As the relatively long half-life (10.64 h) of $^{212}$Pb and high energy (8.78 MeV) of alpha particle emitting from $^{212}$Po, the public exposure to $^{220}$Rn progeny might be significant in those regions. On the other hand, as the uneven distribution of indoor $^{220}$Rn, it is difficult to estimate the equilibrium-equivalent $^{220}$Rn concentration (EETC) from $^{220}$Rn concentrations and an equilibrium factor. Several methods have been developed for measuring $^{220}$Rn progeny. However, convenient and low-cost methods are still few. In this paper, four simple methods of evaluating EETC are introduced. Because of their simplicity and low-cost, they are considered as applicable to the surveys of indoor EETC.

## 2. Materials and methods

The exhalation rate of $^{220}$Rn ($E_T$) from walls is measured with an accumulator and the scintillation cell method [1]. Based on the theoretical study [2], the EETC in Bq·m$^{-3}$ can

---

\* Corresponding author. 4-9-1 Anagawa, Inage-ku, Chiba 263-8555, Japan. Tel.: +81 43 206 3111; fax: +81 43 206 4098.

*E-mail address:* whzhuo@nirs.go.jp (W. Zhuo).

0531-5131/ © 2004 Elsevier B.V. All rights reserved.
doi:10.1016/j.ics.2004.10.001

be derived from $E_T$ in $Bq \cdot m^{-2} \cdot s^{-1}$ with the following formula under several assumptions:

$$EETC = 3.35 E_T S V^{-1}, \tag{1}$$

where $S$ is the surface area of walls from which $^{220}$Rn exhales, and $V$ is the room volume. The main assumptions are: (1) the attached coefficient of free $^{212}$Pb to aerosol particles is 50 h$^{-1}$, (2) the ventilation rate of the room air is 0.5 h$^{-1}$, (3) the deposition rate of attached $^{212}$Pb is 0.2 h$^{-1}$, and (4) outdoor $^{212}$Pb concentration is negligible.

Indoor $^{220}$Rn is measured with the passive monitor [3], and the $E_T$ is estimated from $^{220}$Rn concentrations measured at an appropriate point. Finally, the EETC is evaluated with the formula mentioned above. The best measuring point is at about 50 cm from the source wall. The $E_T$ can be calculated from the $^{220}$Rn concentration at 50 cm ($C_{Tn-50}$) as

$$E_T = 0.0174 C_{Tn-50} \tag{2}$$

Environmental $^{222}$Rn and $^{220}$Rn progeny are simultaneously collected on a filter. After over 6 h of the sampling end, a CR-39 detector is attached on the filter covered with a piece of aluminium foil (15 μm). The registration of alpha particles emitting from $^{212}$Bi and $^{212}$Pb lasted over 24 h. The EETC is calculated by the following equation [4]:

$$EETC = 4.47 N_{net} A / (I_{bc} f), \tag{3}$$

where $N_{net}$ is the net alpha-track density, $A$ is the effective area of filter, $I_{bc}$ is the number of alpha disintegrations of $^{212}$Pb decay registered in Bq·s, and $f$ is the flow rate.

A simple passive monitor using CR-39 as a detector is used to measure the deposition rate of attached $^{212}$Pb [5]. The averaged EETC during the measuring period ($T$) is estimated from the net track density ($N_{net}$) as

$$EETC = 23.7 N_{net} / (v_a T), \tag{4}$$

where $v_a$ is the deposition velocity of attached $^{212}$Pb which was found to be about 0.053 mm·s$^{-1}$ in the general living environment [5].

## 3. Results and discussion

Test of the first method was performed in a model room; the measured EETC (1.1 Bq·m$^{-3}$) agreed well with the estimated value (1.0 Bq·m$^{-3}$). Test of the second method was carried out through a one-year survey in 100 rooms. Averaged $^{220}$Rn concentration measured at 10 cm from the walls was 184 Bq·m$^{-3}$, and the estimated EETC (1.0 Bq·m$^{-3}$) was only about 20% higher than the measured value (0.8 Bq·m$^{-3}$). Tests of the third and fourth methods were performed in 36 cave dwellings in China. The ratio of EETC measured by the two methods was averaged to be 0.95, and the individual value was also confirmed with the commercial monitor (WLx). As the four methods are simple and cost effective, they are considered as applicable to the surveys of indoor EETC.

## References

[1] S. Tokonami, et al., Rev. Sci. Instrum. 73 (2002) 69–72.
[2] A. Katase, et al., Health Phys. 54 (1988) 283–286.
[3] W. Zhuo, et al., Rev. Sci. Instrum. 73 (2002) 2881–2887.
[4] S. Tokonami, et al., Jpn. J. Health Phys. 37 (2002) 59–63.
[5] W. Zhuo, T. Iida, Jpn. J. Health Phys. 35 (2000) 365–370.

# Radon exhalation from a ground surface during a cold snow season

H. Yamazawa[a,*], T. Miyazaki[a], J. Moriizumi[a], T. Iida[a],
S. Takeda[b], S. Nagara[b], K. Sato[b], T. Tokizawa[b]

[a]*Department of Energy Engineering and Science, Nagoya University, Japan*
[b]*Ningyo-Toge Environmental Engineering Center, Japan Nuclear Fuel Cycle Development Institute, Japan*

**Abstract.** Laboratory experiments on the radon diffusion in frozen soils and field measurements of radon exhalation from the snow-covered ground surface were carried out. The experimental results showed that, when soils were frozen, the effective diffusion coefficient decreased by a factor of about 2. The existence of snow cover considerably reduced the radon exhalation to a few percent of that of no-snow seasons. This reduction can be ascribed to wetting of the snow–soil interface. The 1.2-m-deep snow itself reduced the flux density only by 20% to 30%. © 2004 Published by Elsevier B.V.

*Keywords:* Ground surface radon exhalation; Diffusion in snow, frozen soil; Uranium mine disposal soil

## 1. Introduction

This study focuses on effects of a snow cover and freezing of soil water to the radon exhalation from the ground surface. Effective diffusion coefficient of radon in frozen and unfrozen soils and radon flux density over the snow surface were measured.

## 2. Methods

The experiments on radon diffusion in soil were conducted with temperature-controlled three-story cylindrical vessels, with the middle one containing 13-cm-deep sample soils and being separated from the other vessels with mesh screens. The radon concentration in the bottom vessel was kept constant at about 30 to 100 kBq m$^{-3}$. The diffusion coefficient was derived from the temporal changes of the concentration in the upper vessel. Radon flux density over snow surface was measured with an accumulator method. Radon

---

* Corresponding author. Tel.: +81 52 789 5134; fax: +81 52 789 3782.
 *E-mail address:* yamazawa@nucl.nagoya-u.ac.jp (H. Yamazawa).

0531-5131/ © 2004 Published by Elsevier B.V.
doi:10.1016/j.ics.2004.11.153

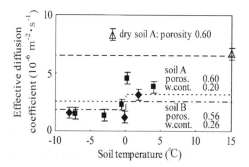

Fig. 1. Experimentally determined effective diffusion coefficients (marks) compared with empirically estimated ones (lines) as a function of porosity and water content.

concentration in snow pore air was measured with a Under Ground Radon Monitoring System [1] during 4-week period from 2nd February 2004.

## 3. Results and discussions

The experimentally determined effective diffusion coefficients are shown in Fig. 1. The coefficient of the moist soils showed clear decrease by a factor of about 2 when the soil temperature decreased below the freezing point while that of the dry soil remained rather constant. The decrease is much larger than expected by the empirical formula by Rogers and Nielson [2] predicts.

The radon flux density at the snow surface was measured to be $9.1 \pm 2.8$ mBq m$^{-2}$ s$^{-1}$ where the depth and average density of snow were 1.2 m and 0.37 g cm$^{-3}$. This value is considerably smaller than that of the no-snow seasons, 360 to 680 mBq m$^{-2}$ s$^{-1}$.

Radon concentrations in snow pore were highly variable depending on the wind conditions and had values between 0.5 to 4.0 kBq m$^{-3}$ at the middle part of the 1.22-m-deep snow. The typical values observed in the 48-h period from the evening of 2nd February 2004 were 1.7 kBq m$^{-3}$ at depth of 0.82 m and 2.2 kBq m$^{-3}$ at 1.07 m.

A one-dimensional numerical radon diffusion model was used to simulate radon transport in snow. The model used the diffusion coefficient which varied with depth depending on the measured snow density profile. With the boundary condition of the snow–soil interface flux density of 11.7 mBq m$^{-2}$ s$^{-1}$, calculated values of the radon exhalation flux density at the snow surface and the concentrations at the above-mentioned depths agreed well with the observation results. The simulation also showed that the snow layer itself reduced the radon exhalation only by 22%. The large reduction of radon exhalation in the snow season can be attributed to the wetting of snow–soil interface due to snow melt water.

## References

[1] S. Yamamoto, et al., Development of a continuous radon concentration monitoring system in underground soil, Trans. Nucl. Sci. 48 (2001) 391–394.
[2] V.C. Rogers, K.K. Nielson, Correlations for predicting air permeabilities and $^{222}$Rn diffusion coefficients of soils, Health Phys. 61 (1991) 225–230.

# The migration of U-238 in the system "soil–plant" and its effect on plant growth

A.A. Kasianenko[a,*], G.A. Kulieva[a], A.N. Ratnikov[b], T.L. Jigareva[b], V.A. Kalchenko[c]

[a]*Ecological Faculty, Russian Peoples' Friendship University, Moscow, Russia*
[b]*Russian Institute of Agricultural Radiology and Agroecology, Obninsk, Russia*
[c]*Scientific research institute of genetics, Moscow, Russia*

**Abstract.** This manuscript estimates the influence of depleted uranium on barley growth and its cytogenetic effect. © 2004 Elsevier B.V. All rights reserved.

*Keywords:* Depleted uranium; Migration; Influence on plants; Cytogenetic effect

## 1. Introduction

The development of nuclear industry, the usage of phosphorus fertilizers in agriculture and utilization of radioactive substances in ammunition lead to anthropogenic pollution of the environment by uranium. Depleted uranium penetrators were used for the first time by the U.S. Army during the Gulf War in Iraq in 1991. For the second time NATO used DU ammunition in Yugoslavia in 1999. About 70% of penetrators turn into aerosol at stroking the armour and DU dust fires. A part of DU oxidizes in $U_3O_8$, $U_3O_7$ and $UO_2$ while exploding. In November 2000 the UNEP commission was directed to Kosovo. A conclusion presented in the report states that the concentration of U-238 in plants was insignificant, and radioactive contamination by U-238 was not harmful to people's health or the environment [1].

Radio nuclides are capable of migrating from soil into plants and then into human organism. The studies on uranium behaviour in the system "soil–plant" are insufficient [2]. Thus, the purpose of this investigation is to study U-238 effect on agricultural plant growth.

---

\* Corresponding author. Tel.: +7 95 438 54 50; fax: +7 95 433 15 11.
*E-mail address:* aakasianenko@tochka.ru (A.A. Kasianenko).

0531-5131/ © 2004 Elsevier B.V. All rights reserved.
doi:10.1016/j.ics.2004.12.006

## 2. Material and methods

In April 2003 we experimented by having a plant growing in vessels of 5 k each. The plant was barley, sort "Elf". The soil was soddy-podzolic light loam. Nutritious substances were brought into the soil in the form of salty solutions ($NH_4$, $NO_3$, $KH_2PO_4$, $K_2SO_4$). The NPK concentration was chosen accordingly for this soil. The experiment was repeated 4 times. The combination of $UO_2(NO_3)_2$ was added to the soil. The additional concentrations of DU in soil were 50, 100, 150 mg/kg. The vegetation period was observed at every stage of the plants' growth: from coming into tubes to ripening. In 2004 the same experiment was again repeated in the same vessels. Spectrometric analysis was made for the determination of uranium concentration in the soil, stems and grains.

Simultaneously, we were engaged in studying uranium influence on plants at genetic level. With this aim we put dry barley grains into a solution of $UO_2(NO_3)_2$. uranium concentration varied from 0 to 200 mg/l.

## 3. Results

As a result, uranium influence on plants' growth in vessels was revealed. The difference in height of stalks was from 5 to 10 sm. The weight of grains recalculated in 1000 pieces was 46.4, 42.8, 41.8 and 39.8 g. In the 2004 experiment uranium concentrations in the soil were 19, 368, 735 and 1042 Bq/kg. The uranium concentrations in grains were 0, 1.28, 2.58, 5.69 Bq/kg. The uranium concentrations in straw were 0, 3.70, 6.69, 14.60 Bq/kg.

It was found that cells' quantity with aberrations had changed from $0.78 \pm 0.28\%$ up to $1.80 \pm 0.35\%$ with the change of uranium concentration in the solution from 0 to 200 mg/l.

## 4. Conclusion

This study has proved that depleted uranium affects plant growth. So, uranium concentration in crops increases with the increase of U-238 in soil. The uranium influence at a genetic level is also confirmed.

## References

[1] Depleted Uranium in Kosovo. Post-conflict environmental assessment. UNEP. 2001. -184 p.
[2] Kasyanenko A.A., Kulieva G.A. Ecological consequences of depleted uranium use in ammunition // Bulletin of Russian Peoples' Friendship University: Series Ecology and Life Safety.-Moscow, Publishing House of RPFU, 2002, #6, p. 90–99.

# $^{210}$Po and $^{210}$Pb content in environmental and human body samples in the Ramsar area, Iran

H. Samavat[a,*], M.R.D. Seaward[b], S.M.R. Aghamiri[c], A. Shabestani Monfared[d]

[a] *Department of Medical Physics, Hamadan University of Medical Sciences, Hamadan, Iran*
[b] *Department of Environmental Sciences, University of Bradford, Bradford, UK*
[c] *Department of Medical Physics, Shahid Beheshti University, Tehran, Iran*
[d] *Department of Medical Physics, Babol University of Medical sciences, Babol, Iran*

**Abstract.** $^{210}$Po and $^{210}$Pb concentrations in air, drinking water and human tissues in a High Level Natural Radiation Area (HLNRA) in Iran have been measured. According to this work, the main intake route of $^{210}$Po and $^{210}$Pb into the human body is via ingestion. Absorption of $^{210}$Po into the blood from the digestive system was 30–40%. The highest concentrations of $^{210}$Po and $^{210}$Pb were absorbed into the skeleton and hair. The mean values in residents and control groups show that the concentrations of $^{210}$Po in Ramsar residents are almost three times higher than those of the control samples. © 2004 Published by Elsevier B.V.

*Keywords:* Radionuclide in human body; Radionuclide in diet; Ramsar; Annual radiation dose

## 1. Introduction

$^{210}$Po and $^{210}$Pb are the main components of natural radioactivity, and α-radiation of $^{210}$Po is an important source of radiation dose to the human body. $^{222}$Rn, the main source of $^{210}$Po and $^{210}$Pb in environment, comes from the earth's crust. The physical half-life of $^{210}$Po is 138.4 days; since the physical half-life $^{210}$Pb is 22 years, the $^{210}$Po content in the environment is replenished by the parent $^{210}$Pb [1]. Hence, the $^{210}$Po content in the environment depends on the content and destiny of $^{210}$Pb.

## 2. Materials and methods

Samples of drinking water, foodstuffs, ground-level air and biological samples of residents in Ramsar area of Iran were collected and bioassayed for their $^{210}$Po and $^{210}$Pb by

---

* Corresponding author. Tel.: +98 9188110828; fax: +98 8118256774.
  *E-mail address:* samavat@yahoo.co.uk (H. Samavat).

0531-5131/ © 2004 Published by Elsevier B.V.
doi:10.1016/j.ics.2004.11.174

Table 1
Concentrations of $^{210}$Po and $^{210}$Pb in ground-level air in the Ramsar area

| Month of sampling | Concentration±S.D. ($\times 10^{-3}$ pCi/m$^3$) | | $^{210}$Po/$^{210}$Pb |
|---|---|---|---|
| | $^{210}$Pb | $^{210}$Po | |
| June | 16.7±1.6 | 1.4±0.2 | 0.09±0.02 |
| October | 9.6±0.9 | 3.2±0.4 | 0.34±0.08 |
| Mean | 13.2±3.5 | 2.3±1.2 | 0.22±0.07 |

the radiochemical methods, α-scintilation and γ-spectroscopy. The biological samples such as human blood, urine, faeces, teeth and hair were collected from residents, and the bone samples were obtained from adult autopsies [2]. Unfortunately, the smoking habits of the subjects were unknown.

## 3. Results and discussion

### 3.1. Air, drinking water and foodstuffs

The results of $^{210}$Po and $^{210}$Pb measurements in air in two different seasons are shown in Table 1; the $^{210}$Po/$^{210}$Pb ratio of 0.22 characterizes the residence time of $^{210}$Pb in air. The mean residence time of $^{210}$Pb in air is 45 days. By assuming the daily inhalation rate of air is 20 m$^3$, the mean inhalations of $^{210}$Po and $^{210}$Pb are 0.23 and 0.05 pCi, respectively. The mean concentrations of $^{210}$Po and $^{210}$Pb in drinking water in the Ramsar area are 0.095 and 0.048 pCi/l, respectively; therefore, the $^{210}$Po/$^{210}$Pb ratio is 0.5. Assuming that the daily consumption of water is 2.2 l [3], the intakes of $^{210}$Po and $^{210}$Pb can be calculated as 0.21 and 0.11 pCi, respectively. The mean daily intake of $^{210}$Pb in adult residents in Ramsar area is 6.5±0.8 pCi, and intake of $^{210}$Po is 4.4±0.8 pCi, giving a $^{210}$Po/$^{210}$Pb ratio of 0.65.

### 3.2. Biological samples

The highest concentrations of $^{210}$Po and $^{210}$Pb were found in bone and hair, being almost 10 times higher than other organs. The concentration in hair is greater than in the skeleton. The $^{210}$Po/$^{210}$Pb ratios in skeleton and hair are 0.6 and 2.3, respectively. Thus, $^{210}$Po is deposited into the hair directly, but not as a result of $^{210}$Pb decay. Hair may serve as a good indicator of the human body burden [2]. Data for other biological material were analysed.

### 3.3. Excretion from the body

The main route of excretion of natural $^{210}$Po and $^{210}$Pb from the human body via faeces, which is 15 times more than urine. The daily total excretion from the body in Ramsar residents was three times higher than the control group. According to the results of different studies [3,4], it can be estimated that the daily absorbance of $^{210}$Po and $^{210}$Pb into the human body is 30–40% of total daily intake.

## References

[1] H. Samavat. High level background radiation areas in Iran. Ph.D. thesis, University of Bradford, UK, 2002.
[2] H. Samavat, M.R.D. Seaward, $^{210}$Po concentration in urine of residents in a high level background radiation area in Iran, International Journal of Low Radiation 1 (3) (2004) 279–284.
[3] M. Sohrabi, N. Alirezazadeh, H. Tajik Ahmadi, A survey of $^{222}$Rn concentrations in domestic water supplies of Iran, Health Physics 75 (4) (1998) 417–421.
[4] R.L. Blanchard, Concentrations of $^{210}$Pb and $^{210}$Po in human soft tissue, Health Physics 13 (1967) 625–632.

# High resolution analyses of temporal variations of airborne radionuclides

K. Komura[a,]*, Y. Yamaguchi[a], N. Muguntha Manikandan[a], Y. Murata[a], M. Inoue[a], T. Iida[b]

[a]Low Level Radioactivity Laboratory, K-INET, Institute of Nature and Environmental Technology, Kanazawa University, Wake, Tatsunokuchi, Ishikawa 923-1224, Japan
[b]Graduate School of Engineering, Nagoya University, Furo, Chikusa, Nagoya 464-8603, Japan

**Abstract.** Temporal variations of $^{210}$Pb, $^{212}$Pb and $^{7}$Be were measured at 1–2 h of intervals simultaneously at Tatsunokuchi, Shishiku Highland (640 m above sea level) and Hegura Island located 50 km from Noto Peninsula. It was recognized that variations of these nuclides occured in a short time depending on meteorological circumstances such as the passage of typhoon, snow fall and even in calm conditions. © 2004 Elsevier B.V. All rights reserved.

*Keywords:* Airborne nuclides; $^{210}$Pb; $^{212}$Pb; $^{7}$Be; Diurnal variation; Low-level gamma spectrometry; Underground laboratory

## 1. Introduction

Concentrations of airborne $^{210}$Pb (half-life: 22.3 years) and $^{7}$Be (53.3 days) are expected to vary hourly or within a shorter time period as observed for $^{222}$Rn [1]; however, no measurements have been performed mainly due to difficulty in measurement because concentrations of $^{210}$Pb and are only in the range of 0.1–10 mBq m$^{-3}$. This problem could be overcome by using ultra low-background Ge detectors in Ogoya Underground

---

* Corresponding author. Tel.: +81 761 51 4440; fax: +81 761 51 5528.
 E-mail address: komura@yu.incl.ne.jp (K. Komura).

0531-5131/ © 2004 Elsevier B.V. All rights reserved.
doi:10.1016/j.ics.2004.11.155

Laboratory (OUL) [2,3]. In this paper, we report examples of high resolution measurements of airborne $^{210}$Pb, $^{212}$Pb and $^{7}$Be.

## 2. Experimental

### 2.1. Sampling points

Airborne radionuclides were collected on a filter paper at Tatsunokuchi, Shishiku Highland (640 m above sea level) and Hegura Island located 50 km from Noto Peninsula (Fig. 1).

### 2.2. Sampling of airborne radionuclides

Airborne nuclides were collected on a silica-fiber filter (Advantec QR-100) by high volume air sampler (Shibata HV-1000 F) [4]. Sampling aimed for the analysis of seasonal variations of $^{210}$Pb and $^{7}$Be was performed at 1–2 days of intervals at LLRL and Shishiku and at 1-week intervals at Hegura. On the other hand, the sampling aimed to investigate diurnal variations of $^{210}$Pb, $^{212}$Pb and $^{7}$Be was made at 1–2 h of intervals at either only at Tatsunokuchi or simultaneously at 2 or 3 points, i.e., Tatsunokuchi–Hegura, Tatsunokuchi–Shishiku and Tatsunokuchi–Shishiku–Hegura.

### 2.3. Measurements

After the sampling, one-half or one-third portion of each filter was taken for gamma spectrometer of $^{210}$Pb, $^{212}$Pb ($T_{1/2}$=10.64 h) and $^{7}$Be. Counting sources of 35 or 21 mm$\phi$ were prepared using 10 tons of hydraulic press and subjected to gamma spectrometry by 11 Ge detectors at OUL. Each sample was measured two times, i.e., short time measurement aimed for $^{212}$Pb and long one for $^{210}$Pb and $^{7}$Be.

## 3. Results and discussion

### 3.1. High resolution analysis of variations of airborne $^{210}$Pb and $^{7}$Be

Main purpose of present study is to investigate temporal variations of $^{210}$Pb and $^{7}$Be at 1–2 h of intervals. More than 10 experiments have been performed under various meteorological conditions. Fig. 2 shows an example performed on November 7–12, 2002. As shown in Fig. 2, both $^{210}$Pb and $^{7}$Be vary in a very short time as expected prior to experiment. However, their variation patterns are different from that of $^{222}$Rn, which

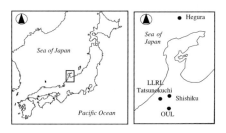

Fig. 1. Locations of monitoring points and OUL.

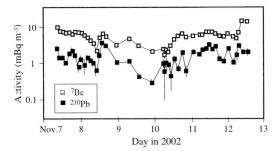

Fig. 2. Variations of airborne $^{210}$Pb and $^{7}$Be at Tatsunokuchi.

shows maximum in the early morning and minimum in the early afternoon. The observation can be explained by vertical and horizontal mixing of air mass caused by sunlight.

### 3.2. Variation of airborne $^{212}$Pb at Tatsunokuchi and Shishiku in winter season

Concentration of short-lived $^{212}$Pb is expected to be low at high altitude, particularly, in winter season because exhalation of 55.6 s $^{220}$Rn from the ground surface is suppressed when the ground surface is frozen due to air temperature goes down to sub-zero levels. Results of experiment on January 10–12, 2004 is shown in Fig. 3, which shows that $^{212}$Pb activity begins to decrease from the late night of January 10 to afternoon of next day, when cold front passed over this area and air temperature at Shishiku went down to −5 °C.

### 3.3. Simultaneous measurement at Tatsunokuchi and Hegura

As described, Hegura Island is very small with only 0.75 km$^2$ of total area. Therefore, concentration of airborne $^{212}$Pb is expected to be lower than at Tatsunokuchi. The $^{212}$Pb at Hegura is composed of the fraction produced from $^{220}$Rn exhaled from the ground surface and the fraction transported from the main island of Japan. As known from Fig. 4, the concentration of $^{212}$Pb at Hegura is 2–10 times lower than at Tatsunokuchi, while the activity levels of both $^{210}$Pb and $^{7}$Be are almost the same. It is noted that the peak activity

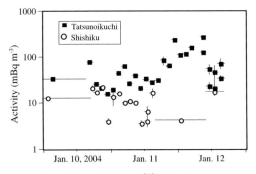

Fig. 3. Variations of airborne $^{212}$Pb in winter season.

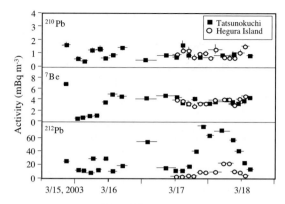

Fig. 4. Variations of airborne $^{210}$Pb, $^{212}$Pb and $^7$Be at Hegura and Tatsunokuchi.

of $^{212}$Pb appeared 4–6 h later than that at Tatsunokuchi, suggesting air-mass transfer from the main island to Hegura.

## References

[1] M. Shimo, KEK Proc. 2003-11, 2003, pp. 1–9.
[2] Y. Hamajima, K. Komura, Applied Radiation and Isotopes 61 (2004) 185–189.
[3] K. Komura, Y. Hamajima, Applied Radiation and Isotopes 61 (2004) 165–169.
[4] Y. Yamaguchi, et al., KEK Proc. 2004-8, 2004, pp. 137–144.

# Naturally accumulated radiation doses and dating of archaeologically burnt materials using luminescence from white minerals

T. Hashimoto[a,*], Y. Nakata[b], T. Yawata[b]

[a]*Department of Chemistry, Faculty of Science, Niigata University, Ikarashi-ninocyo, Niigata 950-2181, Japan*
[b]*Graduated School, Niigata University, Niigata, Japan*

**Abstract.** Radiation-induced luminescence, including thermoluminescence (TL) and optically stimulated luminescence (OSL), was employed to the evaluation of naturally accumulated doses. Quartz grain samples as luminescent mineral were extracted from six parts of a roof tile from Shin-Yakushiji temple in Nara-city, Japan, which has been built about 1300 years ago. All quartz grains revealed the highly sensitive red-TL (RTL) property in addition to moderately OSL-sensitive nature. The naturally accumulated doses from a single aliquot regenerative-dose (SAR) method showed 4.2–3.1 and 2.5–1.1 Gy from the RTL- and OSL-measurements for the middle part, respectively. One of the former results gave an age close to a known age. © 2004 Elsevier B.V. All rights reserved.

*Keywords:* Luminescence dating; Radiation-induced luminescence; Quartz grains; Archaeologically burnt material; Red-TL; Optically stimulated luminescence (OSL); Single-aliquot regenerative-dose (SAR)

## 1. Introduction

The blue-thermoluminescence (BTL) had been widely applied to the evaluation of accumulated radiation-doses and TL-dating before the red TL (RTL)-property had been reported on the quartz grains from volcanic ash layers and archaeologically burnt materials in our laboratory [1,2]. The RTL-property from the quartz grains showed well-linear response and stable long-life preferable for dating over a period of 1 Ma [2]. Subsequently, the causes of RTL have been found to attribute to the rapidly cooling process after thermal heating beyond 867 °C (phase inversion temperature of β-quartz/tridymite), together with the Al-impurity content over 100 ppm. Recently, an optically stimulated luminescence

---

* Corresponding author. Tel.: +81 25 262 6169; fax: +81 25 262 6116.
  *E-mail address:* thashi@curie.sc.niigata-u.ac.jp (T. Hashimoto).

0531-5131/ © 2004 Elsevier B.V. All rights reserved.
doi:10.1016/j.ics.2004.10.013

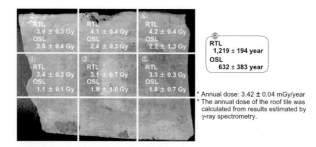

Fig. 1. Naturally accumulated doses estimated from RTL and OSL of quartz aliquots and their age-results (Shin-Yakushiji temple roof tile).

(OSL) of quartz grains has been developed using a single-aliquot regenerative-dose (SAR) technique for the evaluation of paleo-doses or naturally accumulated doses.

In this paper, the naturally accumulated doses and the age-results were compared between RTL- and OSL-methods using quartz aliquots from an archaeological roof tile.

## 2. Experimental

A roof tile from Shin-Yakushiji temple in Nara-city, Japan, which has been built about 1300 years ago, was cut into six parts according to its differently overlapping patterns. Quartz grain samples as luminescent mineral were extracted from each tile piece. On the TL-color images (TLCIs), all quartz grains revealed the highly sensitive RTL-property in addition to moderately OSL-sensitive nature. The luminescence measurements of purified quartz grains were carried out by means of a new automated TL/OSL-reader system, which has been developed initially for both the SAR-method of RTL and OSL.

## 3. Results and discussion

As expected from the production process of roof tile, all quartz grains have proved RTL-property because of the heating treatment beyond 1000 °C using an ancient kiln. The RTL-glow curves always showed a single peak at higher temperature side. The naturally accumulated doses were evaluated by interpolating natural RTL-intensities into the dose/RTL response curve, which were obtained by correcting sensitivity changes. In the OSL-analysis, the procedures similar to the description in a preceding paper was employed for the determination of accumulated doses [2]. All experimental results are summarized in Fig. 1. The RTL-results gave the natural doses amounted to 4.2–2.9, whereas the OSL-doses offers relatively lower ones, giving 2.7–0.7 Gy, respectively. The most preferable age was obtained from the highest naturally accumulated dose using annual dose, estimated from radioactivity analysis of roof tile powder.

## References

[1] T. Hashimoto, K. Yokosaka, H. Habuki, Emission properties of thermoluminescence from natural quartz—blue and red TL response to absorbed dose, Nucl. Tracks Radiat. Meas. 13 (1987) 57–66.
[2] T. Hashimoto, T. Yawata, M. Takano, Preferential use of red-thermoluminescence (RTL)-dating for quartz extracts from archaeologically burnt pottery-comparison of RTL and BTL (blue-TL) measurements using single-aliquot regenerative-dose (SAR) method, Ancient TL 21 (2003) 85–91.

# Case-control study of residential radon and childhood leukemia in Japan: results from preliminary analyses

S. Yoshinaga[a,*], S. Tokonami[a], S. Akiba[b], H. Nitta[c], M. Kabuto[c]

JCCSG[1]

[a]*National Institute of Radiological Sciences, Chiba, Japan*
[b]*Kagoshima University, Kagoshima, Japan*
[c]*National Institute for Environmental Studies, Tsukuba, Japan*

**Abstract.** There is no convincing evidence that exposure to radon increases cancer risks other than lung cancer. We conducted a case-control study of residential radon exposure and childhood leukemia. A total of 255 incident cases of acute lymphoblastic and myeloid leukemia that were diagnosed at age <15, and 248 controls were included in analyses. Relative risks for radon level of 20–49, 50–99, and 100 Bq/m$^3$ or more were 1.00 (95% CI, 0.62 to 1.62), 1.57 (95% CI, 0.47 to 5.22), and 2.05 (95% CI, 0.18 to 23.4), compared to less than 20 Bq/m$^3$ ($p$ for trend: 0.52). The present study provides no evidence for an association between residential radon exposure and childhood leukemia. © 2004 Elsevier B.V. All rights reserved.

*Keywords:* Case-control study; Radon; Childhood leukemia; Epidemiology

## 1. Introduction

There is convincing evidence that exposure to radon increases the risk of lung cancer. However, the carcinogenic effects of radon on other types of malignant tumors have not been well studied. Several ecological studies showed a significant positive association between residential radon and childhood leukemia, but results were inconsistent [1]. We

---

* Corresponding author. 4-9-1, Anagawa, Inage-ku, Chiba, 263-8555, Japan. Tel.: +81 43 206 3108; fax: +81 43 251 6089.
  *E-mail address:* yosinaga@nirs.go.jp (S. Yoshinaga).
[1] Japan Childhood Cancer Study Group.

0531-5131/ © 2004 Elsevier B.V. All rights reserved.
doi:10.1016/j.ics.2004.09.050

conducted a case-control study of residential radon and childhood leukemia, as a part of nationwide study of childhood cancer in Japan.

## 2. Methods

A total of 791 incident cases of acute lymphoblastic leukemia (ALL) and acute myeloid leukemia (AML) that were diagnosed at age <15 between 1999 and 2002 were enrolled in the study. Out of the 791, 312 cases were eligible in the study. One to three controls for each case were selected with matching of sex, age, and population size of their addresses, resulting 603 controls. Radon measurement was carried out for 255 cases and 248 controls in bedrooms at their houses for 6 months using track-etch detectors. Information on potential confounding factors, including parents' smoking, maternal medical X-ray exposure during pregnancy, were obtained by interview survey. In addition, magnetic field and gamma ray radiation were measured in each subject's home.

Subjects were classified into four categories, i.e.,<20, 20–49, 50–99, 100+ $Bq/m^3$ for radon concentration. Another categorization based on quartiles was also done for radon concentration. A conditional logistic model was used to estimate relative risks (RRs) of leukemia according to radon concentration, adjusted for potential confounders.

## 3. Results and discussion

Mean radon concentrations were similar between cases and controls (17 and 18 $Bq/m^3$, respectively). Adjusted RRs of ALL and AML combined were 1.00 (95% confidence interval, 0.62 to 1.62), 1.57 (0.47 to 5.22), and 2.05 (0.18 to 23.4) for radon concentrations of 20–49, 50–99, and 100 $Bq/m^3$ or more compared to less than 20 $Bq/m^3$ (Table 1). This trend was not statistically significant ($p$ for trend: 0.52). Separate analyses of ALL or AML also showed no significant trend.

There have been limited number of case-control studies on residential radon and childhood leukemia. Our results are consistent with those from the previous studies in that there was no significant positive association. The radon level in our study was, however, much lower than the other studies [2,3] despite the comparable number of cases, which could limit the statistical power.

Table 1
Number of subjects, relative risk (RR), and 95% confidence interval (CI) for acute lymphoblastic and myeloid leukemia according to radon concentration

| Radon concentration ($Bq/m^3$) | No. of cases[a] | No. of controls[a] | Crude RR (95% CI) | Adjusted RR[b] (95% CI) |
|---|---|---|---|---|
| <20 (reference) | 166 | 170 | 1.00 | 1.00 |
| 20–49 | 51 | 55 | 0.94 (0.60–1.50) | 1.00 (0.62–1.62) |
| 50–99 | 8 | 5 | 1.67 (0.54–5.15) | 1.57 (0.47–5.22) |
| 100+ | 2 | 1 | 2.00 (0.18–22.1) | 2.05 (0.18–23.4) |
| Total | 227 | 231 | $p$ for trend=0.57 | $p$ for trend=0.52 |

[a] Twenty-eight cases and 17 controls were excluded from the matched analysis.
[b] Adjusted for magnetic field level in subject's bedroom, maternal education background, and maternal medical X-ray exposure during pregnancy.

## 4. Conclusion

The present study provides no evidence for an association between residential radon exposure and childhood leukemia.

## References

[1] D. Laurier, M. Valenty, M. Tirmarche, Radon exposure and the risk of leukemia: a review of epidemiological studies, Health Phys. 81 (2001) 272–288.
[2] J.H. Lubin, et al., Case-control study of childhood acute lymphoblastic leukemia and residential radon exposure, J. Natl. Cancer Inst. 90 (1998) 294–300.
[3] M. Steinbuch, et al., Indoor residential radon exposure and risk of childhood acute myeloid leukaemia, Br. J. Cancer 81 (1999) 900–906.

# The risk of lung cancer in HBR area in India—a case control study

V.S. Binu[a,*], P. Gangadharan[a], P. Jayalekshmi[a], R.R.K. Nair[a], M.K. Nair[a], B. Rajan[a], S. Akiba[b]

[a]*Regional Cancer Centre, Trivandrum, Karunagappally Cancer Registry, Puthenthura P.O., Neendakara, Kollam, Kerala 691588, India*
[b]*Faculty of Medicine, Kagoshima University, Kagoshima, Japan*

**Abstract.** Karunagappally in Kerala, India is known for the radiation-emitting sands, more than 100,000 people are exposed to higher levels of natural radiation. The risk of lung cancer with exposure to natural radiation is investigated in this study. An increasing nonsignificant trend of risk with radiation levels was obtained. © 2004 Published by Elsevier B.V.

*Keywords:* Lung cancer; Environmental radiation; Odds Ratio (OR)

## 1. Introduction

Lung cancer is the leading cancer among males in almost all developed as well as developing countries. The coastal areas of Kollam district of Kerala, India provides a unique opportunity to conduct studies to investigate the effect of chronic exposure to low-level ionizing radiation on human beings. In this paper, a preliminary analysis was attempted to investigate the risk of lung cancer among males in relation to exposure to the external radiation present in Karunagappally Taluk (administrative unit), a coastal area in Kollam district of Kerala, India.

## 2. Material and methods

A complete enumeration on sociodemographic and life style factors of the people residing in the study area along with radiation level measurements were undertaken by the population-based cancer registry functioning in this area from 1990. In the survey, 54,544 males were above 34 years of age, and this constitutes the study population. Cases were incident males with lung cancer reported by the registry during 1991–2001 in the study

---

\* Corresponding author. Tel.: +91 476 2685203; fax: +91 471 2447454.
*E-mail address:* qln_nbrrkply@sancharnet.in (V.S. Binu).

0531-5131/ © 2004 Published by Elsevier B.V.
doi:10.1016/j.ics.2004.11.156

Table 1
Distribution and Odds Ratio (OR) and corresponding 95% Confidence interval (CI) for cancer of lung by selected characteristics

| Characteristics | | Cases (%) | Controls (%) | OR | 95% CI |
|---|---|---|---|---|---|
| Age (year) | 30–39 | 3 (1.5) | 125 (20.3) | 1 | |
| | 40–49 | 17 (8.3) | 212 (34.5) | 3.3 | 0.9–11.5 |
| | 50–59 | 58 (28.3) | 122 (19.8) | 16.3 | 4.9–54.1 |
| | 60–69 | 79 (38.5) | 94 (15.3) | 29.1 | 8.7–97.0 |
| | ≥70 | 48 (23.4) | 62 (10.1) | 27.1 | 7.9–92.7 |
| Religion | Hindu | 135 (65.9) | 442 (71.9) | 1 | |
| | Muslim | 55 (26.8) | 108 (17.6) | 2.0 | 1.3–3.1 |
| | Christian | 15 (7.3) | 65 (10.6) | 0.8 | 0.4–1.5 |
| Smoking | Nonsmokers | 15 (7.3) | 157 (25.5) | 1 | |
| | Ever smokers | 190 (92.7) | 458 (74.5) | 3.2 | 1.7–5.8 |
| Outside radiation (mGy/Yr) | <1 mGy | 25 (12.2) | 89 (14.5) | 1 | |
| | 1≥mGy<2 | 49 (23.9) | 205 (33.3) | 0.9 | 0.5–1.6 |
| | 2≥mGy<5 | 90 (43.9) | 222 (36.1) | 1.5 | 0.9–2.7 |
| | 5≥mGy<10 | 27 (13.2) | 63 (10.2) | 1.6 | 0.8–3.3 |
| | ≤10 mGy | 14 (6.8) | 36 (5.8) | 2.3 | 0.9–5.7 |

Similar results were obtained for microscopically confirmed primary cases also.

population, and there were only 205 such cases. Three controls per case were randomly selected using systematic sampling technique from the study population who were free of any cancer. Among 205 male cases, 105 (51%) were microscopically confirmed primary lung cancer cases. The analysis was done on these 205 and 105 cases separately. Odds Ratio and 95% confidence interval (CI) were computed using unconditional multiple logistic regression models which include age, education, religion, smoking habit and outside house radiation measured in milliGray per year (mGy/Yr) using the GMBO software in Epicure.

## 3. Results

Table 1 gives the distribution, Odds Ratio (OR) and 95% confidence interval (CI) for all lung cancer cases by age, religion, smoking habit and outside house radiation measured in milliGray per year (mGy/Yr). The OR for all age groups (except ≥70) shows an increasing trend with age ($p$ value<0.001). An increasing nonsignificant trend of lung cancer risk with different radiation levels was obtained ($p$ value for trend=0.154).

## 4. Conclusion

The investigated population has been living in the natural background radiation areas of Karunagappally Taluk for many decades, and the lack of any significant increase in the incidence of lung cancer either indicates (a) the existence of no association with the radiation levels present in this area or (b) very low levels of elevated or reduced risk, which were not detectable by means of simple analysis applied and due to the small number of cases under study or (c) other factors, like diet, which may confound or compete with the effects of radiation exposure. Further in-depth studies are thus necessary to investigate [1] the interaction of risk factors acting together with radiation [2] risk in relation to cumulative dose and [3] the risk among Muslims.

## References

[1] Sources and effects of ionizing radiation—UNSCEAR 1994 report.
[2] L.X. Wei, High background radiation area—an important source of exploring the health effects of low dose ionizing radiation, in: L.X. Wei (Ed.), High Levels of Natural Radiation 1996: Radiation Dose and Health Effects, Elsevier, 1997.
[3] P. Gangadharan, et al., Natural background radiation cancer registry, Technical report 1990–1999.

# Improvement of the radon situation at former uranium mining and milling sites in East Germany

P. Schmidt*, J. Regner

*WISMUT GmbH, Department of Environmental Monitoring and Radiological Protection, Jagdschänkenstrasse 29, D-09117 Chemnitz, Germany*

**Abstract.** Uranium mining in East Germany was accompanied by piling up of more than 300 Mio $m^3$ of radioactive waste rock material, often adjacent to settlements. This resulted at some places in mean outdoor radon concentrations of 1000 $Bq/m^3$ and more. In 1991, the national corporation WISMUT was charged with rehabilitation of the sites. For waste rock piles, WISMUT developed the concept of in situ remediation involving reshaping and covering of the piles. The optimised 1 m thick cover allows to reduce the radon exhalation rate by a factor greater than 10. Reshaping and covering of the piles guarantees that the remediation target of 1 mSv of annual effective dose for the local population will be met. © 2004 Published by Elsevier B.V.

*Keywords:* Uranium mining; Radon concentration; Remediation; Effective dose

## 1. Introduction

From 1946 to 1990, the Soviet-German WISMUT produced 231,000 tons of uranium and became the world's third largest in cumulative uranium production during that time. The mining was accompanied by piling up of more than 300 million cubic metres of radioactive waste rock material, often adjacent to settlements. This resulted (and still continues to result) at some places in mean annual outdoor radon concentrations locally in excess of 1000 $Bq/m^3$ equivalent to an annual effective dose >20 mSv for persons living permanently at the sites.

In 1991, the national corporation WISMUT was charged with the decommissioning of the uranium production facilities and with the rehabilitation of the sites. Since then, the WISMUT Rehabilitation Project has emerged as an international reference project for state-of-art remediation of uranium mining sites. Improvement of the radon situation at former mining sites is one of the central tasks within the project.

---

\* Corresponding author. Tel.: +49 371 8120 176; fax: +49 371 8120 107.
*E-mail address:* p.schmidt@wismut.de (P. Schmidt).

0531-5131/ © 2004 Published by Elsevier B.V.
doi:10.1016/j.ics.2004.09.041

## 2. The Schlema-Alberoda radon case study

Compared to other sites of former uranium mining and milling in East Germany, the highest radon concentrations are found in Schlema-Alberoda, in Saxony. At this site, stockpiling of subgrade ore and waste rock having mean specific activities of Ra-226 from 0.4 to 2.0 Bq/g resulted in 20 big waste rock piles. The rock material was mainly piled up at slopes around the village which itself is in a valley location. At the end of uranium mining, the piles contained more than 47 million cubic metres of waste rock and covered a surface area of 342 ha.

Since relocation of the waste rock at Schlema-Alberoda was not feasible due to technical and financial reasons, WISMUT developed the concept of in situ, i.e. in-place remediation involving reshaping and covering of the piles. In order to optimise the cover construction, soil air transport processes in piles and cover layers were investigated (radon exhalation rate and gas permeability measurements, tracer gas experiments). It was found that radon exhalation is dominated mainly by convection and not by radon diffusion. The convection process is driven by temperature differences between the soil air within the pile and the outside air, and the resulting pressure differences. As a consequence, the pattern of radon exhalation rate as well as of the radon concentration free in air show daily and seasonal fluctuations.

An optimised cover construction was developed to be built of inert material of 1 m thickness. The cover allows to reduce the radon exhalation rate by a factor greater than 10. The effect of placement of a cover on a waste rock pile adjacent to a residential building is shown in Fig. 1.

## 3. Results and conclusion

In recent years, the radon situation at WISMUT sites has continuously improved. This comes in particular true for the Schlema-Alberoda site, where the exposure of the local public is dominated by inhalation of radon and its decay products. Reshaping of and placement of covers of optimised thickness on waste rock piles have resulted in low radon concentrations so that the remediation target of 1 mSv/year as reference level for the residual induced effective dose above the background is met.

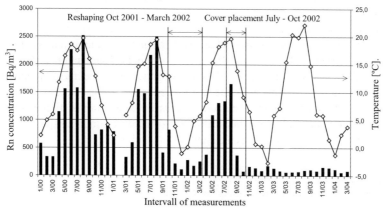

Fig. 1. Development of the radon concentration free in air at the toe of a waste rock pile before and after placement of a cover.

# Investigation and reduction of personnel exposure levels in Bavarian water supply facilities

S. Körner*, M. Trautmannsheimer, K. Hübel

*Bavarian Environmental Protection Agency, D-86177 Augsburg, Germany*

**Abstract.** Since 2001, the protection against natural radiation is implemented in German law. The limit of the annual effective dose is 20 mSv. But the annual effective dose due to radon exposure is only limited at certain work places, e.g., mines and water supply facilities. The federal states are responsible for the observation of this law. Consequently, all Bavarian water supply facilities were investigated with regard to radon concentrations in indoor air as well as radon exposure of the staff working in these buildings. In 0.7% of all water supply facilities, the staff were exposed to levels that exceed the limit of 20 mSv/a. In 1.7% of the water supply facilities, the personnel exposure level was between 6 and 20 mSv/a. There are several remedial actions that have proven to be successful: it is inexpensive to reduce the radon exposure by reducing the time spent in the units; and the buildings can be ventilated with stationary or mobile devices by blowing fresh air directly to the working place. In one case, a combination of an effective ventilation system with a separating wall between the water purification basins and the water supply control centre reduced the annual effective dose of the process controller from 100 to about 6 mSv. © 2004 Elsevier B.V. All rights reserved.

*Keywords:* Radon; Working place; Track-etch detector; Effective dose; Remediation measure

## 1. Introduction

The protection of the health of workers and the general public against the dangers arising from natural radiation entered the German radiation protection ordinance [1] in 2001. The limit of the annual effective dose is 20 mSv. Therefore, the European Council directive 96/29/Euratom [2] was implemented in German law. This directive was based on awareness that high exposure to radon and its progenies through inhalation can cause lung cancer. High radon concentrations that can lead to increased radon exposure have also

---

\* Corresponding author. Bayerisches Landesamt für Umweltschutz, D-86177 Augsburg, Germany. Tel.: +49 821 9071 5334; fax: +49 821 9071 5554.

*E-mail address:* simone.koerner@lfu.bayern.de (S. Körner).

0531-5131/ © 2004 Elsevier B.V. All rights reserved.
doi:10.1016/j.ics.2004.11.071

been measured at other work places, such as visitor caves and mines, radon spas, and water supply facilities [3].

## 2. Radon mearurements and results

In Germany, the federal states, such as Bavaria, are responsible for the observation of the radiation protection ordinance [2]. Consequently, all 2550 Bavarian water supply facilities were investigated with regard to radon concentrations in indoor air as well as the radon exposure of the staff working in these buildings.

To estimate the radon exposure level, the processing plant workers had to wear a personal track-etch detector for 3 months. When not in use, the personal detector was stored near a reference detector at a place with low radon concentration. To obtain the mean room concentration, track-etch detectors were exposed for a period of 2 weeks in mainly reservoirs and purification units.

The exposure levels of 2000 personnel and the indoor concentrations of 5000 rooms were measured. Personnel exposure levels of up to 400 mSv/a and room concentrations of up to 1000 kBq/$m^3$ were found. In 0.7% of all water supply facilities, the exposure level of the staff exceeded the limit of 20 mSv/a. In 1.7% of all water supply facilities, the exposure level of the staff was between 6 and 20 mSv/a. In both cases, the exposure level of the staff is constantly monitored and remediation measures are in progress.

## 3. Reduction of personnel exposure levels

It is inexpensive to reduce the radon exposure by reducing the time spent in the units. To achieve this raised awareness amongst staff, information about radon and its characteristics was distributed.

Another very effective way to reduce the radon concentration inside buildings is to blow fresh air directly into the working place of the staff with stationary or mobile devices. In one specific water purification building, active ventilation reduced the indoor radon concentration from an average of 20 kBq/$m^3$ (without ventilation) to about 1 kBq/$m^3$. This method can also be used for the annual cleaning of the basins and reservoirs as well as actions in badly ventilated pits. In one reservoir, the radon concentration was reduced from 400 to 8 kBq/$m^3$ during cleaning. It is also possible to install an effective ventilation system and a separating wall between the water purification basins and the control centre. In one specific case, these measures reduced the annual effective dose of the process controller from 100 to about 6 mSv.

An alternative way to reduce the indoor air concentration is to avoid any transfer of radon polluted air exhausted from purification tanks to indoor air. In one case, the indoor air concentration in the purification building was reduced from 4.6 to 0.5 kBq/$m^3$.

## References

[1] Verordnung über den Schutz vor Schäden durch ionisierende Strahlung vom 20. Juli 2001, Bundesgesetzblatteil 1 (2001) 1713–1848.
[2] Council Directive 96/29/Euratom, Basic safety standards for the protection of the health of workers and the general public against the dangers arising from ionizing radiation, Off. J. 159 (1996) 1–114.
[3] M. Trautmannsheimer, W. Schindlmeier, K. Börner, Radon concentration measurements and personnel exposure levels in Bavarian water supply facilities, Health Phys. 84 (2003) 100–110.

www.ics-elsevier.com

# The ENVIRAD project: a way to control and to teach how to protect from high indoor radon level

A.M. Esposito[a,b,*], M. Ambrosio[b], E. Balzano[c], L. Gialanella[b], M. Pugliese[b,c], V. Roca[b,c], M. Romano[b,c], C. Sabbarese[a,b], G. Venoso[b,c]

[a]*Dipartimento di Scienze Ambientali, Seconda Università di Napoli, Caserta Italy*
[b]*Istituto Nazionale di Fisica Nucleare, Napoli, Italy*
[c]*Dipartimento di Scienze Fisiche, Università Federico II, Napoli, Italy*

**Abstract.** The environmental radioactivity (ENVIRAD) project involves students of Italian secondary schools with the aim to make them acquainted with the radon measurement techniques. The first step of the project has consisted of the introduction of the participants into some themes not treated in usual curricula to supply the student with the cultural skill to carry out survey about radon concentrations in the school and eventually in other buildings. The second ENVIRAD's part is the creation of a network for the continuous on-line measurements of radon exhalation from soil for geophysical applications. The feasibility of the project has been verified by 1-year work with a pilot school during the season 2002/2003. In the following year, 18 schools will join the project. © 2004 Published by Elsevier B.V.

*Keywords:* Radon; Environmental radioactivity; School; Network

## 1. Introduction

The Campania is a volcanic region whose territory is characterised by the high indoor radon concentrations [1,2]. In spite of a mean value of radon indoor concentration of 97 Bqm$^{-3}$, several rooms were found to have concentrations higher than 1000 Bqm$^{-3}$ [3]. These values could imply health problems. The possible mitigation of these problems can not rely only a technical/medical intervention, but they have to be accompanied by an

---

* Corresponding author. Dipartimento di Scienze Ambientali, Seconda Università di Napoli, Caserta Italy. Tel.: +39 823 055946; fax: +39 823 274605.
   *E-mail address:* aesposit@na.infn.it (A.M. Esposito).

0531-5131/ © 2004 Published by Elsevier B.V.
doi:10.1016/j.ics.2004.11.084

educational program started in the school training. To reach this goal, the involvement of teachers and students is mandatory; this is, the environmental radioactivity (ENVIRAD) mission.

## 2. The layout of the ENVIRAD project

The ENVIRAD Project, financially supported by the Istituto Nazionale di Fisica Nucleare (INFN), is a cooperation between university and secondary schools [4]. The project started at the end of 2002 and has a time span of 3 years. The 1st year was spent in collaboration with a pilot school to set-up and optimize the working methodology to propose and realise our objective with the young students in a context where, usually, the proposed theme is not treated. After this general training period, a 2-year working program with other schools has started. In the 1st year, participants learn the fundamental concepts of the radioactivity and the properties of the ionising radiations, the characteristics of the radon, the motivation of the monitoring and measurement techniques. To better understand the last topics, they performed a lot of laboratory work and carried out preliminary measurements, having the aim to take confidence with experimental techniques and with the behaviour of radon. This objective was reached with the design and the realization of series of 1/2 weeks of measurements and the discussion of the results. Charcoal canisters and E-perm detectors have been used to perform fast measurement series in various conditions; this has allowed students to gain expertise. After that, students were able to sketch the long-term measurements (organised on two consecutive semesters) suited to estimate the annual mean effective dose. LR115 detectors have been used with an integration time of 6 months; thus, the first results will be available in October 2004. During the 2nd year, besides the analysis and the discussion of data of the long-term survey, the second purpose of the experiment will be developed: the monitoring of the radon levels in soil using the RaMonA equipment. It is a system based on a silicon $\alpha$-detector coupled with an electrostatic collection chamber for radon detection and on three sensors to control the environmental parameters [5]. It will be used for continuous measurement of radon exhalation and environmental parameters and for the real time transmission of data to a database. Specimens of the system will be installed in the participant schools. They will be connected via Ethernet in an online network, and they will be managed completely by the students. In this year, for each school, new groups of students will begin the same activity to start a virtuous cycle that will guarantee the activity in the following years; teachers and "expert" students will organise the training sessions for the new groups. At the end of the 2nd year, a network database will be available, containing both indoor and in-soil data.

## 3. Preliminary results and conclusion

Eighteen schools from Campania region have accomplished the first step of the project. Satisfying results in terms of understanding of the scientific approach have been obtained especially on a selected student's group for each school. Although the preliminary measurements with typical short-term detectors were used for didactic purposes, they allowed to get useful information about the school environment. In particular, the radon level in the computer and laboratory rooms was higher than 500 $Bqm^{-3}$ in the 12% of rooms, while no classrooms showed radon levels as high as 500 Bqm$^{-3}$, and 3% of them had concentrations higher than 400 Bqm$^{-3}$. These results will be compared with the long-term measurements. In conclusion, this project has posed the bases for the realization of a network that will allow to map the environmental radioactivity in Campania region and, more importantly, that will contribute to diffusion of a more complete scientific culture about radioactivity.

## References

[1] F. Bochicchio, et al., Health Phys. 71 (5) (1996) 741–748.
[2] C. Sabbarese, et al., Radiat. Prot. Dosim. 48 (3) (1993) 257–262.
[3] C. Sabbarese, et al., 5th Int. Conf. on High Levels of Natural Radiation and Radon Areas, 4–7 September, Munich, Germany, 2000.
[4] M. Pugliese, First Int. Meeting on Appl. Ph. Badjoz, Spain, Oct. 2003.
[5] V. Roca, et al., Appl. Radiat. Isotopes 61 (2–3) (2004) 243–247.

# Radiation hazard and protection for the nuclear weapon terrorism

Jun Takada*

Medical School, Sapporo Medical University, Sapporo, Japan

**Abstract.** A radiation hazard is predicted in case of nuclear weapon terrorism on the surface. An effective radiation protection, in which 70% of victims may survive, is proposed. © 2004 Elsevier B.V. All rights reserved.

*Keywords:* Nuclear weapon terrorism; Dosimetry; Radiation protection; Semipalatinsk test site

## 1. Introduction

The possibility of nuclear weapon attack by terrorists (NWAT) occurring after September 11th 2001 may be considered as the most dangerous radiation hazard in the 21st century [1]. It is difficult for us to believe that the risk of nuclear attack, unrelated to war, may occur in a city of a developed nation. Emergency measures for radiation protection are one of the main topics regarding nuclear terrorism.

For a 3-year period, we especially focused on the assessment of hazard and radiation protection in case of a terrorist attack with nuclear weapons. We studied field investigations of ground zero (GZ) of small weapons in Semipalatinsk Nuclear Test Site (SNTS). Here, we report dose fore-assessment and propose a method of radiation protection in case of 1 kt NWAT in a city.

## 2. Method

Scientific information of the present report is based on our field research of nuclear weapon testing [1] in and out of ground zero and on the US reports on the effects of nuclear explosion [2]. We have radiation data on GZ of a small nuclear weapon which was exploded on the surface in SNTS. This provided us with information on residual radiation

---

\* Tel./fax: +81 11 644 2001.
 *E-mail address:* juntakada@sapmed.ac.jp.

0531-5131/ © 2004 Elsevier B.V. All rights reserved.
doi:10.1016/j.ics.2004.11.089

hazards after the NWAT. Dose evaluation due to fallout was carried out in case of outdoors with no protection and in case of indoor sheltering and then evacuation with subway by using the USA scheme and certain parameters [2].

The small crater size in SNTS suggests the yield of NWAT less than 1 kt equivalent of TNT. In our evaluation of disasters of NWAT, we deal with a 1 kt weapon.

## 3. Result and discussion

A nuclear explosion of 1 kt yield instantly becomes a fire ball with a diameter of about 70 m with a temperature of several million degrees by a nuclear fission chain reaction of plutonium. It occurs in a shock wave, a bomb blast and radiation such as heat radiation, gamma ray and neutrons at the same time. This may happen in a building such as a hotel in case of NWAT.

A range of 500 m radius from GZ will be affected severely by shock wave. People in the 260 m zone will receive the third thermal injury. The initial nuclear radiation within a minute from the fire ball extends wider than heat radiation. People near ground zero in the 800 m zone will receive dose of level-A (lethal more than 50%).[1]

A huge amount of radioactive concrete particles containing fission products becomes a pillar several kilometers high above GZ. Consequently, the most dangerous impact for the population will be radioactive fallout in the case of NWAT. Areas of level-A, level-B and level-C may extend to 2, 4 and 12 km, respectively, in case of an hour outdoor stay in the fallout area. Therefore, this fallout will produce a very severe nuclear disaster and severe radiation hazard will remain in the lee area.

Evacuation by subway following indoor sheltering as a protection method is very effective. Seventy percent of victims potentially with level-A radiation will be saved in this manner. Short-term nuclear hazard will remain at ground zero and cause late recovery work after several months or a year. Some measure should be carried out for long-term nuclear hazard such as Cs-137, Sr-90 as well as plutonium.

## References

[1] Jun Takada, Nuclear Weapon Attack in Tokyo by Terrorists, Kodansha Tokyo, 2004 (in Japanese).
[2] S. Glasstone, P.J. Dolan, The Effects of Nuclear Weapons, United States Department of Defense and the Energy Research and Development Administration, Washington, DC, 1977.

---

[1] Dose level is classified into six grades, from A to F. Level-A with risk of lethality more than 50% (more than 4 Sv), level-B with risk of acute radiation sickness (more than 1 Sv), level-C with risk of embryo and late health effects, level-D with some safety zone, level-E safety zone (less than 1 mSv), level-F in relief.

# Posters

# Effects of radon and thermal therapy on osteoarthritis

Kiyonori Yamaoka*, Fumihiro Mitsunobu, Katsumi Hanamoto, Takahiro Kataoka, Yoshiro Tanizaki

*Medical Radioscience, Okayama University Medical School, 2-5-1 Shikata-cho, Okayama 700-8558, Japan*

**Abstract.** We examined the temporal changes in antioxidants, immune-, vasoactive- and pain-associated substances in human blood by therapy to elucidate the mechanism of osteoarthritis in which radon therapy is used as a treatment. Results showed that radon inhalation enhanced the antioxidation and immune function, and that the changes in vasoactive and pain-associated substances indicate increases in tissue perfusion brought about by radon therapy. © 2004 Published by Elsevier B.V.

*Keywords:* Radon and thermal therapy; Osteoarthritis; Active oxygen diseases; Antioxidants; Immunity; Pain-associated substance; Vasoactive substance

Radon therapy uses radon ($^{222}$Rn) gas, which mainly emits α-rays and induces a small amount of active oxygen in the body. We examined the temporal changes in antioxidant- (the activities of superoxide dismutase (SOD) and catalase which inhibit lipid peroxidation, the levels of lipid peroxide (Fig. 1A) and total cholesterol), immune- (concanavalin A (ConA)-induced mitogen response, and the percentage of CD4 positive cells, which is the marker of helper T cells, and the percentage of CD8 positive cells, which is the common marker of killer T cells and suppressor T cells, in the white blood cell differentiation antigen (CD8/CD4)), vasoactive- (the levels of α atrial natriuretic polypeptide (αANP) and vasopression) and pain- (the levels of βendorphin (Fig. 1B) and adrenocorticotropic hormone (ACTH)) associated substances in human blood by therapy to elucidate the mechanism of osteoarthritis in which radon therapy is used as a treatment. Results showed that radon inhalation enhanced the antioxidation and immune function, and the findings suggest that radon therapy contributes to the prevention of osteoarthritis related to peroxidation reactions and immune depression. Moreover, the

---

* Corresponding author. Tel./fax: +81 86 235 6852.
  *E-mail address:* yamaoka@md.okayama-u.ac.jp (K. Yamaoka).

0531-5131/ © 2004 Published by Elsevier B.V.
doi:10.1016/j.ics.2004.12.013

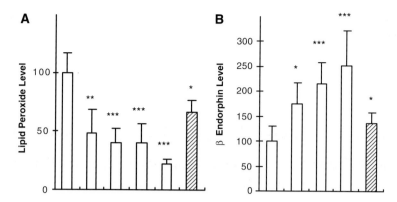

Fig. 1. Temporal changes (%) in antioxidant-associated substance (A) lipid peroxide and pain-associated substance (B) βendorphin of blood of patient with osteoarthritis after first radon and thermal therapy. Each value represents the mean±S.E.M. Normal is the mean of 20 healthy persons living in Misasa area. The number of patients in experiment was 20 (11 men and 9 women). Significantly different from the no therapy control (before) group at *$P<0.05$, **$P<0.01$ and ***$P<0.001$, respectively.

changes in vasoactive and pain-associated substances indicate increases in tissue perfusion brought about by radon therapy, suggesting that radon inhalation plays a role in alleviating pain. The findings suggest that an appropriate amount of active oxygen is produced in the body after radon inhalation and this contributes to the alleviation of the symptoms of active oxygen diseases such as osteoarthritis [1].

On the other hand, the radioactive and thermal effects of radon hot spring were biochemically compared under a sauna room or hot spring condition with a similar chemical component, using the parameters that are closely involved in the clinic for radon therapy. The results showed that the radon and thermal therapy enhanced the antioxidation functions, such as the SOD activity and the total cholesterol level produced in the body. Moreover, the therapy enhanced ConA-induced mitogen response, and increased the percentage of CD4 positive cells and decreased the percentage of CD8 positive cells in the white blood cell differentiation antigen assay. Furthermore, the therapy increased the levels of αANP, βendorphin and ACTH, and it decreased the vasopression level. The results were on the whole larger in the radon group than in the thermal group. The findings suggest that radon therapy contributes more to the prevention of life-style-related diseases related to peroxidation reactions and immune suppression than to thermal therapy. Moreover, these indicate what may be a part of the mechanism for the alleviation of osteoarthritis brought about more by radon therapy than by thermal therapy [2]. Furthermore, we have studied on the other effects of radon inhalation on biological function [3–5].

## References

[1] K. Yamaoka, et al., J. Pain 5 (2004) 20–25.
[2] K. Yamaoka, et al., J. Radiat. Res. 45 (2004) 83–88.
[3] F. Mitsunobu, et al., J. Radiat. Res. 44 (2003) 95–99.
[4] K. Yamaoka, et al., Neurosciences 20 (1994) 17–22.
[5] K. Yamaoka, et al., Arch. Biochem. Biophys. 302 (1993) 37–41.

# Induction of micronuclei in rat tracheal epithelial cells following radon exposure at air–liquid interface culture

K. Fukutsu*, Y. Yamada, W. Zhuo, S. Tokonami, A. Koizumi

*Radon Research Group, National Institute of Radiological Sciences, 4-9-1, Anagawa, Inage-ku, Chiba 263-8555, Japan*

**Abstract.** The micronuclei for rat tracheal epithelial cells were examined by in vitro exposure of radon. The rat tracheal epithelial cells were exposed in air–liquid interface culture (ALI culture), which was developed for the purpose of simulating in vivo conditions. The micronuclei induction gradually increased when the radon concentration was over $10^3$ Bq/m$^3$. The absorbed dose in Gy was tried to convert from exposure concentration in Bq/m$^3$ by using alpha spectrum data. As a result, the micronucleus induction significantly increased when the total absorbed dose was over 100 mGy.
© 2004 Elsevier B.V. All rights reserved.

*Keywords:* Radon; Micronucleus; Air–liquid interface culture; Tracheal epithelial cell

## 1. Introduction

The conventional cell culture methods are covered with medium on surfaces of cells. In this condition, the cells were irradiated through thick medium. On the other hand, radon exposes epithelial cells through thin mucus in vivo. Therefore, exposure conditions may cause differences between in vivo and in vitro. Air–liquid interface culture (ALI culture) has been reported for a new culture method for epithelial cells [1,2]. In this culture, the apical surface of epithelial cells is in contact with air as same as in vivo, and the morphology of epithelial airway cells is similar to in vivo. For the dose response of X-rays and alpha rays, rat tracheal epithelial cells (RTE cells) in ALI culture showed the same response as those for in vivo and ex vivo [3].

* Corresponding author. Tel.: +81 43 206 3096; fax: +81 43 206 4097.
*E-mail address:* fukutsu@nirs.go.jp (K. Fukutsu).

0531-5131/ © 2004 Elsevier B.V. All rights reserved.
doi:10.1016/j.ics.2004.11.039

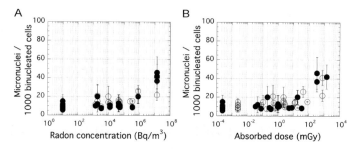

Fig. 1. Micronuclei induction of RTE cells with ALI culture. Panel (A) shows the relationship between radon concentration and micronuclei. Panel (B) shows the relationship between absorbed dose and micronuclei. Closed circle shows 3-day exposure, and open circle shows 7-day exposure.

In this paper, we made conversion from the exposure concentration in $Bq/m^3$ to the absorbed dose in Gy by using alpha spectrum data. The results for micronucleus induction of RTE cells were shown as relationship of total absorbed dose in Gy.

## 2. Materials and methods

The RTE cells were kept by ALI culture following a previous paper [3]. Briefly, cells were grown on a cell culture insert coated with collagen gel with serum-free medium. For micronucleus assay, cells were incubated for 96 h with 3 μg/ml Cytochalasin B and then fixed in methanol and stained using 0.1% acridine orange solution. The number of micronuclei observed per 1000 binucleated cells was recorded.

Radon source with high stability [4] was used for exposure of cells. Radon concentration for exposure ranged from $2.0 \times 10^3$ to $1.4 \times 10^7$ $Bq/m^3$.

## 3. Results and discussion

In order to calculate the absorbed dose in Gy, alpha energy spectrum was measured in the incubator as the same condition as cell exposure. On the basis of this spectrum, flux density was calculated when the cross-section of cell was assumed to be 100 $\mu m^2$. Furthermore, absorbed dose was calculated when the mass of cell was assumed to be 1500 $\mu m^3$. For example, the absorbed dose was 6.2 mGy/day with lid when the radon concentration was $8.1 \times 10^5$ $Bq/m^3$. In this paper, the RTE cells were exposed during 3 or 7 days at each level of radon concentrations. Fig. 1A shows the relationship between radon concentration and micronucleus induction. The micronucleus induction gradually increased when the concentration was over 1000 $Bq/m^3$. Fig. 1B shows the relationship between total absorbed dose and micronucleus induction. The micronucleus induction was almost the same as the control up to the total absorbed dose of 10 mGy and increased significantly when the total absorbed dose was over 100 mGy.

## References

[1] C.B. Robinson, R. Wu, Culture of conducting airway epithelial cells in serum-free medium, J. Tissue Cult. Methods 13 (1991) 95–102.
[2] L. Kaartinen, et al., Rat tracheal epithelial cell differentiation in vitro, In Vitro Cell. Dev. Biol. 29A (1993) 481–492.
[3] K. Fukutsu, Y. Yamada, M. Shimo, Dose response of tracheal epithelial cells to ionizing radiation in air–liquid interface cultures, High Levels of Natural Radiation and Radon Areas: Radiation Dose and Health Effects, vol. II, 2002, pp. 475–477.
[4] A. Koizumi, et al., Radon emission rate of ceramic radon source, Proceeding of IRPA 2000.

# Simplification of analysis for comparison of radioactive characteristics and its application to some minerals for radon therapy

Katsumi Hanamoto*, Kiyonori Yamaoka

*Department of Radiological Technology, Okayama University Medical School, 2-5-1 Shikata-cho, Okayama 700-8558, Japan*

**Abstract.** To simplify the analysis of the radioactive characteristics of the minerals for the radon therapy, we developed a simple method to calculate the ratio of the uranium series to the thorium series (U/Th) without using activity. The ratios calculated by this method agreed well with the ratios calculated using the activity. We applied this method to some minerals for the radon therapy. As a result, the U/Th's for the sludge of Misasa, the hokutolite, the mineral of Badgastein, and the two kinds of commercially available minerals were 22, 50, 0.94, 0.26, and 0.20, respectively. © 2004 Elsevier B.V. All rights reserved.

*Keywords:* Radon therapy; Misasa; Badgastein; Radioactive mineral; High-purity germanium detector

## 1. Introduction

Many attempts have been made to elucidate the mechanism of the radon therapy [1–4]. It is meaningful for these studies to analyze the radioactive characteristics of the minerals for the radon therapy. To simplify this analysis, we developed a method to calculate the ratios of the uranium series to the thorium series (U/Th) of the minerals without using the activity and applied this method to some radioactive minerals.

## 2. Materials and methods

We prepared the sludge of Misasa and a commercially available mineral in the shape of sand. Gamma rays of radioelements in these specimens were measured by a high-purity

---

* Corresponding author. Tel.: +81 86 235 6878; fax: +81 86 222 3717.
 *E-mail address:* hana@md.okayama-u.ac.jp (K. Hanamoto).

germanium detector. The specimens were enclosed in a standard plastic container (U-8 container) with same geometry and the detection efficiency was calibrated for this geometry. The U/Th can be expressed as

$$U/Th = \frac{A_U}{A_{Th}} = \frac{N_U}{N_{Th}} \frac{\varepsilon_{\gamma Th}}{\varepsilon_{\gamma U}} \frac{\varepsilon_{effiTh}}{\varepsilon_{effiU}}, \qquad (1)$$

where $A$ is an activity of a radioelement, $N$ a net count rate of a γ ray of the specimen, $\varepsilon_\gamma$ a branching ratio of a γ ray [5], $\varepsilon_{effi}$ a detection efficiency of the detector, and subscripts U and Th show the symbols of the uranium series and the thorium series, respectively. The U/Th can be calculated using the activities $A_U$ and $A_{Th}$ when the detection efficiency $\varepsilon_{effi}$ is correctly calibrated. When energies of γ rays from the radioelements of both series are mutually similar, the U/Th can be calculated without using $\varepsilon_{effi}$ because $\varepsilon_{effiTh}/\varepsilon_{effiU}$ is nearly equal to unity. The U/Th's were calculated using the activities and compared with the values calculated without using the activity. We employed four pairs of the photo peaks with the similar energy between the uranium series and the thorium series; $^{214}$Pb(241.981 keV) and $^{212}$Pb(238.632 keV), $^{214}$Pb(295.213 keV) and $^{212}$Pb(300.087 keV), $^{214}$Pb(351.921 keV) and $^{228}$Ac(338.322 keV), $^{214}$Bi(609.312 keV) and $^{208}$Tl(583.191 keV). The method to calculate the U/Th without using the activities was applied to the hokutolite, the mineral of Badgastein, and another commercially available mineral in the shape of pellet.

## 3. Results and discussion

The specific activities and the U/Th calculated using activity (U/Th1) and without using activity (U/Th2) were shown in Table 1. For the three specimens with non-calibrated efficiency containing geometry, the specific activities were calculated by assuming the detection efficiency. For the sludge of Misasa and the mineral in the shape of sand, the values of the U/Th1 are in good agreement with the values of the U/Th2. This suggests that the calibration of the detection efficiency may successfully be correct and this method is available to calculate the U/Th. The sludge of Misasa and the hokutolite extremely include the larger radioelements of the uranium series than those of the thorium series. The mineral of Badgastein includes the radioelements of the both series with almost the same ratio. The commercially available minerals for radon therapy include the larger radioelements of the thorium series than those of the uranium series.

Table 1
The ratios of the uranium series to the thorium series (U/Th)

| Specimen | Activity of U (Bq g$^{-1}$) | Activity of U (Bq g$^{-1}$) | U/Th1 | U/Th2 |
|---|---|---|---|---|
| Sludge of Misasa | 0.58±0.02 | 0.026±0.003 | 22±3 | 22±2 |
| Mineral (sand) | 16±0 | 79±2 | 0.20±0.01 | 0.20±0.00 |
| Hokutolite | (38±1) | (0.75±0.01) | (51±1) | 50±2 |
| Mineral of Badgastein | (0.029±0.001) | (0.029±0.002) | (0.99±0.07) | 0.94±0.04 |
| Mineral (pellet) | (4.2±0.1) | (16±0) | (0.26±0.01) | 0.26±0.0 |

The U/Th1 and U/T2 show the ratios calculated with and without using the activity, respectively. Activities in parentheses stand for the presence of ambiguity.

# References

[1] P. Deetjen, Epidemiology and biological effects of radon, in: H.G. Pratzel, P. Deetjen (Eds.), Radon in derkurmedizin, Verlag Geretsried, ISMH, 1997, pp. 32–38.
[2] K. Yamaoka, et al., Effects of radon inhalation on biological function—lipid peroxide, SOD activity and membrane fluidity, Arch. Biochem. Biophys. 302 (1993) 37–41.
[3] K. Yamaoka, et al., Changes in biogenic amine neurotransmitters in rabbits brain by inhalation of radon spring, Neurosciences 20 (1994) 17–22.
[4] K. Yamaoka, Y. Komoto, Experimental study of alleviation of hypertension, diabetes and pain by radon inhalation, Physiol. Chem. Phys. 28 (1996) 1–5.
[5] R.B. Firestone, V.S. Shirley, Table of Isotopes, 8th ed., John Wiley and Sons, New York, 1996.

# Bystander cellular effects in normal human fibroblasts irradiated with low-density carbon ions

Masao Suzuki*, Chizuru Tsuruoka, Nakahiro Yasuda, Kenichi Matsumoto, Kazunobu Fujitaka

*International Space Radiation Laboratory, National Institute of Radiological Sciences, Chiba 263-8555, Japan*

**Abstract.** We have studied bystander lethal and mutagenic effects in normal human fibroblasts irradiated with low-density carbon ions following four different methods: (1) irradiate with all cells; (2) irradiate with half of cells; (3) mix with two cell populations (1) and (2); (4) treat with the inhibitor of cell–cell communication to half-cell irradiated group. The results of both cell killing and mutation induction showed that there observed in the same effects between (1) and (2). Furthermore, both effects for (4) showed the same level with (3). There is evidence that bystander effects of cell killing and mutation induction occurred in unirradiated half-cells of the irradiation method (2), and cell–cell communication may play an important role of inducing the bystander effects. © 2004 Elsevier B.V. All rights reserved.

*Keywords:* Bystander effects; Normal human fibroblasts; *HPRT* locus; Carbon ions

## 1. Introduction

It is well known that high-LET radiations are more effective in causing biological effects in vivo and in vitro than low-LET radiations. However, almost all of the reports for high-LET radiations are using high doses with acute irradiation. It should therefore be important for estimating radiation risk in natural radiation from the earth or outer space to study biological effects using low-density irradiation of high-energy and charged particles. In recent years, there has been evidence that low fluences of alpha particles induce biological effects in the cell population close to, but not directly hit by alpha particles. These studies suggest that some signals induced by the alpha particles in directly hit cells

---

\* Corresponding author. Tel.: +81 43 206 3238; fax: +81 43 251 4531.
*E-mail address:* m_suzuki@nirs.go.jp (M. Suzuki).

Table 1
Surviving fraction for four different irradiation methods

| Irradiation methods | Surviving fraction |
| --- | --- |
| (1) Irradiate with all cells on Mylar film | 0.64 |
| (2) Irradiate with half of cells on Mylar film | 0.67 |
| (3) Mix with all irradiated cells and unirradiated control cells | 0.88 |
| (4) Treated with the inhibitor of cell–cell communication | 0.89 |

might be transmitted to unirradiated neighbours [1–3] and might induce such "bystander effects". In this study, we studied bystander cellular responses in normal human fibroblasts irradiated with low-density carbon-ion beams.

## 2. Materials and methods

### 2.1. Cell

Normal human skin fibroblasts were distributed by the RIKEN Cell Bank (Cell No. RCB0222) in Japan. The cells were cultured in an Eagle's minimum essential medium (MEM), supplemented with 10% fetal bovine serum in a 5% $CO_2$ incubator at 37 °C.

### 2.2. Irradiation

Irradiation dishes were constructed by drilling a 6.5-mm hole in the center of the 35-mm-diameter tissue culture dishes, and 2.5-μm-thick Mylar film was epoxied over the bottom of the hole. The cells were inoculated onto the Mylar film in the dishes and then irradiated with 6 MeV/n carbon ions at the fluence density of $1.5 \times 10^5$ particles/cm$^2$ generated by the Medium Energy Beam Course at the Heavy Ion Medical Accelerator in Chiba (HIMAC), following four different methods: (1) irradiate with all cells on Mylar film; (2) irradiate with half of cells on Mylar film; (3) mix with all irradiated cells and unirradiated control cells (1:1); (4) irradiate with half of cells treated with the specific inhibitor of cell–cell communication (40 μM of Lindane).

## 3. Results and discussion

The cell killing effect (Table 1) showed that the surviving fraction of all-cell irradiated group (1) was almost the same with half-cell irradiated group (2). Furthermore, it was a similar level between half-cell irradiated group treated with the inhibitor of cell–cell communication (4) and mixed population group (3). Also, mutation induction at *HPRT* locus showed the same results with the cell killing effect. Our results indicated that there is evidence that bystander effects of both cell killing and mutation induction occurred in unirradiated half-cells of the irradiation method (2). And gap-junction mediated cell–cell communication may play an important role of inducing the bystander effects.

## References

[1] H. Zhou, G. Randers-Pehrson, C.A. Waldren, et al. Induction of a bystander mutagenic effect of α-particles in mammalian cells. Proc. Natl. Acad. Sci. U. S. A. 97 (200) 2099–2104.
[2] C. Mothersill, C.B. Seymour, Cell–cell contact during gamma irradiation is not required to induce a bystander effect in normal human keratinocytes: evidence for release during irradiation of a signal controlling survival into the medium, Radiat. Res. 149 (1998) 256–262.
[3] R. Iyer, B.E. Lehnert, Factors underlying the cell growth-related bystander responses to α particles, Cancer Res. 60 (2000) 1290–1298.

www.ics-elsevier.com

# Effects of low dose-rate long-term gamma-ray irradiation on DNA damage in mouse spleen

Kensuke Otsuka*, Kazuo Sakai

*Low Dose Radiation Research Center, Central Research Institute of Electric Power Industry, 2-11-1 Iwado-Kita, Komae-shi, 201-8511, Japan*

**Abstract.** The effects of low dose-rate γ-irradiation on the induction of DNA damage were investigated in the spleen of C57BL/6N mice. A modified single-cell gel electrophoresis (comet assay) was employed to measure DNA damage. Mice were irradiated with 137Csγ-rays at 1.2 mGy/h for 23 days (0.5 Gy total). The amount of DNA damage was compared to that in the mice irradiated with the same dose of X-rays at 96 Gy/h (high dose-rate). When the mice irradiated at the low dose-rate were then exposed to 1 Gy of X-rays, the initial amount of DNA damage was less than that in mice, which received the 1 Gy only. These results indicated that damage was repaired during irradiation period and that the low dose-rate irradiation induced an adaptive response in terms of DNA damage. © 2004 Published by Elsevier B.V.

*Keywords:* Low dose-rate; DNA damage; Comet assay; Mouse spleen

## 1. Introduction

The dose-rate in the high background natural radiation area is, from a viewpoint of radiation biology, very low. Damage on DNA molecules, which are most important targets for biological effects of ionizing radiation [1], has been studied at high dose-rate, and mostly in vitro. To assess the health effects of high background radiation, in vivo study at low dose-rate should be necessary for appropriate interpretation of epidemiological data. We carried out low dose-rate whole body irradiation of mice and analyzed DNA damage by a sensitive technique.

## 2. Materials and methods

C57BL/6N female mice (6 weeks old) were irradiated with $^{137}$Cs gamma rays at 1.2 mGy/h for 23 days (0.5 Gy total). To compare with low dose-rate, the other mice were exposed to

---

* Corresponding author. Tel.: +81 3 3480 2111; fax: +81 3 3480 3113.
*E-mail address:* ohken@criepi.denken.or.jp (K. Otsuka).

0531-5131/ © 2004 Published by Elsevier B.V.
doi:10.1016/j.ics.2004.11.037

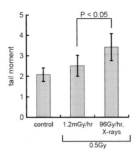

Fig. 1.

0.5 Gy of X-rays (300 kV, 10 mA, 1 mm Al+0.5 mm Cu filter) at 96 Gy/h as high dose-rate. After each irradiation, spleen was removed and prepared for following assay. We modified a comet assay described in Ref. [2] for sensitive detection of DNA damage. A comet "tail moment" was calculated to estimate the amount of DNA damage.

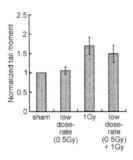

Fig. 2.

## 3. Results and discussion

DNA damage by 0.5 Gy at the low dose-rate was less severe than that at high dose-rate, and no statistically significant difference was found between the control and the low dose-rate group (Fig. 1). The damage induced by the low dose-rate irradiation was possibly repaired during the irradiation period.

The amount of DNA damage by 1 Gy of X-rays given after the low dose-rate irradiation was less than that by 1 Gy only (Fig. 2), indicating that an adaptive response was induced by the low dose-rate irradiation.

It was unlikely that the enhancement of DNA repair mechanisms were involved under the conditions used, because the amount of DNA damage was analyzed immediately after the 1 Gy irradiation. Rather, other biological responses, such as the increase in antioxidants and the exclusion of sensitive cells, may possibly contribute to the reduction of DNA damage.

## References

[1] J.H. Hoeijmakers, Genome maintenance mechanisms for preventing cancer, Nature 411 (2001) 366–374.
[2] R.S. Malyapa, et al., Detection of DNA damage by alkaline comet assay after exposure to low-dose gamma radiation, Rad. Res. 149 (1998) 396–400.

# Detection of radiation-induced mutations in the liver after partial hepatectomy using HITEC transgenic mice

A. Ootsuyama*, F. Kato, T. Norimura

*Dept. of Radiat. Biol. and Health, Univ. of Occup. and Environm. Health., Japan,
1-1 Iseigaoka Yahatanishi-ku Kitakyushu, 807-8555, Japan*

**Abstract.** We investigate the dynamics of radiation-specific mutations in a mammalian differentiated cell of the liver in the transgenic mouse called Hypersensitive In vitro Test of Carcinogenecity (HITEC), carrying the *rpsL* gene as a reporter for mutation detection. In this study, we performed a partial hepatectomy (PH) to induce cell proliferation and increase the detection of mutation frequency. Prior to PH, the mice were exposed to 5 Gy gamma rays at 2 days, 10 days, 1 month or 2 months. The remnant liver was then resected at 3 or 10 days after PH, and the mutation frequencies (MFs) for both resected livers and remnant livers were analyzed. In mice where PH was done at 2 or 10 days after irradiation, the MF of the remnant livers of them increased approximately two times compared with that of the livers in which only irradiation was done. However, in mice where PH was done at 1 or 2 months after irradiation, the increase in MF was not seen. These results show that HP is effective in increasing the MF, and the repair system is clearly working even in mammalian differentiated cell. © 2004 Published by Elsevier B.V.

*Keywords:* Mutation; Mice; Radiation; Hepatectomy; Liver

## 1. Introduction

In stem and embryonic cells, repairing the radiation-induced mutations or removing the injury cells from tissues prevents the diffusion of mutation cells and disorders produced by these abnormal cells. In differentiated cells, it is thought DNA damage cannot be repaired. Those cells that have the radiation-induced mutations leave from tissues according to their homeostatic destiny without repair of the mutation. It looks like a reasonable mechanism, but the mechanism means that the differentiated cells having a mutation that does not kill could

---

\* Corresponding author. Tel.: +81 93 691 7433; fax: +81 93 692 0559.
*E-mail address:* aootsuya@med.uoeh-u.ac.jp (A. Ootsuyama).

0531-5131/ © 2004 Published by Elsevier B.V.
doi:10.1016/j.ics.2004.11.123

Table 1
Experimental protocols and results

| Group | Total dose (Gy) | Dose rate (Gy/min) | Days between radiation and hepatectomy | Days between hepatectomy and sacrifice | Mutation frequency ($\times 10^{-5}$) Resected liver | Mutation frequency ($\times 10^{-5}$) Remnant liver | P-value |
|---|---|---|---|---|---|---|---|
| A | 5 | 1.04 | 2 | 3 | 1.8 | 3.9 | <0.01 |
| B | 5 | 1.04 | 2 | 10 | 3.0 | 6.1 | <0.01 |
| C | 5 | 1.04 | 10 | 3 | 2.4 | 6.9 | <0.01 |
| D | 5 | 1.04 | 10 | 10 | 2.9 | 4.6 | <0.05 |
| E | 5 | 1.04 | 30 | 3 | 3.8 | 4.8 | – |
| F | 5 | 1.04 | 60 | 3 | 2.7 | 3.9 | – |

remain in the tissue for some period. It could be one of the causes in the induction of gene instability. We analyzed it using the transgenic mouse whether such a phenomenon actuary occurred in differentiated cells after exposure to radiation.

## 2. Materials and methods

### 2.1. Mutation frequency

For the mutation assay, heterozygous transgenic female mice [1–3] carrying the *rpsL* gene in the pSSW shuttle vector were utilized. Mice were randomized into seven groups (Table 1). The mice were exposed to 5 Gy of gamma-rays before the partial hepatectomy (PH). After irradiation, all mice had half of the liver resected at different time points. After PH, the mice were sacrificed at different time points (Table 1). Genomic DNA was extracted from the resected and the remnant liver. Mutation analysis was performed as described by Gondo et al. [1].

### 2.2. Sequence of mutation in rpsL gene

We analyzed a spectrum of mutation in the *rpsL* gene extracted from mutant *E. coli* colonies obtained from the resected and a remnant liver DNA.

## 3. Results and discussions

In Groups A, B, C and D mice, the mutation frequency (MF) in the remnant liver was two to three times greater than that of the resected liver. In contrast, in Groups E and F mice, there were no significant differences in the MF between the resected and the remnant liver (Table 1).

The ratio of multiple-base deletions, including long deletions, of the remnant liver was about three times that of the resected liver. When the damage is fixed to mutations during cell proliferation after PH, it may be prone to develop the multiple-base or large deletion-type mutation.

These results show that PH is effective in increasing the MF when carried out with appropriate timing relative to radiation exposure, and the mutation repair system is clearly working even in differentiated cell.

## References

[1] Y. Gondo, et al., A novel positive detection system of in vivo mutations in rpsL (strA) transgenic mice, Mutat. Res. 360 (1996) 1–14.
[2] S. Muto, et al., Inhibition of benzo[a]pyrene-induced mutagenesis by (−)-epigallocatechin gallate in the lung of rpsL transgenic mice, Carcinogenesis 20 (1999) 421–424.
[3] Y. Shioyama, et al., Different mutation frequencies and spectra among organs by *N*-methyl-*N*-nitrosourea in rpsL (strA) transgenic mice, Jpn. J. Cancer Res. 91 (2000) 482–491.

# Role of *p53* gene in genetic instability induced by ionizing radiation

Kazuyuki Igari*, Yuka Igari, Hiroyo Kakihara, Fumio Kato, Akira Ootsuyama, Toshiyuki Norimura

*Department of Radiation Biology and Health, School of Medicine, University of Occupational and Environmental Health, Japan, 1-1 Iseigaoka, Yahatanishi-ku, Fukuoka 807-8555, Japan*

**Abstract.** To know the role of *p53* gene for delayed somatic mutation, we analysed T-cell receptor (TCR) mutation in X-irradiated mice. Wild-type *p53*(+/+) and heterozygous *p53*(+/−) mice were given a whole-body acute exposure of 3 Gy X-ray at 8 weeks of age. TCR assays were done at various time intervals between 9 days and 40 weeks after exposure. In *p53*(+/−) mice, TCR mutation frequency (TCR-MF) increased again 36 weeks after irradiation. However, in *p53*(+/+) mice, TCR-MF did not increase again. These findings show that the loss of p53 function becomes relevant to the acquisition delayed mutations. © 2004 Elsevier B.V. All rights reserved.

*Keywords:* p53; Ionizing radiation; TCR; Delayed mutation

## 1. Introduction

The *p53* gene plays a key role in the cellular response to genotoxic stress. Although the loss of p53 function is associated with radiation-induced delayed mutation in vitro, it is not clear in vivo. To clarify the in vivo role of p53 in the delayed somatic mutation, we analysed mutation frequency (MF) at the T-cell receptor (TCR) locus in wild-type *p53*(+/+) and heterozygous *p53*(+/−) mice over 40 weeks after X-irradiation with 3 Gy.

## 2. Materials and methods

### 2.1. Mouse strains

C57BL/6N mice were used as wild-type *p53*(+/+) mice. Nonfunctional *p53* gene were derived by homologous recombination of an embryonic stem cell line from 129/SvJ mice

---

\* Corresponding author. Tel.: +81 93 603 1611; fax: +81 93 692 0559.
*E-mail address:* kigari@apricot.ocn.ne.jp (K. Igari).

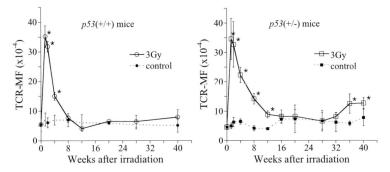

Fig. 1. Time course of the TCR-MF in $p53(+/+)$ and $p53(+/-)$ mice after X-ray exposure to 3 Gy. Each point indicates the mean±S.D. of 3 to 10 mice. The statistical significance was determined by the Student's $t$ test ($*p<0.05$).

[1]. Heterozygous $p53(+/-)$ mice were obtained by crossing the male $p53(-/-)$ mice with the female $p53(+/+)$ mice.

2.2. TCR mutation assay and radiation treatments

The in vivo T-cell receptor (TCR) assay detects somatic mutation with a high sensitivity. Mutant $CD4^+$ lymphocytes are found to be $CD3^-CD4^+$ by flow cytometry. The number of these mutant cells increases in humans [2] and mice [3] following exposure to radiation. At 8 weeks of age, $p53(+/+)$ and $p53(+/-)$ mice were given a whole-body acute exposure of 3 Gy (450 mGy/min) X-rays. TCR assays were done at various time intervals between 9 days and 40 weeks after exposure.

3. Results and discussion

In $p53$ $(+/+)$ mice, the TCR-MF reached a peak level 9 days after irradiation and gradually decreased to background levels within 8 weeks (Fig. 1). In $p53(+/-)$ mice, the TCR-MF reached a peak level 9 days after irradiation and gradually decreased to background levels within 16 weeks. However, TCR-MF increased again 36 weeks after irradiation (Fig. 1). These findings show that the loss of p53 function becomes relevant to the acquisition delayed mutations.

References

[1] Y. Gondo, et al., Gene replacement of the *p53* gene with the lacZ gene in mouse embryonic stem cells and mice by using two steps of homologous recombination, Biochem. Biophys. Res. Commun. 202 (2) (1994) 830–837.
[2] S. Kyoizumi, et al., Frequency of mutant T lymphocytes defective in the expression of the T-cell antigen receptor gene among radiation-exposed people, Mutat. Res. 265 (2) (1992) 173–180.
[3] F. Kato, et al., Role of *p53* gene in apoptotic repair of genotoxic tissue damage in mice, J. Radiat. Res. (Tokyo) 43 (2002) S209–S212 (Suppl.).

# Adaptive response of bone marrow stem cells induced by low-dose rate irradiation in C57BL/6 mice

Kazuniori Shiraishi[a,*], Akira Tachibana[b], Morio Yonezawa[a], Seiji Kodama[a]

[a]*Research Institute for Advanced Science and Technology, Osaka Prefecture University, Gakuen-chyo 1-2, Sakai city, Osaka 599-8570, Japan*
[b]*Department of Mutagenesis, Radiation Biology Center, Kyoto University, Yoshidakonoe-chyo, Skyo-ku, Kyoto, Japan*

**Abstract.** We previously showed that whole body pre-irradiation with 0.5 Gy of X-rays enhanced the survival of C57BL/6 mice after the subsequent sub-lethal dose of radiation. In the present study, we examined whether the similar radio-adaptive response of bone marrow stem cells can be observed when the mice are exposed to a low dose-rate (1 mGy/min) of pre-irradiation of Co-60 gamma-rays. On 10th day after the low dose-rate irradiation with 0.5 Gy, C57BL/6 mice were exposed to 6 Gy (1 Gy/min) of gamma-rays, and the numbers of spleen colonies were scored to assess radio-adaptive response of bone marrow stem cells on 12th day after irradiation. The result demonstrated that the number of spleen colonies in the pre-irradiated mice increased two fold as compared with those in the unirradiated mice. This finding suggests that pre-irradiation at a low dose-rate is also effective on the induction of radio-adaptive response of bone marrow stem cells. © 2004 Published by Elsevier B.V.

*Keywords:* Adaptive response; Low-dose rate; Mouse

## 1. Introduction

To know the risk of low dose-rate radiation, whole-body irradiation studies using laboratory animals have a significant implication in addition to epidemiological studies [1,2]. We previously showed that whole body pre-irradiation with 0.5 Gy of X-rays enhanced the survival of C57BL/6 mice that were subsequently received with a sub-lethal dose (6.75 Gy) of X-rays [3]. In the present study, we examined whether the similar radio-adaptive response of bone marrow stem cells was observed when the mice were exposed to a low dose-rate (1 mGy/min) of pre-irradiation.

---

\* Corresponding author. Tel.: +81 72 254 9860; fax: +81 72 254 7938.
*E-mail address:* shiraish@riast.osakafu-u.ac.jp (K. Shiraishi).

0531-5131/ © 2004 Published by Elsevier B.V.
doi:10.1016/j.ics.2004.12.008

## 2. Materials and methods

### 2.1. Animals

Specific pathogen free (SPF) C57BL/6N mice of 7 week old (Oriental Yeast, Tokyo) were used. They were maintained in a clean conventional environment at $24\pm1$ °C in humidity of $60\pm10\%$.

### 2.2. Irradiation

Eight-week-old mice were irradiated continuously in a plastic chamber under the above described conditions. A low dose-rate of pre-irradiation (0.5–1.5 Gy) was performed with a Cs-137 gamma-ray source at a dose-rate of 1 mGy/min in the Radiation Biology Center, Kyoto University. The mice were subsequently irradiated with a sub-lethal (challenge) dose of X-rays using a X-ray machine (RadioFlex, Tokyo) at a dose rate of 1 Gy/min immediately or 10 days after the pre-irradiation.

### 2.3. Spleen colony formation assay

To assess the radiosensitivity of bone marrow stem cells, the number of spleen colonies was scored. To know the shrinkage of spleen by radiation, the ratio of spleen weight to whole-body weight was determined.

## 3. Results and discussion

We examined the effect of the pre-irradiation of 1.5 Gy at a low-dose rate (1 mGy/min) on changes of the spleen weight and the number of spleen colonies caused by the immediate irradiation (5 Gy) at a high dose rate on 10th day after the irradiation and compared with the case exposed to only the acute irradiation (6.5 Gy). The result indicated that the low dose-rate of pre-irradiation (1.5 Gy) gave no effect on the shrinkage of spleen.

Because our former study indicated that radio-adaptive response to bone marrow cell death by pre-irradiation at a high dose-rate required 12–14 day-intervals between the pre-irradiation and the challenge irradiation [4], we examined whether the similar time interval for the induction of radio-adaptive response was observed by the pre-irradiation at a low dose-rate. The result demonstrated that time-intervals required for the induction of radio-adaptive response by the pre-irradiation of low dose-rate were 10–12 days, showing the similar interval observed at a high dose-rate of pre-irradiation. In summary, the present study indicates that whole body pre-irradiation at a low dose-rate induces radio-adaptive response effectively in C57Bl/6N mice.

## References

[1] S. Wolf, Aspects of the adaptive response to very low doses of radiation and other agents, Mutat. Res. 358 (2) (1996) 135–142.
[2] M. Yonezawa, A. Takeda, J. Misonoh, Acquired radioresistance after low dose x-irradiation in mice, J. Radiat. Res. 31 (3) (1990) 256–262.
[3] K. Horie, K. Kubo, M. Yonezawa, p53 dependency of radio-adaptive responses in endogenous spleen colonies and peripheral blood-cell counts in C57BL mice, J. Radiat. Res. 43 (4) (2002) 353–360.
[4] M. Yonezawa, et al., Increase in endogenous spleen colonies without recovery of blood cell counts in radioadaptive survival response in C57BL/6 mice, Radiat. Res. 161 (2) (2004) 161–167.

# The life saving role of radioadaptive responses in long-term interplanetary space journeys

S.M.J. Mortazavi[a,*], J.R. Cameron[b], A. Niroomand-Rad[c]

[a]*Medical Physics Department, School of Medicine, Rafsanjan University of Medical Sciences (RUMS), Rafsanjan, Iran*
[b]*Departments of Medical Physics, Radiology, and Physics, University of Wisconsin, Madison, WI, USA*
[c]*Department of Radiation Medicine, Georgetown University, LL Bles Building, 3800, Reservoir Road NY Washington DC, 20007-2197, USA*

**Abstract.** It has been estimated that exposure to unpredictable extremely large solar particle events would kill the astronauts without massive shielding in interplanetary space. The high LET component of space radiation, neutrons especially, makes the major contribution to the genetic risks. Recent findings concerning the induction of adaptive response by neutrons and high cumulative doses of gamma radiation in human cells have opened a new horizon for possible implications of radiation adaptive response in human. Screening the candidates of long-term space missions by in vitro adaptive response studies help to identify the individuals with low radiation susceptibility and high radioadaptive response. In these selected individuals, chronic exposure to elevated levels of space radiation during a long-term mission can considerably decrease their radiation susceptibility and protect them against the unpredictable exposure to relatively high radiation levels caused by solar activity. © 2004 Elsevier B.V. All rights reserved.

*Keywords:* Adaptive response; Space radiation; Deep space missions

## 1. Introduction

We have shown that in high background radiation areas (HBRAs), cultured human lymphocytes of the inhabitants whose cumulative radiation doses were as much as 170 times more than those of a nearby control area (2550 mSv and 15 mSv, respectively) were significantly more radioresistant to chromosomal damage compared to the residents of the

---

\* Corresponding author. Vice Chancellor for Academic Affairs' Office, Central Building of the Rafsanjan University of Medical Sciences, Rafsanjan, Iran. Tel.: +98 391 822 0097; fax: +98 391 822 0097.
*E-mail address:* jamo23@lycos.com (S.M.J. Mortazavi).

0531-5131/ © 2004 Elsevier B.V. All rights reserved.
doi:10.1016/j.ics.2004.12.019

control area when subjected to 1.5 Gy challenge dose [1,2]. Adaptive response is an increased radioresistance in cells or organisms exposed to a high challenging dose after exposure to a low adapting dose. The adaptive response of the residents of HBRAs was more pronounced at higher cumulative doses except for 2 residents whose cumulative doses were much higher than the others, that is increased dose from natural radiation decreased the radiation sensitivity of the cells. On the other hand, we and other investigators have reported that some individuals do not show an adaptive response and even in some individuals an exaggerated sensitivity to subsequent high dose can be observed after irradiation with a low dose [3].

## 2. Adaptive response and deep space missions

The four major components of space radiation are gamma rays and electrons, high-energy protons, high-energy heavy ions and neutrons [4]. The high LET component of space radiation, especially neutrons, is the major contributor to genetic risks. Adaptive response, that is an increased radioresistance in cells or organisms exposed to a high challenging dose after exposure to a low adapting dose, can considerably reduce the radiation susceptibility of individuals. It has recently been shown that neutrons can induce adaptive response in Chinese hamster V79 cells or human lymphocytes. These findings, if confirmed by similar adaptive response experiments with high energy protons and heavy ions, can reduce the risk of long-term stay of human in space.

## 3. Selection of candidates

Astronauts are irradiated with high levels of radiation. Solar activity is currently unpredictable and especially space-walking astronauts and astronauts who participate in long-term space missions may receive high doses of radiation in a short time without massive shielding. Recent findings concerning the induction of adaptive response by neutrons and high cumulative doses of gamma radiation in human cells have opened a new horizon for possible implications of adaptive response in radiation protection. Screening the candidates of long-term space missions by in vitro adaptive response studies identifies the individuals who show low radiation susceptibility and demonstrate a high magnitude of radioadaptive response. In these individuals, chronic exposure to elevated levels of space radiation during a long-term mission can considerably decrease their radiation susceptibility and protect them against the unpredictable exposure to relatively high radiation levels caused by solar activity.

## References

[1] S.M.J. Mortazavi, et al., How should governments address high levels of natural radiation and radon? Lessons from the Chernobyl nuclear accident, Risk: Health Saf. Environ. 13/1.2 (2002) 31–45.
[2] M. Ghiassi-nejad, et al., Very high background radiation areas of Ramsar, Iran: preliminary biological studies, Health Phys. 82 (2002) 87–93.
[3] S.M.J. Mortazavi, T. Ikushima, H. Mozdarani, Variability of chromosomal radioadaptive response in human lymphocytes, Iran. J. Radat. Res. 1 (1) (2003) 55–61.
[4] A.A. Edwards, RBE of radiations in space and the implications for space travel, Phys. Med. 17 (2001) 147–152.

# Biosorption of chromium(III) by new bacterial strain (NRC-BT-2)

M. Rabbani[a], H. Ghafourian[a], S. Sadeghi[b], Y. Nazeri[b],*

[a]*Nuclear Research Center, AEOI, Tehran, Iran*
[b]*Islamic Azad University, North branch of Tehran, Chemistry Department, Iran*

**Abstract.** Biosorption of $Cr^{3+}$ has been investigated by 17 various bacterial strains isolated from Ramsar warm springs, Iran. A strain of gram-positive cocobacilli bacteria is highly capable for biosorption of $Cr^{3+}$. The effect of initial concentration of the metal solution in the range of 10–250 ppm has been studied, and the maximum removal occurs in 10 ppm, in pH 4 and 1 h of contact time (100%). Sodium acid, high temperature and γ-ray did not affect the biosorption. Presence of the other cations such as $Ni^{2+}$, $Zn^{2+}$, $Cu^{2+}$ and $Pb^{2+}$, reduce removal efficiency. Desorption of Cr from biomass by nitric acid was revealed in 31.45% of biosorbed chromium. © 2004 Elsevier B.V. All rights reserved.

*Keywords:* Biomass; Bacteria; Chromium; Biosorption; Removal

## 1. Introduction

The removal of chromium, which is present in different types of industrial effluents such as electroplating and metallurgy, traditionally made by chemical precipitation. The conventional techniques are extremely expensive and not completely feasible to reduce the chromium concentration to levels required by environmental legislation [1–3]. The purpose of this study is the biological uptake of $Cr^{3+}$ and optimizing $Cr^{3+}$ uptake conditions.

## 2. Materials and methods

All chemicals have been supplied by Merck.
The best strain has been chosen from 17 new various isolated and purified strains. Each strain was added to 150 ml of cultivation medium (GMS), after autoclaving and cooling. They were shaken at 150 rpm (Heidolph-unemax1010) for 72 h in 25 °C. The biomass was

---

* Corresponding author. Tel.: +98 21 2237937; fax: +98 21 802 1412; mobile: +98 912 159 9129.
 *E-mail address:* yasmin2802@yahoo.com (Y. Nazeri).

0531-5131/ © 2004 Elsevier B.V. All rights reserved.
doi:10.1016/j.ics.2004.11.125

isolated by centrifugation (12,000 rpm) (Heraeus) for 5 min in 20 °C. To prepare the suspended solution, distilled water was added to the biomass in a volume to the extent of 10 times more than biomass weight. A total of 5 ml of the suspension was mixed with 50 ml of $Cr^{3+}$ solution and shaken for 60 min at 100 rpm and 25 °C. The solution was separated from biomass by centrifugation. The $Cr^{3+}$ concentration of separated solution was determined by AAS (GBC 932 plus).

## 3. Results and discussion

The $Cr^{3+}$ removal in initial concentrations of 10, 30, 50, 100, 150, 200 and 250 ppm are, respectively, 100%±0.009, 54.35%±0.041, 39.84%±0.033, 25.00%±0.037, 20.58%± 0.023, 17.98%±0.039, 15.77%±0.052 in pH 4.5 and 60 min of contact time.

The removal percentages for 5, 15, 30, 60, 90, 120 and 150 min, constant concentration of 150 ppm and pH 4.5 were 8.70%±0.044, 13.08%±0.053, 17.56%±0.040, 18.00%±0.029, 18.64%±0.028, 19.08%±0.043 and 19.09%±0.037, respectively. So the equilibrium contact time is 30 min.

In pH 1, 2, 3 and 4, the results were, respectively, 3.00%±0.015, 4.61%±0.021, 21.05%±0.043 and 24.58%±0.038. In pH 1 and 2, the removal percentage is very low and, by increasing pH, it has been increased.

The $Cr^{3+}$ removal by nonliving biomass for 10, 30, 50, 100, 150, 200 and 250 ppm are, respectively, 95.15%±0.020, 35.51%±0.036, 26.72%±0.052, 22.41%±0.037, 17.50%± 0.058, 12.94%±0.063 and 10.68%±0.036.

By biomass contacted with sodium acid (10 mM), the removal percentages for 10, 30, 50, 100, 150, 200 and 250 ppm in pH 4.5 for 60 min are, respectively, 96.82%±0.017, 40.89%± 0.054, 21.02%±0.039, 10.58%±0.062, 9.61%±0.053, 7.63%±0.057 and 6.31%±0.049.

$Cr^{3+}$ removal by biomass with the absorbed dose of 0, 100, 200, 300 and 400 Gy of gamma ray are, respectively, 16.70%±0.045, 15.50%±0.030, 14.87%±0.051, 13.88%±0.039 and 13.82%±0.059.

In the presence of $Cu^{2+}$, $Pb^{2+}$, $Zn^{2+}$, $Ni^{2+}$ and $Na^+$, chromium removal percentages are, respectively, 21.64%±0.051, 11.95%±0.058, 0.00%±0.011, 9.37%±0.022, 14.96%± 0.041 and 20.88%±0.035. The initial concentration of each cation was 150 ppm.

By 50 ml of $HNO_3$ (1 N), $Na_2CO_3$ (0.1 N) and EDTA (1 N), the chromium desorption percentages are, respectively, 31.45%±0.056, 4.66%±0.037 and 3.41%±0.049.

## 4. Conclusion

The results show that the uptake mechanism is biosorption and pH-dependent. Even nonliving biomass takes up $Cr^{3+}$. Sodium acid and γ-ray have no effect on $Cr^{3+}$ removal but the presence of other cations except $Na^+$ reduce it.

## References

[1] A.M. Zayed, N. Terry, Chromium in the environment: factors affecting biological remediation, Plant Soil 249 (2003) 139–156.
[2] C. Ebner, T. Pumpel, M. Gmper, Biosorption of Cr(III) by the cell wall of *Mucor hiemalis*, Eur. J. Miner. Process. Environ. Prot. 2 (3) (2002) 168–178.
[3] E.S. Cossich, C.R.G. Tavares, T.M.K. Ravagnani, Biosorption of chromium(Ø) by *Sargassum* sp. Biomass, Electron. J. Biotechnol. 5 (2) (2002) (ISSN: 0717-3458).

# Uptake and removal of nickel by new bacterial strain (NRC-BT-1)

H. Ghafourian[a], M. Rabbani[a], Y. Nazeri[b,*], S. Sadeghi[b]

[a]Nuclear Research Center, AEOI, Tehran, Iran
[b]Islamic Azad University, North branch of Tehran, Chemistry Department, Iran

**Abstract.** Nickel removal has been investigated by new 16 various bacterial strains isolated from Ramsar Warm Springs, Iran. Results showed that a strain of gram-negative cocobacilli bacteria is highly capable to take up nickel in optimum pH about 6. The $Ni^+$ initial concentrations of 20–200 ppm have been studied. Uptake capacity of bacterial biomass regarding concentrations below 150 ppm is nearly constant; it is decreased from 150 to 200 ppm and at 200 ppm, is not any nickel taken up by bacterial biomass. After 60 min of contact time, nickel uptake reaches maximum by 53%. $Ni^+$ sorption mechanism was demonstrated mainly as bioaccumulation and biosorption was very limited. The presence of the other cations such as $Zn^{2+}$, $Cu^{2+}$ and $Pb^{2+}$ did not affect $Ni^+$ uptake. Nickel taken up by biomass can be easily recovered by $HNO_3$ (0.1 M). © 2004 Elsevier B.V. All rights reserved.

*Keywords:* Biomass; Bacteria; Nickel; Uptake; Removal

## 1. Introduction

Industrial processes release heavy and toxic metals to the environment. Microbial processes were considered as an appropriate solution in industrial waste treatment [1,2]. These methods are inexpensive, with high efficiency and feasibility to metal recovery [3,4]. The purpose of this investigation is the biological uptake of nickel and optimizing $Ni^+$ uptake conditions.

## 2. Materials and methods

All chemicals have been supplied by Merck.
The best strain has been chosen from 16 new various isolated and purified bacterial strains. Each strain was added to 150 ml of cultivation medium (GMS), after autoclaving

---

* Corresponding author. Tel.: +98 21 2237937; fax: +98 21 802 1412.
*E-mail address:* yasmin2802@yahoo.com (Y. Nazeri).

0531-5131/ © 2004 Elsevier B.V. All rights reserved.
doi:10.1016/j.ics.2004.11.124

and cooling, and then shaken at 100 rpm for 72 h. Bacterial biomass was isolated by centrifugation (12,000 rpm) for 5 min at 20 °C. To prepare a suspended solution, distilled water was added to the biomass in a volume to the extent of 10 times more than biomass weight. A total of 5 ml of the suspension was mixed with 50 ml of $Ni^{2+}$ solution $(Ni(NO_3)_2 \cdot 6H_2O)$ in pH 5 and shaken for 60 min at 100 rpm and 25 °C. The solution was separated from biomass by centrifugation. The nickel concentration of isolated solution was determined by AAS (GBC 932 plus).

## 3. Results and discussion

$Ni^+$ removal percentages in initial solution concentrations of 20, 50, 100, 150 and 200 ppm in pH 5 and 60 min of contact time are 50.815±0.039%, 49.658±0.021%, 45.056±0.05%, 43.553±0.016% and near 0%, respectively.

The removal percentages in 5, 15, 30, 40, 50, 55, 60 and 65 min, pH 5 and 150 ppm were 10.221±0.035%, 12.795±0.046%, 19.768±0.028%, 25.657±0.032%, 35.650±0.021%, 45.787±0.016%, 51.800±0.013% and 53.450±0.016%, respectively.

Removal percentage for initial concentration of 150 ppm and 60 min, in pH 1, 2, 3, 4 and 5, increases gradually (6.944±0.032%, 14.479±0.041%, 18.245±0.025%, 18.940±0.013% and 19.650±0.021%, respectively). The maximum uptake took place in pH 6 and the removal percentage is 43.864±0.034%.

Presence of the other cations such as $Zn^{2+}$, $Cu^{2+}$ and $Pb^{2+}$, with the concentration of 150 ppm, pH 5 and 60 min, causes a decrease in nickel uptake from 43.55% to 34.12±0.023%, 32.78±0.026% and 31.20±0.031%, respectively.

Uptake by living biomass in 20 ppm is 50.815±0.028% and by nonliving biomass is 5.300±0.032%.

By $HNO_3$ (0.1 M) nickel desorption percentage is 60±0.05%.

The study of gamma ray effect shows that the Ni removal by biomass with 6 Gy absorbed dose has been reduced.

## 4. Conclusion

This investigation shows that $Ni^+$ uptake by this biomass depends on metabolism and is pH-dependent. The equilibrium time will be reached within about 60 min. The uptake is not considerably reduced by the presence of the other cations, so it can be used as a moderate selective biomass for nickel uptake. Nickel could be desorbed very easily from biomass by using $HNO_3$ (0.1 M).

## References

[1] A. Esposito, F. Pagnanelli, F. Veglio, pH-related equilibria models for biosorption in single metal systems, Chem. Eng. Sci. 73 (2002) 307–313.
[2] K. Tsekova, G. Petrov, Removal of heavy metals from aqueous solution using *Rhizopus delemar* mycelia in free and polyurethane, Bulg. Acad. Sci. 57 (2002) 629–633 (bond from).
[3] F. Pagnanelli, et al., Pergamon, metal specification and pH effect on Pb, Cu, Zn and Cd biosorption on to sphaezotilusnatans: langmuir type empirical model, Water Res. 37 (2003) 627–633.
[4] K. Sekhar Chandra, et al., Removal of heavy metals using a plant biomass with reference to environmental control, Elsevier 68 (2003) 37–45.

# Biosorption of Cs$^+$ by new bacterial strain (NRC-BT-2)

Y. Nazeri[a,*], S. Sadeghi[a], M. Rabbani[b], H. Ghafourian[b]

[a]*Islamic Azad University, North branch of Tehran, Chemistry Department, Iran*
[b]*Nuclear Researchc Center, AEOI, Tehran, Iran*

**Abstract.** Biosorption of Cs$^+$ has been investigated by new bacterial strain of Gram-positive cocobacilli bacteria isolated from Ramsar warm springs, Iran. In the initial concentration of 100 ppm the binding sites are occupied considerably, and the removal efficiency is 19.954%. The maximum biosorption capacity is obtained in pH 6 and 60 min of contact time. Presence of the other cations such as Na$^+$, K$^+$, Sr$^{2+}$ and Co$^{2+}$ reduced the biosorption capacity, but sodium acid, high temperature and gamma ray are ineffective. Biosorbed Cs, successfully recovered by oxalic acid (0.1 M), nitric acid (0.1 M), sodium carbonate (0.1 M) and EDTA (0.025 M), is over 95%. © 2004 Elsevier B.V. All rights reserved.

*Keywords:* Biomass; Cesium; Biosorption; Removal

## 1. Introduction

The term biosorption describes the physicochemical interaction between dissolved ionic species and binding sites in cell walls and exopolymers [1]. Uptake of heavy metals by microorganisms is done through metabolism-dependent or -independent mechanisms [2,3]. The purpose of this investigation is biological uptake of Cs$^+$ and optimizing its conditions.

## 2. Materials and methods

All chemicals have been supplied from Merck.

The strain was added to 150 ml of cultivation medium of GMS after autoclaving and cooling. Then it was shaken at 150 rpm (Heidolph-unemax1010) for 72 h in 25 °C. The biomass was isolated by centrifugation (12,000 rpm) (Heraeus), for 5 min in 20 °C. To

---

\* Corresponding author. Tel.: +98 21 2237937; fax: +98 21 8021412.
*E-mail address:* yasmin2802@yahoo.com (Y. Nazeri).

0531-5131/ © 2004 Elsevier B.V. All rights reserved.
doi:10.1016/j.ics.2004.11.049

prepare the suspended solution, distillated water was added to the biomass in a volume to the extent of 10 times more than biomass' weight. About 5 ml of the suspension was mixed with 50 ml of $Cs^+$ solution and shaken for 60 min at 100 rpm and 25 °C. The solution was separated from biomass by centrifugation (13,000 rpm) at 10 °C. The $Cs^+$ concentration was determined by AAS (Varian AA-20).

## 3. Results and discussion

The results of cesium removal in initial concentrations of 10, 50, 100, 150, 200 and 250 ppm are 39.850%, 25.560%, 19.954%, 14.213%, 10.871%, 8.845%, respectively, in pH 6 and 60 min. The reduction of $Cs^+$ removal by increasing initial concentration is the result of occupation of available binding sites on cell walls.

The removal percentage for 5, 60, 180, 360 and 960 min of contact time for 100 ppm and pH 6 were 20.238%, 23.254%, 20.754%, 18.845% and 16.160% respectively. Removal of $Cs^+$ has been increased up to 60 min, so the equilibrium contact time is 60 min.

In pH 1, 3, 5, 6, 7, 8 and 10 in initial concentration of 100 ppm for 60 min, the results were respectively 0.010%, 0.020%, 16.582%, 18.825%, 18.425%, 17.300% and 15.730%. In pH 1 and 3 the removal percentage revealed to 0, and by increasing pH it has been increased because of the winning of other cations in the competition with $H^+$ to bind on the sites.

$Cs^+$ removal percentages by living and nonliving biomass in pH 6 for initial concentration of 100 ppm and 60 min are 23.53% and 23.25%, respectively.

$Cs^+$ removal percentages by biomass contacted with sodium azid (10 mM) and living biomass are 23.18% and 23.53%, respectively, for initial concentration of 100 ppm and 60 min in pH 6.

The removal percentages for 100 ppm, pH 6 and 60 min by biomass with the absorbed dose of 0, 100, 200, 300 and 400 Gy from gamma ray are respectively 19.75%, 18.16%, 17.04%, 16.25% and 16.10%.

In the presence of, $Na^+$, $K^+$, $Sr^{2+}$ and $Co^{2+}$, cesium removal percentages are respectively 4.378%, 3.276%, 4.030% and 1.736%. Initial concentrations of Cs and other cations were 100 ppm and the study has been done in pH 6 for 60 min.

By oxalic acid (0.1M), $HNO_3$ (0.1M), $Na_2CO_3$ (0.1M), EDTA (0.025M), the cesium desorption percentages are respectively 99.30%, 99.29%, 97.12% and 95.69%.

## 4. Conclusion

According to the obtained results the uptake mechanism is biosorption and pH dependent. Even nonliving biomass takes up $Cs^+$. Sodium acid and gamma ray have no effect on removal but the presence of the other cations reduces it.

## References

[1] T. Pumpel, K.M. Paknik, Bioremediation technologies for metal containing wastewaters using metabolically active microorganisms, Adv. Microbiol. 48 (2000) 135–169.
[2] A.M. Marques, et al., Removal of uranium by an exopolysacharide from Pseudomonas sp., Appl. Microbiol. Biotechnol. 34 (1990) 429–431.
[3] K.J. Blackwell, I. Sngletn, J.M. Tobin, Biosorption of uranium by a spergillus fumigates, Biotechnol. Tech. 13 (1995) 695–699.

www.ics-elsevier.com

# The effect of low dose-rate γ-radiation on the chemiluminescence of blood serum at chronic inflammation in rats

N. Klimenko[a], M. Onyshchenko[a,*], N. Dikij[b], E. Medvedeva[b]

[a]*Kharkov State Medical University, Lenin Ave. 4, Kharkov 61022, Ukraine*
[b]*NSC Kharkov Institute of Physics and Technology, Academic Str. 1, Kharkov 61108, Ukraine*

**Abstract.** Chemiluminescence assay was used for the indirect estimation of DNA damage at low dose-rate γ-irradiation under chronic inflammation in rats. A good correlation between blood serum chemiluminescence intensity and dose was found. © 2004 Published by Elsevier B.V.

*Keywords:* Chronic inflammation; Chemiluminescence; Low dose-rate γ-irradiation

## 1. Introduction

Inflammation has long been associated with the increased risk of cancer. Over the last decade, it has been clearly demonstrated that chronic inflammation significantly contributes to neoplastic transformation [1]. At the same time, a lot of statistical data concerning radiation carcinogenesis have been accumulated, as well as molecular mechanisms of DNA damage and repair from ionising radiation have been examined by the present moment of time [2,3]. However, little is known about the interaction of low intensive irradiation and chronic inflammation in mutagenesis. This problem is especially important for Ukraine, Belarus and Russia, where the large populations of people suffer from the consequences of the Chernobyl accident.

## 2. Materials and methods

The carrageenan-induced air-pouch-type granulomatous inflammation served as a model of chronic inflammation [4]. Twelve milliliters of sterile air was injected s.c. on the back of the rat to make an air pouch oval in shape. Thirty hours later, 2 ml of a 0.5% (w/v) delivered solution in PBS

---

\* Corresponding author. Tel.: +7 38 057 3352895; fax: +7 38 057 3353564.
*E-mail address:* nionish@mail.ru (M. Onyshchenko).

0531-5131/ © 2004 Published by Elsevier B.V.
doi:10.1016/j.ics.2004.11.038

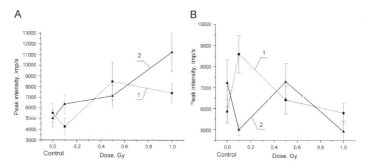

Fig. 1. (A) Dose dependence of the chemiluminescence peak intensity. Irradiation was performed to the 3rd day after inflammation induction. Curve 1: immediately after irradiation; curve 2: 7th day after inflammation induction. (B) Dose dependence of the chemiluminescence peak intensity. Irradiation was performed to the 7th day after inflammation induction. Curve 1: immediately after irradiation; curve 2: 14th day after the inflammation induction.

was injected into the air pouch. Gamma-irradiator OB-6 ($^{137}$Cs, 20 Ci, 14.3 µGy/s on the distance 1 m) was used. Doses of 0.1, 0.5 and 1.0 Gy were delivered within 4.8, 24 and 48 h correspondingly. For the chemiluminescence assay, 1 ml of blood serum was injected into the thermostatically controlled chamber, placed above the photomultiplier tube.

## 3. Results and discussion

As can be seen from Fig. 1A, radiation response to the 3rd day of chronic inflammation was not found out immediately after irradiation (curve 1) but was revealed 4 days after irradiation (curve 2). Possibly, the accumulation of chemical products of radiation influence, such as products peroxidate oxidation of lipids, and the structural damage of cells occur to this moment of time, which define radiation effect in the absence of irradiation.

As can be seen from Fig. 1B, radiation response at inflammation on 7th day is distinct from that at inflammation on 3rd day. First, the inflammation on the 7th day radiation response was immediate and, apparently, was defined by the immediate action of the ionizing radiation, or it was mediated by the some short-living molecules. Second, a higher sensitivity to the action of ionizing radiation at dose of 0.1 Gy and the absence of the immediate response were found out at big doses (0.5 and 1.0 Gy; Fig. 1B, curve 1). It is possible to explain the last phenomenon by failure of antioxidative protective system at small doses. Noted bimodal character of dose-dependence curve at the second group of animals (Fig. 1B, curve 2) was previously described by Burlakova et al. [5] in irradiated initially intact animals; however, in our case, the effect at doses of 0.1 and 1.0 Gy, on the contrary, was less than in the control and at dose of 0.5 Gy, which is probably caused by inflammation, which independently activates peroxidate oxidation and alters radiation response.

## References

[1] S. Murthy, J.D. Winkler, Inflammation and oncogenesis, Inflamm. Res. 51 (2002) 76.
[2] A. Dorozynski, Chernobyl damaged health, says study, BMJ 309 (1994) 1321.
[3] Chernobyl 10 years on, BMJ 312 (1996) 1052–1053.
[4] Q. Jiang, B.C. Blount, B.N. Ames, 5-Chlorouracil, a marker of DNA damage from hypochlorous acid during inflammation, J. Biol. Chem. 278 (35) (2003) 32834–32840.
[5] E.B. Burlakova, et al., Radiation effects at small doses, Rad. Biol. and Radioecol. 36 (4) (1996) 610–632.

# An intercomparison exercise for thoron gas measurement

T. Ishikawa[a,*], A. Cavallo[b], S. Tokonami[a]

[a]Radon Research Group, National Institute of Radiological Sciences, 4-9-1 Anagawa Inage-ku, Chiba, 263-8555, Japan
[b]Applied Physics, Environmental Measurements Laboratory, USA

**Abstract.** Four devices for measuring thoron gas were compared using an exposure test chamber. Average concentrations measured with these devices ranged from 650 to 1000 Bq m$^{-3}$. It is needed to make intercomparison exercises at higher thoron concentrations. © 2004 Elsevier B.V. All rights reserved.

*Keywords:* Thoron; Measurement; Intercomparison; Chamber

## 1. Introduction

Several types of detection techniques have been developed for measuring thoron gas [1]. Intercomparison studies on measurement methods are useful to check the accuracy of measurements. Although many intercomparison studies have been conducted for radon, those for thoron have been scarcely done. In the present study, four methods for measuring thoron gas were compared using the Environmental Measurements Laboratory (EML; New York) environmental chamber. The results of the intercomparison exercise are presented.

## 2. Materials and methods

Four types of devices were used: a two-filter tube, an ionization chamber with a pump, a radon/thoron discriminative monitor using a silicon semiconductor detector, and nuclear truck detectors. A brief description of each device is as follows: (1) the two-filter tube [2]; thoron concentration was estimated from gross alpha counting of filtered samples. (2) The ionization chamber; this device was originally developed for measuring radon but can be

\* Corresponding author. Tel.: +81 43 206 3099; fax: +81 43 206 4097.
*E-mail address:* tetsuo_i@nirs.go.jp (T. Ishikawa).

0531-5131/ © 2004 Elsevier B.V. All rights reserved.
doi:10.1016/j.ics.2004.09.030

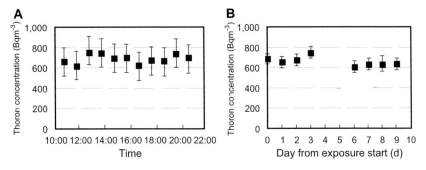

Fig. 1. Variation of thoron concentration in the exposure chamber: hourly variation (A) and daily variation (B).

made sensitive to thoron by sucking air with a pump. Thoron concentration was estimated from a difference in measured concentration between the pumping and diffusion modes. (3) The radon/thoron discriminative monitor; this monitor employs a silicon semiconductor detector, which enables to discriminatively measure radon and thoron using alpha spectrometry. (4) The nuclear track detectors; two sets of detectors (20 detectors each) were exposed in the chamber.

The detail of the EML environmental chamber is described elsewhere [3]. It is a walk-in chamber that has a volume of 30 m$^3$. Thoron was generated from a $^{228}$Th source located in the chamber. The intercomparison was divided into two phases. First, the two-filter tube, the ionization chamber, and the radon/thoron discriminative monitor were compared. Secondly, two sets of nuclear track detectors were exposed for several days, measuring thoron concentration with the radon/thoron discriminative monitor.

## 3. Results and discussion

Variation of thoron concentration is shown in Fig. 1A and B (hourly and daily variation, respectively). There was no significant difference in thoron concentration. As a result, the average thoron concentration measured with the radon/thoron monitor was not significantly different between the first and second phases. Thus, the results for the nuclear truck detectors could be compared with those for other three devices.

Average concentrations measured with four devices were as follows: 650 Bq m$^{-3}$ (radon/thoron discriminative monitor), 800 Bq m$^{-3}$ (ionization chamber), 870 Bq m$^{-3}$ (two-filter tube), and 1000 Bq m$^{-3}$ (nuclear track detectors). An average of four values was about 830 Bq m$^{-3}$. For some devices, counting errors were large due to the relatively low thoron concentration. It is needed to make intercomparison exercises at higher thoron concentrations.

## References

[1] C. Nuccetelli, F. Bochicchino, The thoron issue: monitoring activities, measuring techniques and dose conversion factors, Radiat. Prot. Dosim. 78 (1998) 59–64.
[2] E.O. Knutson, et al., EML thoron gas measurements, Radiat. Prot. Dosim. 56 (1994) 263–266.

# Particle size measurement of radon decay products using MOUDI and GSA

S. Tokonami*, K. Fukutsu, Y. Yamada, Y. Yatabe

*Radon Research Group, National Institute of Radiological Sciences, Japan*

**Abstract.** In order to assess the dose due to inhalation of radon decay products, it is important to characterize their particle size distribution in various environments. For this purpose, a cascade impactor method and a graded screen array (GSA) method were used in the present study. This paper describes their performance of these two techniques and several data obtained in the environment. © 2004 Elsevier B.V. All rights reserved.

*Keywords:* Radon decay product; Dose assessment; Particle size distribution; Underground mine; Radon chamber

## 1. Introduction

It is well known that particle size distribution of radon decay products depends on the dose conversion factor. When assessing the dose, two different approaches are used: dosimetric and epidemiological approaches. However, two approaches provide two different estimates. The particle size distribution can be regarded as one of important physical parameters to facilitate their correspondence. From this point of view, both home and mine aerosols needed to be sufficiently characterized. In order to evaluate the particle size distribution with a wide range, two sampling techniques were used in the present study. This paper describes their performance of these two techniques and several data obtained in the environment.

## 2. Materials and methods

Fig. 1A shows sectional views of the graded screen array (GSA) holder [1]. A cascade impactor is commercially named Micro-Orifice Uniform Deposit Impactor (MOUDI) [2].

---

\* Corresponding author. Anagawa 4-9-1, Inage, Chiba 263-8555. Tel.: +81 43 206 3105; fax: +81 43 206 4098.
*E-mail address:* tokonami@nirs.go.jp (S. Tokonami).

0531-5131/ © 2004 Elsevier B.V. All rights reserved.
doi:10.1016/j.ics.2004.09.049

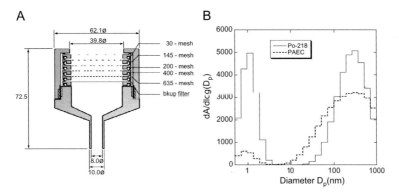

Fig. 1. (A) Sectional view of GSA sampler. (B) Activity-weighted particle size distribution (Po-218 and PAEC) measured in an underground mine.

The GSA holder consists of five stainless steel wire screens (30-, 145-, 200-, 400-, 635-mesh) and a backup filter (Whatman GF/F). The flow rate of MOUDI is optimally designed for air sampling with 30 L/min. The flow rate of the GSA is set at around 24 L/min with an effective flow area of 11.2 cm$^2$. After a 5-min sampling, alpha particles on 10 impaction stages (backup filter included) or five screens and filter are simultaneously counted using proper number of ZnS scintillation detecting systems. The counting data are analyzed by the Thomas method. The analyzed data are unfolded to obtain an activity weighted particle size distribution by the Expectation–Maximization algorithm.

## 3. Results and discussion

One measurement was made in well-controlled environment (e.g., NIRS radon chamber) in comparison with the ambient particle distribution using the Scanning Mobility Particle Sizer (SMPS). Another measurement was carried out in an underground mine located in Gifu prefecture. Fig. 1B exemplifies the activity-weighted particle size distribution (Po-218 and PAEC) measured in the underground mine. Table 1 summarizes typical data obtained by these techniques.

## 4. Conclusion

Two techniques for particle size distribution measurement were shown in the present study. The performances were investigated in various environments. From the data obtained in an underground mine, their particle sizes for unattached and attached modes

Table 1
Examples of particle size distribution of radon decay products (expressed by GMD and GSD)

| Site | Po-218 | PAEC | Remarks |
|---|---|---|---|
| Underground mine | 0.9 nm (GSD: 1.5) | 0.8 nm (GSD: 1.5) | |
| | 222 nm (GSD: 2.2) | 162 nm (GSD: 3.1) | |
| NIRS radon chamber | 101 nm (GSD: 1.4) | 107 nm (GSD: 1.3) | Carnauba wax aerosols generated. |

SMPS measurement results for radon chamber GMD=90 nm, GSD=1.4.

were consistent with other studies. Further data should be accumulated with these techniques for more accurate dose assessment.

## References

[1] S. Tokonami, Health Phys. 78 (2000) 74–79.
[2] S. Tokonami, Radiat. Prot. Dosim. 81 (1999) 285–290.

www.ics-elsevier.com

# Practicality of the thoron calibration chamber system at NIRS, Japan

## Y. Kobayashi, S. Tokonami*, H. Takahashi, W. Zhuo, H. Yonehara

*Radon Research Group, National Institute of Radiological Sciences, Anagawa 4-9-1, Inage, Chiba 263-8555, Japan*

**Abstract.** A thoron calibration chamber system has been set up at NIRS in Japan. Another exposure condition in relative humidity (approximately 7%, 40% and 70%) was examined in this study. Thoron and radon concentrations were continuously measured with commercial measuring device (RAD7, Durridge, USA) throughout the exposure period. The present study mainly describes another exposure conditions and its relevant result from the viewpoint of the practicality. © 2004 Elsevier B.V. All rights reserved.

*Keywords:* Thoron; Calibration; Continuous measurement

## 1. Introduction

When radon concentration is measured with passive methods, thoron gas can be functioned as a noise for passive radon monitors. Thus, thoron sensitivity of passive radon monitor is required to evaluate the actual value of radon. In order to examine thoron sensitivity for passive radon detectors, Tokonami et al. [1,2] set up the thoron chamber system and devised the prompt measurement system with a scintillation cell. This thoron chamber system has been used with a continuous measuring device "RAD7" based on the electrostatic collection method, and practicality and stability of the chamber system were examined.

## 2. Materials and methods

The thoron calibration chamber system consists of four parts: the thoron gas generator, the exposure chamber, the environmental monitor and the RAD7. In addition, the prompt

---

* Corresponding author. Tel.: +81 43 206 3105; fax: +81 43 206 4098.
  *E-mail address:* tokonami@nirs.go.jp (S. Tokonami).

0531-5131/ © 2004 Elsevier B.V. All rights reserved.
doi:10.1016/j.ics.2004.09.052

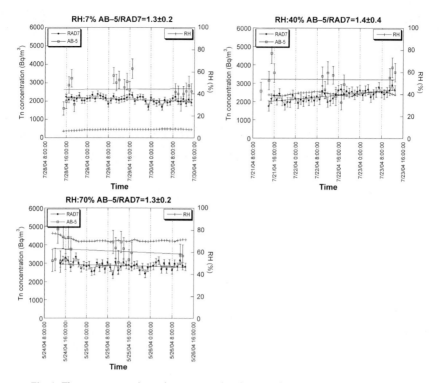

Fig. 1. Thoron concentration values measured under several relative humidity conditions.

measurement system with the Pylon model AB-5 is used for a calibration of the RAD7. The flow rate of gas supply is 20 L/min and measurement period is 48 h. The experiments were carried out under following three relative humidity conditions: (1) 7% as dry air, (2) 40% as indoor air and (3) 70% as humid air.

## 3. Results

The thoron concentrations measured with two type measuring devices under each relative humidity condition are shown in Fig. 1. The calibration coefficient of the RAD7 under each condition is roughly 1.3, which is nearly constant.

## 4. Conclusion

Thoron concentration can be uniquely controlled with relative humidity. It was checked that not only the thoron concentrations but also the calibration coefficients of the RAD7 under each relative humidity condition are nearly constant. The stability and practicality of the thoron chamber system with a continuous measuring device were obtained in this study.

### References

[1] S. Tokonami, M. Yang, T. Sanada, Health Phys. 80 (2001) 612–615.
[2] S. Tokonami, et al., Rev. Sci. Instrum. 73 (2002) 69–72.

# Measurements of radon, thoron and their progeny in a dwelling in Gifu prefecture, Japan

C. Németh[a,b,*], S. Tokonami[a], T. Ishikawa[a],
H. Takahashi[a], W. Zhuo[a], M. Shimo[c]

[a]*Radon Research Group, National Institute of Radiological Sciences, Japan*
[b]*University of Veszprém, Hungary*
[c]*School of Health Sciences, Fujita Health University, Japan*

**Abstract.** Radon and thoron and their progeny were measured using different methods in a dwelling in Gifu prefecture, Japan. The results indicate a considerably higher background level concerning radon and thoron compared to the average in Japan. © 2004 Elsevier B.V. All rights reserved.

*Keywords:* Radon; Thoron; High background natural radiation; Dose assessment

## 1. Introduction

Due to the rocky neighborhood, consisting of mostly granite with high radium content, elevated radon concentration was suspected in a territory of Gifu prefecture situated in the middle of Japan. First, radon concentrations in water collected here were measured and it was found considerably high. Suspecting relatively high indoor concentrations of radon and radon progeny, their measurements were conducted. In addition, thoron and thoron progeny were measured as well. Alpha track detectors were used for 3- and 1-month periods, and short-term continuous measurements were also conducted. Using the results thus obtained the annual dose contribution was assessed.

## 2. Methods, results, and discussion

The radon concentration measured in the spring-water at the site was $2638 \pm 86$ Bq/L. Nuclear track detectors (Radopot [1,2]) were placed at five rooms for a 1- and 3-month-long measurements. The average was 43 Bq m$^{-3}$.

---

* Corresponding author. Radon Research Group, National Institute of Radiological Sciences, Anagawa 4-9-1, Inage, Chiba 263-8555, Japan. Tel.: +81 43 206 3111; fax: +81 43 206 4098.
E-mail address: csaba@fml.nirs.go.jp (C. Németh).

Continuous measurements were executed by two AlphaGuards for 17 h long on 24 September, 2003. The averages from the measurements of AlphaGuards I and II (AlphaGuards) were $79\pm25$ Bq m$^{-3}$ (with range 13–129) and $69\pm20$ Bq m$^{-3}$ (with range 19–98), respectively, which are in good accordance with the 1-month-long track detector result. The AlphaGuards I and II were put 5 and 50 cm away from the wall, respectively.

The equilibrium equivalent radon and thoron concentrations (EERC and EETC, respectively) were measured by a Pylon AB-5 equipment. The average for EERC is 24.9 and 6.8 Bq m$^{-3}$ for EETC. The EERC–EETC ratio was 3.6. Using the EERC data, the equilibrium factor ($F$) was estimated at $0.33\pm0.12$.

Another short-term measurement conducted at the same site from June 7 to 10, 2004. Two RAD7s were used for these measurements. The RAD7 I and II were put 5 and 50 cm away from the wall, respectively.

The average values of the EERC and EETC measured by the Pylon AB-5 were 25.7 (range: 0–101) and 10.2 (range: 0.2–14) Bq m$^{-3}$, respectively. The ratio of EERC/EETC was 2.5. During this period the EERC and EETC were determined by alpha spectrometry (grab-filter method) as well. The averages of EERC and EETC were 53.9 (range: 19–137) and 18.4 (range: 2–47) Bq m$^{-3}$, respectively. These values were about two times higher than those measured by the Pylon AB-5. The ratio of ERC/EETC was 2.9 in the case of alpha spectrometry measurements.

The 3-month-long measurements allow us to do a rough dose assessment; however, the dose calculated using these results would be a preliminary one. If the information on seasonal variation in radon/thoron concentrations at the site is obtained, more precise calculation could be made. So, we used a common conversion of average radon concentration to dose equivalent, which is 1.7 mSv (annual effective dose) per 100 Bq m$^{-3}$ (average radon concentration at home). This is based on an occupation factor of 0.8 (this means 7000 h/year) and an F value of 0.4 (ICRP-65) [3]. Using these data and the average of the 3-month-long measurements (43 Bq m$^{-3}$), the annual dose contribution due to radon was calculated at 0.73 mSv. If we choose the dose calculation model given in UNSCEAR 2000 Report [4] the dose can be calculated as follows: annual effective dose=43 Bq m$^{-3}\times0.4\times7000$ h$\times9$ nSv (Bq h m$^{-3}$)$^{-1}$=1.1 mSv.

The dose due to thoron progeny was derived using the average value of 10 Bq m$^{-3}$ (roughly assessed) for EETC (UNSCEAR 2000): 10 Bq m$^{-3}\times7000$ h$\times40$ nSv (Bq h m$^{-3}$)$^{-1}$=2.8 mSv.

## 3. Conclusion

Based on the results, it can be assumed that this territory presents a higher background radiation area than the average Japanese places. A rough estimation of the population dose resulting from radon and thoron in indoor air was made using different calculation methods. It can be concluded that the dose contribution from thoron seems to be about three times higher than the dose due to radon, emphasizing that thoron could play an important role in exposure from natural radiation.

## References

[1] S. Tokonami, et al., Radiat. Prot. Dosim. 106 (2003) 71–75.
[2] W. Zhuo, et al., Rev. Sci. Instrum. 73 (2002) 2877–2881.
[3] ICRP Publication No. 65: 1994.
[4] UNSCEAR 2000, Sources and effect of ionizing radiation, Vol. I: sources; (p. 107).

# Soil radon flux and outdoor radon concentrations in East Asia

Weihai Zhuo[a,*], Masahide Furukawa[a], Qiuju Guo[b], Yoon Shin Kim[c]

[a]*Radon Research Group, National Institute of Radiological Sciences, Chiba 263-8555, Japan*
[b]*Department of Technical Physics, School of Physics, Peking University, Beijing 100871, China*
[c]*Institute of Environmental and Industrial Medicine, Hanyang University, Seoul 133-791, South Korea*

**Abstract.** Soil $^{222}$Rn flux and outdoor $^{222}$Rn concentrations were continuously measured at 20 sites in China, Japan and Korea. Annual averaged $^{222}$Rn flux and outdoor $^{222}$Rn concentrations ranged from 6.9 to 59.6 mBq m$^{-2}$ s$^{-1}$ and 5.3 to 17.0 Bq m$^{-3}$ among those sites. Seasonal variations of both $^{222}$Rn flux and outdoor $^{222}$Rn were observed at most of the sites. A significant positive correlation exists between the soil $^{222}$Rn flux and outdoor $^{222}$Rn concentrations. © 2004 Elsevier B.V. All rights reserved.

*Keywords:* Radon-222 flux; Outdoor radon-222; Radium-226; Terrestrial gamma radiation

## 1. Introduction

Soil $^{222}$Rn flux is an important parameter for validation of global chemistry and transport computer models whenever $^{222}$Rn is used as a tracer. On the other hand, soil $^{222}$Rn flux is also a main parameter for identifying regions likely to have a strong potential of radon exposure. In this study, both soil $^{222}$Rn flux and outdoor $^{222}$Rn were continuously measured at the 20 typical sites in East Asia throughout a year. The correlations among soil $^{222}$Rn flux, outdoor $^{222}$Rn concentration, soil $^{226}$Ra content and terrestrial gamma radiation were analysed.

## 2. Materials and methods

An aluminium box (100×80×80 cm$^3$) with an opened bottom was embedded 20 cm into soil for accumulating $^{222}$Rn. The accumulator has a nearly stable air exchange rate of 0.25 h$^{-1}$ throughout measuring periods. $^{222}$Rn accumulated in the box was measured by the passive $^{222}$Rn and $^{220}$Rn discriminative monitors [1]. The soil $^{222}$Rn flux was

\* Corresponding author. Tel.: +81 43 206 3111; fax: +81 43 206 4098.
*E-mail address:* whzhou@nirs.go.jp (W. Zhuo).

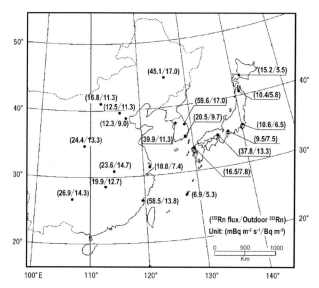

Fig. 1. Annual averaged soil $^{222}$Rn flux and outdoor $^{222}$Rn concentrations at 20 selected sites in East Asia.

calculated from the $^{222}$Rn concentration, air exchange rate and the height of the box. The passive monitor was also set outside the box for measuring of outdoor $^{222}$Rn.

Taking the geographical and meteorological distributions into account, the 20 sites shown in Fig. 1 were selected for the measurements. The measurements were made in about 3-month integrating intervals in the fiscal year of 2003. Soil $^{226}$Ra content and terrestrial gamma radiation were also measured for each site.

## 3. Results and discussion

Annual averaged soil $^{222}$Rn flux and outdoor $^{222}$Rn concentration in those sites were 24.0 mBq m$^{-2}$ s$^{-1}$ and 10.7 Bq m$^{-3}$, respectively. Averaged soil $^{222}$Rn fluxes and outdoor $^{222}$Rn varied from site to site (see Fig. 1). The ratios of the maxima to minima are about 8.6 and 3.2 for soil $^{222}$Rn flux and outdoor $^{222}$Rn, respectively. Furthermore, seasonal variations of soil $^{222}$Rn flux and outdoor $^{222}$Rn were observed at most of the sites. These phenomena can be explained as the inhomogeneous distributions of soil $^{226}$Ra content, soil texture and meteorological conditions among those sites.

Correlation between the annual averaged soil $^{222}$Rn flux and the soil $^{226}$Ra content among those sites was significant ($R=0.8394$). It indicates that the annual soil $^{222}$Rn flux can be estimated from the soil $^{226}$Ra content. On the other hand, even though the correlation between outdoor $^{222}$Rn concentrations and soil $^{226}$Ra contents was weak, a significant correlation ($R=0.7874$) was found between the annual outdoor $^{222}$Rn concentration and the annual soil $^{222}$Rn flux. It suggests that outdoor $^{222}$Rn concentrations can be roughly estimated from soil $^{222}$Rn fluxes.

## Reference

[1] W. Zhuo, et al., A simple passive monitor for integrating measurements of indoor thoron concentrations, Rev. Sci. Instrum. 73 (2002) 2877–2881.

# Estimation of radon-222 exhalation rate and control of radon-222 concentration in ventilated underground space

J. Moriizumi[a,*], M. Mori[a], E. Sasao[b], H. Yamazawa[a], T. Iida[a]

[a]*Graduate School of Engineering, Nagoya University, Chikusa-ku, Nagoya 464-8603, Japan*
[b]*Tono Geoscience Center, Japan Nuclear Cycle Development Institute, Toki City, Gifu Pref. 509-5212, Japan*

**Abstract.** Representative radon-222 exhalation rates into a tunnel with diverse and rugged surface have been evaluated by measurements of radon-222 concentration increases along airflow. Using these results, a numerical analysis has been conducted to optimize distribution of airflow in the tunnel and to minimize lung doses to personnel in it. © 2004 Elsevier B.V. All rights reserved.

*Keywords:* Radon-222; Exhalation rate; Airflow control; Numerical analysis; Lung dose

## 1. Introduction

Closed underground spaces have a tendency to accumulate radon-222 ($^{222}$Rn) gas emitted from its inner surface. As a case study, representative $^{222}$Rn exhalation rates, which are essential to control $^{222}$Rn concentrations in any closed space, have been estimated in a mine cavity tunnel with diverse and rugged surface. On the basis of these estimations, optimum distribution of airflow in the tunnel has been numerically analyzed to minimize lung doses to personnel/users in it.

## 2. Measurements of representative radon-222 exhalation rates

The tunnel studied was 1.8 m in height, 2.4 m in cross sectional width and 1300 m in total length, and had a relatively complicated structure: two airflow intakes, main horizontal tunnel which branches into two parallel tunnels and some short dead-end branches. A blower fan placed at the confluence of the parallel tunnels continuously discharged inner air and created one directional airflow in the tunnel. Wide variation in its surface condition, i.e., bare base rock which was sometimes plunked with concrete or

---

* Corresponding author. Tel.: +81 52 789 5134; fax: +81 52 789 3782.
*E-mail address:* j-moriizumi@nucl.nagoya-u.ac.jp (J. Moriizumi).

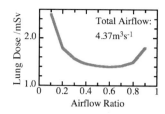

Fig. 1. The calculation of annual (2000 h) lung dose for the airflow ratio between two parallel tunnels [2].

coated with mortar or synthetic resin, and rugged shape made it difficult to obtain valid and representative $^{222}$Rn exhalation rates by closed chamber methods. Under steady state, increase of $^{222}$Rn concentration $\Delta C$ [Bq m$^{-3}$] between a given interval along airflow in the tunnel can be expressed by an equation $\Delta C=ES/Q$, where $E$, $S$ and $Q$ are exhalation rate, surface area and airflow rate in the interval, respectively. Mean $^{222}$Rn concentrations have been measured from 12 December 2002 to 25 January 2004 at 13 intervals through the tunnel with cup-type passive $^{222}$Rn monitors with CR-39 SSNTD [1]. The intervals were decided in consideration of shapes and surface conditions of the tunnel. Airflow rates were calculated from wind velocity profiles on the cross sections of the tunnel. Total airflow rate was 4.37 m$^3$ s$^{-1}$, and ratio of flow rates between the parallel tunnels was 0.35:0.65. Radon-222 exhalation rates, calculated by using these observations, were higher in the intervals with bare base rock surface than those with coated surface.

## 3. Discussions—numerical analysis of radon-222 concentration profiles

A compartment model based on the 13 intervals was developed to predict $^{222}$Rn concentration profile in the tunnel by numerical analysis. Stationary solutions of $^{222}$Rn concentrations of simultaneous differential equations for the 13 intervals were calculated by Runge–Kutta method. The calculation by using estimated exhalation rates and airflow rates agreed well with the observed $^{222}$Rn concentration profile.

Since the calculated $^{222}$Rn concentration profile in the tunnels varied according to the airflow distribution, the dependency of lung dose on airflow rate has been estimated. In the case that one stays uniformly in every horizontal compartment, his/her lung dose is expected to be minimized at an airflow ratio of 0.6:0.4 between the parallel tunnels as shown in Fig. 1, assuming that residence time in each compartment can be weighted by the ratio of length of the compartment to the total length of the tunnel.

## 4. Conclusion

The method of evaluating representative exhalation rates has been developed for ventilated underground space with diverse and rugged surface. From the observations of exhalation rates and airflow rates, $^{222}$Rn concentration profiles were successfully calculated by numerical analysis for various airflow conditions. Optimal airflow condition can be predicted for various operations in the underground space.

## References

[1] T. Iida, R. Nurishi, K. Okamoto, Passive integrating $^{222}$Rn and $^{220}$Rn cup monitors with CR-39 detectors, Environ. Int. 22 (1) (1996) S641–S647.
[2] UNSCEAR, Sources and Effects of Ionizing Radiation, vol. 1, United Nations, New York, 2000, p. 107.

# Measurements of concentrations and its ratio of radon decay products in rainwater by gamma-ray spectrometry with a low background germanium detector

M. Takeyasu[a,*], T. Iida[b], T. Tsujimoto[c], K. Yamasaki[d]

[a]*Tokai Works, Japan Nuclear Cycle Development Institute, 4-33 Muramatsu, Tokai-mura, Naka-gun, Ibaraki 319-1194, Japan*
[b]*Graduate School of Engineering, Nagoya University, Nagoya, Japan*
[c]*Electron Science Institute, Japan*
[d]*Research Reactor Institute, Kyoto University, Kyoto, Japan*

**Abstract.** The concentrations of $^{214}$Pb and $^{214}$Bi in rainwater were measured by a low background Ge detector. The concentration of $^{214}$Pb had an inverse relation with rainfall rate in some rainfall events, and the concentration ratio of $^{214}$Bi to $^{214}$Pb had a weak inverse relation with rainfall rate in all rainfall events. These relations were the same as those obtained by the removal model of radon decay products from stratiform clouds. © 2004 Elsevier B.V. All rights reserved.

*Keywords:* Concentration; Ratio; $^{214}$Pb; $^{214}$Bi; Rainwater; Gamma-ray spectrometry; Removal model

## 1. Introduction

Rainfall brings radon ($^{222}$Rn) decay products ($^{214}$Pb and $^{214}$Bi) in the atmosphere to the ground by scavenging effects, and causes the increase of environmental gamma radiation near the ground, which makes it difficult to evaluate precisely the radiation originating from a nuclear facility.

Knowledge of the relation between rainfall and environmental radiation has been provided by numerous studies [1]. The increase of radiation is explained by the concentrations of $^{214}$Pb and $^{214}$Bi in rainwater or rainfall rate. However, few studies have

---

* Corresponding author. Tel.: +81 29 282 1111; fax: +81 29 282 3838.
 E-mail address: take@tokai.jnc.go.jp (M. Takeyasu).

0531-5131/ © 2004 Elsevier B.V. All rights reserved.
doi:10.1016/j.ics.2004.11.172

discussed the concentrations of $^{214}$Pb and $^{214}$Bi, whose contributions to the increase of radiation dose were different.

In this paper, the concentrations of $^{214}$Pb and $^{214}$Bi in rainwater were measured by gamma-ray spectrometry with a low background Ge detector, and the results were compared with the model that had been reported by Ikebe et al. [2].

## 2. Measurement

The concentrations of $^{214}$Pb and $^{214}$Bi in rainwater were measured on the campus of the Research Reactor Institute of Kyoto University from March 1992 to July 1993. Rainwater sample was collected with a funnel sampler, and subjected to gamma-ray spectrometry with a low background Ge detector. The details of the measurement method had been described in a previous paper [3].

## 3. Results and discussion

The concentrations of $^{214}$Pb and $^{214}$Bi in rainwater varied with time, together with rainfall rate, during one rainfall event. But the concentration of each seemed a power function of rainfall rate, which was represented by the equation, $A=xP^y$, where $A$ is the concentration of $^{214}$Pb and $^{214}$Bi (Bq cm$^{-3}$), $P$ is the rainfall rate (mm h$^{-1}$), and $x$ and $y$ are constants.

The distribution of the values of $x$ of $^{214}$Pb, which was obtained from all the measured rainfall events, ranged from 0.29 to 3.84, and that of $y$ of $^{214}$Pb had two peaks around $-0.5$ and 0. This means that the concentration of $^{214}$Pb has an inverse relation with rainfall rate in some rainfall events, and that the concentration was constant and independent of rainfall rate in other events. The concentration ratio of $^{214}$Bi /$^{214}$Pb was represented by the same equation. The distributions of the values of $x$ and $y$ of the ratio of $^{214}$Bi/$^{214}$Pb have one peak around about 0.75 and $-0.25$, respectively. This means that the ratio has a weak inverse relation with rainfall rate in all rainfall events.

Ikebe et al. had proposed the model for the removal of $^{222}$Rn decay products from stratiform clouds. The model shows the same relations as those from the measurement, except for the relation that the concentration of $^{214}$Pb is constant and independent of rainfall rate in some rainfall events. This reason might be that the condition of clouds adopted by the model is different from that at those events. From the measured value of $x$ of the concentration of $^{214}$Pb and the model, $^{222}$Rn concentration in the cloud was evaluated to be 0.5–6.5 Bq m$^{-3}$, whose value was almost same as that measured at a height of a few kilometers around central Japan by aircraft [4].

## References

[1] T. Nishikawa, et al., Analysis of the time variation of environmental gamma radiation due to the precipitation, Appl. Radiat. Isotopes 46 (1995) 603–604.
[2] Y. Ikebe, et al., A model for precipitation scavenging of short-lived Rn-222 daughters, Res. Lett. Atmos. Electr. 5 (1985) 17–21.
[3] M. Takeyasu, et al., Radon daughters in rain water, in: M. Shimo, T. Tsujimoto (Eds.), Environmental Radon, Electron Science Institute, Japan, 1992, pp. 279–286 (in Japanese).
[4] T. Iida, et al., Vertical profile of radon concentrations in the atmosphere, in: J. Inaba, H. Tsukada, A. Takeda (Eds.), Radioecology and Environmental Dosimetry, Institute for Environmental Sciences, Japan, 2003, pp. 409–415.

www.ics-elsevier.com

# Radon exhalation rate monitoring in/around the closed uranium mine sites in Japan

## Y. Ishimori*, Y. Maruo

*Japan Nuclear Cycle Development Institute, Okayama, Japan*

**Abstract.** The radon exhalation rates have been measured regularly in/around closed uranium mine sites in Japan based on the closed can method with scintillation cells. The measurement results of radon exhalation rate had a large fluctuation depending on the measurement points, and the averages of radon exhalation rates in the sites were approximately from 0.1 to 1 Bq m$^{-2}$ s$^{-1}$, while those at controls were around 0.02 Bq m$^{-2}$ s$^{-1}$. © 2004 Elsevier B.V. All rights reserved.

*Keywords:* Radon exhalation rate; Radon flux; Closed can method; Uranium mine; Waste rock

## 1. Introduction

The Japan Nuclear Cycle Development Institute has the closed uranium mine sites around Ningyo-toge at the boundary of Okayama and Tottori Prefectures in the western part of the Honshu Island in Japan, such as waste rock sites, a mill tailing dam and so on. In this study, the results of radon exhalation rate (radon flux) monitoring in/around closed uranium mine sites are illustrated.

## 2. Methods

Based on the closed can method with scintillation cells, the radon exhalation rates have been measured regularly by the grab sampling method since 1989, and done as a supplemental monitoring by the continuous monitor since 1993 [1].

## 3. Results and discussion

The radon exhalation rate had a large fluctuation depending on the measurement points and the averages of radon exhalation rate in the sites were approximately from 0.1 to 1 Bq m$^{-2}$ s$^{-1}$, while those at controls were around 0.02 Bq m$^{-2}$ s$^{-1}$, as shown in Table 1. The higher

---

\* Corresponding author. Tel.: +81 868 44 2211; fax: +81 868 44 2851.
*E-mail address:* yuu@ningyo.jnc.go.jp (Y. Ishimori).

0531-5131/ © 2004 Elsevier B.V. All rights reserved.
doi:10.1016/j.ics.2004.11.026

Table 1
Average radon exhalation rates evaluated from all data obtained in/around closed uranium mine sites by the grab sampling method

| Site name | Size, m$^2$ | Radon exhalation rate average (min–max), Bq m$^{-2}$ s$^{-1}$ | Number of measurement points |
|---|---|---|---|
| Mill tailing dam | 25,228 | 1.05 (0.042–3.50) | 11 |
| Waste rock site | | | |
| Yotsugi | 1750 | 0.27 (0.009–1.82) | 21 |
| Nakatsugo | 10,650 | 0.15 (0.010–0.93) | 60 |
| Choja | 410 | 0.98 (0.089–2.59) | 3 |
| Kan'nokura-1 | 8450 | 0.44 (0.23–0.61) | 3 |
| Kan'nokura-2 | 7670 | 1.32 (1.23–1.42) | 2 |
| Asabatake-1 | 2300 | 0.73 (0.11–2.87) | 12 |
| Asabatake-2 | 3100 | 0.68 (0.062–2.33) | 30 |
| Asabatake-2B | 250 | 0.31 (0.10–0.80) | 4 |
| Asabatake-3 | 1100 | 0.49 (0.080–2.13) | 7 |
| Katamo-1 | 2200 | 0.56 (0.087–1.39) | 11 |
| Katamo-2 | 3250 | 0.64 (0.065–2.71) | 11 |
| Katamo-S2 | 570 | 0.36 (0.30–0.41) | 2 |
| Katamo-3 | 270 | 0.24 (0.26–0.23) | 2 |
| Ayumidani | 330 | 0.17 (0.10–0.41) | 5 |
| Control | | | |
| Kamisaibara-son | | 0.017 (0.015–0.019) | 2 |
| Misasa-cho | | 0.018 (0.017–0.019) | 2 |
| Togo-cho | | 0.016 (0.013–0.022) | 3 |
| Kurayoshi-shi | | 0.025 (0.024–0.025) | 2 |

radon exhalation rate was generally observed around Ningyo-toge in summer season (see Fig. 1). The radon exhalation rate was affected on by rainfall strongly, and it had a characteristic recovery curve within several days after the end of rain [1]. Because the monitoring of radon exhalation rate by the grab sampling method is not performed in rainy days, the environmental impact evaluated with the monitoring result would be rather conservative as average.

## References

[1] Y. Ishimori, K. Ito, S. Furuta, Environmental effects of radon and its progeny from uranium waste rock piles: part 1 and 2, in: A. Katase, M. Shimo (Eds.), Proc. 7th Tohwa Univ. Int. Sym. on Radon and Thoron in the Human Environment, 23–25 October 1997, World Scientific, Fukuoka, Japan, 1998, pp. 282–293.

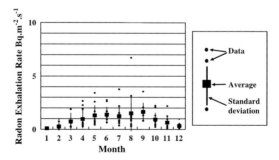

Fig. 1. An example of the seasonal variation of radon exhalation rate around Ningyo-toge.

www.ics-elsevier.com

# Radon-222 concentrations and the environmental radiation dose rates at Misasa spa districts, Japan

M. Inagaki[a,*], S. Nojiri[a], T. Takemura[a], T. Koga[a], H. Morishima[a], Y. Tanaka[a], M. Mifune[b], I. Kobayashi[c]

[a]*Kinki University, 3-4-1, Kowakae, Higashi-osaka, Osaka, Japan*
[b]*Former Okayama University., Tottori, Japan*
[c]*Nagase Landauer Ltd., Tokyo, Japan*

**Abstract.** We have been carrying out continuously the measurements of Rn-222 concentrations in air with a set of indoors and outdoors using the convenient and highly sensitive Pico-rad detectors with active carbon adsorber for several years at Misasa spa districts. Rn-222 concentrations in air on geologically formed granite layers were higher than those basaltic layers. And the mean $\gamma$ radiation dose rates at Misasa spa districts measured by means of optically stimulated luminescence (OSL) dosemeters for 3 months, they showed the similar tendency to those of Rn-222 on geological map.
© 2004 Elsevier B.V. All rights reserved.

*Keywords:* Radon-222; Liquid scintillation spectrometer; Pico-rad detector; Misasa, In air; Granite; OSL; $\gamma$ radiation dose rate

## 1. Introduction

There are a several kinds of sampling devices, such as filter trapping method, electrostatic trapping, active carbon, etc. We have adopted a simple and highly sensitive radon measurement method named the Pico-rad system using active carbon adsorption vials, and a liquid scintillation spectrometer manufactured by Packard Misasa spa on Tottori pref. is one of the most highly radioactive spa in Japan which contains mainly Rn-222. We have measured the radon concentrations in air and the environmental radiation dose rates by optically stimulated luminescence (OSL) dosemeters, it was named Luxel badge manufactured by Nagase Landauer. In this paper, we reported the distribution

---

* Corresponding author. Tel.: +81 6 6721 2332; fax: +81 6 6721 3743.
*E-mail addresses:* masayo.inagaki@itp.kindai.ac.jp (M. Inagaki), taekoga@msa.kindai.ac.jp (T. Koga).

0531-5131/ © 2004 Elsevier B.V. All rights reserved.
doi:10.1016/j.ics.2004.10.012

characteristics of Rn-222 concentrations and the environmental radiation dose rates on Misasa spa districts.

## 2. Materials and methods

### 2.1. Sampling

Hot springs gush out at Misasa and Yamada along both sides of the Mitoku river, the branch of Tenjin river flowing to the north in the middle of the Tottori pref. We sampled the environmental air with a set of indoors and outdoors along both banks of the Mitoku river, Kamo river, Katani river and Tenjin river.

### 2.2. Measurements of radon and $\gamma$ radiation dose rates

The measurement of radon in environmental air using Pico-rad detectors and a liquid scintillation spectrometer have been carried out. Packard Pico-rad detector is a convenient detecting device containing an active carbon adsorber. (1) The Pico-rad vials are exposed for 24 h at the sampling points, collected, and then 10 ml of scintillator (Insta-fluor) is poured into each vials shaken for 15 s and over. These sampling vials are brought to our laboratory, and the radioactivity of Rn-222 and these decay products become equilibrium to measure with a liquid scintillation spectrometer(TRI-CARB 2250CA Type). (2) On the measurement of $\gamma$ radiation dose rate, we set optically stimulated luminescence(OSL) dosemeters manufactured by Nagase Landauer for 3 months and measured.

## 3. Results and discussions

The detection limit of Pico-rad system is 1.7 $Bq/m^3$ in air with accuracy of about 10% on 200 min counting by a liquid scintillation spectrometer, Packard. Their concentration ranges are 19–163 $Bq/m^3$ indoors, 7–50 $Bq/m^3$ outdoors at Misasa spa district, and those of the control Rn-222 at Higashi-osaka are ND-18 $Bq/m^3$ indoors, ND-9.6 $Bq/m^3$ outdoors. [1–3]. Those concentrations made a wide fluctuation on the ventilated conditions, and it showed a seasonal tendency that those in air were fluctuated low level on summer and rain seasons and high on winter. Rn-222 concentrations in air on geologically formed granite layers were higher than those basaltic layers.

The mean $\gamma$ radiation dose rates at Misasa spa districts measured by means of optically stimulated luminescence (OSL) dosemeters for 3 months, they showed the similar tendency to those of Rn-222 on geological map. The minimum detectable limit using the OSL dosemeter was 35 μSv with accuracy of 16% on accumulated dose for 3 months. It was possible to measure natural radiation dose rates, 0.08 μSv/h at Osaka as control area using the OSL dosemeter.

## References

[1] T. Koga, et al., Radon measurement using a liquid scintillation spectrometer, Annual report of Kinki University Atomic Energy Research Institute 29 (1992) 17–24.
[2] T. Koga, et al., Measurement of radon in air using the PICO-RAD detector and a liquid scintillation spectrometer (2), Annual report of Kinki University Atomic Energy Research Institute 31 (1994) 7–13.
[3] T. Koga, et al., Measurement of radon on Misasa district, Tottori pref. using the PICO-RAD detector and a liquid scintillation spectrometer (3), Annual report of Kinki University Atomic Energy Research Institute 33 (1996) 11–23.

# Distribution characteristics of natural radionuclides at some spa districts in Japan

T. Koga[a,*], T. Takemura[a], S. Nojiri[a], H. Morishima[a], M. Inagaki[a], Y. Tanaka[a], M. Mifune[b]

[a]*Kinki University Atomic Energy Research Institute, 3-4-1, Kowakae, Higashi-Osaka, Osaka 577-8502, Japan*
[b]*Former Okayama Univ., Japan*

**Abstract.** In Japan, there are many high-level natural radiation areas such as Masutomi spa (Yamanashi pref.), Misasa spa (Tottori pref.), Ikeda spa (Shimane pref.), and Tamagawa spa (Akita pref.), etc. The Uranium and thorium decay series and K-40 are widely distributed as the sources of terrestrial radiation. Their concentrations are influenced by geological features and are nonuniform. We carried out a study of the behavior and distribution of natural radionuclides in the surrounding soils of high-level natural radiation spa areas by γ-ray energy spectrometry using a Ge(INT) diode detector. It was found that the Pb-212(Th)/Pb-214(U) ratios fluctuated widely in the same areas, of which Tamagawa spa and Masutomi spa were typical. © 2004 Elsevier B.V. All rights reserved.

*Keywords:* Natural radionuclide; Masutomi spa; Tamagawa spa; Soil; γ-Ray spectrometry; Pb-212(Th)/Pb-214(U) ratio

## 1. Introduction

In Japan, there are many high-level natural radiation areas such as Masutomi spa (Yamanashi pref.), Misasa spa (Tottori pref.), Ikeda spa (Shimane pref.), and Tamagawa spa (Akita pref.), etc. The Uranium and thorium decay series and K-40 are widely distributed as the sources of terrestrial radiation. Their concentrations are influenced by geological features and are non-uniform. We carried out a study of the behavior and distribution of natural nuclides in the environmental soils at these areas by γ-ray energy spectrometry using a Ge diode detector.

## 2. Materials and methods

We measured the Rn-222 concentrations in the air and water at spa areas in Japan using the Pico-rad detectors and a liquid scintillation spectrometer. Air samples were taken over a

---

* Corresponding author. Tel.: +81 6 6721 2332x4407; fax: +81 6 6721 3743.
 *E-mail address:* taekoga@msa.kindai.ac.jp (T. Koga).

0531-5131/ © 2004 Elsevier B.V. All rights reserved.
doi:10.1016/j.ics.2004.11.085

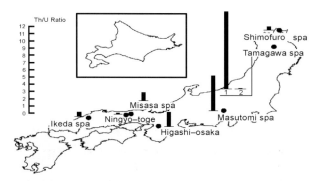

Fig. 1. Th/U concentration ratio of soils at some spa districts in Japan.

24-h period and then mixed with a scintillator to measure the concentration of Rn-222, while water samples were mixed directly with an equal amount of scintillator to measure the same.

Soil samples were collected from the spa areas, dried, sieved and enclosed in 100-ml plastic containers (U-8). These containers were sealed, kept for one month to attain radioactive equilibrium between various decay products in the Th-232 and U-238 series and measured for approximately $10^5$ s, using a Ge(INT) diode detector system. The spectrum was analysed with reference to the following photo-peaks: 239 keV of Pb-212(Th), 352 keV of Pb-214(U), and 1461 keV of K-40.

## 3. Results and discussions

The Pico-rad method for measuring concentrations of Rn-222 in air and water is suitable for both the measuring of many samples, and drawing a concentration distribution map. The maximum Rn-222 concentrations in spring water and air were 20 kBq/l and 3.3 kBq/m$^3$ indoors at Masutomi, 9.7 kBq/l and 690 kBq/m$^3$ outdoors at Ikeda and 4.8 kBq/l and 4.4 kBq/m$^3$ indoors at Misasa, respectively.

A total of 166 soil samples from spa areas were analysed. These concentrations of Pb-212(Th) and Pb-214(U) were calculated in Bq/kg. The concentrations of some samples collected at control areas in Osaka pref. are nearly the same as the world average [1]. The maximum Pb-212(Th) concentration of soil samples showed 33.5 kBq/kg and Pb-214(U) 2.7 kBq/kg in the Tamagawa spa area, and Pb-212 8.8 kBq/kg and Pb-214 1.5 kBq/kg in the Masutomi spa area, respectively. The world averages of Th-232 and U-238 are 30 and 35 Bq/kg, respectively [1]. These concentrations in the Tamagawa and Masutomi spa areas were remarkably higher than those in the control areas. Of the mean Pb-212(Th)/Pb-214(U) ratios in the Tamagawa spa areas, one was 0.018 in which a large number of U series nuclides were contained, while that of the other was 11 (Fig. 1). It was typically found that the Pb-212(Th)/Pb-214(U) ratios in the Tamagawa spa and Masutomi spa areas fluctuated widely. There are areas made up of the accumulate-piled soils of aqueous uranium and radium compounds.

## Reference

[1] UNSCEAR 2000 Report, II Terrestrial Radiation (2000).

www.ics-elsevier.com

# Optimization of measuring methods on size distribution of naturally occurring radioactive aerosols

K. Yamasaki[a],*, Y. Oki[a], Y. Yamada[b], S. Tokonami[b], T. Iida[c]

[a]*Research Reactor Institute, Kyoto University, Japan*
[b]*Radiation Safety Research Center, National Institute of Radiological Sciences, Japan*
[c]*Graduate School of Engineering, Nagoya University, Japan*

**Abstract.** Several domestic intercomparison experiments of the size distributions of $^{222}$Rn progeny by using four different types of low-pressure cascade impactors have been conducted in the KUCA reactor room of Kyoto University, of Japan. There were serious discrepancies between some impactors. The cause of these discrepancies originated mainly from the differences of the jet nozzle to the collecting plate distances and selection of the collecting substrates of the impactors. © 2005 Elsevier B.V. All rights reserved.

*Keywords:* Intercomparison; Size; Distribution; $^{222}$Rn; Progeny; Impactor; Jet; Collecting plate; Distance; Substrate

## 1. Introduction

Size distribution of inhaled aerosol-attached $^{222}$Rn progeny is one of the important parameters in determining lung dose, as size depends on the deposition at the lung [1]. Several researchers have applied some types of cascade impactors (Andersen, Berner, MOUDI, etc.) to measure the activity size distribution of $^{222}$Rn progeny in the environment [2,3]. The reliability of the measured activity size distribution has not been established owing to lack of standard calibration method and facility. In these situations, Japanese domestic intercomparison experiments on size distribution of $^{222}$Rn progeny were carried out in the KUCA reactor room of Kyoto University, using combustion aerosol of incense sticks as a carrier aerosol. This paper describes some countermeasures for improving reliability of size distribution data measured by cascade impactors.

* Corresponding author. Tel.: +81 724 51 2377; fax: +81 724 51 2603.
  E-mail address: yamasaki@rri.kyoto-u.ac.jp (K. Yamasaki).

0531-5131/ © 2005 Elsevier B.V. All rights reserved.
doi:10.1016/j.ics.2004.10.028

Table 1
Typical intercomparison result in the building of KUCA for various collecting substrates

| Impactor | Nov. 27, 2002 | | Nov. 28, 2002 | |
| --- | --- | --- | --- | --- |
| | AMD (μm) | GSD (-) | AMD (μm) | GSD (-) |
| MOUDI-Model 110 | 0.25[a] | 1.58[a] | 0.26[a] | 1.60[a] |
| LP-20RS | 0.26[b] | 1.66[b] | 0.28[b] | 1.63[b] |
| LP-20RPS47 | 0.20[b] | 2.16[b] | 0.35[c] | 2.10[c] |
| LP-2015J | 0.34[b] | 2.44[b] | 0.50[c] | 1.95[c] |

[a] Grease-coated aluminum foil; [b] Grease-coated stainless steel plate; [c] Teflon binder filter (T60A20).

Table 2
The influence of jet to plate distances of low pressure cascade impactors

| Impactor | Jet to plate distance (mm) | AMD (μm) | GSD (-) |
| --- | --- | --- | --- |
| LP-2015J | 3.5 | 0.33 | 2.19 |
| | 2.0 | 0.43 | 1.49 |
| LP-20RS | 2.0 | 0.34 | 1.41 |

## 2. Experimental

Instruments used for intercomparison experiments were 4 cascade impactors listed as follows, a Micro Orifice Uniform Deposit Impactor (MOUDI, Model 110, MSP, USA) and 3 Andersen low pressure impactors (LP-20RS, LP-2015J and LP-20RPS47, Tokyo Dylec, Japan). MOUDI consists of eight stages and a back up filter, and sucks air at 30 lpm [4]. LP-20RS, LP-2015J and LP-20RPS47 are products by the same maker and consist of 12 stages, 15 stages, 12 stages and a back up filter, and sucks air at 22.9, 21.4 and 25.7 lpm, respectively. Experiments were carried out at 400–600 Bq m$^{-3}$ of $^{222}$Rn concentration and $1 \times 10^4$–$2 \times 10^4$ cm$^{-3}$ of aerosol concentration by combustion of incense sticks.

## 3. Results and discussion

Table 1 shows a typical intercomparison result in the building of KUCA for various collecting substrates, such as grease-coated aluminum (stainless steel) plate and Teflon binder filter. The data in Table 1 shows that MOUDI and LP-20RS agreed quite well, but LP-2015J and LP-20RPS47 gave a different tendency. By using LP-20RS and LP-2015J, causes of these differences are discussed below. The distance between jet nozzle and collecting plate of each impactor is 2.0 mm for LP-20RS and 3.5 mm for LP-2015J. Comparing size and clearness of the aerosol spots deposited on the collecting plate, LP-20RS gave small and clear spots, but LP-2015J gave larger and less clear spots than LP-20RS. In the case of the same distance of both impactors by putting a spacer of 1.5 mm thickness between jet and collecting plate of LP-2015J, these impactors gave the same spots. Table 2 shows that the distance between jet nozzle and collecting plate of the impactor gives a serious influence on the shape of measured size distribution.

## References

[1] International Commission on Radiological Protection, Human Respiratory Tract Model for Radiological Protection, ICRP Publ., vol. 66, Pergamon Press, Oxford, England, 1994.
[2] J. Porstendörfer, J. Aerosol Sci. 25 (1994) 219–263.
[3] E.O. Knutson, K.W. Tu, Environ. Int. 22 (Suppl. 1) (1996) S617–S632.
[4] V.A. Marpl, K.L. Rubow, S.M. Behm, Aerosol Sci. Technol. 14 (1991) 434–446.

# Measurement of radon concentration in water using direct dpm method of liquid scintillation counter

Y. Yasuoka[a,*], T. Ishii[b], T. Sanada[c], W. Nitta[c], Y. Ishimori[d],
Y. Kataoka[e], T. Kubo[f], H. Suda[g], S. Tokonami[h],
T. Ishikawa[h], M. Shinogi[a]

[a]*Kobe Pharmaceutical University, Hyogo, Japan*
[b]*University of Yamanashi, Yamanashi, Japan*
[c]*Japan Chemical Analysis Center, Chiba, Japan*
[d]*Japan Nuclear Cycle Development Institute, Okayama, Japan*
[e]*Keio University, Tokyo, Japan*
[f]*Oita University, Oita, Japan*
[g]*Kagawa University, Kagawa, Japan*
[h]*National Institute of Radiological Sciences, Chiba, Japan*

**Abstract.** For measuring radon ($^{222}$Rn) concentration in water with a direct method, a two-layer sample is measured by a liquid scintillation counter (LSC). A calibration factor for an LSC is generally evaluated by a comparison measurement of a radium ($^{226}$Ra) standard. However, it is difficult to use the $^{226}$Ra standard in Japan. Thus, we studied a direct dpm method based on the integral counting method for a two-layer sample. The optimized window values of three efficiency tracing points for the direct dpm method were determined to be 50–2000, 75–2000, and 100–2000 keV. It was determined that a calibration factor of 5 could be used with an LSC. © 2004 Elsevier B.V. All rights reserved.

*Keywords:* Radon; Groundwater; Liquid scintillation counter; Integral counting method

## 1. Introduction

It is popular to measure radon ($^{222}$Rn) concentration in water with a liquid scintillation counter (LSC). It is accurate and simple that a direct method is designated by the American Society for Testing and Materials. In the direct method, a two-layer sample, which is 10

---

\* Corresponding author. 4-19-1, Motoyamakita-machi, Higashinada-ku, Kobe, Hyogo 658-8558, Japan. Tel.: +81 78 441 7519; fax: +81 78 441 7519.
 *E-mail address:* yasuoka@kobepharma-u.ac.jp (Y. Yasuoka).

0531-5131/ © 2004 Elsevier B.V. All rights reserved.
doi:10.1016/j.ics.2004.11.023

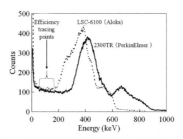

Fig. 1. The pulse height spectra of $^{222}$Rn and its decay products measured with the LSC-6100 and the 2300TR.

mL of water in a vial containing 10 mL of toluene scintillator, is measured by an LSC. A calibration factor of the LSC is generally evaluated by a comparison measurement of a radium ($^{226}$Ra) standard. However, it is difficult to use the $^{226}$Ra standard solution in Japan. Thus, we studied a direct dpm method based on the integral counting method for a two-layer sample. For measuring $^{222}$Rn with the direct dpm method, we show optimized window values of three efficiency tracing points and a calibration factor for an LSC. The following criteria of the Environmental Protection Agency (EPA) were adopted as our acceptance limit. The proposed acceptance limit for a single sample is ±5%. The proposed limit for triplicate analyses at the 99th percent confidence interval is ±9%.

## 2. Methods and results

The optimized window values of the three efficiency tracing points, after a two-layer sample is measured by an LSC, were determined to be 50–2000 keV, 75–2000 keV, and 100–2000 keV (Fig. 1). We used some major LSCs from PerkinElmer Life and Analytical Sciences and Aloka [1].

The $^{222}$Rn concentration Ct(Bq/kg) in the toluene scintillator is expressed as follows:

$$Ct = N_0/(F Wt_m)$$

where $N_0$ is the total net counting rate in cps, $Wt_m$ is the toluene weight in the counting vial in kg, and $F$ is a calibration factor for the LSC. While this factor was theoretically 5 considering three alpha rays and two beta rays, the factor was proven to be 4.7 according to our comparison experiments with a $^{226}$Ra standard. This 6% difference in calibration factors is small enough to have negligible effect on the calculations for the proposed acceptance limit for triplicate analyses.

## 3. Conclusion

For measuring the $^{222}$Rn concentration in groundwater with the direct dpm method, the optimized window values of the three efficiency tracing points were determined to be 50–2000, 75–2000, and 100–2000 keV for the major LSCs in Japan. It was determined that a calibration factor of 5 could be used with an LSC.

## Reference

[1] Y. Yasuoka, et al., Determination of radon concentration in water using liquid scintillation counter, Radioisotopes 53 (3) (2004) 123–131.

# Airborne and waterborne radon concentrations in houses with the use of groundwater

T. Ishikawa*, S. Yoshinaga, S. Tokonami

*National Institute of Radiological Sciences, Japan*

**Abstract.** A preliminary survey of airborne and waterborne radon concentrations was conducted in an area where groundwater is used as a source of public water supplies. An average of waterborne radon concentrations was 58.4 Bq $l^{-1}$ for public water supplies and that of airborne radon concentrations was 18 Bq $m^{-3}$ for 10 houses. © 2004 Elsevier B.V. All rights reserved.

*Keywords:* Radon; Groundwater; Release; Ingestion; Inhalation

## 1. Introduction

In Japan, radon concentrations in public water supplies are generally low, because their source is surface water in most cases. However, high radon concentrations can occur in water supplied from groundwater. Water supplies in a part of Nijo town (Fukuoka prefecture, Japan) are such cases. In some areas in this town, groundwater is used as a source of public water supplies. Radon in such water supplies presents a possible risk to the population in two pathways: ingestion of water and inhalation of radon (progeny) released from water [1]. Radon-in-water could be a public health concern in this town because of potential exposures to a large number of population ($n=13,000$). Thus, a preliminary survey of airborne and waterborne radon concentrations was conducted in Nijo town.

## 2. Materials and methods

Radon concentrations in drinking water were surveyed for 19 houses in Nijo town. Some houses use private well water as well as public water supplies. In such cases, water samples were taken from sources mainly used for drinking. Radon-in-water concentrations were measured with an electrostatic-type monitor, which employs a silicon semiconductor

---

\* Corresponding author. 4-9-1 Anagawa Inage-ku, Chiba, 263-8555, Japan. Tel.: +81 43 206 3099; fax: +81 43 206 4097.

*E-mail address:* tetsuo_i@nirs.go.jp (T. Ishikawa).

0531-5131/ © 2004 Elsevier B.V. All rights reserved.
doi:10.1016/j.ics.2004.09.022

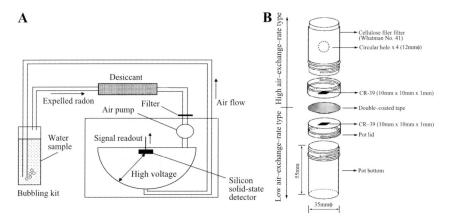

Fig. 1. Schematics of the devices for measuring waterborne radon (A) and airborne radon (B).

detector (Fig. 1A). For 10 out of 19 houses, nuclear track detectors (Fig. 1B) were placed in kitchens, living rooms and bedrooms to estimate airborne radon concentrations. The detectors were exposed for about 2 months.

## 3. Results and discussion

For public water supplies, nine water samples were collected. Radon concentrations for these samples ranged from 0.6 to 175 Bq $l^{-1}$ with an average of 58.4 Bq $l^{-1}$. For private well water, 16 samples were collected. Radon concentrations for these samples ranged from 2.6 to 287 Bq $l^{-1}$ with an average of 81.7 Bq $l^{-1}$. US EPA proposed regulations on radon concentration in drinking water: a maximum contaminant level (MCL: 11 Bq $l^{-1}$) and an alternative maximum contaminant level (AMCL: 148 Bq $l^{-1}$) [1]. For 4 samples out of 25, radon concentrations were over the AMCL. For 14 samples, radon concentrations were between the MCL and the AMCL.

Airborne radon concentrations for ten houses ranged from 9 to 27 Bq $m^{-3}$ with an average of 18 Bq $m^{-3}$. The value was not so different from a geometric mean of indoor radon concentrations obtained from a nation-wide survey in Japan [2]. For these houses, there was no significant increase of indoor radon concentration due to radon released from water. Average radon concentrations in living rooms, bed rooms and kitchens are 21, 17 and 17 in Bq $m^{-3}$, respectively.

The dose due to ingestion of radon was calculated using a dose coefficient of UNSCEAR (3.5E−9 Sv/Bq [3]). The calculation showed that a daily ingestion of 0.6-l water [1] with a radon concentration of 70 Bq $l^{-1}$ causes an annual effective dose of 0.05 mSv. The dose could be much smaller than the total dose from natural radiation.

## References

[1] National Research Council, Risk Assessment of Radon in Drinking Water, Academy Press, Washington, DC, 1999.
[2] T. Sanada, et al., Measurement of nationwide indoor radon concentration in Japan, J. Environ. Radioact. 45 (1999) 129–137.
[3] UNSCEAR, Sources and Effects of Ionizing Radiation, United Nations, New York, 2000.

# Emanating power of $^{220}$Rn from a powdery radiation source

T. Iimoto*, T. Nagai, N. Sugiura, T. Kosako

*Research Center for Nuclear Science and Technology, The University of Tokyo, 2-11-16 Yayoi, Bunkyo-ku, Tokyo 113-0032, Japan*

**Abstract.** The emanating power (Fr) of $^{220}$Rn from a powdery sample was experimentally estimated. The powdery sample used in this study consisted of natural mineral encrustations left by a radioactive hot spring. The emanating power of thoron was analyzed based on both the theoretical emanation model of UNSCEAR and the result of the emanation test using the delayed-coincidence and time-analysis method with a flow-through scintillation flask. Finally, the maximum value of Fr for thoron was calculated to be approximately 0.5. © 2004 Elsevier B.V. All rights reserved.

*Keywords:* $^{220}$Rn; Emanating power; Powdery sample; Scintillation flask; Delayed-coincidence and time-analysis method

## 1. Introduction

Some materials emanating $^{220}$Rn become more important radiation sources for general public. NORM (naturally occurring radioactive materials) and TENORM (technologically enhanced naturally occurring radioactive materials) are those of the representative materials. Both external and internal exposures due to these sources must be considered, of course. Especially an emanation of thoron from these materials has largely been taken interest on because of the lack of data on the related parameters.

In this study the emanating power (Fr) of thoron from a powdery sample was experimentally estimated. The Fr parameters for $^{222}$Rn were defined in the UNSCEAR 2000 report [1]. Fr is the radon fraction formed that enters the pores. We defined the parameter of Fr for thoron in the same way. The powdery sample must show the biggest Fr value among common materials because the gap between the particles is large. The primary objective of this study was to determine the maximum value of an effective Fr for the general thoron source.

---

\* Corresponding author. Tel.: +81 3 5841 2915; fax: +81 3 3813 2010.
   *E-mail address:* iimoto@rcnst.u-tokyo.ac.jp (T. Iimoto).

## 2. Materials and methods

The powdery sample used in this study consisted of natural mineral encrustations left by a radioactive hot spring (Tamagawa hot springs) located in Akita Prefecture in northeastern Japan. The $^{228}$Ra content in the powder, which was approximately 6 Bq g$^{-1}$, was estimated using a pure Ge semiconductor detector. A simple thoron chamber was constructed using a plastic glove box with an internal effective volume of 0.1 m$^3$. In this chamber, an emanation test of thoron was performed with 2.4 kg of a powdery source material (30 g pack$^{-1}$×81 packs). Two fans were installed and operated in the chamber in order to stir the inside air and achieve a uniform thoron concentration. The thoron concentration in the chamber was estimated by the delayed-coincidence and time-analysis (DC-TA) method [2] with a flow-through scintillation flask (300 ml) [3]. This method is based on the use of the short decay half-life of 145 ms of $^{216}$Po in order to distinguish between $^{220}$Rn–$^{216}$Po pair counts and random coincidence pulses (i.e., chance coincidence). Three time-windows in the DC-TA unit, whose opening durations were 150, 300 and 450 ms, were designed to start counting delayed pulses in each window from a trigger pulse. We can analyze the decay curve of $^{216}$Po by the accumulated counts in the three time-windows. Using this result, the emanating power of thoron was analyzed based on the following emanation model of UNSCEAR, developed in 1982 [4].

$$R = \lambda_{Tn} Fr\, X_{Ra}(\Delta_k/\lambda_{Tn})^{0.5}$$

$R$=emanation rate of thoron [Bq m$^{-2}$ s$^{-1}$], $\lambda_{Tn}$=decay constant of thoron [s$^{-1}$], $X_{Ra}$=$^{228}$Ra content in the source material [Bq m$^{-3}$], $\Delta_k$=diffusion coefficient [m$^2$ s$^{-1}$].

## 3. Results, discussion, and conclusion

Finally, the effective maximum value of Fr for thoron was calculated to be approximately 0.5. In this calculation, $\Delta_k$ was assumed to be $5\times10^{-6}$ m$^2$ s$^{-1}$ based on the representative value for the standard soil. According to UNSCEAR [1], the Fr for radon ranges widely from 0.05 to 0.7 depending on various factors. On the other hand, in this study, the upper limit value of the Fr for thoron was estimated to be much smaller than that of radon. This means that the Fr for thoron ranges from 0.05 to 0.5. The reason for this difference can be explained primarily by the half-lives of radon (3.8 days) and thoron (56 s). As is well known, thoron emanates from the extremely surface layer of source, not from a deep position. Therefore, using some assumptions and measurement data, such as the maximum Fr=0.5 and the $^{228}$Ra content in the source material of X Bq m$^{-3}$, a plausible maximum concentration of thoron can easily be estimated in the air close to the natural source. This effective maximum value of Fr can also be adopted for the pre-estimation of thoron concentration before the continuous monitoring of thoron concentration in both the natural environment and special artificial facilities, such as nuclear cycle facilities involving the radiation safety control.

## References

[1] Sources and effects of ionizing radiation, United Nations Scientific Committee on the Effects of Atomic Radiation UNSCEAR 2000 Report to the General Assembly, with scientific annexes, volume I: SOURCES.
[2] T. Iimoto, R. Kurosawa, Environ. Int. 22 (Suppl. 1) (1996) S1139–S1145.
[3] T. Iimoto, et al., Radiat. Prot. Dosim. 77 (3) (1998) 185–190.
[4] Ionizing radiation: sources and biological effects, United Nations Scientific Committee on the Effects of Atomic Radiation, 1982 Report to the General Assembly, with annexes.

www.ics-elsevier.com

# The environment dose measurement using the OSL dosimeter

Ikuo Kobayashi[a,*], Hiroshi Sekiguchi[a], Fumiaki Tatsuta[a], Hironobu Komori[a], Taeko Koga[b], Hiroshige Morishima[b]

[a]Nagase Landauer, Ltd., Tokyo 103-8487, Japan
[b]Atomic Energy Research Institute, Kinki University, Osaka 577-8502, Japan

**Abstract.** The OSL (Optically Stimulated Luminescence) dosimeter using $Al_2O_3{:}C$ as a detection material has been commercialized for monitoring personal radiation exposure. We have obtained nationwide ambient doses from control badges for compensating natural environments in the case of personal radiation doses. They were used from August 2001 to July 2002. The nationwide annual average dose for 30 days is 0.9 mSv. The lowest dose is 0.59 mSv of Okinawa Prefecture and the highest is 1.16 mSv of Ishikawa Prefecture. The doses during the month of May are the highest for all prefectures. © 2004 Elsevier B.V. All rights reserved.

*Keywords:* Optically Stimulated Luminescence; Personal monitor; Ambient dose

## 1. Introduction

Ambient radiation doses are measured using active dosimeters or TLD. Nagase Landauer provides nationwide services to approximately 170,000 individuals. The measurement interval is normally one month. Control badges are sent to the customers in order to compensate for the natural environments. They are placed in an environment without any artificial radiation sources. We measured environmental radiation doses for these badges. These data are continuously obtained from every prefecture throughout the year. Measurement is conducted by Nagase Landauer and calibrated with the same standard. The data shown here are the results of measurement using our personal dosimeters for each prefecture.

---

\* Corresponding author. Tel.: +81 3 3665 3682; fax: +81 3 3662 9518.
E-mail address: kobayashi@nagase-landauer.co.jp (I. Kobayashi).

0531-5131/ © 2004 Elsevier B.V. All rights reserved.
doi:10.1016/j.ics.2004.11.191

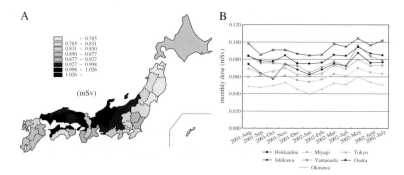

Fig. 1. Panel A shows Annual Environmental Dose from August 2001 to July 2002. Panel B shows changes of the monthly ambient doses at some typical prefectures.

## 2. Materials and methods

The doses are read using a green LED or a CW-green Laser as a stimulating light [1,2].

We obtained a number of samples for all badges and read the initial values of Eb doses when they are first worn. One month later, the Em doses of the returned control badges are read. The environmental dose will be obtained by subtracting Eb from the Em dose. The exposure time of each badge differs depending on the return date. We calculate the 1-day dose from the measurement date. The environmental dose $E$ for 30 days is calculated using the following equation:

$$E = (Em - Eb) \times 30/Dm$$

where Dm is defined as the days passed from when the badge was first worn to the measurement date. A total of 197,044 data are obtained.

## 3. Results and discussion

Fig. 1A shows the annual dose of each prefecture from August 2001 to July 2002. The total average annual dose is 0.90 mSv. The highest dose of 1.16 mSv is recorded at Ishikawa prefecture while the lowest dose of 0.59 mSv is at Okinawa. The western part of Japanese Main Island shows higher doses. Fig. 1B shows seasonal variations of the environmental doses. May shows the largest doses for the year. Slightly larger doses are also observed in November. January shows the lowest dose. Yamanashi shows the largest variation of 0.031 mSv in a year, while Miyagi shows the smallest variation of 0.016 mSv.

The differences in the annual doses arise from the geophysical structures. High dose prefectures are abundant in granite on the ground surface. In this case, we gathered control badges, which were placed indoors. These seasonal variations are often discussed in relation to weather conditions such as rainfall. However, in this case, no relation was found with the monthly amount or days of rainfall.

## References

[1] I. Kobayashi, H. Morishima, Development of optically stimulated luminescence dosimeter systems of green cw-laser, Proc. Advanced Radiation Application Symp., Japanese Soc. of Appl. Phys. Osaka, Japan, Aug. 8–10, 2001, pp. 1–4, 2001 (in Japanese).
[2] H. Yasuda, I. Kobayashi, Optically stimulated luminescence from $Al_2O_3$:C irradiated with relativistic heavy ions, Radial. Prot. Dosim., 95 (2001) 339–343.

# Eolian dust may be an effective expander of HBRA: a case of China–Ryukyu connection

M. Furukawa[a,*], W. Zhuo[a], S. Tokonami[a], N. Akata[b], Q. Guo[c]

[a]Radon Research Group, National Institute of Radiological Sciences(NIRS), Chiba 263-8555, Japan
[b]Department of Radioecology, Institute for Environmental Sciences, Aomori 039-3212, Japan
[c]Department of Technical Physics, School of Physics, Peking University, Beijing 100871, China

**Abstract.** Extensive survey for the natural radiation was carried out in the East Asia. The results strongly suggest that the natural radiation level in the Ryukyu Islands, southwestern part of Japan, has been enhanced due to the deposition of eolian dusts mainly transported from the high background radiation area (HBRA) in China. © 2004 Elsevier B.V. All rights reserved.

*Keywords:* Eolian dust; HBRA in China; Ryukyu Islands of Japan; Enhanced natural radiation level

## Introduction

In the Ryukyu Islands, subtropical region of Japan, peculiar red soils are deposited on the Pleistocene limestone mainly composed of coral reef. The recent pedological studies concluded that the mother material of the red soils is not residuals from the limestone, but is the East Asian eolian dust "*Kosa* (yellow sand)" (e.g. Ref. [1]). In order to estimate the change of natural radiation level with the dust deposition in the Ryukyu Islands and to research the origin area of the dust, a review was performed in this study based on the results of extensive survey for the natural radiation in the East Asia by the Radon Research Group of NIRS and some Japan–China cooperative research teams (e.g. Ref. [2]).

## Data

The following new and reported data were mainly used in this study (see Table 1).

(1) Dose rates in air due to terrestrial gamma radiation measured in the Ryukyu Islands (Okinawa, Miyako [3], Kitadaito and Minamidaito Islands).

---

* Corresponding author. Tel: +81 43 206 3104; fax: +81 43 206 4098.
*E-mail address:* m_furu@nirs.go.jp (M. Furukawa).

0531-5131/ © 2004 Elsevier B.V. All rights reserved.
doi:10.1016/j.ics.2004.09.046

Table 1
Mean values of the dose rate in air due to terrestrial gamma radiation, the contents of radioactive elements (U-238, Th-232) and the outdoor radon (Rn-222) concentration mainly obtained by our research

| Locality | Material (nGy/h) | Dose rate (Bq/m$^3$) | Rn-222 (Bq/kg) | U-238 (Bq/kg) | Th-232 (Bq/kg) |
|---|---|---|---|---|---|
| *Japan* | | | | | |
| Okinawa, Ryukyu Is. | Red soil | 71 | – | – | – |
| | Limestone | 19 | | | |
| Miyako, Ryukyu Is. | Red soil | 95 | 14 | 176 | 67 |
| | Limestone | 29 | – | 8 | 8 |
| Kitadaito, Ryukyu Is. | Red soil | 114 | – | 69 | 88 |
| | Limestone | 20 | – | | |
| Minamidaito, Ryukyu Is. | Red soil | 118 | – | 61 | 59 |
| | Limestone | 10 | – | – | – |
| Nationwide | – | 50 [5] | 6 [6] | – | – |
| *China* | | | | | |
| Taklamakan Desert | Sand | 54 | 27 | 25 | 33 |
| Loess Plateau | Loess | 66 | – | 32 | 45 |
| Tibet Heights | Alpine soil | 109 | – | 48 | 74 |
| Fuzhou (a part of HBRA) | Red soil | 157 | – | 101 | 127 |

(2) Concentrations of radioactive elements (U-238 and Th-232) for the soil samples from the Ryukyu Islands (Miyako [3], Kitadaito and Minamidaito Islands) and the wide area of China.
(3) Outdoor radon (Rn-222) concentrations detected in Miyako Island [4].

## Results and discussion

The Ryukyu Islands presents relatively high background radiation level in comparison with the others of Japan (see Table 1). Correlation of the gamma radiation dose rate and geology in the Ryukyu Islands showed that the high-rate areas (>100 nGy/h) and the distribution of the red soils overlap each other. On the other hand, the low dose rates (<30 nGy/h) are mainly found in the outcrops of Pleistocene limestone. These results strongly suggest that the background radiation level in the Ryukyu Islands has been enhanced due to the deposition of the mother materials of the red soils, i.e. the East Asian eolian dust, since the last glacial epoch at the latest.

From viewpoints of the contents of radioactive elements (see Table 1) and the geographic arrangement, the high background radiation area (HBRA) and the Tibet Heights in the southeast and southwest of China, respectively, are considered to be the key regions for the origin of the red soils deposited in the Ryukyu Islands. These findings also indicate that the eolian dust may be an effective expander of the HBRA.

## References

[1] K. Inoue, et al., Quat. Res. (Tokyo) 32 (1993) 139–155 (in Japanese with English abstract).
[2] S. Tokonami, et al., Radiat. Res. 162 (2004) (in press).
[3] M. Furukawa, et al., Jpn. J. Health Phys. 36 (2001) 195–206 (in Japanese with English abstract).
[4] M. Furukawa, et al., Radioisotopes 53 (2004) 141–147 (in Japanese).
[5] M. Furukawa, J. Geogr. 102 (1993) 868–877 (in Japanese with English abstract).
[6] S. Oikawa, et al., J. Environ. Radioact. 65 (2003) 203–213.

# Radon and thoron in Yongding Hakka Earth Building in Fujian Province of China

M. Wei[a], W. Zhuo[b,*], Y. Luo[a], M. Furukawa[b], X. Chen[a], Y. Yamada[b], D. Lin[a], F. Jian[a]

[a]*Fujian Centre for Prevention and Control of Occupational Diseases and Chemical Poisoning, PR China*
[b]*Radon Research Group, National Institute of Radiological Sciences, Chiba 263-8555, Japan*

**Abstract.** A 1-yr survey of indoor $^{222}$Rn, $^{220}$Rn and its progeny was performed in a typical earth building in Fujian Province, China. Annual arithmetic mean concentrations of indoor $^{222}$Rn, $^{220}$Rn and its progeny were 25, 232 and 0.7 Bq m$^{-3}$, respectively. Although seasonal variation of indoor $^{222}$Rn was observed, variations of indoor $^{220}$Rn and its progeny were not obvious. Indoor $^{220}$Rn progeny concentration might be roughly estimated from measuring indoor $^{220}$Rn. © 2004 Elsevier B.V. All rights reserved.

*Keywords:* Radon; Thoron; Equilibrium-equivalent thoron concentration; Gamma dose rate

## 1. Introduction

Yongding Hakka Earth Building, a traditional building mainly made of raw soils, has a history of over 1000 years. More than 20,000 of these still exist in Yongding Prefecture, Fujian Province, China. Earth buildings with an area over 200 m$^2$ and three stories account for more than 80%. Usually a whole family of many generations lives together in it. As high contents of $^{238}$U and $^{232}$Th in soil were reported in the area [1], it was expected that high levels of radon and thoron might exist in the building. In this study, a survey of natural radiation levels was carried out in the buildings, and their results are reported.

## 2. Materials and methods

Indoor $^{222}$Rn and $^{220}$Rn were measured by the passive integrating $^{222}$Rn and $^{220}$Rn discriminative monitor [2]. The monitors were set at about 10 cm from a wall in a bedroom.

---

* Corresponding author. Tel.: +81 43 206 3111; fax: +81 43 206 4098.
*E-mail address:* whzhuo@nirs.go.jp (W. Zhuo).

The equilibrium-equivalent thoron concentration was estimated from measuring the deposition rate of its progeny [3]. The deposition rate monitors were set in the middle of the same room. Indoor and outdoor gamma radiation levels were measured by a NaI(Tl) scintillation radiation meter at 1 m above the ground. Measurements of $^{222}$Rn, $^{220}$Rn and its progeny were made from December 2002 to January 2004 in 6-month intervals. A total of 100 buildings were randomly selected from 10 villages. One set of the $^{222}$Rn, $^{220}$Rn and the deposition rate monitors was used in each room.

## 3. Results and discussion

Indoor $^{222}$Rn, $^{220}$Rn and equilibrium-equivalent thoron concentrations in the earth buildings are summarized in Table 1. Annual arithmetic mean concentrations of indoor $^{222}$Rn and $^{220}$Rn are 25 and 232 Bq m$^{-3}$, respectively. $^{220}$Rn was detected in 96% of the surveyed rooms. The abundant indoor $^{220}$Rn can be explained as the high soil $^{232}$Th content and the uncoated walls. However, indoor $^{222}$Rn is somewhat lower than expected, as the soil $^{238}$U content was also reported to be high [1]. This may be explained by poor airtightness of the soil walls as well of wooden doors and windows.

The mean $^{222}$Rn concentration in the second period was 70% of that in the first period. However, both $^{220}$Rn and its progeny concentrations were nearly the same in the two periods. It is considered as the different ventilation rates of the rooms in the two periods. The second survey performed in the summer and autumn, the ventilation rate is expected to be higher than that of the first survey covered the winter and spring. While the decay constant of $^{220}$Rn is much larger than the ventilation rate, the change of ventilation does not affect $^{220}$Rn concentrations.

The averaged indoor and outdoor $\gamma$-dose rates are $141\pm18$ and $120\pm15$ nGy h$^{-1}$, respectively, in the 100 rooms. The relatively high value is also considered as the high contents of $^{238}$U and $^{232}$Th in soil. No significant correlation was found between indoor $\gamma$-dose rates and indoor $^{222}$Rn or $^{220}$Rn concentrations. On the other hand, even though the direct correlation between the concentrations of indoor $^{220}$Rn and its progeny was weak, a better correlation ($R=0.7641$) was found by modified $^{220}$Rn concentrations with the ratios of source areas to volumes of rooms. It suggests that indoor concentrations of $^{220}$Rn progeny may be roughly estimated from the measurements of indoor $^{220}$Rn.

Table 1
Results of indoor $^{222}$Rn, $^{220}$Rn and its progeny concentrations (Bq m$^{-3}$) in Yongding Hakka Earth Building

| Period | Rn-222 | | Rn-220 | | Equilibrium-equivalent thoron | |
|---|---|---|---|---|---|---|
| | Mean*±S.D. | Range | Mean±S.D. | Range | Mean±S.D. | Range |
| 2002.12–2003.06 | 28±19 | 6–118 | 225±288 | ND$^a$–1540 | 0.69±0.57 | 0.08–3.41 |
| 2003.07–2004.01 | 21±15 | 5–109 | 239±271 | ND–1163 | 0.71±0.55 | 0.12–3.36 |

$^a$ Refers to lower than detection limit; * Refers to arithmetic mean.

## References

[1] J. Chen, et al., Chin. J. Radiol. Med. Prot. 8 (Suppl. 2) (1988) 60–66 (in Chinese).
[2] W. Zhuo, et al., Rev. Sci. Instrum. 73 (2002) 2881–2887.
[3] W. Zhuo, T. Iida, Jpn. J. Health Phys. 35 (2000) 365–370.

# Natural radiation levels in Fu'an city in Fujian Province of China

Y. Luo[a], W. Zhuo[b,*], M. Wei[a], S. Tokonami[b], W. Wang[a], Y. Yamada[b], J. Chen[a], M. Chen[a]

[a]*Fujian Centre for Prevention and Control of Occupational Diseases and Chemical Poisoning, P.R. China*
[b]*Radon Research Group, National Institute of Radiological Sciences, Chiba 263-8555, Japan*

**Abstract.** A survey of natural radiation levels was performed in an area with the highest mortality of lung cancer in Fujian Province of China. Even though relatively high terrestrial gamma radiation was observed, the concentrations of indoor $^{222}$Rn, $^{220}$Rn and its progeny are near the same as the provincial averages. Annual arithmetic mean concentrations of indoor $^{222}$Rn, $^{220}$Rn and its progeny were 29, 184 and 0.8 Bq m$^{-3}$, respectively. The public exposure to radon and thoron is estimated to be 1.2 mSv year$^{-1}$, similar to the world average. © 2004 Elsevier B.V. All rights reserved.

*Keywords:* Radon; Thoron; Equilibrium-equivalent thoron concentration; Gamma dose rate

## 1. Introduction

Fu'an city populated with about 600,000 people locates in the eastern part of Fujian Province in China. Based on the Cancer Registry of Fujian Province, although the total cancer mortality in Fu'an city was not high, the mortality of lung cancer was the highest in the past decades. The reason is still unknown so far. On the other hand, very high content of $^{232}$Th in soil was reported in the area [1], it is expected that high thoron might exist in the area. In this study, a survey of the natural radiation levels was carried out, their results are reported.

## 2. Materials and methods

Indoor $^{222}$Rn and $^{220}$Rn were measured by the passive integrating $^{222}$Rn and $^{220}$Rn discriminative monitor [2]. The monitors were set at about 10 cm from a wall in a bedroom. The equilibrium-equivalent thoron concentration was estimated from measuring the deposition rate of its progeny [3]. The deposition rate monitors were set in the middle

---

\* Corresponding author. Tel.: +81 43 206 3111; fax: +81 43 206 4098.
 *E-mail address:* whzhuo@nirs.go.jp (W. Zhuo).

0531-5131/ © 2004 Elsevier B.V. All rights reserved.
doi:10.1016/j.ics.2004.09.048

Table 1
Results of indoor $^{222}$Rn, $^{220}$Rn and its progeny concentrations (Bq m$^{-3}$) in Fu'an city of Fujian Province, China

| Period | Rn-222 | | Rn-220 | | Equilibrium-equivalent thoron | |
|---|---|---|---|---|---|---|
| | Mean*±S.D. | Range | Mean±S.D. | Range | Mean±S.D. | Range |
| 2002.11–2003.05 | 33±25 | 4–151 | 182±290 | ND**–1589 | 0.85±0.77 | 0.12–4.26 |
| 2003.06–2003.12 | 24±19 | 4–92 | 186±326 | ND–1678 | 0.72±0.67 | 0.12–3.83 |

\* Refers to arithmetic mean.
\*\* Refers to lower than detection limit.

of the same room. Indoor and outdoor gamma radiation levels were measured by a NaI(Tl) scintillation radiation meter at 1 m above the ground.

Indoor $^{222}$Rn, $^{220}$Rn and its progeny were measured from Nov. 2002 to Dec. 2003 in 6-month intervals. A total of 100 rooms were randomly selected from 10 villages. One set of the $^{222}$Rn, $^{220}$Rn and the deposition rate monitors was used in each room.

## 3. Results and discussion

Indoor $^{222}$Rn, $^{220}$Rn and equilibrium-equivalent of thoron concentrations measured in the 100 rooms are summarized in Table 1. The annual arithmetic mean concentrations of indoor $^{222}$Rn, $^{220}$Rn and equilibrium-equivalent thoron are 29, 184 and 0.8 Bq m$^{-3}$, respectively. They are nearly the same as the averages of the whole province [4]. The mean $^{222}$Rn concentration in the second period was about 73% of that in the first period. However, both $^{220}$Rn and equilibrium-equivalent thoron concentrations were nearly the same in the two periods. It is considered as the different ventilation rates of the rooms in the two periods. The second survey performed in the summer and autumn, the ventilation rate is expected to be higher than that of the first survey covered the winter and spring. While the decay constant of $^{220}$Rn is much larger than the ventilation rate, the change of ventilation does not affect $^{220}$Rn concentrations.

Indoor and outdoor gamma dose rates are averaged to be 184±26 and 188±26 nGy h$^{-1}$ in the 100 rooms. $^{220}$Rn was detected out in about 90% of the rooms. The abundant indoor $^{220}$Rn and the high gamma dose rate are considered as the high content of $^{232}$Th in the soil [1].

Based on the results from this survey and another study [4], public exposure to radon and thoron in Fu'an city are estimated to be 0.90 and 0.32 mSv year$^{-1}$ by using the conversion factors provided by the UNSCEAR report [5]. The total exposure to radon and thoron is about 1.2 mSv year$^{-1}$. It is nearly the same as the world average [5]. Public exposure to radon and thoron in Fu'an city is not significant. Other studies are still needed to explore reasons of high mortality rate of lung cancer in the area.

## References

[1] J. Chen, et al., Radionuclide contents in soil and their contribution to public dose in Fujian, Chin. J. Radiol. Med. Prot. 8 (Suppl. 2) (1988) 60–66 (in Chinese).
[2] W. Zhuo, et al., A simple passive monitor for integrating measurements of indoor thoron concentrations, Rev. Sci. Instrum. 73 (2002) 2881–2887.
[3] W. Zhuo, T. Iida, Estimation of thoron progeny concentrations in dwellings with their deposition rate measurements, Jpn. J. Health Phys. 35 (2000) 365–370.
[4] W. Zhuo, T. Iida, X. Yang, Environmental radon and thoron progeny concentrations in Fujian Province of China, Radiat. Prot. Dosim. 87 (2000) 137–140.
[5] UNSCEAR, Sources and Effects of Ionizing Radiation, New York, United Nations, 2000.

ELSEVIER

www.ics-elsevier.com

# A pilot survey on indoor radon and thoron progeny in Yangjiang, China

## Guo Qiuju[a,]*, Chen Bo[a], Sun Quanfu[b]

[a]*Department of Technical Physics, School of Physics, Peking University, Beijing, 100871, China*
[b]*National Institute for Radiological Protection, China CDC, Beijing, China*

**Abstract.** A pilot survey on indoor radon and thoron progeny concentration was carried out in five villages in HBRA of Yangjiang and one village in the controlled area(CA) in March 2004. Totally 26 adobe houses and 29 brick houses in HBRA were investigated. A portable 24-h integrating monitor with CR-39 detectors was adopted for progeny measurements, and the average equilibrium equivalent $^{222}$Rn and $^{220}$Rn concentrations (EEC$_{Rn}$ and EEC$_{Tn}$) were obtained. Rather high levels of EEC$_{Rn}$ and EEC$_{Tn}$ with an average of 57.1±33.3 and 12.6±9.5 Bq·m$^{-3}$, respectively, were observed in adobe houses. In addition, $^{222}$Rn/$^{220}$Rn concentrations and exhalation rate were also measured by an Electrostatic-Radon-Sampler (ERS-2) monitor. © 2004 Published by Elsevier B.V.

*Keywords:* Radon; Thoron; Progeny; High Background Radiation Area; Yangjiang, China

## 1. Introduction

Yangjiang, located in the very southern part of China mainland, is well-known as a high background radiation area (HBRA), and an epidemiological study has been carried out for several decades. It is noticed according to previous research that the $^{232}$Th content of soil in HBRA is rather high; the average value is 206±92 Bq kg$^{-1}$ [1]. Therefore, the evaluation of the contribution from thoron ($^{220}$Rn) progenies should be necessary and important from the viewpoint of internal exposure. Measurements on both radon ($^{222}$Rn) and $^{220}$Rn gas have been performed, but data of $^{222}$Rn and $^{220}$Rn progenies was very limited until now. For the purpose of the evaluation of dose contribution from those progenies, direct and integrating measurements are desirable.

---

\* Corresponding author. Tel.: +86 10 6275 5403; fax: +86 10 6275 1615.
*E-mail address:* qjguo@pku.edu.cn (G. Qiuju).

0531-5131/ © 2004 Published by Elsevier B.V.
doi:10.1016/j.ics.2004.10.005

Table 1
Indoor EEC$_{Rn}$ and EEC$_{Tn}$ in HBRA (Bq m$^{-3}$)

| Dwelling types | Dwelling numbers | EEC$_{Rn}$ | | | EEC$_{Tn}$ | | |
|---|---|---|---|---|---|---|---|
| | | Range | Mean | S.D. | Range | Mean | S.D. |
| Adobe houses | 26 | 1.2–128.2 | 57.1 | 33.3 | 1.2–37.0 | 12.6 | 9.5 |
| Brick houses | 29 | 3.6–88.8 | 41.8 | 26.0 | 0.6–29.6 | 4.7 | 5.5 |

## 2. Materials and methods

### 2.1. $^{222}Rn/^{220}Rn$ progeny integrating measurements

To simultaneously measure both $^{222}$Rn and $^{220}$Rn progeny concentrations, a new type of portable integrating monitors with etched track detectors (CR-39) was developed by Dr. Zhou [2] and adopted in our survey. The monitor sampled ambient air at a flow rate around 1 L min$^{-1}$ for 24 h continuously, and then the average equilibrium-equivalent $^{222}$Rn and $^{220}$Rn concentrations (EEC$_{Rn}$ and EEC$_{Tn}$) during the sampling intervals were obtained through etching and calculating.

### 2.2. Measurements of $^{222}Rn$ and $^{220}Rn$ gas concentrations and exhalation rates

A set of Electrostatic-Radon-Sampler (ERS-2, Tracerlab Instruments, Germany) was adopted to measure both $^{222}$Rn and $^{220}$Rn gas concentrations as well as exhalation rates from the surface of typical walls and ground.

## 3. Results and discussions

A pilot survey of 64 dwellings in five villages of the HBRA and one village in the control area was conducted in a typical spring season, March 2004. The main results of HBRA are shown in Table 1.

EEC$_{Tn}$ seemed more sensitive to ventilation conditions than that of EEC$_{Rn}$. Indoor EEC$_{Rn}$ and EEC$_{Tn}$ also showed a good correlativity with field $\gamma$ dose rate. The ranges of $^{222}$Rn and $^{220}$Rn exhalation rate from the surface of adobe wall were 22.5–42.9 and 2111–4012 mBq/m$^2$ s, respectively. For the surface of brick wall, the ranges of $^{222}$Rn and $^{220}$Rn exhalation rate were 11.1–26.4 and 317–1022 mBq/m$^2$ s, respectively. Both $^{222}$Rn/$^{220}$Rn concentrations were measured with 1-h cycle in five adobe and brick houses; the range of the average concentrations of each measured room was 152–412 Bq m$^{-3}$ for $^{222}$Rn and 82–403 Bq m$^{-3}$ for $^{220}$Rn (5 cm from the ground and far from the wall).

## 4. Conclusion

Rather high levels of EEC$_{Rn}$ and EEC$_{Tn}$ with a wide and varied range were found in both adobe and brick houses as well. High $^{220}$Rn exhalation rates from the surface of adobe walls suggested that the adobe walls were the sources of the high EEC$_{Tn}$ concentrations indoors. Investigations in detail were needed in the future.

## References

[1] L.-X. Wei, T. Sugahara, High background radiation area in chain, J. Radiat. Res. 41 (2000) 1–76 Suppl.
[2] W. Zhuo, T. Iida, An instrument for measuring equilibrium-equivalent $^{222}$Rn and $^{220}$Rn concentrations with etched track detector, Health Phys. 77 (5) (1999) 584–587.

ELSEVIER

www.ics-elsevier.com

# High background radiation valley formed by Peitou Hot Spring

## Ching-Jiang Chen*, Pei-Hou Lin, Ching-Chung Huang

*Radiation Monitoring Center, Atomic Energy Council, Kaohsiung, Taiwan, 83305, Republic of China*

**Abstract.** In 1905, the high radioactivity crystallized stone so-called "hokutolite" was discovered in the so-called "hell valley" in Peito. Its terrestrial gamma dose rate, determined by in situ measurement, was around 180 nGy/h, which was about three times higher than the radiation levels in the surrounding area. The deposited soil samples were measured by gamma spectrometry. Data shows that the $^{228}$Ra and its progeny is about 270 Bq/kg, while $^{226}$Ra and its progeny is about 44 Bq/kg only. The $^{228}$Ra and its progeny dominate the dose contribution. But the radioactivity of hokutolite is mainly from $^{226}$Ra and its progeny. This difference might be caused by different deposition processes and physical half-life of $^{226}$Ra and $^{228}$Ra. © 2004 Elsevier B.V. All rights reserved.

*Keywords:* Peito hot spring; Hokutolite; Radiation; Radium; Radioactivity

## 1. Introduction

Peito hot spring in north Taiwan is famous for its high temperature and low pH value. This area is also famous for "hokutolite," or Peito stone, containing high-level radium. There were many geologists and radiologists who have studied the characteristic of hokutolite since it was discovered by Okamoto in 1905 [1]. This heavy crystal was formed by barium sulphate and lead sulphate. The thickness of this glassy crystal is normally less than 10 mm. Radium is dissolved into geothermal water with low pH value (1.2–1.6) at high temperature (>90 °C). On reaching the surface of a dormant volcano, deposition and crystallization of dissolved radium occurs in the valley and its downstream regions where hokutolite is formed. The area of this volcanic valley is about 3500 m$^2$. Tourists used to boil eggs in this valley and called it "hell valley" owing to the foggy steam all year long especially in winter season and cloudy days. The gross α and β activities of this hot spring are 0.59–4.0 and 1.25–2.0 Bq/l, which is more than 15 times higher than other river water in Taiwan.

---

\* Corresponding author. Tel.: +886 73709206x200; fax: +886 73701660.
 *E-mail address:* jiang@trmc.aec.gov.tw (C.-J. Chen).

0531-5131/ © 2004 Elsevier B.V. All rights reserved.
doi:10.1016/j.ics.2004.11.168

## 2. Experimental

### 2.1. In situ γ dose rate measurement

The in situ γ dose rate was measured with a 25.4φ×25.4 mm NaI(Tl) survey meter, which can measure γ energy above 50 keV. A standard calibration method published by Japan Science and Technology was applied. Another survey meter using issue-equivalent plastic scintillation detector with heavy metal admixtures, model AT1121, was used to measure γ dose rate at 1 m above ground. A 76.2φ×76.2 mm L NaI(Tl) in situ γ spectrum method was also used to measure the γ dose rate. The spectra were unfolded to true photon energy flux by a response function [2].

### 2.2. HPGe γ spectrometry for deposit samples and hokutolite

Three deposit samples and a hokutolite sample were measured by HPGe γ spectrometry to identify the natural radionuclides from $^{232}$Th and $^{238}$U series.

## 3. Results and discussion

The γ dose rates at 1 m above ground in the valley measured by survey meters were between 161 and 192 nGy/h. If we put the survey meter on ground, some spots with high dose rates showed the radiation levels over 400 nGy/h. Dose rate measured by 3″ φ NaI(Tl) spectrometry showed an average of 180 nGy/h, which is about three times higher than the radiation levels in the surrounding areas. Spectrum showed that the $^{232}$Th progeny dominated the dose contribution. Radionuclides in deposits measured by HPGe γ spectrometry were $^{232}$Th progeny $^{208}$Tl, $^{212}$Pb, $^{212}$Bi, $^{228}$Ac and $^{238}$U progeny $^{210}$Pb, $^{214}$Pb, $^{214}$Bi, and $^{226}$Ra. The appearance of $^{228}$Ac showed the deposition of its parent nuclide $^{228}$Ra. But the absence of $^{234m}$Pa suggested no significant deposition of $^{238}$U and $^{234}$Th. The radioactivity of the deposits mainly came from the deposition of $^{228}$Ra, $^{224}$Ra, and $^{226}$Ra. The activity of $^{228}$Ra and its progeny was about 270 Bq/kg, while that of $^{226}$Ra and its progeny was only 44 Bq/kg which is close to the mother rock content. Gamma spectrometry of a hokutolite sample showed that the $^{226}$Ra progeny were dominant, while $^{234}$Pa and $^{228}$Ac were negligible. Because the hokutolite samples were collected before the 1970s, most $^{228}$Ra($T_{1/2}$=5.76y) has decayed to stable $^{208}$Pb.

## 4. Recovery of hokutolite ecology

Since the discovery of hokutolite in 1905, it has been known to be a precious stone even more expensive than gold. Many people collected hokutolite illegally, and the ecology of Peito stream was destroyed seriously. Hokutolite could not be found anymore since the early 1970s. The reconstruction of this district to revive hokutolite started in 1997. Hokutolites were exhibited in Peito Hot Spring Museum, which was opened on October 11, 1998. The dormant volcano is now covered with 0.5–1.0 m deep hot water to revive hokutolite.

## References

[1] H. Hamaguchi, Y.T. Lee, H.S. Cheng, Study of the radioactivity of hokutolite, J. Chin. Chem. Soc. Ser. II 9 (1) (1962).
[2] Y.M. Lin, et al., Natural radiation of interest in Taiwan, R.O.C., Nucl. Sci. J. 21 (4) (1984) 226–232.

# Natural radionuclide distribution in soil samples around Kudankulam Nuclear Power Plant Site (Radhapuram taluk of Tirunelveli district, India)

G.M. Brahmanandhan, D. Khanna, J. Malathi, S. Selvasekarapandian*

*Solid State and Radiation Physics Laboratory, Bharathiar University, Coimbatore-641046, India*

**Abstract.** A study on the concentration of natural radionuclides in soil samples collected around forthcoming Kudankulam Nuclear Power Plant Site (Radhapuram taluk of Tirunelveli district, India) has been undertaken. In total, 145 soil samples have been collected. NaI(Tl) gamma ray spectrometer has been used for the measurement of radio nuclide's activity. The dose contribution due to $^{40}$K and $^{232}$Th varies in the range of 0.35–69.15 nGy/h and 13.02–5750.71 nGy/h, respectively. The activity concentration of $^{238}$U lies in the below detectable limit (8.5 Bq kg$^{-1}$) in all the samples, so the dose due the $^{238}$U is negligible. The annual effective dose due to these primordial radionuclides is 0.26 mSv/year. © 2004 Elsevier B.V. All rights reserved.

*Keywords:* Primordial radionuclides; NaI(Tl) gamma ray spectrometer; Dose rate; Effective dose

## 1. Introduction

Measurement of natural background radiation around nuclear power plant site provides the knowledge of the distribution pattern of both natural and anthropogenic radionuclides. Indian Government, with the help from Russia, is now building a 2000-MW nuclear power plant in Kudankulam. A base line data regarding radiation in the surrounding areas of Kudankulam is precious for cross checking the radiation levels after the commission of nuclear reactors in 2007. So a detailed systematic study has been carried out in the soil samples of the surrounding areas of Kudankulam, i.e., Radhapuram taluk of Tirunelveli district, India (Fig. 1A).

* Corresponding author. Tel.: +91 422 2422222 422; fax: +91 422 2422387.
*E-mail address:* sekarapandian@yahoo.com (S. Selvasekarapandian).

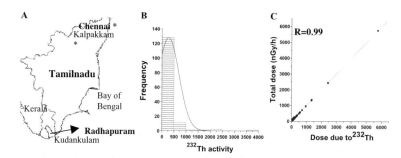

Fig. 1. (A) Study area map. (B) Frequency distribution of the activity concentration of $^{232}$Th. (C) Correlation between the dose due to $^{232}$Th and total dose.

## 2. Materials and methods

According to Selvasekarapandian et al. [1], the ambient gamma measurements have been carried out using a portable NaI(Tl) scintillometer. The samples have been analysed using 3×3 in. NaI(Tl) based detector.

## 3. Results and discussions

The ambient gamma dose rate has been found to vary from 45 to 810 nGy/h. The highest ERDM value has been found in the Thilaivanamthoppu sample (810 nGy/h). It is due to the presence of the monazite content in the soil. From the NaI (Tl) gamma ray spectrometer, the average activity concentration of $^{40}$K has been found to vary from 8.5 to 1608.3 Bq kg$^{-1}$ with an average value of 348.49 Bq kg$^{-1}$. The highest value (1608.3 Bq kg$^{-1}$) of $^{40}$K has been found in Paarivillai, which is an important agricultural area situated on the bank of "Tamiraparani" river. The activity concentration of $^{238}$U has been found to be below detectable limit (8.5 Bq kg$^{-1}$) in all the samples. The activity concentration of $^{232}$Th has been found to vary from 20.91 to 9230.68 Bq kg$^{-1}$ with the average of 334.36 Bq kg$^{-1}$. Three samples have been collected at 20–30, 100–150 and more than 500 m away from the sea. Gamma spectroscopic measurement show sample collected within 20–30 m from the sea has the highest value of $^{232}$Th than the other samples; so the deposition of monazites from the sea is the sole reason for the higher value. The highest value of $^{232}$Th has been observed in Kuthankuli-II sample and the value is 9230.68 Bq kg$^{-1}$. The Kuthankuli-II is situated along the Bay of Bengal and monazite deposition from the sea is the reason for the high thorium value. Thorium has been observed in all the samples and this may be due to the existence of igneous and granitic nature of rock and soil present in this area. The geological survey already proved the existence of rich deposits of red garnet sands in the beds of Uvari river, which is flowing across this area. Frequency distribution of $^{232}$Th is illustrated in the Fig. 1B. The annual effective dose due to the three primordial radionuclides has been calculated [2], and it is found to be 0.26 mSv/year, which is within the permissible limit. Good correlation ($R$=0.99) has been observed between the total dose due to the three primordial radionuclides and dose due to $^{232}$Th and it is shown in Fig. 1C. Taken together, it is very clear that this area has natural gamma radiation dose primarily due to thorium content and the dose due to potassium and uranium is negligible.

## References

[1] S. Selvasekarapandian, et al., Journal of Radio Analytical and Nuclear Chemistry 252-2 (2002) 429–435.
[2] J. Sannappa, et al., Radiation Measurements 37 (2003) 55–65.

# Measurement of activity concentrations of $^{40}$K, $^{238}$U and $^{232}$Th in soil samples of Agastheeswaram taluk, Kanyakumari district, India

D. Khanna, J. Malathi, G.M. Brahmanandhan, S. Selvasekarapandian*

*Solid State and Radiation Physics Laboratory, Bharathiar University, Maruthamalai Road, Coimbatore 641046, India*

**Abstract.** Southern coastal region of Kanyakumari district, Tamilnadu state, India is one of the high natural background radiation areas. A total of 41 soil samples have been collected in Agastheeswaram taluk of Kanyakumari district. By γ-ray spectrometry using NaI(Tl) detector, average activity concentrations of $^{40}$K, $^{238}$U and $^{232}$Th have been measured for soil samples and are found to be in the range between Below-Detectable-Level (BDL) to 613.24 Bq kg$^{-1}$, BDL to 229.86, and 32.03 to 567.76 Bq kg$^{-1}$, respectively. The total absorbed dose rate due to the distribution of these elements is 112.97 nGy/h. © 2004 Elsevier B.V. All rights reserved.

*Keywords:* Primordial radionuclides; Activity concentration; Effective dose rate

## 1. Introduction

Natural radiation, a ubiquitous feature of our earth since its origin, is composed of the cosmogenic radionuclides such as $^3$H, $^7$Be and $^{14}$C and primordial radionuclides such as $^{40}$K, $^{232}$Th and $^{238}$U. According to UNSCEAR report 2000, the greatest contribution to exposure of mankind comes from natural background radiation due to presence of these primordial radionuclides. These elements occur in minerals such as monazites and zircons. Significant quantities of monazite ore are known to occur along the coastal line and traces in inland areas of Agastheeswaram taluk, Kanyakumari district. Many studies have been carried out to measure the radiation levels along the coastlines there [1] and none inland. Hence, this study has been undertaken to measure the activity concentrations of $^{40}$K, $^{232}$Th and $^{238}$U in the soil samples collected at Agastheeswaram taluk. The present study gains

---

\* Corresponding author. Tel.: +91 422 2422222x422; fax: +91 422 2422387.
*E-mail address:* sekarapandian@yahoo.com (S. Selvasekarapandian).

0531-5131/ © 2004 Elsevier B.V. All rights reserved.
doi:10.1016/j.ics.2004.10.018

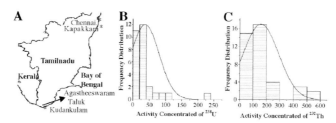

Fig. 1. (A) Area map, (B) frequency distribution of the activity concentration of $^{238}$U, (C) frequency distribution of the activity concentration of $^{232}$Th.

significance due to the construction of Kudankulam Nuclear Power Plant (20 km from the study area) and will form baseline data for the pre-operational survey.

## 2. Materials and methods

Forty-one soil samples have been collected from undisturbed areas through the taluk (Fig. 1A). Collection and processing of soil samples have been carried out following the procedure described elsewhere [2]. Ambient γ-dose has been measured using NaI(Tl) scintillometer (SM141D, ECIL). γ-ray spectrometry analysis has been carried out using a 3″×3″ NaI(Tl) crystal detector-based γ-ray spectrometer. The activity concentration of the primordial radionuclides $^{40}$K, $^{238}$U and $^{232}$Th are measured from photo peaks of $^{40}$K (1.46 MeV), $^{214}$Bi (1.76 MeV) and $^{208}$Tl (2.614 MeV), respectively.

## 3. Results and discussion

Activity concentration of $^{40}$K has been found to vary from BDL (Below-Detectable-Level) to 613.24 Bq kg$^{-1}$ with an average of 69.71 Bq kg$^{-1}$. In 25 places, $^{40}$K concentration has been found to be BDL. The highest value for $^{40}$K is found in Kottaram, which is an agricultural area. Activity concentration of $^{238}$U is also found to vary from BDL to a maximum value of 229.56 Bq kg$^{-1}$ with 12 places having BDL value. It has an average value of 26.45 Bq kg$^{-1}$; frequency distribution of $^{238}$U is given in Fig. 1B.

The value of activity of $^{232}$Th has been found to vary from a minimum of 32.03 to 567.76 Bq kg$^{-1}$. Interestingly, no BDL is observed for $^{232}$Th. Lowest concentration of thorium is found in Marungoor about 10 km from the seashore. The highest concentration is found in Andivillai, which is 2 km away from the seashore. This may be due to the transportation of $^{232}$Th by the wind from the seashore where significant pockets of monazite ore are known to occur. This is supported by the fact that Kanyakumari is a high windy place throughout the year. The Western Ghats, a source of monazite ore [2], which ends in this taluk, may also be the reason for the omnipresent distribution of thorium throughout the taluk with an average of 158.4 Bq kg$^{-1}$. Thorium frequency distribution is given in Fig. 1C. Average absorbed dose rates due to $^{40}$K, $^{238}$U and $^{232}$Th are found to be 8.78, 15.97 and 98.68 nGy/h, respectively. Average absorbed dose rate due to the distribution of these elements is 112.97 nGy/h. Major contribution comes from thorium. Annual effective dose rate for the region is 138.64 μSv/y assuming 20% outdoor occupancy factor.

## References

[1] K.S. Lakshmi, PhD thesis, Bharathiar University, Coimbatore, India, (2001).
[2] S. Selvasekarapian, et al., Applied radiation, Isotope 52 (2000) 299.

# Study of primordial radionuclide distribution in sand samples of Agastheeswaram taluk of Kanyakumari district, India

J. Malathi, G.M. Brahmanandhan, D. Khanna, S. Selvasekarapandian*

*Solid State and Radiation Physics Laboratory, Bharathiar University, Coimbatore 641046, India*

**Abstract.** The activity concentrations of the three primordial radionuclides have been measured in the sand samples of Agastheeswaram taluk using NaI(Tl) gamma ray spectrometer. The average activities of $^{232}$Th, $^{238}$U and $^{40}$K have been found to be 5787.15, 1082.90 and <13.85 Bq kg$^{-1}$ (BDL), respectively. The calculated annual effective dose is 4.78 mSv year$^{-1}$ and the results are discussed in this paper. © 2004 Elsevier B.V. All rights reserved.

*Keywords:* Gamma ray spectrometer; Primordial radionuclide; Effective dose

## 1. Introduction

It is a well-known fact that west coast of Southern Peninsula of India has the richest deposits of monazite, a thorium-bearing mineral. It is also known that some regions of east coast have sparse distribution of monazites. A 2000 MW Nuclear Power Plant is also under construction in Kudankulam, located in south east coast of India. It would be of interest to evaluate primordial radionuclide distribution in Kudankulam and its surrounding areas and their impact on the background radiation levels. Such an evaluation would also provide the baseline data for estimating the change in the environment. So a measurement of the activity concentration of the primordial radionuclides in the sand samples of the Agastheeswaram taluk of Kanyakumari district, which is situated within radius of 30–50km from the Kudankulam Nuclear Power Plant Site, has been carried out.

---

\* Corresponding author. Tel.: +91 422 2422222x422; fax: +91 422 2422387.
*E-mail address:* s.sekarapandian@rediffmail.com (S. Selvasekarapandian).

0531-5131/ © 2004 Elsevier B.V. All rights reserved.
doi:10.1016/j.ics.2004.11.119

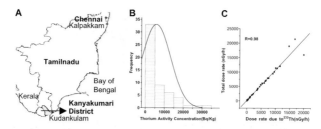

Fig. 1. (A) Area map. (B) Frequency diagram of $^{232}$Th activity concentration. (C) Correlation graph of $^{232}$Th vs. total dose.

## 2. Materials and methods

In total, 55 sand samples are collected along the coastal areas of Agastheeswaram taluk and processed as per the procedure given by Selvasekarapandian et al. [1]. A typical NaI(Tl) scintillator has been used to measure the ambient gamma dose. The activity concentrations of the primordial radionuclides have been measured using a NaI(Tl) gamma ray spectrometer. The peaks corresponding to $^{40}$K (1.460 MeV), $^{214}$Bi (1.764 MeV) and $^{208}$Tl (2.614 MeV) have been considered for the estimation of the primordial radionuclides $^{40}$K, $^{238}$U and $^{232}$Th, respectively.

## 3. Results and discussion

The ambient gamma dose rate varies from 90 to 9000 nGy/h with the average value of 1691 nGy/h. From the ERDM value, it is inferred that the study area is one of the high background radiation areas in India. Activity concentration of $^{232}$Th lies in the range of 32.18–32147.22 Bq kg$^{-1}$ with geometric mean of 1717.71 Bq kg$^{-1}$. Activity concentration of $^{238}$U found to lie in the range of 11.9–5030.81 Bq kg$^{-1}$ with the geometric mean of 425.06 Bq kg$^{-1}$, while the activity concentration of $^{40}$K found to lie in below detectable limit (<13.85 Bq kg$^{-1}$). Frequency distribution pattern of the activity concentration of $^{232}$Th is shown in Fig. 1B.

The gamma dose rate due to $^{232}$Th lies between 20.04 and 20027.71 nGy/h, with the mean value of 3605.33 nGy/h. The gamma dose rate due to $^{238}$U is found to lie between 5.08 and 2148.15 nGy/h, its mean value being 395.91 nGy/h. Good correlation exists between the total dose and the dose due to the $^{232}$Th (Fig. 1C), which indicates that this area has sizeable thorium content. The annual effective dose has been calculated by using the UNSCEAR 2000 report [2]. The annual effective dose is found to be 4786.78 μSv year$^{-1}$ (4.78 mSv year$^{-1}$). Thus, the dose due to the three primordial radionuclides in the study area is 4.7 times higher than the normal level [2]. From the geological survey results, it is very clear that the high background radiation is due to the presence of the monazites and other heavy minerals in the sediments of the beach places. Enhancement of monazite content in the sand is due to cyclic subsidence of the coast as a result of wind and wave action. The huge variations observed in the activity concentrations are due to the discontinuous distribution of Monazites in sand.

## References

[1] Selvasekarapandian, et al., Gamma radiation dose from radionuclides in soil samples of Udagamandalam (Ooty)in India, Radiation Protection Dosimetry 82–83 (1999) 225–228.
[2] UNSCEAR, Report on "Sources and Effects of Ionizing radiation of General assembly with Scientific Annexes", United Nations, New York, 2000.

# Primordial radionuclides concentrations in the beach sands of East Coast region of Tamilnadu, India

K.S. Lakshmi[a], S. Selvasekarapandian[b],*, D. Khanna[b], V. Meenakshisundaram[c]

[a]*Meenakshi College for Women, Kodambakkam, Chennai-600024, India*
[b]*Solid State and Radiation Physics Laboratory, Bharathiar University, Coimbatore-641046, India*
[c]*Radiation Safety System Division, Indira Gandhi Centre for Atomic Research, Kalpakkam-603102, India*

**Abstract.** The Eastern coastal region of Tamil Nadu (8°11′ N to 13°04′ N latitude, 77°29′ E to 80°17′ E longitude) India is known to have some of regions of significant distribution of monazite. A total of 66 sand samples have been collected from 33 locations situated 25–30 km apart. By γ-ray spectrometry using NaI(Tl) detector, activity concentrations of primordial radionuclides such as $^{40}K$, $^{232}Th$, and $^{238}U$ have been determined as 15 to 524 Bq kg$^{-1}$, 13 to 3576 Bq kg$^{-1}$ and BDL to 254 Bq kg$^{-1}$. Depth profile study has been also carried out. Abnormal values of thorium concentrations have been observed at Kanyakumari (1315 Bq kg$^{-1}$), Kudankulam (2860 Bq kg$^{-1}$), Uvari (1225 Bq kg$^{-1}$), Tharangampadi (3576 Bq kg$^{-1}$) and Pondicherry (1641 Bq kg$^{-1}$). The absorbed γ-dose rate above 1 m due to the presence of these primordial radionuclides has been found to be in the range of 12–2337 nGy/h. Subsequently, the annual effective dose has been found to vary from 0.014 to 2.87 mSv/y. © 2004 Elsevier B.V. All rights reserved.

*Keywords:* Monazite ore; Activity concentration; Depth profile study; Effective dose

## 1. Introduction

Background radiation levels are from a combination of terrestrial (from the $^{40}K$, $^{232}Th$, $^{226}Ra$, etc.) and cosmic radiation (photons, muons, etc.). The level is fairly constant over the world (0.8–1.5 nGy/h), although there are some areas with sizable populations that have high background radiation levels, the highest being found primarily in Brazil, India, and China [1]. Higher radiation levels are due to high concentrations of radioactive minerals in soil. One, monazite, is a highly insoluble rare earth mineral that occurs in beach sand together with the mineral ilmenite. In Brazil, the monazite sand deposits are found along certain beaches. On the SW coast of India, the monazite deposits are larger

* Corresponding author. Tel.: +91 422 2422222 422; fax: +91 422 2422387.
 *E-mail address:* sekarapandian@yahoo.com (S. Selvasekarapandian).

0531-5131/ © 2004 Elsevier B.V. All rights reserved.
doi:10.1016/j.ics.2004.10.022

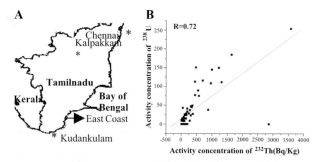

Fig. 1. (A) Area map. (B) Correlation between $^{232}$Th and $^{238}$U activities.

than those in Brazil. Our study has been undertaken to estimate the concentration of primordial radionuclides along the East coast of Tamilnadu, India.

## 2. Materials and methods

East coast of Tamil Nadu is a little more than 1000 km in length (Fig. 1A). Sixty-six sand samples have been collected from 33 sites at a distance of 25–30 km apart. Ambient γ-dose rate has been measured by using a NaI(Tl) scintillometer (SM 141D model) by ECIL, India. γ-ray spectrometry measurement has been carried out with NaI(Tl) crystal detector. Activity concentration of the primordial radionuclides $^{40}$K, $^{238}$U, and $^{232}$Th are measured from photo peaks of $^{40}$K (1.46 MeV), $^{214}$Bi (1.76 MeV), and $^{208}$Tl (2.614 MeV).

## 3. Results and discussion

Highest $^{232}$Th activity of 3576 Bq kg$^{-1}$ has been observed in beach sand collected at Tharangampadi shore, while Manora beach sand has the lowest $^{232}$Th activity. Activity concentration of $^{238}$U varies from below-detectable-level (BDL) to 254 Bq kg$^{-1}$. Highest activity of $^{238}$U is in Tharangampadi. BDL activity has been observed at 22 locations. $^{40}$K has been found to be 524 Bq kg$^{-1}$ in the sand sample collected at Tuticorin and the lowest activity of 15 Bq kg$^{-1}$ has been observed in the sand samples collected at 5 sites. Range is much broader for $^{232}$Th series than for $^{238}$U series. This is understandable from the very large fluctuations in the activity of $^{232}$Th series from site to site in this study, as abnormal values of thorium concentrations have been observed at Kanyakumari (1315), Kudankulam (2860), Uvari (1225), Tharangampadi (3576), and Pondicherry (1641 Bq kg$^{-1}$, respectively). The arithmetic mean of the ratio of activity of $^{232}$Th to $^{238}$U is 8.9 varying from 1.3 to 3.1, which indicates the significant presence of $^{232}$Th bearing monazite ore in the beach sands of the East coast. For correlation graph see Fig. 1B. Depth profile study indicates that the highest activity of primordial radionuclides is confined to the upper 0–10 cm layer of the soil. Both $^{232}$Th and $^{238}$U concentrations have been found to decrease with depth in sand samples indicating that the deposition of radionuclides at different time in the past is not uniform whereas that of $^{40}$K is more or less uniform [2]. Absorbed γ-dose rate above 1 m due to the presence of these primordial radionuclides is found to be in the range of 12–2337 nGy/h with a mean of 160.85 nGy/h. Subsequently, the annual effective dose is found to vary from 0.014 to 2.87 mSv/y with a mean value of 0.197.

## References

[1] V. Kannan, M.P. Rajan, et al., Applied Radiation and Isotopes 57 (2002) 109–119.
[2] A. Baeza, et al., Journal of Environmental Radioactivity 23 (1994) 19–37.

www.ics-elsevier.com

# Measurement of primordial radionuclide distribution in the soil samples along the East coast of Tamilnadu, India

V. Meenakshisundaram[a], K.S. Lakshmi[b],
J. Malathi[c], S. Selvasekarapandian[c],*

[a]*Radiation Safety System Division, IGCAR, Kalpakkam-603102, India*
[b]*Meenakshi College for Women, Kodambakkam, Chennai-600024, India*
[c]*Solid State and Radiation Physics Laboratory, Bharathiar University, Coimbatore- 641046, India*

**Abstract.** Thorium, uranium and potassium activity concentration in soil samples along the East coast of Tamilnadu is measured using NaI(Tl) gamma ray spectrometer. In all, 66 places were chosen for the study. $^{232}$Th activity has been found to vary from 6 to 17693 Bq kg$^{-1}$. $^{238}$U activity has been found to vary from BDL to 1114 Bq kg$^{-1}$, $^{40}$K activity has been found to vary from 15 to 447 Bq kg$^{-1}$. Depth profile study has also been conducted. The average dose due to the primordial radionuclide is found to be 125.6 nGy h$^{-1}$, which gives an annual effective dose of 115.2 μGy h$^{-1}$. The gross α-activity has been found to be in the range from 30 to 7200 Bq kg$^{-1}$ and that of the gross β-activity between 185 and 29,252 Bq kg$^{-1}$. © 2004 Published by Elsevier B.V.

*Keywords:* Gamma ray spectrometer; Gross alpha; Gross beta; Effective dose

## 1. Introduction

The operating experience of nuclear power plant has well established that the radiation dose received by the population from nuclear industry is far less than that from natural sources of radiation. Notwithstanding this, concern is being expressed frequently about the possible impact of nuclear installations on the environment through increase of ambient radiation levels attributable to reactor operations. A detailed survey of natural background radiation fields would provide a proper reference base for assessing any increase in radiation exposure arising from reactor operations [1]. Hence, systematic investigation has been

---

* Corresponding author. Tel.: +91 422 2422222, +91 422 2422422; fax: +91 422 2422387.
*E-mail addresses:* s.sekarapandian@rediffmail.com, sekarapandian@yahoo.com (S. Selvasekarapandian).

0531-5131/ © 2004 Published by Elsevier B.V.
doi:10.1016/j.ics.2004.11.118

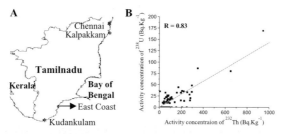

Fig. 1. (A) Area map. (B) Frequency distribution of annual indoor dose rates.

undertaken to study the background radiation along the entire east coast of Tamilnadu where a nuclear power plant (Madras Atomic Power Station) is under operation in an area near the northern end and another being constructed (Kudankulam Nuclear Power Plant) at a site near the southern end of this coastline.

## 2. Materials and method

In all, 66 soil samples were collected from 33 sites along the east coast of Tamilnadu, covering a length of 1000 km. Distance between any two site is ± 25-30 km. A typical NaI(Tl) scintillator has been used to measure the ambient γ-dose. For gross α- and gross β-measurements ZnS(Ag), based α-counter and GM counter based low β-counting system, respectively, have been used.

## 3. Results and discussion

$^{232}$Th activity has been found to vary from 6 to 17,693 Bq kg$^{-1}$. $^{238}$U activity has been found to vary from BDL to 1114 Bq kg$^{-1}$, and $^{40}$K activity varies from 15 to 447 Bq kg$^{-1}$. Abnormal thorium values have been found in Kanyakumari (17,693 Bq kg$^{-1}$), Uvari (1300 Bq kg$^{-1}$) and Mamallapuram (1124 Bq kg$^{-1}$). Wide range of variation is due to the wide variety of lithological components existing in the study area [2]. Analysis of depth profile study has been limited to 30 cm. Both $^{232}$Th and $^{238}$U concentrations are found to decrease with depth in soil. This result indicates that the deposition of radioactive heavy minerals at different time duration in the past is not uniform, whereas that of $^{40}$K is more or less uniform. Null value of skewness obtained for $^{40}$K activity concentration shows that its distribution is symmetrical. Positive and higher values of skewness obtained for other two radionuclides ($^{238}$U and $^{232}$Th) show that their distributions are asymmetric. Average dose due to these primordial radionuclide is found to be 125.6 nGy h$^{-1}$, which gives an annual effective dose of 115.2 çGy h$^{-1}$. Correlation graph between the $^{232}$Th and $^{238}$U activities have been drawn (Fig. 1B). Absence of a very good correlation ($R=0.83$) between $^{232}$Th and $^{238}$U is possibly due to differential transportation of radionuclides in soil. Gross α-activity in the soil samples has been found to be 307200 Bq kg$^{-1}$, and that of the gross β-activity has been found to be in the range of 18529,252 Bq kg$^{-1}$. The higher β-activity compared to α is definitely due to the very high β (1.314 MeV) as well as high γ energy (1.460 MeV) lines of $^{40}$K, in addition to strong β-γ emitting daughter products of $^{232}$Th and $^{238}$U.

## References

[1] D.L. Donohue, R. Zeisler, Analytical Chemistry 65 (1993) 359A–368A.
[2] A. Baeza, et al., Journal Environmental Radioactivity 23 (1994) 19–37.

www.ics-elsevier.com

# Indoor gamma dose measurement along the East coast of Tamilnadu, India using TLD

S. Selvasekarapandian[a,*], K.S. Lakshmi[b], G.M. Brahmanandhan[a], V. Meenakshisundaram[c]

[a]*Solid State and Radiation Physics Laboratory, Bharathiar University, Coimbatore-641046, India*
[b]*Meenakshi College for Women, Kodambakkam, Chennai-600024, India*
[c]*Radiation Safety System Division, IGCAR, Kalpakkam-603102, India*

**Abstract.** Indoor gamma dose rate measurement using TLD has been performed in the East Coast of Tamilnadu. Totally, 56 places have been chosen for the study. Four TLD measurements have been performed throughout a year (July 1994–November 1995), covering four different seasons to find the annual effective dose. The average annual gamma dose has been found to be 1332.4 $\mu Gy \cdot y^{-1}$. The annual effective dose to the public residing along the east coast region of Tamilnadu is 1.3 $mSv \cdot y^{-1}$. The calculated gamma dose from TLD and the observed gamma dose from the environmental survey meter have been compared. © 2004 Elsevier B.V. All rights reserved.

*Keywords:* Indoor gamma dose; TLD; Environmental survey meter; Annual effective dose

## 1. Introduction

Virtually, all materials and environments in our planet have been exposed to ionizing radiation. The primordial radionuclides ($^{232}$Th, $^{238}$U and $^{40}$K) found in the Earth's crust is giving an exposure up to 0.5 mSv/year. Some places in the Earth have very high radiation levels due to the increase in the primordial radionuclide content. In India, the west coast area is a very high background radiation area due to the monazite content in the sand [1]. The east coast also has a sparse distribution of monazite content. The east coast of Tamilnadu is around 1000 km in length, and a nuclear power plant (Madras Atomic Power Station) is under operation near the northern end, while another one is being constructed (Kudankulam Nuclear Power Plant) near the southern end of this coastline. So a detailed systematic investigation throughout the east coast of Tamilnadu will give a finite data about

---

* Corresponding author. Tel.: +91 422 2422222x422; fax: +91 422 2422387.
*E-mail address:* sekarapandian@yahoo.com (S. Selvasekarapandian).

0531-5131/ © 2004 Elsevier B.V. All rights reserved.
doi:10.1016/j.ics.2004.10.026

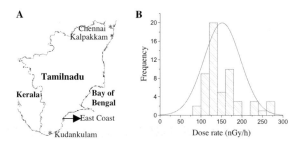

Fig. 1. (A) Area map. (B) Frequency distribution of annual indoor dose rates.

the human exposure to radiation. Majority of the people spend 80% of their time in indoors. Hence, it is important to have knowledge of the indoor radiation level in the living areas.

## 2. Materials and methods

To measure indoor gamma dose distribution along the southeast coast of India (Fig. 1A) in Tamilnadu, TLD technique has been used. Totally, 56 locations are chosen and in every location; two to three TLDs are used. TLD measurements have been made for four periods, i.e., (i) July 1994–November 1994, (ii) November–March 1994–1995, (iii) March 1995–August 1995 and (iv) August 1995–November 1995, covering 484 days. $CaF_2$ powder is used as the dosimetric material. TL measurements have been performed using a computerized TL reader from room temperature to 350 °C at a heating rate of 300 °C per minute. The TLDs were hung 70 to 100 cm below the house roof. Spot reading of the ambient radiation levels has been obtained using the 'scintillometer' every time while keeping fresh dosimeters. The scintillometer consists of NaI(Tl) crystal coupled with a PMT.

## 3. Results and discussion

TLD measurements for the four periods have been carried out, and the mean gamma dose has been found to be 158, 165, 136 and 145 $nGy \cdot h^{-1}$. The average annual gamma dose has been found to be 1332.4 $\mu Gy \cdot y^{-1}$, which is 2.5 times higher than the world average (520 $\mu Gy \cdot y^{-1}$). The average is taken, leaving an extremely high value of 8838.83 $\mu Gy \cdot h^{-1}$ recorded at one site (Kanyakumari). Fig. 1B shows the frequency distribution of annual indoor dose rates measured by TLD. The annual effective dose to the public residing along the east coast region of Tamilnadu is 1.3 $mSv \cdot y^{-1}$ (except Kanyakumari), which is higher than the world average value (1 $mSv \cdot y^{-1}$) [2]. In the present study, out of 56 locations, 12 were in range of 500–700 $\mu Gy \cdot y^{-1}$, 25 locations were in the range of 701–1000 $\mu Gy \cdot y^{-1}$ and the remaining 19 locations were more than 1001 $\mu Gy \cdot y^{-1}$. The calculated gamma dose from TLD and the observed gamma dose from the scintillometer are compared, and the ratio (scintillometer/TLD) was found to vary from 0.54 to 0.91. This study has shown that, as the dose increases, the values of the ratio also increases and approaches to 1.

## References

[1] K.S.V. Nambi, et al., Radiation Protection Dosimetry 18–1 (1987) 31–38.
[2] UNSCEAR, Report on "Sources and Effects of Ionizing Radiation of General Assembly with Scientific Annexes", United Nations, New York, 2000.

# Study of background radiation from soil samples of Udumalpet Taluk of Coimbatore district, India

R. Sarida[a], G.M. Brahmanandhan[a], J. Malathi[a],
D. Khanna[a], S. Selvasekarapandian[a,*], R. Amutha[a],
V. Meenakshisundaram[b], V. Gajendran[b]

[a]*Solid State and Radiation Physics Laboratory, Bharathiar University, Coimbatore-641046, India*
[b]*Radiation Safety System Division, IGCAR, Kalpakkam-603102, India*

**Abstract.** Ninety-seven soil samples collected in Udumalpet Taluk are analysed by a gamma ray spectrometer. The activity concentrations of $^{232}$Th, $^{238}$U, and $^{40}$K have been found to be varying from 15.8 to 335.04 Bq kg$^{-1}$, BDL to 41.96 Bq kg$^{-1}$, and from 114.21 to 1155.54 Bq kg$^{-1}$, respectively. The absorbed gamma dose of $^{232}$Th, $^{238}$U, and $^{40}$K are in the range of 9.84 to 208.72 nGy/h, BDL to 17.91 nGy/h, and 5.4 to 54.65 nGy/h, respectively. The doses due to these three primordial radionuclides have been found to be varying from 15.24 to 281.28 nGy/h. The annual effective dose due to the primordial radionuclides is 0.11 mSv/y. © 2004 Elsevier B.V. All rights reserved.

*Keywords:* Gamma ray spectrometer; Activity concentration; Annual effective dose

## 1. Introduction

Human beings are exposed to ionizing radiation from soil, rocks, and building materials. The intensity of exposure depends upon the type of emission involved or the duration of exposure. These are very important in determining the risk of the harmful biological effects that can result from exposure to radiation arising from many sources both natural and artificial. Udumalpet Taluk is one of the famous summer resorts of India surrounded by mountains on three sides. The limited amount of information available in the literature indicates that the monazite deposits in the coastal area of Kerala are formed due to weathering of rocks in the nearby Western Ghats. Udumalpet Taluk is at the foothills of Anaimalai Hills, a part of Western Ghats, for which no background radiation data are

---

\* Corresponding author. Tel.: +91 422 2422222 422; fax: +91 422 2422387.
*E-mail address:* s.sekarapandian@rediffmail.com (S. Selvasekarapandian).

0531-5131/ © 2004 Elsevier B.V. All rights reserved.
doi:10.1016/j.ics.2004.11.117

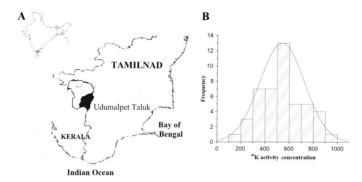

Fig. 1. (A) Area map; (B) frequency distribution of $^{40}$K activity concentration.

available. Hence, a systematic study has been carried out in Udumalpet Taluk to measure the natural radiation levels in soil.

## 2. Materials and methods

Totally, 97 samples have been collected in the Udumalpet Taluk (Fig. 1A) and processed by the procedure given elsewhere [1]. A typical NaI(Tl) scintillometer has been used to measure the ambient gamma dose. The activity concentration of the primordial radionuclides has been measured using a NaI(TI) gamma ray spectrometer.

## 3. Results and discussion

The activity concentrations of $^{232}$Th, $^{238}$U, and $^{40}$K have been found to vary from 15.8 to 335.04 Bq kg$^{-1}$, BDL to 41.96 Bq kg$^{-1}$, and from 114.21 to 1155.54 Bq kg$^{-1}$, respectively. The soil type in this study area is "alluvial type", due to which this Taluk is one of the important agricultural areas in Tamil Nadu. This alluvial soil contains major portion of $^{40}$K. The activity concentration of $^{40}$K has been found to be high in this Taluk. This also may be the due to the fact that the potassium concentration in soil can appreciably be altered by agricultural inputs, particularly through the application of potassium bearing fertilizers which have been absorbed to increase the potassium content in the soil under cultivation for 20 years by an order of a magnitude [2]. The frequency distribution of $^{40}$K is shown in Fig. 1B. The absorbed gamma dose rate of $^{232}$Th, $^{238}$U, and $^{40}$K are found to vary from 9.84 to 208.72 nGy/h, BDL to 17.91 nGy/h, and from 5.4 to 54.65 nGy/h, respectively. It has been found that the dose due to thorium is high. The mean percentage contributions to the external dose rate from the radionuclides are 32.5% (29.55 nGy/h) from $^{40}$K, 66.8% (60.73 nGy/h) from $^{232}$Th, and 0.01% (0.83 nGy/h) from $^{238}$U. The total average dose rate is 90.81 nGy/h. Frequency distribution of $^{40}$K is illustrated in the Fig. 1B. The annual effective dose due to the three primordial radionuclides has been calculated, and it is found to be 0.11 mSvy$^{-1}$, which is within the permissible limit.

## References

[1] Selvasekarapandian, et al., Gamma radiation dose from radionuclides in soil samples of Udagamandalam (Ooty) in India, Radiation Protection Dosimetry 82–83 (1999) 225–228.
[2] A.W. Kelment Jr., Hand-Book of Environmental Radiation, CRC press, 1982, pp. 15–21.

www.ics-elsevier.com

# Study of background radiation from soil samples of Pollachi taluk, Tamilnadu, India

R. Amutha[a], G.M. Brahmanandhan[a], J. Malathi[a], D. Khanna[a],
S. Selvasekarapandian[a,*], R. Sarida[a],
V. Meenakshisundaram[b], V. Gajendran[b]

[a]Solid State and Radiation Physics Laboratory, Bharathiar University, Coimbatore-641046, India
[b]Radiation Safety System Division, IGCAR, Kalpakkam-603102, India

**Abstract.** Pollachi is one of the nine taluks in Coimbatore district, India. Totally 100 soil samples have been collected from different places of Pollachi taluk. It is found, from the gamma-ray spectra, that activity concentrations of $^{232}$Th, $^{238}$U and $^{40}$K are varying from 3.63 to 263.39, 4.31 to 32.27 and 36.68 to 1821.76 Bq kg$^{-1}$, respectively. The effective dose rates of $^{232}$Th, $^{238}$U and $^{40}$K, estimated from soil samples, are ranging from 2.77 to 201.38, 2.19 to 16.46 and 1.93 to 96.12 µSv year$^{-1}$, respectively.
© 2004 Elsevier B.V. All rights reserved.

*Keywords:* Natural radioactivity; Gamma radiation; Absorbed dose; $^{40}$K; $^{238}$U; $^{232}$Th

## 1. Introduction

The imperceptible, everlasting natural background radiation levels are composed of Cosmogenic and Primordial radionuclides. Natural environmental radiation depends on geological and geographical conditions. West coast of India, China and certain beaches in Brazil are the areas well known for their high background radiation [1]. Higher background radiation levels are associated with igneous rocks such as granite and sedimentary rocks. The specific levels due to terrestrial background radiation are related to the types of rock from which the soil originates. Literature survey indicates that the monazite deposits in the coastal area are formed due to the weathering of rocks in the Western Ghats [2]. The present study has been undertaken to investigate the activity of primordial radionuclides in the Pollachi taluk, a taluk lying at the foothills of Western Ghats.

---

* Corresponding author. Tel.: +91 442 2422222 422; fax: +91 422 2422387.
*E-mail address:* sekarapandian@yahoo.com (S. Selvasekarapandian).

0531-5131/ © 2004 Elsevier B.V. All rights reserved.
doi:10.1016/j.ics.2004.10.025

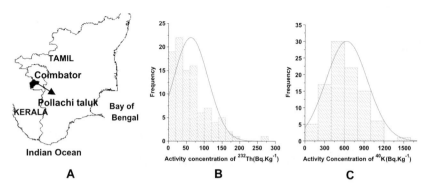

Fig. 1. (A) Study area map. (B) Frequency distribution of the activity concentration of $^{232}$Th. (C) Frequency distribution of the activity concentration of $^{238}$U.

## 2. Materials and Methods

Totally 100 soil samples have been collected from the study area (Fig. 1A) at a distance of 20–30 km apart. The ambient gamma dose rate has been measured by using a NaI(Tl) Scintillometer. A. NaI(Tl) based detector is used for the Gamma ray Spectrometry measurements. For the estimation of the primordial radionuclides $^{40}$K, $^{238}$U and $^{232}$Th, the peaks corresponding to $^{40}$K(1.460MeV), $^{214}$Bi(1.764MeV) and $^{208}$Tl(2.614MeV) have been considered respectively.

## 3. Results and discussion

The ambient gamma dose rate has been found to vary from 40 to 210 nGy h$^{-1}$. The activity concentration of $^{232}$Th is found to vary from BDL to 263.39 Bq kg$^{-1}$, the highest activity of $^{232}$Th has been observed as 263.39 Bq kg$^{-1}$ at Makkinampalayam, which is in beneath of the Western Ghats and BDL value is observed in two samples. The frequency distribution of $^{232}$Th is shown in Fig. 1B. The activity concentration of $^{238}$U varies from BDL to 32.27 Bq kg$^{-1}$. The highest activity of $^{238}$U is in Meenakshipuram (32.27 Bq kg$^{-1}$) and BDL value has been observed in 78 samples. The frequency distribution of $^{238}$U is shown in Fig. 1C. $^{40}$K activity is in the range from 36.68 to 1821.76 Bq kg$^{-1}$. The arithmetic mean of the ratio of activity of $^{232}$Th to $^{238}$U is 6.13, varying from 0.85 to 15.67, indicating the significant presence of $^{232}$Th in the soil samples. A wide range of variation in the soil dose rate due to $^{232}$Th, $^{238}$U and $^{40}$K radionuclides have been observed. The range of variation in the dose rate is much broader for $^{232}$Th than $^{40}$K and $^{238}$U, confirming the abundance of $^{232}$Th in the soil samples. The total absorbed dose rate ranges from 3.52 to 242.43 nGy h$^{-1}$, with the mean value of 64.04 nGy h$^{-1}$. The annual effective dose rate is found to be varying from 4.32 to 297.48 μSv year$^{-1}$. The International Commission on Radiation Protection recommends 1 mSv year$^{-1}$ to be the effective dose limit for the public; so, the dose due to the three primordial radionuclides in this study area (Pollachi taluk) is within the permissible limit [3].

## References

[1] A.W. Klement Jr., Hand-Book of Environmental Radiation, CRC Press, 1982, pp. 15–21.
[2] S. Selvasekarapandian, et al., Radiation Protection Dosimetry 82-.3 (1999) 225–228.
[3] J. Sannappa, et al., Radiation Measurements 37 (2003) 55–65.

# Enrichment of natural radionuclides in monazite areas of coastal Kerala

Y. Narayana*, P.K. Shetty, K. Siddappa

*Department of Studies in Physics, Mangalore University, Mangalagangagotri, India*

**Abstract.** Detailed investigations were carried out on the distribution and enrichment of natural radinuclides in the high background areas of coastal Kerala. The activity of $^{232}$Th and $^{226}$Ra at all beach sands was found to be maximum at a distance of 20 m away from the water line except for Karunagapalli beach. The activity of natural radionuclides was determined in different size fractions of sand to study the enrichment pattern. The highest activity of $^{232}$Th and $^{226}$Ra was found in 125–63 μ particle size fraction. The $^{40}$K activity in beach sand samples was below detectable level in the most of the samples. The results of these systematic investigations are presented and discussed in this paper. © 2004 Elsevier B.V. All rights reserved.

*Keywords:* High background radiation area; Monazite; Gamma; $^{232}$Th

## 1. Introduction

There are regions in the world known as high background radiation areas due to local geology and geochemical effects that cause enhanced levels of terrestrial radiation [1,2]. In India, there are quite a few high background radiation areas along its long coastline due to the presence of monazite placer deposits. One of the areas along southwest coast of India where high radiation level has been reported was from coastal Kerala.

## 2. Materials and methods

Sand samples were collected from Karunagappally, Chavara, Neendakara and Kollam beaches of coastal Kerala. Samples were collected from sea waterline, 20 m away from

---

\* Corresponding author. Tel.: +91 824 2287363; fax: +91 824 2287364.
*E-mail address:* narayanay@yahoo.com (Y. Narayana).

0531-5131/ © 2004 Elsevier B.V. All rights reserved.
doi:10.1016/j.ics.2004.11.163

waterline and 40 m away from waterline in each sampling locations. Standard techniques were followed in the collection of samples. The dried sand samples were passed through different sieves to obtain four different particle size fractions, namely 1000–500, 500–250, 250–125 and 125–63 μ. The concentration of natural radionuclides in the samples were determined employing high efficiency 5″ ×5″ NaI (Tl) γ-ray spectrometer [3].

## 3. Results and discussion

The highest $^{232}$Th activity of 446,026 Bq kg$^{-1}$ was observed in sand samples at waterline in grains of size 125–63 μ along Chavara beach and a minimum of 29 Bq kg$^{-1}$ at 40 m away from waterline in grains of size 1000–500 μ in Kollam beach sand sample. Similarly, the highest $^{226}$Ra activity 12323 Bq kg$^{-1}$ was found at waterline in grains of size 125–63 μ along Chavara beach and a lowest of 29 Bq kg$^{-1}$ at waterline in grains of size 1000–500 μ in Kollam beach sand sample. The $^{40}$K activity in beach sand samples was below detectable level in most of the sampling stations. The highest $^{40}$K activity of 1389 Bq kg$^{-1}$ was observed at Kollam beach sand with grain size of 125–63 μ.

From the results, it was found that the highest activity of $^{232}$Th and $^{226}$Ra were found in top layer of sand at a distance of 20 m away from the waterline. This may be due to the continuous wave action, as the waves reaches up to about 20 m from the waterline during high tide and results in the fresh deposition of heavy minerals along the seashore. The activity of $^{232}$Th and $^{226}$Ra was found to be minimal in 1000–500 μ particle size fraction of sand. The activity of these radionuclides found to increase in the subsequent (500–250 and 250–125 μ) fractions. However, the increase of activities in these two fractions is rather narrow. The highest activity was found confined in 125–63 μ particle size fraction. The highest activity in 125–63 μ fractions indicates the selective enrichment of radioactive mineral in this particle size fraction.

## References

[1] B.G. Bennett, Exposure to natural radiation worldwide, Fourth International Conference on High Levels of Natural Radiation: Radiation Doses And Health Effects, Beijing, China, 1996, Elseveir, Tokyo, 1997, pp. 15–23.
[2] United Nations Scientific Committee on the Effects of Atomic Radiation (UNSCEAR). Forty-second session of UNSCEAR. A/AC.82/R.526 United Nations, New York, 1993.
[3] M.C. Abani, Methods for processing of complex gamma ray spectra using computers, Refresher Course in Gamma Ray Spectrometry, BARC, June 27–July 1, 1994.

# Thoron ($^{220}$Rn) levels in dwellings around normal and high background areas in India

T.V. Ramachandran*, K.P. Eappen, R.N. Nair, Y.S. Mayya

*Environmental Assessment Division, Bhabha Atomic Research Centre, Mumbai 400 085, India*

**Abstract.** The paper presents the thoron levels in Indian dwellings. Results are compared with the literature values available. © 2004 Elsevier B.V. All rights reserved.

*Keywords:* Indoor $^{220}$Rn; Dwellings; Passive dosimeter; LR-115 film

## 1. Introduction

Ever since studies on uranium miners established the presence of a positive risk coefficient for the occurrence of lung cancer due to elevated levels of $^{222}$Rn and its progeny, there was a great upsurge of interest in the measurement of $^{222}$Rn in the environment. As a result of these, huge data is generated on $^{222}$Rn levels in the indoor environment. However, data on $^{220}$Rn (thoron) are quite scare due to the general perception that its level is negligible. $^{220}$Rn levels in the environment are governed by its emanation from the surface layer of soil or building materials containing $^{232}$Th. Available estimates show a world average value of $^{232}$Th as 40 Bq kg$^{-1}$ on par with that of $^{238}$U in soil [1].

## 2. Measurement procedure

Measurement of $^{220}$Rn were initiated along with $^{222}$Rn using a twin cup dosimeter developed at this center, with Kodak LR 115 Type II 12-μm-thick strippable Solid State Nuclear Track Detector (SSNTD) films placed on the two compartments as detectors. Each compartment of the dosimeter has a length of 4.5 cm and a radius of 3.1 cm. SSNTD placed in the membrane compartment measures only $^{222}$Rn, which diffuses into the cup from ambient air through a semipermeable membrane sandwiched between glass fiber filter. SSNTD placed on the other compartment having a glass fiber filter paper barrier allows both

---

\* Corresponding author. Tel.: +91 22 25593231; fax: +91 22 25505151.
*E-mail address:* tvr@apsara.barc.ernet.in (T.V. Ramachandran).

0531-5131/ © 2004 Elsevier B.V. All rights reserved.
doi:10.1016/j.ics.2004.09.021

$^{222}$Rn and $^{220}$Rn gas to diffuse in and hence, the tracks registered on the SSNTD detector in this chamber are related to both $^{222}$Rn and $^{220}$Rn gases [2]. These dosimeters were deployed into the field on a quarterly cycle of 3 months covering all the seasons of a calendar year. In all, about 1425 houses of different construction types over 45 locations in this country have been surveyed; while in the high background radiation region, measurements were carried out in three villages covering nearly 300 houses of different types of construction. After the exposure, the SSNTDs were processed under standard protocols and were scanned under a spark counter to get the total track densities recorded in the filter and membrane compartment. From the total tracks recorded, $^{220}$Rn levels were estimated using the sensitivity factor derived from the controlled experiments. Only $^{220}$Rn levels are presented in this paper.

## 3. Results

Mean value of $^{220}$Rn varied from 5.7 Bq m$^{-3}$ around Kalpakkam, Tamil Nadu state (southern region) to 42.4 Bq m$^{-3}$ around Digboi oil field around Assam (northeastern region) state of the country with an overall geometric mean concentration (GM) of 12.2 Bq m$^3$ (GSD 3.22). High $^{220}$Rn levels are recorded in locations having high $^{232}$Th content in the soil. Estimated $^{232}$Th content in soil around normal background region varied from 3.5 to 24.7 Bq kg$^{-1}$ with a mean value of 18.4 Bq kg$^{-1}$ [3]. Some locations around the world have been classified as high background radiation areas either due to heavy deposits of monazite or uranium. One such area is located in the southern parts of India with high $^{232}$Th content in soil. $^{232}$Th content in the soil of this region varied from 75 to 9070 Bq kg$^{-1}$ with a mean concentration value of 827.0 Bq kg$^{-1}$ [4]. This works out to be 56 times higher than the estimated national average value of 18.4 Bq kg$^{-1}$ for the country as a whole around normal background region. Results of a sample survey carried out in nearly 300 dwellings of different types of construction spread over four electoral wards of this region (two outside the monazite belt and two near to the monazite belt) shows that the indoor $^{220}$Rn levels varied from 5.0 to 69.6 Bq m$^{-3}$ with a median value of 8.3 Bq m$^{-3}$ in dwellings belonging to the normal background region and from 5.0 to 420.0 Bq m$^{-3}$ with a mean value of 24.0 Bq m$^{-3}$ in the monazite belt region [5]. The estimated inhalation dose rate due to thoron and its progeny in Indian dwellings around normal background region varied from 0.047 to 0.39 mSv year$^{-1}$ with a mean of 0.11 mSv year$^{-1}$ (GSD 1.36); while the mean value around the high background region works out to be 0.22 mSv year$^{-1}$, about two times higher than that recorded for the country from normal background region. Results are comparable with those reported in literature.

## References

[1] United Nations Scientific Committee on the Effect of Atomic Radiation Mishra, Report to the United Nations General Assembly, New York, 2000.
[2] T.V., Ramachandran, et al., Radon-thoron levels and inhalation dose distribution patterns in Indian dwellings, BARC Report No. BARC/2003/E/026(2003).
[3] U.C. Mishra, S. Sadasivan, Natural radioactivity levels in Indian soil, J. Sci. Ind. Res. 30 (1971) 59–62.
[4] V.K. Shukla, S.J. Sartendel, T.V. Ramachandran, Natural radioactivity levels and exposure to population in a high background area of Kerala, Radiat. Prot. Environ. 24 (2001) 437–439.
[5] K.P. Eappen, et al., A study of inhalation dose estimates at high background radiation area of Kerala, Proc. 9th Nat. Symp. on Envir., Bangalore, 2000, pp. 173–176.

# Background radiation exposure levels: Indian scenario

## T.V. Ramachandran*, K.P. Eappen, Y.S. Mayya

*Environmental Assessment Division, Bhabha Atomic Research Center, Mumbai 400 085, India*

**Abstract.** This paper presents the contributions of natural background radiation exposure levels from various sources to the Indian population. © 2004 Elsevier B.V. All rights reserved.

*Keywords:* Natural background radiation; Sources; Indian scenario

## 1. Introduction

Man has been exposed to radiation emitted by naturally occurring species like $^{238}$U, $^{232}$Th, and $^{40}$K ever since his existence on Earth. He has been exposed to this radiation until recent times, when growth of nuclear energy created additional sources of exposure, like fallout from weapon tests, radioactive releases from reactor operations and accidents, exposure due to radioactive waste deposals, and other industrial, medical, and agricultural uses of radioisotopes. Still, the major contribution to background radiation exposure arises from natural sources. Natural sources of exposures are due to: (a) external sources of extraterrestrial origin (cosmic rays); (b) sources of terrestrial origin (radioactive nuclides present in the Earth's crust, atmosphere, and building materials); (c) internal sources of exposure from naturally occurring radionuclides through ingestion of food materials, etc.; and (d) lung irradiation due to radon and thoron and their daughter products inhaled through air. Some of these exposures are relatively constant and uniform throughout the world, while other exposures vary widely depending on the location and due to elevated levels of naturally occurring radioactive elements like $^{238}$U and $^{232}$Th and potassium ($^{40}$K) in specific localized areas. Extent of exposure to natural radiation depends on the occupation, type of dwelling, location of habitation, and lifestyle of the population. Natural radiation is of importance because it accounts for the largest contributor to the collective dose to the world population from all sources. Cosmic rays consist of protons (85%), alpha particles

---

\* Corresponding author. Tel.: +91 22 25593231; fax: +91 22 25505151.
*E-mail address:* tvr@apsara.barc.ernet.in (T.V. Ramachandran).

0531-5131/ © 2004 Elsevier B.V. All rights reserved.
doi:10.1016/j.ics.2004.11.169

(14%), and nuclei with atomic numbers between 4 and 26 (about 1%). Cosmogenic nuclides like $^{3}$H, $^{7}$Be, $^{14}$C, and $^{22}$Na are produced by cosmic rays of heavily charged particles, coming from outer space and interacting with the atmosphere. Inhalation of those radionuclides contributes to internal exposure to natural radiation. Annual effective doses from cosmic ray radiation around the world are estimated to range from 0.26 to 2.00 mSv/year [1]. Radionuclides existing on the Earth's crust since its formation are known as primordial radionuclides. These important nuclides are $^{40}$K and $^{87}$Rb and the isotopes of $^{238}$U and $^{232}$Th present in rocks and soil, giving rise to external as well as inhalation and ingestion exposures. Levels of terrestrial radiation differ from place to place around the world, as the concentrations of these nuclides in the Earth's crust vary considerably.

## 2. External and internal exposures

A countrywide survey of outdoor natural gamma radiation in India using a thermoluminescent dosimeter has yielded a national average value of 0.734 mSv/year as external gamma radiation consisting of cosmic ray (0.355 mSv/year) and the rest from terrestrial components (0.379 mSv/year) [2]. Out of the terrestrial component, 48.7% is from $^{40}$K and the rest is from $^{232}$Th (33.6%) and $^{238}$U series (17.7%) in India [2]. Internal exposures arise from the ingestion of primordial radionuclides and $^{40}$K through dietary intake. Total dose received through the ingestion pathway of the dietary intake of long-lived radionuclides of $^{238}$U and $^{232}$Th series, as well as $^{40}$K to the Indian population is estimated to be 0.315 mSv/year [2] as against the global estimate of 0.310 mSv/year [1]. Annual effective dose contribution from cosmogenic radionuclides through internal exposure is estimated to be 0.015 mSv/year. Internal exposure via inhalation arises from inhalation of $^{222}$Rn and $^{220}$Rn, rare gases formed in the decay chain of $^{238}$U and $^{232}$Th series. Typical worldwide outdoor levels of $^{222}$Rn and $^{220}$Rn are estimated to be about 10 Bq/m$^3$ while that of indoor $^{222}$Rn and $^{220}$Rn are 40 and 10 Bq/m$^3$, respectively, giving a total inhalation dose rate of 1.275 mSv/year. A countrywide survey of indoor radon and thoron using solid-state nuclear track detector-based radon–thoron dosimeters has yielded a national average of 1.235 mSv/year [3] for the inhalation dose due to $^{222}$Rn and $^{220}$Rn and their progenies (both indoor and outdoor).

## 3. Conclusion

Total annual effective doses from natural radiation sources to the Indian population from normal background areas are estimated to be 2.3 mSv/year as against the global value of 2.4 mSv/year. We all are in the midst of a radiation environment. However low it may be, exposure from natural sources cannot be avoided altogether. Therefore, a constant endeavor to control the radiation component due to $^{222}$Rn and $^{220}$Rn and their progenies is needed to keep the exposure as low as reasonably achievable.

## References

[1] United Nations Scientific Committee on the Effect of Atomic Radiation, Report to the United Nations General Assembly, New York, 2000.
[2] K.S.V. Nambi, V.N. Bapat, M. David, V.K. Sundaram, C.M. Sunta, S.D. Soman, Natural background radiation and population dose distribution in India, Health Physics Division, BARC Report (1986).
[3] T.V. Ramachandran, et al., Estimation of inhalation dose due to radon, thoron and their progeny in Indian dwellings, Rad. Prot. Environ. 26 (2003) 139–142.

# Radiological impact of utilization of phosphogypsum and fly ash in building construction in India

## V.K. Shukla, T.V. Ramachandran*, S. Chinnaesakki, S.J. Sartandel, A.A. Shanbhag

*Environmental Assessment Division, Bhabha Atomic Research Center, Mumbai 400 085, India*

**Abstract.** The paper presents the results of gamma spectrometric analysis of enhanced natural radioactivity levels in construction materials. Enhanced radioactivity levels in these materials are due to the use of phosphogypsum ($CaSO_4 \cdot 2H_2O$), by-product formed during production of phosphoric acid in fertilizer industry and fly ash, a by-product of coal fired thermal power plant. © 2004 Elsevier B.V. All rights reserved.

*Keywords:* NORM; Phosphogypsum; Building material; Fly ash

## 1. Introduction

Natural radioactive sources are the highest contributors of dose to the population. Main nuclides contributing to this are $^{238}U$, $^{232}Th$ and their daughter products as well as $^{40}K$. External exposure is due to the presence of nuclides in construction material and internal exposure is due to inhalation due to $^{222}Rn$, $^{220}Rn$ and thoron progeny. Use of bricks made of cement and gypsum with phosphogypsum and fly ash has been increased in construction. Enhancement of radioactivity levels in these materials thus results in increased radiation exposure inside a building.

## 2. Measurement procedures

Samples of these materials were collected, powdered, homogenized and filled in plastic containers of 6.5 cm in diameter×7.5 cm in height and sealed to make them airtight for 1

---

* Corresponding author. Tel.: +91 22 25593231; fax: +91 22 25505151.
*E-mail address:* tvr@apsara.barc.ernet.in (T.V. Ramachandran).

month to ascertain establishment of secular equilibrium between $^{226}$Ra and $^{228}$Th with their daughters [1]. For the analysis, high-resolution gamma ray spectrometric technique was used [2].

## 3. Results

Radioactivity levels due to $^{238}$U in rock phosphates varied from 25.0 to 1750 Bq/kg [1,2]. During the production of phosphoric acid, 95% of uranium from rock phosphate remains in phosphoric acid, while 90% of $^{226}$Ra is precipitated with phosphogypsum. Usage of these materials increases the radioactivity levels of $^{226}$Ra in cement and in gypsum bricks made from phosphogypsum. $^{226}$Ra levels in phosphogypsum and bricks made of it as a component varied from 340 to 1250 Bq/kg. $^{238}$U, $^{232}$Th, $^{40}$K in cement and in gypsum bricks varied from 54 to 94, 8 to 32 and 63 to 270 Bq/kg and 640 to 952, 57 to 101 and 113 to 424 Bq/kg, respectively Fly ash is a by-product from a coal fired thermal power plant. Coal contains naturally occurring primordial radionuclides. Its combustion results in concentration of noncombustible mineral matter containing most of the radionuclides in the ash or gaseous residues. Composition of raw material used in the manufacture of fly ash brick is fly ash (55%), sand (36%), cement (9%) or fly ash (50%), sand (38%), cement (10%) and gypsum (2%). $^{238}$U, $^{232}$Th, $^{40}$K in coal are ranging from 23 to 60, 30 to 83 and 63 to 280 Bq/kg, respectively, while in fly ash it varied from 81 to 206, 110 to 305 and 208 to 515 Bq/kg, respectively. Radioactivity levels due to $^{238}$U, $^{232}$Th, $^{40}$K in fly ash bricks samples varied from 47 to 195, 66 to 142 Bq/kg and from 112 to 299 Bq/kg [2].

## References

[1] S.J. Sartandel, et al., Measurement and dose assessment due to use of fly ash bricks and granite in construction materials, Proc. of 15th Nat. Symp. on Rad. Phys. (NSRP-15), 2003, pp. 333–337.
[2] V.K. Shukla, et al., Assessment of gamma radiation exposure inside a newly constructed building and a proposed regulatory guideline for exposure control from natural radioactivity in future buildings, Radiat. Prot. Dosim. 59 (1995) 127–133.

# Indoor and outdoor radon levels and their diurnal variations in the environs of southwest coast of India

N. Karunakara*, H.M. Somashekarappa,
K.M. Rajashekara, K. Siddappa

*University Science Instrumentation Centre, Mangalore University Mangalagangotri-574199, Mangalore, India*

**Abstract.** Indoor and outdoor $^{222}$Rn concentrations and their diurnal variations with atmospheric air pressure, humidity and temperature were studied in the southwest coast region of India using AlphaGUARD PQ-2000PRO and accessories. The indoor $^{222}$Rn concentration varied in the range 7–64 Bq m$^{-3}$ with a mean value of 32 Bq m$^{-3}$, and the outdoor concentration varied in the range 3–35.2 Bq m$^{-3}$ with a mean value of 15.7 Bq m$^{-3}$. The mean annual effective doses were 0.81 and 0.1 mSv for indoor and outdoor, respectively. © 2004 Elsevier B.V. All rights reserved.

*Keywords:* $^{222}$Rn concentration; Annual effective dose; Diurnal variation; Southwest coast of India

## 1. Introduction

Detailed studies on background radiation levels prevailing in Coastal Karnataka, Kaiga and Goa regions (12°51′ N, 74°49′ E to 15°45′ N, 74°20′ E) of southwest coast of India were carried out. Geographically, the region is flanked by the Arabian Sea on its western side and the world famous Western Ghats on the east. This paper presents the results of studies on indoor and outdoor $^{222}$Rn and its progeny concentrations.

## 2. Method

The concentration of $^{222}$Rn in indoor and outdoor air of major towns/cities were measured using a AlphaGUARD PQ-2000PRO system (Genitron Instruments, Germany). The $^{222}$Rn progeny concentration was measured using a Thomson and Nielson Radon WL

---

* Corresponding author. Tel.: +91 824 2287671; fax: +91 824 2287367.
 *E-mail address:* karunakara_n@yahoo.com (N. Karunakara).

0531-5131/ © 2004 Elsevier B.V. All rights reserved.
doi:10.1016/j.ics.2004.11.025

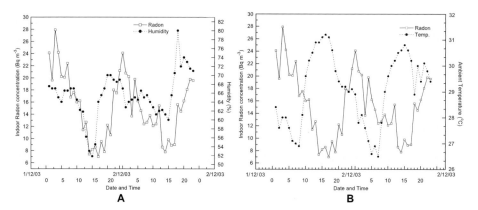

Fig. 1. Diurnal variation of $^{222}$Rn concentration with atmospheric (a) humidity and (b) temperature.

meter (Thomson and Nielsen Electronics, Canada). Soil samples collected from $^{222}$Rn measurement stations were analysed for the $^{226}$Ra activity using an HpGe (Canberra, USA) detector with a relative efficiency of 41.3%. The details of the measurement techniques were discussed elsewhere [1].

## 3. Results and discussion

The indoor $^{222}$Rn concentration varied in the range 7–64 Bq m$^{-3}$ with a mean value of 32 Bq m$^{-3}$, and the outdoor concentration varied in the range 3–35.2 Bq m$^{-3}$ with a mean value of 15.7 Bq m$^{-3}$ in the study region. These are comparable to the worldwide mean values (40 Bq m$^{-3}$ for indoor and 10 Bq m$^{-3}$ for ouodor $^{222}$Rn concentrations) reported in UNSCEAR [2]. The indoor and outdoor annual effective doses to the population of the region due to the inhalation of $^{222}$Rn and its progeny were evaluated using the equilibrium factors of 0.4 for indoors and 0.6 for outdoors and the dose coefficients given in UNSCEAR [2]. The mean annual effective doses for indoor and outdoor were 0.81 and 0.1 mSv, respectively. The initial results of indoor $^{222}$Rn progeny concentration show that the equivalent equilibrium concentration varied in the range 0.5–46.0 Bq m$^{-3}$ with a mean value of 11.3 Bq m$^{-3}$. The equilibrium factor for indoor environment was found to have a mean value of 0.35, and this is comparable to the value (0.4) reported in UNSCEAR [2].

Fig. 1A and B shows typical diurnal variation of indoor $^{222}$Rn along with atmospheric humidity and temperature. The $^{222}$Rn concentration is higher in the early morning hours and lower during early afternoon hours. The concentration decreases with decreasing humidity and vice versa (Fig. 1A). The concentration showed a similar variation with atmospheric pressure also. However, it is in inverse relation with ambient temperature (Fig. 1B).

## Acknowledgements

The work was carried out under the research project sponsored by Board of Research in Nuclear Science (BRNS), Department of Atomic Energy, Government of India, and the authors gratefully thank the BRNS for encouraging research studies. The authors thank Dr.

S. Sadasivan, Former Head, EAD, BARC and Coordinating Scientist of the BRNS project and Sri M Raghavayya, Former Head, Health Physics Unit, RMP, Mysore for many useful suggestions.

## References

[1] N. Karunakara, et al., $^{226}$Ra, $^{232}$Th and $^{40}$K concentrations in soil samples of Kaiga of south west coast of India, Health Physics 80 (5) (2001) 470–476.
[2] United Nations Scientific Committee on the Effects of Atomic Radiation (UNSCEAR). Report to the General Assembly, vol. 1, Annex B, 2000.

# Study of indoor gamma radiation in Coimbatore City, Tamilnadu, India

J. Malathi, A.K. Andal Vanmathi, A. Paramesvaran, R. Vijayshankar, S. Selvasekarapandian*

*Solid State and Radiation Physics Laboratory, Bharathiar University, Coimbatore-641046, India*

**Abstract.** Indoor gamma radiation measurement using TLD has been performed in the Coimbatore City. The indoor gamma radiation dose of Coimbatore City is in the range from 115 to 190.6 nGy/h with the geometric mean value of 157.3 nGy/h. The gamma dose in Coimbatore City has been found to be lesser than that observed at beach side houses (276.25 nGy/h) and higher than that in inland houses (142.68 nGy/h) at Kalpakkam. The indoor gamma radiation in Coimbatore City is more or less equal to that observed along east coast of Tamil Nadu (153.31 nGy/h). The annual effective dose has been calculated as 772.3 µSv/year. © 2004 Elsevier B.V. All rights reserved.

*Keywords:* Indoor gamma radiation; TLD; Annual effective dose

## 1. Introduction

Natural radioactivity is present in the environment right from the time of formation of the universe. Mankind has been compulsorily exposed to ionizing radiations of natural origin at every place. Ionizing radiation should have played a great role in the evolution of all the components, living and non-living of earth as we see it today. The effective dose due to this ionizing radiation for members of the public varies substantially depending on where they live, occupation, personal habits, diet, building type and house utilization pattern. The single largest contributor to the collective dose to human population is the natural background radiation inside residential places. So, a detailed systematic investigation has been undertaken in Coimbatore City.

## 2. Materials and methods

Coimbatore City, situated at latitude of 11.00N and a longitude of 77.00E is the second largest city in Tamil Nadu. It is exposed to the Palghat gap of Western Ghats. The

---

\* Corresponding author. Tel.: +91 422 2422222 422; fax: +91 422 2422387.
*E-mail address:* sekarapandian@yahoo.com (S. Selvasekarapandian).

0531-5131/ © 2004 Elsevier B.V. All rights reserved.
doi:10.1016/j.ics.2004.11.048

Fig. 1. A study area map.

Coimbatore City has a number of textile mills and industrial factories; so the major population is residing in a congested atmosphere (Fig. 1). The technique of TLD has been chosen for monitoring the ambient gamma radiation in Coimbatore City. $CaSO_4$: Dy powder is used as the dosimetric material. Preparation of the TLD material is done according to Yamashita et al. [1]. TL measurements have been performed using a computerized TL reader (Model: Nucleonix TL reader) from room temperature to 300 °C at a heating rate of 300 °C/min. The peak shape method is used for the calculation of the dose due to environmental radiation. Approximately 30 mg of the TLD powder filled in brass capsule is kept in a plastic container to avoid any atmospheric effect. These TLDs are kept approximately 1 m below the roof of the houses. Bedrooms in the houses are chosen for the placement of TLDs because one spends 80% of his time in that room. The houses are chosen with great care to have uniform distribution in the Coimbatore City. After 90 days, these TLDs are taken and gamma dose is measured. In every location, the ambient gamma ray dose is measured using an environmental survey meter.

## 3. Results and discussion

The indoor gamma radiation dose of Coimbatore City is in the range from 115 to 190.6 nGy/h. The calculated geometric mean value is 157.3 nGy/h. Lowest value is recorded in a concrete roof house, while the maximum value has been recorded in an asbestos house. The gamma dose in Coimbatore is found to be lesser than that observed at beachside houses (276.25 nGy/h) and higher than that in inland houses (142.68 nGy/h) at Kalpakkam Nuclear Power Plant Site [2]. The indoor gamma radiation in Coimbatore City is more or less equal to that observed along east coast of Tamil Nadu (153.31 nGy/h). The indoor gamma radiation dose calculated from the TLD has been compared with environmental survey meter reading and a good correlation is obtained. The annual effective dose is calculated as 772.3 μSv by using UNSCEAR 2000 Report.

## References

[1] Yamashita, et al., Calcium sulphate activated by thulium or dysprosium for thermo luminescence dosimetry, Health Physics 21 (1971) 295–300.
[2] K.S.V. Nambi, et al., Countrywide environmental radiation monitoring using the dosimeters, Radiation Protection Dosimetry 18-1 (1987) 31–38.

ELSEVIER

www.ics-elsevier.com

# Natural radioactivity in South West Coast of India

N. Karunakara[a,*], H.M. Somashekarappa[a], K. Siddappa[b]

[a]*University Science Instrumentation Centre, Mangalore University, Mangalagangotri-574199, Mangalore, India*
[b]*Department of Physics, Mangalore University, India*

**Abstract.** Soil samples from South West Coast region of India were analysed for their $^{226}$Ra, $^{232}$Th, and $^{40}$K activities by gamma spectrometry. The $^{226}$Ra concentration varied in the range of 8.2–68.4 Bq kg$^{-1}$ with a mean of 30.6 Bq kg$^{-1}$, $^{232}$Th in the range of 5.9–77.2 Bq kg$^{-1}$ with a mean of 38.2 Bq kg$^{-1}$, and $^{40}$K in the range of 14.6–344.9 Bq kg$^{-1}$ with a mean of 152.2 Bq kg$^{-1}$ in 0–5 cm soil profile. Depth profile studies showed a uniform distribution of these radionuclides throughout the soil depth in most of the sampling stations. The $^{226}$Ra and $^{232}$Th activities show significant positive correlation with the organic matter content of the soil. © 2004 Elsevier B.V. All rights reserved.

*Keywords:* $^{226}$Ra activity; $^{232}$Th activity; $^{40}$K activity; South West Coast of India; Gamma spectrometry

## 1. Introduction

Detailed studies were carried on the background radiation levels and radionuclides concentration in different environmental matrices of Coastal Karnataka, Kaiga, and Goa regions (12°51′N, 74°49′E to 15°45′N, 74°20′E) of South West Coast of India. This region is witnessing a rapid industrialization with setting up of many major industries. The study has established the baseline data on radionuclides distribution in different environmental matrices and this would help to assess, in the future, the impact of operation of major industries on the environment. The paper discusses the results of $^{226}$Ra, $^{232}$Th, and $^{40}$K activities, their vertical profiles in soil and dependence on physico-chemical properties of the soil.

## 2. Method

Soil samples were collected from 0 to 5, 5 to 10, 10 to 15, and from 15 to 20 cm depths from 32 sampling stations, processed and physico-chemical parameters, such as grain size, clay and silt contents, organic matter content, pH, and density, were determined employing standard methods. The activities of $^{226}$Ra, $^{232}$Th, and $^{40}$K were determined by gamma

\* Corresponding author. Tel: +91 824 2287671; fax: +91 824 2287367.
*E-mail address:* karunakara_n@yahoo.com (N. Karunakara).

0531-5131/ © 2004 Elsevier B.V. All rights reserved.
doi:10.1016/j.ics.2004.11.088

spectrometry using an HpGe detector. Intercomparison measurements were carried out with the Environmental Assessment Division, Bhabha Atomic Research Centre, Mumbai to ensure the reliability of measurements. Details of the measurements technique and inter comparison measurements were published elsewhere [1].

## 3. Results and discussion

The activity of $^{226}$Ra varied in the range of 8.2–68.4 Bq kg$^{-1}$ with a mean value of 30.6 Bq kg$^{-1}$, $^{232}$Th in the range of 5.9–77.2 Bq kg$^{-1}$ with a mean value of 38.2 Bq kg$^{-1}$, and that of $^{40}$K in the range of 14.6–344.9 Bq kg$^{-1}$ with a mean value of 152.2 Bq kg$^{-1}$ in the surface soils of the study region. These values are comparable to those reported for other normal background regions of India and the world [2]. Results of depth profile studies show that the distribution of these radionuclides is more or less uniform along the soil depth. Correlation studies show a significant positive correlation between the organic matter content of the soil and the $^{226}$Ra ($y=2.0x+9.1$, $r=0.72$ for $n=29$) and $^{232}$Th activities ($y=2.5x+11.1$, $r=0.63$ for $n=29$). Although a positive correlation was observed between the organic matter content and the $^{40}$K activity, the coefficient was lower when compared to that of $^{226}$Ra and $^{232}$Th. These radionuclides did not show significant correlations with other soil parameters, such as silt, clay, and sand contents. This suggests that the organic matter plays an important role in the fixation of $^{226}$Ra and $^{232}$Th to the surface soil.

From the results of $^{226}$Ra, $^{232}$Th, and $^{40}$K activities in the soil, the gamma dose rates in air were calculated using the dose coefficients given in United Nations Scientific Committee on the Effects of Atomic Radiation (UNSCEAR) [2]. The absorbed dose from $^{226}$Ra subseries varied in the range of 3.8–31.6 nGy h$^{-1}$ with a mean value of 14.4 nGy h$^{-1}$, $^{232}$Th series in the range of 3.6–46.6 nGy h$^{-1}$ with a mean value of 23.0 nGy h$^{-1}$, and that due to $^{40}$K varied in the range of 0.6–14.4 nGy h$^{-1}$ with a mean value of 6.3 nGy h$^{-1}$. The dose delivered by these radionuclides, in total, ranged from 10.2 to 80.5 nGy h$^{-1}$ with a mean value of 43.7 nGy h$^{-1}$.

## 4. Conclusions

The study has established the baseline data on $^{226}$Ra, $^{232}$Th, and $^{40}$K in the soils of South West Coast of India. The activities of these radionuclides in soil and the gamma absorbed dose delivered by these radionuclides to the population of the region are comparable to those reported for other normal background regions.

### Acknowledgements

This work is carried out under the research project sponsored by Board of Research in Nuclear Science (BRNS), Department of Atomic Energy, Govt. of India and the authors gratefully thank the BRNS for supporting research studies. Authors thank Dr. S. Sadasivan, Former Head, EAD, BARC and Coordinating Scientist of the BRNS project for many useful suggestions and Dr. V. K. Shukla, EAD, BARC for intercomparison measurements.

### References

[1] N. Karunakara, et al., $^{226}$Ra, $^{232}$Th and $^{40}$K concentrations in soil samples of Kaiga of South West Coast of India, Health Physics 80 (5) (2001) 470–476.
[2] United Nations Scientific Committee on the Effects of Atomic Radiation (*UNSCEAR*), Report to the General Assembly, vol. 1, 2000 Annex B.

www.ics-elsevier.com

# Transportation of radionuclides from Western Ghats to Arabian sea through some major rivers of South India

K.M. Rajashekara, Y. Narayana*, N. Karunakara, K. Siddappa

*Department of Studies in Physics, Mangalore University, Mangalagangotri 1-574 199, Mangalore, Karnataka, India*

**Abstract.** The main objective of the present investigation is to make a systematic study on the transportation of radionuclides from Western Ghats to Arabian Sea through some major rivers to trace the source and processes that are responsible for the newly discovered monazite deposits near Ullal beach. As part of the programme, the activity concentration of natural radionuclides such as $^{226}$Ra, $^{232}$Th and $^{40}$K in rocks, soil and sediments at different locations in the catchment area of these rivers were analysed to understand the release of these radionuclides by weathering and transportation by rivers. © 2004 Elsevier B.V. All rights reserved.

*Keywords:* Monazite; Radioactivity; $^{226}$Ra; $^{232}$Th

## 1. Introduction

Transportation of radionuclides in the riverine environment is an important pathway since radionuclides released from weathering of rocks in the catchment areas of the river can be carried long distances and subsequently deposited in the river mouth. In the present study, systematic analysis of soil, sediment and rock samples, collected from the riverine environment of two major rivers from coastal Karnataka, was made to understand the transportation and enrichment of natural radionuclides in the environment.

## 2. Materials and methods

Standard techniques were followed in the collection of rock, soils and sediment samples of two major rivers of coastal Karnataka, namely Netravathi and Kali [1]. The

---

* Corresponding author. Tel.: +91 824 2287363; fax: +91 824 2287367.
  *E-mail address:* narayanay@yahoo.com (Y. Narayana).

0531-5131/ © 2004 Elsevier B.V. All rights reserved.
doi:10.1016/j.ics.2004.11.164

concentrations of primordial radionuclides in the samples were determined employing high efficiency 5"×5" NaI (Tl) gamma ray spectrometer [2].

## 3. Results

In sediments of Netravathi and Kali rivers, the mean activity concentration of $^{226}$Ra is 39.0 and 37.0 Bq kg$^{-1}$, respectively. The mean concentration of $^{232}$Th is found to be 57.4 and 26.4 Bq kg$^{-1}$ and that of $^{40}$K is 455.0 and 391.0 Bq kg$^{-1}$, respectively. The $^{226}$Ra activity in the soil samples in the banks of the river Netravathi is 59.4 Bq kg$^{-1}$ and Kali river 44.0 Bq kg$^{-1}$. The activity of $^{232}$Th is 100.0 and 40.0 Bq kg$^{-1}$. The mean activity of $^{40}$K is 564.0 and 423.0 Bq kg$^{-1}$. The mean activity in rocks for $^{226}$Ra, $^{232}$Th and $^{40}$K are 27.4, 39.0 and 450.0 Bq kg$^{-1}$ in Netravathi river and 23.0, 37.0 and 499.0 Bq kg$^{-1}$ for Kali river.

The concentration of $^{232}$Th, $^{226}$Ra and $^{40}$K in the Netravathi river sediment is higher than the Kali river sediment. The concentration of $^{232}$Th is more than twice that of $^{226}$Ra in both of the rivers. The activity of $^{226}$Ra in soil is almost constant except for a few samples of both of the rivers.

## References

[1] Herbert L. Volchok, Gail de Planque (Eds.), Procedure EML Manual, 26th edn., Environmental Measurement Laboratory, 1983.
[2] M.C. Abani, Methods for processing of complex gamma ray spectra using Computers, Refresher Course in Gamma Ray Spectrometry, BARC, June 27–July 1, 1994.

# Application of well-type NaI(Tl) detector for indoor radon measurements

Darwish Al-Azmi*

*Department of Applied Sciences, College of Technological Studies,
Public Authority for Applied Education and Training, Shuwaikh, P.O. Box: 42325, Code 70654, Kuwait*

**Abstract.** Measurements of indoor radon gas concentration levels were performed using a well-type NaI(Tl) detector accommodating vials containing 10 g of activated charcoal for the detection of radon gas. After the exposure of the vials and the elapse of 3 h, gamma-ray spectrometry was performed. A large energy window of about 150 to 1360 keV was used to study the relative increase in the number of counts due to radon progeny. Comparison of the method was made with an active radon monitor. The work presents the feasibility of use of a small amount of activated carbon along with a well-type NaI detector for the measurement of radon concentration. © 2004 Elsevier B.V. All rights reserved.

*Keywords:* Well-type NaI(Tl) detector; Radon gas measurements; Gamma-ray spectrometry; Large energy window

## 1. Introduction

It is common to measure radon gas concentrations in dwellings by using charcoal canisters of a typical size of 4-in. (containing 70 g of charcoal) and utilizing routine gamma-ray measurements. Flat-type NaI(Tl) detectors have been used satisfactorily for this purpose due to their high detection efficiency and low cost. Also, well-type NaI(Tl) detectors have been used for radon gas measurements: precipitation [1] and calibration [2]. Such well-type detectors are useful when the samples are available in small sizes or quantities.

## 2. Materials and method

The work presented here utilizes a standard well-type NaI(Tl) detector (7.6×7.6 cm) which can accommodate vials of capacity of 20 ml containing small amounts (10 g) of activated charcoal for use in indoor radon detection. Applying the same procedure as with

---

* Tel.: +965 9622019; fax: +965 4838196.
 *E-mail addresses:* dalazmi@paaet.edu.kw, dalazmi@yahoo.co.uk.

0531-5131/ © 2004 Elsevier B.V. All rights reserved.
doi:10.1016/j.ics.2004.10.016

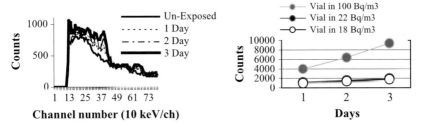

Fig. 1. (Left) Energy spectra of un-exposed vial (background) and vials exposed for different durations in the basement with an average radon concentration level of 100 Bq/m³. (Right) Variation of counts (background subtracted) in the energy region of 150–1360 for different exposure time and radon concentration levels.

the standard 4-in. charcoal canisters, the vials are left open in the location for exposure (1–3-day period). After the collection, and after the elapse of 3 h for allowing radioactive equilibrium, the vials were counted for 10,000 s using the well-type NaI (Tl) gamma-ray spectrometer. Three locations were chosen in a typical house in Kuwait constructed of cement and concrete, in the basement (air-conditioning turned-off to obtain a location with high radon concentration level as possible), and two air-conditioned rooms in the first and second floors. An active radon monitoring system (AlphaGuard) was in use to provide direct data on the radon concentration levels.

## 3. Results and discussion

From Fig. 1(left), it can be seen that there is an increase in the count rate indicating the radon detection and the differences between the spectra are proportional to the exposure time. The counts as seen from the spectra of the three exposed vials are clearly higher than the background level in the energy region up to 352-keV photopeak due to the high detection efficiency of the detector for the low photon energies from the radon progeny. The other main 609-keV photopeak is also observed. However, it is possible to utilize this spectral information and use a large energy window of about 150 to 1360 keV to study the relative increase in the number of counts due to the detection of the radon gas. This energy window includes the photon emissions produced in the decay of $^{222}$Rn daughters and their sum peaks [2]. The overall counts (background subtracted) from the 150–1360-keV energy window are compared with the readings from the AlphaGuard active radon monitor in Fig. 1(right). It is evident from the figure that the number of collected counts increases for longer exposure time and radon concentration level.

## Acknowledgements

This work is part of the project of the "Study of Radon Level Concentrations in Houses in Kuwait" supported by the Public Authority of Environment in Kuwait.

## References

[1] T. Nishikawa, et al., Automatic measuring instrument for radon daughters concentration of precipitation, Journal of Nuclear Science and Technology 23 (11) (1986) 1001–1007.
[2] P. De Felice, Xh Myteberi, The $^{222}$Rn reference measurement system developed at ENEA, Nuclear Instruments & Methods in Physics Research. Section A, Accelerators, Spectrometers, Detectors and Associated Equipment 369 (1996) 445–451.

# Assessment of exposure to the population of Russia from radon

A.M. Marenny[a], S.M. Shinkarev[b,c,*], A.V. Penezev[a], M.N. Savkin[b], M. Hoshi[c]

[a]*Research and Testing Center of Radiation Safety, Ministry of Health, Moscow, Russia*
[b]*State Research Center-Institute of Biophysics, Ministry of Health, Moscow, Russia*
[c]*International Radiation Information Center, Research Institute for Radiation Biology and Medicine, Hiroshima University, Hiroshima, Japan*

**Abstract.** A model has been developed to assess the dose to the population of Russia from radon. Only measurements of radon concentration in dwellings carried out by integral track method were used in the calculations. Population-weighted average radon concentration in dwellings determined on the basis of the results of indoor surveys for the whole Russian population is estimated to be about 52 Bq m$^{-3}$. According to the conditions of radon exposure, all the Russian population was divided into two categories: (1) living in high radon-prone zone (10% of total) and (2) living in the moderately hazardous or safe zone (90% of total). Assuming the value of dose conversion factor is equal to 6.1 nSv (Bq h m$^{-3}$)$^{-1}$, the annual per caput effective doses in the high and low exposed categories are estimated to be 3.2 and 0.63 mSv, respectively. An estimate of annual collective effective dose to the population in Russia (148,100,000 people) is about 130,000 manSv. The annual per caput effective dose is estimated to be 0.88 mSv. About 1 million people receive annual dose above 10 mSv, including approximately 200,000 people with doses above 20 mSv. © 2004 Elsevier B.V. All rights reserved.

*Keywords:* Radon; Collective effective dose; Model; Population; Russia

## 1. Introduction

The purpose of the paper is to present the currently available estimates of effective doses from radon received by the Russian population. The collective dose, per caput doses, and numbers of persons in high dose groups are presented.

* Corresponding author. Postal address: State Research Center-Institute of Biophysics, Zhivopisnaya 46, Moscow 123182, Russia. Tel.: +7 095 190 9687; fax: +7 095 190 3590.
*E-mail address:* sshinkarev@atom.ru (S.M. Shinkarev).

0531-5131/ © 2004 Elsevier B.V. All rights reserved.
doi:10.1016/j.ics.2004.10.019

## 2. Materials and methods

A model has been developed to assess exposure of the population in Russia from radon [1]. According to the model, the Russian population was divided into two categories depending upon the geological formation type of the residential territory (a high radon-prone zone or a moderately hazardous (or safe) zone). In addition, the Russian population was divided into several groups depending upon: (a) type of the settlement (urban or rural); (b) floor of living in a residential building (the ground floor, the first floor, or higher than the first floor). It was assumed that for each population group the distribution of indoor radon concentration measurements was representative and satisfactorily fitted by a lognormal curve with the values of the parameters (a geometric mean and a geometric standard deviation) derived from the results of measurements related to a given group. Only measurements of radon concentration in dwellings carried out by integral track method were used to characterize exposure to the population from radon in each group.

## 3. Results and discussion

Population-weighted average radon concentration in dwellings determined on the basis of the results of indoor surveys for the whole Russian population is estimated to be about 52 Bq m$^{-3}$. Comparison of this estimate with those published in the UNSCEAR 2000 report [2] shows that it is close to that in Germany (50 Bq m$^{-3}$), higher than in Japan (16 Bq m$^{-3}$) and the US (46 Bq m$^{-3}$) and lower than in Sweden (108 Bq m$^{-3}$) and Finland (120 Bq m$^{-3}$). It was estimated that 90% of the population lives in a moderately hazardous (or safe) zone and 10% lives in a high radon-prone zone. Assuming the value of dose conversion factor is equal to 6.1 nSv (Bq h m$^{-3}$)$^{-1}$ [3], the annual per caput effective dose in the high and low exposure categories are 3.2 and 0.63 mSv, respectively. The annual per caput effective dose is estimated to be 0.88 mSv. About 1 million people receive annual dose above 10 mSv, including approximately 200,000 people with doses above 20 mSv. An estimate of annual collective effective dose for the population in Russia (148,100,000 people) is about 130,000 manSv. Comparison of radon exposure with exposure to radioactive fallout from the Chernobyl accident for the Russian population (1,980,000 people) living in the areas where the $^{137}$Cs deposition density exceeds 37 kBq m$^{-2}$ shows that the average effective dose from external and internal exposures due to the Chernobyl accident for that population (living in the moderately hazardous and/or safe zone), estimated to be 6.8 mSv for 10 years (1986–1995) [4], is close to 6.3 mSv received by that population from radon for the same period of time. But, at the present time, an annual exposure to that population from the Chernobyl accident fallout is substantially lower than that from radon.

## References

[1] A.M. Marenny, M.N. Savkin, S.M. Shinkarev, Radiat. Prot. Dosim. 90 (4) (2000) 403–408.
[2] UNSCEAR, Report to the General Assembly. Vol. 1: Sources. Annex B. Exposures from Natural Sources of Radiation, United Nations, New York, 2000.
[3] ICRP Publ. 65. Int. Commission on Radiological Protection. Protection against radon-222 at home and at work. Ann. ICRP 22(2). Pergamon Press, Oxford, 1993.
[4] UNSCEAR, Report to the General Assembly. Vol. 2: Effects. Annex J. Exposures and Effects of the Chernobyl Accident, United Nations, New York, 2000.

# Altai Region of Russia—a high radon potential area

V.P. Borisov*, I.P. Saldan, A.P. Strokov

*Center for Sanitary and Epidemiological Inspections in Altaisky krai of Russia, Russia*

**Abstract.** The Altai Region of Russian Federation is known as the target of extensive Semipalatinsk Scientific Program, in the context of which population exposures due to long-term nuclear tests at Semipalatinsk Test Site have been reconstructed and the effects on public health studied. The region is also known as an increased natural radiation area. Radon surveys conducted by leading national institutions and local radiation protection authorities have revealed that some parts of the region have high radon potential. The geometric means of annual average radon concentration for administrative areas of the Altaisky krai and the Republic Altai vary in the range of 30–250 $Bq \cdot m^{-3}$. The most radon-prone administrative territories in the Republic Altai are Turochak and Choya raions (counties), where mean values exceed 400 $Bq \cdot m^{-3}$. The difference in radon potential of territories is due to the variety in underlying geology with most radon-prone areas situated on uranium-rich granites, soils with high permeability for radon and near regional neotectonic fault spreading along the north face of Altai Mountains. Regional radon programs make provisions for implementation of radon limits established by national radiation protection legislation. © 2004 Published by Elsevier B.V.

*Keywords:* Radon; Radon survey; SSNTD

## 1. Introduction

The Altai Region of Russian Federation is situated in the south of West Siberia. The region includes two administrative territories—Altaisky krai and the Republic of Altai. The attention was drawn to the considerably high levels of indoor radon in some areas of the region during the last decade. Early results were obtained by the Research Center of Spacecraft Radiation Safety (Moscow), and the Institute for Radiation Hygiene (Saint-Petersburg) in the context of Semipalatinsk Scientific Program in which provisions were made for screening radon surveys [1]. The main purpose of the Program was reconstruction of exposures and health consequences for population of the Altai region resulted from long-term nuclear tests at Semipalatinsk Test Site [2]. The present work describes the results of

---

\* Corresponding author. Tel.: +7 3852 248731; fax: +7 3852 249949.
*E-mail address:* radprot@ab.ru (V.P. Borisov).

0531-5131/ © 2004 Published by Elsevier B.V.
doi:10.1016/j.ics.2004.11.177

Table 1
Radon measurements results

| Town or area | GM, Bq·m$^{-3}$ | $\sigma_{LN}$ | Percentage above 200 Bq·m$^{-3}$ | Max $C_{Rn}$, Bq·m$^{-3}$ |
|---|---|---|---|---|
| *Rural settlements* | | | | |
| Mountain area | 134 | 0.66 | 27 | 1016 |
| Steppen area | 109 | 0.51 | 10 | 463 |
| *Cities* | | | | |
| Barnaul | 66 | 0.42 | 0.5 | 141 |
| Biysk | 116 | 0.43 | 10 | 292 |

survey in the Altai region of Russia performed with the purpose to estimate local differences in indoor radon levels and collective exposures to population.

## 2. Materials and methods

Measurements of radon concentration in indoor air were performed by solid-state nuclear track detectors (SSNTDs). Radon radiometers consisted of a plastic case with SSNTDs LR-115 (Dosirad, France) placed inside. Radiometers were exposed for 3 months during warm and cold seasons. More than 300 dwellings were surveyed in 20 townships selected to represent different parts of the region. After the exposure, SSNTDs were etched in a 6N NaOH solution at 50 °C. The etching time complied with the value stated in calibration certificate of radiometers and measuring instrument. Track density was measured by spark counting technique at 500 V after pre-discharges at 900 V. Quality assurance included detectors exposed in radon chamber under control of radon-monitor AlphaGUARD and detectors exposed in buildings side-by-side.

## 3. Results

Distribution parameters of the annual average radon concentration and percentage of dwellings where radon concentration exceeds national reference level for new constructions (100 Bq·m$^{-3}$ effective equivalent concentration—EEC) were calculated. The results are summarized in Table 1.

## 4. Conclusions

The distribution of high radon potential areas across the territory of the Altai region proved to be sufficiently uneven. Most of them are situated in the foot-hill and mountain areas in the south and southeast of the region. Indoor radon levels are even higher in the neighboring administrative territory, the Republic of Altai, where there are settlements with mean radon concentrations exceeding 400 Bq·m$^{-3}$. High radon potential areas of the Altai region of Russia are an appropriate territory for prospective epidemiological study. Radon reduction techniques should be implemented for new constructions in these areas to ensure the compliance with national standards.

## References

[1] A.M. Marenny, A.S. Vorozhtsov, N.A. Nefedov, Results of radon concentration measurements in some regions of Russia, Radiat. Meas. 25 (1995) 649–653.
[2] Y.N. Shoikhet, V.I. Kiselev, V.M. Loborev, et al., Nuclear tests at the Semipalatinsk Test Site, Radiation impact on the Altai Region population. Barnaul, 1999.

# Assessment of exposure to the population of Moscow from natural sources of radiation

A.M. Marenny[a], S.M. Shinkarev[b,c,*], A.V. Penezev[a], A.V. Frolova[a],
Yu.A. Morozov[d], S.E. Okhrimenko[e], M.N. Savkin[b], M. Hoshi[c]

[a]*Research Center of Spacecraft Radiation Safety, Ministry of Health, Moscow, Russia*
[b]*State Research Center–Institute of Biophysics, Ministry of Health, Moscow, Russia*
[c]*International Radiation Information Center, Research Institute for Radiation Biology and Medicine, Hiroshima University, Hiroshima, Japan*
[d]*Ltd. "Moscow Committee on Science and Technology", Moscow, Russia*
[e]*Center of State Sanitary and Epidemiological Supervision, Moscow, Russia*

**Abstract.** A model has been developed to assess exposure to the residents of Moscow, Russia from external (cosmic rays and terrestrial gamma rays) and radon exposures. This model accounts for different factors (demographic, social, materials of buildings, etc.) affecting the exposures considered and is based on contemporary measurement data. Ten administrative districts of Moscow were considered separately with respect to both external and internal exposures to the residents. The population-weighted average radon concentration in dwellings in Moscow, determined on the basis of the results of indoor surveys, is estimated to be about 82 Bq m$^{-3}$. Current estimates of annual collective effective doses to the residents in Moscow city (8,638,100 people) are (1) 6300 man Sv with a 68% confidence interval of 5900–7300 man Sv from external gamma-exposure and (2) 12,000 man Sv with a 68% confidence interval of 8000–16,000 man Sv from radon. The annual per caput effective dose is estimated to be 0.73 mSv from external gamma-exposure and about 1.4 mSv from radon. About 4300 people (0.05% of total) receive an annual effective dose from radon above 5 mSv.
© 2004 Elsevier B.V. All rights reserved.

*Keywords:* Natural radiation; Radon; Model; Collective effective dose; Moscow

## 1. Introduction

The first purpose of this paper is to indicate the methods used to assess exposure to the population in Moscow, Russia from external (cosmic rays and terrestrial gamma rays) and

---

* Corresponding author. State Research Center—Institute of Biophysics, Zhivopisnaya 46, Moscow 123182, Russia. Tel.: +7 95 190 9687; fax: +7 95 190 3590.
  *E-mail address:* sshinkarev@atom.ru (S.M. Shinkarev).

0531-5131/ © 2004 Elsevier B.V. All rights reserved.
doi:10.1016/j.ics.2004.10.020

radon exposures. Then, dose estimates based on currently available information are presented. Data collection is continuing, and the results will be updated at a later time.

## 2. Materials and methods

A model developed has two independent parts related to estimation of exposure to the population of Moscow from (1) external (cosmic rays and terrestrial gamma rays) and (2) radon exposures. This model is based on the general approaches described in UNSCEAR 2000 Report [1]. The model accounts for different factors (demographic, social, materials of buildings, etc.) affecting the exposures considered. Ten administrative districts of Moscow were considered separately. With respect to exposure of the residents to radon, the population was further subdivided to form population subgroups with similar exposure conditions. The following factors were considered: (a) radon-hazard category according to the radio-geochemical survey results; (b) floor of living in a residential building; (c) type of building materials; (d) number of floors in a building; (e) whether the year of building construction was before 1997; (f) peculiarity of ventilation rates; and (g) floor of the buildings where people work, study, etc. It was assumed that the distribution of indoor radon concentration in each group is satisfactorily fitted by a lognormal curve. Values of the parameters of a lognormal curve for each group were derived from the measurements. If no data were available for the groups considered, they were merged to a larger group, trying to form such a group more or less homogeneous with respect to radon exposure. Contemporary direct measurements of outdoor and indoor dose rates in air were used to estimate the external doses. Indoor radon concentrations, measured using the integral track method, were used to estimate the radon doses.

## 3. Results and discussion

The population-weighted average radon concentration in dwellings in Moscow, determined using the indoor survey data, is estimated to be about 82 Bq m$^{-3}$. Current estimates of annual collective effective doses to the residents in Moscow city (8,638,100 people) are (1) 6300 man Sv with a 68% confidence interval of 5900–7300 man Sv from external gamma-exposure and (2) 12,100 man Sv with a 68% confidence interval of 8000–16,000 man Sv from radon, assuming the value of dose conversion factor is equal to 6.1 nSv (Bq h m$^{-3}$)$^{-1}$ [2]. The annual per caput effective dose is estimated to be 0.73 mSv from external gamma-exposure and about 1.4 mSv from radon. The current data suggest that about 4300 people (0.05% of total) receive annual doses from radon that exceed 5 mSv. By the present time, a limited number of measurements (824) of radon concentration in dwellings has been carried out and used to estimate the radon doses. So, the current assessment of exposure to the residents of Moscow from radon is the subject to change. This work is in progress.

### References

[1] UNSCEAR, United Nations Scientific Committee on the Effects of Atomic Radiation, Sources and Effects of Atomic Radiation. Report to the General Assembly, Sources. Annex B. Exposures from Natural Sources of Radiation, vol. 1, United Nations, New York, 2000.
[2] ICRP Publication 65, International Commission on Radiological Protection. Protection against radon-222 at home and at work, Annals of ICRP, vol. 22(2), Pergamon Press, Oxford, 1993.

www.ics-elsevier.com

# Comparison of contemporary and retrospective radon concentration measurement in dwellings in Poland

J. Jankowski[a,b,*], J. Skubalski[b], J. Olszewski[a], P. Szalanski[b], A. Zak[b]

[a]*Institute of Occupational Medicine, Poland*
[b]*Nuclear Physics and Radiation Safety Department, University of Lodz, Poland*

**Abstract.** This paper presents the results of a comparison of the retrospective and contemporary radon concentration in 24 rooms in Kowary city. The average concentration of radon in Kowary city is the highest of any city in Poland. The contemporary radon concentration was estimated using a passive detector (CR-39). The retrospective radon concentration was estimated by measuring alpha activity of $^{210}$Po implanted on glass surfaces. The average radon concentrations estimated by contemporary and retrospective measurements were 375 and 404 Bq/m$^3$, respectively. © 2004 Published by Elsevier B.V.

*Keywords:* Retrospective radon assessment; Alpha spectroscopy; Dwellings

## 1. Introduction

In Poland, the average radon concentration in homes is 50 Bq/m$^3$ [1], although there are homes with much higher concentrations. The concentration depends on the geological formation of the site on which the house is built and also on the materials used to construct the house. In Poland, the Kowary region has a high radon concentration in dwellings. If we assume that these inhabitants spend about 80% time in closed spaces, then the annual effective dose from $^{222}$Rn is about 9.5 mSv. It is four times higher than the average worldwide annual effective dose [2].

In epidemiological studies of radon-related health risk, it is very important to know what concentration of radon there was in the past. Concentration of radon is subject to

---

\* Corresponding author. Radiation Protection Department, The Nofer Institute of Occupational Medicine, Teresy 8, Lodz 90-950, Poland. Tel.: +48 426314547; fax: +48 426314548.

*E-mail address:* jjan@imp.lodz.pl (J. Jankowski).

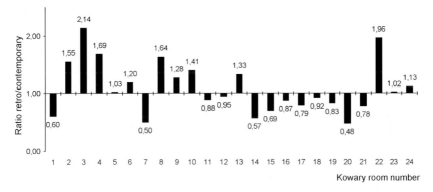

Fig. 1. Ratio of retrospective to contemporary estimation radon concentration in Kawary region.

daily, seasonal and yearly fluctuations; thus, retrospective measurements eliminate influence of these fluctuations.

## 2. Methods

The contemporary radon concentration was measured using a passive detector (CR-39). The measurements were conducted quarterly over a 1-year period. The retrospective radon concentration was estimated by measurement of alpha activity of $^{210}$Po implanted in glass samples [3,4]. For these measurements, an alpha spectrometer with Si-detector was used. Using the alpha particle spectrometer, it was possible to define precisely the energy range ($d_E$) in which the alpha particles were emitted from the glass. Thanks to that, the background emissions of the glass could be significantly reduced.

## 3. Results and conclusions

Using a radon chamber, the calibration coefficient was determined as follows: $k_R = 1.37 \pm 0.04$ kBq·year·m$^{-3}$/Bq·m$^{-2}$. The sensitivity of the alpha spectrometry was calculated on the basis of Currie criterion. The minimal implanted alpha activity which can be registered is $A_{min} = 0.42$ Bq/m$^2$. The average contemporary radon concentration in homes in Kowary was 375 Bq/m$^3$ with a maximum of 648 Bq/m$^3$ and a minimum of 134 Bq/m$^3$. The average retrospective radon concentration was 404 Bq/m$^3$ with a maximum of 1006 Bq/m$^3$ and a minimum of 121 Bq/m$^3$. Fig. 1 shows results comparing the retrospective and contemporary radon concentrations. With the exception of dwelling number 3 and number 20, these values are very similar. The obtained ratios of concentrations are lower than a factor of 2 or higher than a factor of 0.5. The results suggest that an epidemiological study of 1-year contemporary radon measurements can be representative for estimation of average radon concentration over a longer period.

## References

[1] J. Jagielak, et al., Radiological Atlas of Poland, Central Laboratory for Radiation Protection, National Atomic Agency, Warsaw, 1992.
[2] UNSCEAR. Report to the General Assembly with scientific annexes. New York: United Nationals 2000.
[3] C. Samuelson, Retrospective determination of radon in house, Nature 334 (1988) 338–340.
[4] J. Skubalski, et al., Alpha spectrometry for measuring activity of glass-embedded $^{210}$Po isotope, International Journal of Occupational Medicine and Environmental Health 13 (4) (2000) 361–367.

# Radon on underground tourist routes in Poland

## J. Olszewski*, W. Chruścielewski, J. Jankowski

*Department of Radiological Protection Nofer Institute of Occupational Medicine Teresy 8, 90-950 Lodz, Poland*

**Abstract.** In Poland, there are over 40 underground tourist routes, but radon concentrations are measured only in a few of them. The underground tourist routes, such as caves, old mines, building constructions under old towns and dungeons are the environments with potential radon concentrations that may exceed several thousand times the atmospheric concentrations. Among other routes, the following should be singled out: the drift of the former Kowary uranium mine, Niedzwiedzia Cave located near Kletno and close uranium mine in Kletno (the Sudetes). In the former, the mean annual radon concentration is of the order of 400 Bq m$^{-3}$, and in the latter it reaches over 2000 Bq m$^{-3}$. Measurements performed in 2004 in the closed uranium mine in Kletno showed that average radon concentration in that route is 1800 Bq m$^{-3}$. © 2004 Elsevier B.V. All rights reserved.

*Keywords:* Radon concentration; Underground tourist rout

## 1. Introduction

All over the world, tourists are seeking more and more fascinating attractions. Poland offers an extremely rich variety of underground attractions, including over 40 underground tourist routes. They could be grouped into four basic categories: caves, mines, post-military underground structures, cellars and other. Work in underground tourist routes (e.g. as a guide) or just visiting them by the tourists is associated with a considerable health hazard. The hazard is due to natural radioactive elements, particularly radon, present in the underground spaces. Here are results of measurements of radon concentration in some Polish underground tourist' routes (Table 1).

## 2. Methods

Our Nofer Institute of Occupational Medicine in Lodz performs regular measurements of radon concentrations in three underground tourists' routes. Mean long-term radon

---

\* Corresponding author. Tel.: +48 426 314 540; fax: +48 426 314 535.
*E-mail address:* jolsz@imp.lods.pl (J. Olszewski).

0531-5131/ © 2004 Elsevier B.V. All rights reserved.
doi:10.1016/j.ics.2004.10.015

Table 1
Values of average monthly radon concentration in the air of selected underground tourists routes in Poland [1–3]

| Name of object | Radon concentration (kBq m$^{-3}$) | | |
|---|---|---|---|
| | Minimum | Average | Maximum |
| Radochowska Cave | 0.06 | 0.45 | 1.37 |
| Underground tourist route in Chelm (mine) | 0.04 | 0.09 | 0.33 |
| Underground tourist route in Walim (Post-military) | 0.002 | 0.166 | 0.622 |
| Underground tourist route in Sandomierz (cellar) | 0.015 | 0.044 | 0.077 |
| Millennium of the Polish State Underground Tourist Route in Klodzko (Post-military) | 0.07 | 0.29 | 2.21 |
| Gold Mine in Zloty Stok | 0.07 | 1.66 | 18.5 |

concentrations are measured by means of NRPB passive integrating dosimeter with a track detector CR-39 (Pershore). The meters are placed at the monitoring points for a period of 1–3 months.

## 3. Results

Concentration of radon in controlled routes varies considerably. The average concentration in the Niedzwiedzia Cave was 2000±700 Bq m$^{-3}$. The average radon concentration in Kowary Drift was almost five times lower, 400±100 Bq m$^{-3}$. The average radon concentration during the first 6 months of 2004 in the tourist's route in Kletno was 1800±400 Bq m$^{-3}$ (Fig. 1).

## 4. Conclusions

(1) In none of these routes, the maximum admissible annual dose of 20 mSv is likely to be exceeded.
(2) Persons employed in the tourist services should be regarded as those occupationally exposed to ionising radiation and, as such, their exposure has to be monitored.
(3) Tourists visiting both routes are exposed to a dose of the order of several μSv, which in comparison with a dose received by every individual during a year (2.4 mSv) is a negligible, with no effects on tourist health.

## References

[1] T.A. Przylibski, Radon concentration changes in air of Niedzwiedzia Cave in Kletno (Sudety Mountains), Prz. Geol. 44 (9) (1996) 942–944 (in Polish).
[2] T.A. Przylibski, Radon in the millennium of the polish state underground tourist route in Klodzko (Lower Silesia), Arch. Environ. Prot. 24 (1998) 33–42.
[3] T.A. Przylibski, Radon concentration changes in the air of two caves in Poland, J. Environ. Radioact. 45 (1999) 81–94.

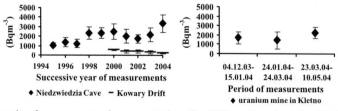

Fig. 1. Results of measurement radon concentrations. For 2004 results, refer to the first 6 months.

# Radioactivity in vine cellars in Hungary and Slovenia

P. Szerbin[a,*], J. Vaupotic[b], I. Csige[c], I. Kobal[b], I. Hunyadi[c], L. Juhász[a], E. Baradács[c]

[a]"Frédéric Joliot-Curie" National Research Institute for Radiobiology and Radiohygiene, Budapest, Hungary
[b]"Jozef Stefan" Institute, Ljubljana, Slovenia
[c]Institute of Nuclear Research of Hungarian Academy of Science, Debrecen, Hungary

**Abstract.** It is well known that Slovenia and particularly Hungary have vine regions, where most inhabitants are involved in vine production. The aim of the study was to identify the radon level differences between cellar types and to estimate the dose to the workers. © 2005 Published by Elsevier B.V.

*Keywords:* Radioactivity; Vine cellar; Radon; Radon progeny; Dose; Worker

## 1. Introduction

In spite of the strong interest in the health effects of radon and radon progeny in underground mines, dwellings, spas, and caves, there are still some unexplored workplaces such as vine cellars, which seems worth of attention.

## 2. Materials and methods

The map shows the location of the Hungarian vine regions. In some cellars sampling was made by Lucas cells, and simultaneous radon and radon progeny measurements were performed (Pylon Electronics Inc. AB-5 monitors, CPRD, AEP-47 detectors, and SARAD EQF 3020 radon monitor). Nuclear track detectors were used in Slovenia. The equilibrium factor was calculated on the basis of continuous measurements for the large cellars, and for the smaller cellars equilibrium factor was

---

\* Corresponding author. NRIRR, POB 101, 1775, Budapest, Hungary.
 E-mail addresses: szerbin@npp.hu, pavel@hp.osski.hu (P. Szerbin).

0531-5131/ © 2005 Published by Elsevier B.V.
doi:10.1016/j.ics.2004.12.063

Fig. 1. Map of Hungary with the most important vine regions.

1. Csongrád
2. Hajós-Baja
3. Kunság
4. Ászár-Neszmély
5. Badacsony
6. Balatonfüred-Csopak
7. Balatonmellék
8. Etyek-Buda
9. Mór
10. Pannonhalma
11. Somló
12. Sopron
13. Dél-Balaton
14. Mecsekalja
15. Szekszárd
16. Villány-Siklós
17. Bükkalja
18. Eger
19. Mátraalja
20. Tokajhegyalja
21. Zala
22. Tolna

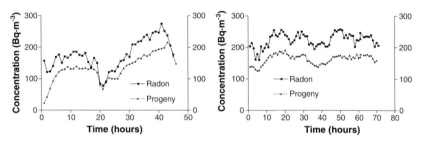

Fig. 2. Radon and radon progeny levels in Hungarovin cellar, Budafok region, sampling points 3, 4.

assumed 0.5 on the basis of our previous measurements of atmosphere of caves [1,2] (Fig. 1).

## 3. Results and discussion

There is the evidence of irregular variations of radon and radon progeny levels indicating the close connection of the cellar atmosphere with the outdoor atmosphere through ventilation shafts in the cellars, necessary to reduce the $CO_2$ produced during the fermentation process. In our previous investigations we found such irregular variations in shallow caves connected with cracks and fissures with the outdoor air [1,2]. In isolated caves the radon levels were very stable. No regular diurnal or seasonal variations have been found so far (Fig. 2).

Table 1
Results of radon measurements in vine cellars in Hungary using Pylon CPRD and AEP-47 detectors (Pylon)

| Sampling point | $C_{Rn}$ (Bq·m$^{-3}$) | F | WLM | Eff. dose (mSv·year$^{-1}$) |
|---|---|---|---|---|
| Hungarovin, Promontorvin cellars cellars | 179–992 | 0.6–1.0 | 0.028–0.139 | 1.7–8.3 |
| Promontorvin cellars (large vinery) | 70–217 | 0.6–1.0 | 0.011–0.034 | 0.5–2.0 |
| Siklós-Villány (small–medium cellars) | 130–10.840 | 0.5 | 0.07–0.119 | 0.4–7.1 |
| Tokaj region (small–medium cellars) | 140–460 | 0.5 | 0.02–0.80 | 0.5–1.3 |
| North–West Slovenia | 42–141 | | | |
| North–East Slovenia | 109–456 | | | |

Occupancy was assessed to be for 6 h day$^{-1}$ for large vineries on the basis of staff interviews, 3 h week$^{-1}$ occupancy was assumed for small cellars and 15 h·week$^{-1}$ assessment was conducted for medium-size cellars. The radon levels are one to two orders of magnitude lower than that of the caves in Hungary and Slovenia [1–3]. In view of lack of occupancy data and low levels, no dose assessment was done for the Slovenian cellars (Table 1).

## References

[1] P. Szerbin, Radon concentrations and exposure levels in Hungarian caves, Health Physics 71 (3) (1996) 362–369.
[2] P. Szerbin, Natural radioactivity of certain spas and caves in Hungary, Environment International 22 (1) (1996) 389–398.
[3] Janja Vaupotiè, et al., Methodology of radon monitoring and dose estimates in Postojna Cave, Slovenia, Health Physics 80 (2001) 142–147.

# Tenorm's around coal fired power plant tailings ponds in Hungary

P. Szerbin[a,*], L. Juhász[a], I. Csige[b], A. Várhegyi[c], J. Vincze[d], T. Szabó[d], F.-J. Maringer[e]

[a]"Frédéric Joliot-Curie" National Research Institute for Radiobiology and Radiohygiene, POB. 101, 1775, Budapest, Hungary
[b]Institute of Nuclear Research, Debrecen, Hungary
[c]Mecsek Ore Environmental Protection Company, Pécs, Hungary
[d]Pannonpower Energetic Company, Pécs, Hungary
[e]Arsenal Research, Vienna, Austria

**Abstract.** During several years of operation of Pannonpower Energetic, hundreds of thousands of tons of ash and slag, containing elevated levels of naturally occurring radioactive materials (NORMs), have been produced. The company decided to launch an environmental restoration program. The main goals are the environmental restoration of the area and to work out recommended options for future industrial, agricultural, and urban utilization. © 2004 Published by Elsevier B.V.

*Keywords:* TENORM; Tailings ponds; Coal fired power plant; Remediation

## 1. Introduction

The tailings ponds comprise approximately $2.7 \times 10^6$ m$^2$. Area A is covered by 30 cm soil layer and it is vegetated by grass. Area B consists of old dumping areas that have been covered by 30 cm soil layer and vegetated by mixed gramineous plants and trees; unrestored dried tailings ponds; and fresh tailings ponds which are currently in use.

## 2. Materials and methods

Soil, water, and ash samples were measured by gamma-spectroscopy, in situ external gamma dose rate and radon exhalation measurements were done. Column and field experiments were carried out to find the optimal option for tailings pond covering, and mathematical modeling was performed to predict radionuclide migration in the soil.

* Corresponding author.
  E-mail addresses: szerbin@npp.hu, pavel@hp.osski.hu (P. Szerbin).

0531-5131/ © 2004 Published by Elsevier B.V.
doi:10.1016/j.ics.2004.12.018

Table 1
Radiation limits for the restored places (on the basis of previous practices [1,2])

| Measurements | Regulatory requirements | Numerical limits |
|---|---|---|
| Rn flux | 0.37 Bq m$^{-2}$ s$^{-1}$ | 0.37 Bq m$^{-2}$ s$^{-1}$ |
| Rn activity concentration (open air) | Bkg$^a$+10 Bq m$^{-3}$ | 20 Bq m$^{-3}$ |
| Gamma dose rate | Bkg+100 nSv h$^{-1}$ | 190 nSv h$^{-1}$ |
| Activity concentration in soil | | |
| $^{226}$Ra in upper 15 cm | Bkg+90 Bq kg$^{-1}$ | 140 Bq kg$^{-1}$ |
| $^{226}$Ra in lower than 15 cm | Bkg+280 Bq kg$^{-1}$ | 380 Bq kg$^{-1}$ |

$^a$ Bkg=background.

Table 2
Radioactivity of the soil, ash, and cover soil in Bq kg$^{-1}$

| | $^{235}$U | $^{234}$Th | $^{226}$Ra | $^{214}$Pb | $^{214}$Bi | $^{210}$Pb | $^{228}$Ac | $^{212}$Pb | $^{212}$Bi |
|---|---|---|---|---|---|---|---|---|---|
| Soil (background) | 2 | 55 | 45 | 38 | 35 | 70 | 53 | 53 | 53 |
| Soil, cover layer/1 | 2 | 45 | 40 | 30 | 30 | 50 | 41 | 40 | 41 |
| Ash | 11–13 | 275–300 | 250–290 | 240–280 | 230–280 | 160–400 | 165–217 | 168–213 | 152–205 |

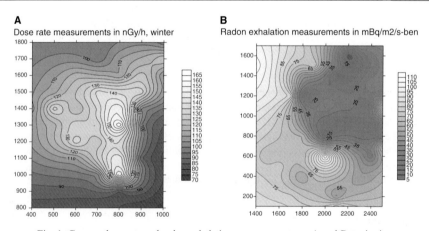

Fig. 1. Gamma dose rate and radon exhalation measurements on A and B territories.

## 3. Results and discussion

According to the measurements, experiments, and modeling, the 30 cm cover layer fulfils the requirements of the radiation criteria system for remediation (Tables 1 and 2; Fig. 1). On long term at least 50 cm layer should be applied to avoid migration of radium to the root zone of gramineous plants, and to reduce the destructive effect of soil erosion. No agricultural activities are allowed in the tailings pond area, although restricted utilization for industrial use is possible.

## References

[1] P. Szerbin, et al., Environmental restoration of uranium mining tailings ponds in Hungary, Proceedings of InternationalSsymposium on Geology and Environment, GEOENV '97, The Chamber of Geological Engineers of UCEAT, vol. 430, 1999, pp. 237–241.
[2] L. Juhász, et al., Results of pilot studies of environmental restoration of uranium mining tailings ponds in Hungary, Science of the Total Environment 272 (2001) 251–252.

ELSEVIER

www.ics-elsevier.com

# Evaluation of the technologically enhanced naturally occurring radioactive material in Hungary

L. Juhász[a,*], P. Szerbin[a], I. Czoch[b]

[a]"Frédéric Joliot-Curie" National Research Institute for Radiobiology and Radiohygiene, Budapest, Hungary
[b]Hungarian Atomic Energy Authority, Budapest, Hungary

**Abstract.** In Hungary, the evaluation of technologically enhanced naturally occurring radioactive materials (TENORM) situation has not been performed in a full scale. So a surveying project was launched for study of TENORM situation. © 2004 Elsevier B.V. All rights reserved.

*Keywords:* TENORM; NORM; Remedial action

## 1. Introduction

A surveying project was launched in order to collect all relevant information about the Hungarian technologically enhanced naturally occurring radioactive materials (TENORM) situation. This project covered a lot of data collection (work activities, disposal places, residue quantities) and in situ and laboratory radiological measurements. Regarding the size of the project, each task was being performed with a stepwise approach. Nevertheless, the Hungarian situation of TENORM definitely differs from that of other countries in the aspect of occurrence forms of natural sources (or in the imported raw materials), in the quantities of exploitation, in the level of the radioactivity and in the applied technological processes.

## 2. Surveying project task

Firstly, those work activities were chosen where a huge amount of residues have been produced. The other criteria were that the activity concentration in the majority of the given residues can be much higher than the average activity concentration of the typical Hungarian soil. After filtering and ranking, the following main activities are left: uranium mining and milling, coal mining, coal firing in power plants, bauxite mining and alumina

---

\* Corresponding author. Tel.: +36 14822008; fax: +36 12291931.
*E-mail address:* juhasz@hp.osski.hu (L. Juhász).

0531-5131/ © 2004 Elsevier B.V. All rights reserved.
doi:10.1016/j.ics.2004.11.092

production. The surveying program was focused on the main processes of a given activity. On TENORM site, firstly, in situ dose rate measurements were carried out in mesh points and then characteristic mixed representative samples were taken according to the distribution of dose rate. Radon gas activity concentration in the soil and radon exhalation rate on the surface were also measured on the TENORM area.

### 2.1. Uranium mining and milling

The waste rock and heap leaching piles were restored at the closed mine near Pécs city. The tailings were discharged into two ponds and the restoration of the ponds is still in progress. On the tailings ponds, the dose rate ranged from 0.4 to 10 $\mu$Sv h$^{-1}$ and radon activity concentration was 40–1100 Bq m$^{-3}$ and radon exhalation rate was 4–8 Bq m$^{-2}$ s$^{-1}$ [1].

### 2.2. Coal mining, coal fired power plant

In the mountain Mecsek and on the territory from the highland Balaton to the mountain Vértes, the radioactivity level of the coals mined on these areas is 10–50 times higher than the worldwide average.

The radioactivity of U-238 series of ash and slag discharged from power plant Ajka was in the range from 200 to 2000 Bq·kg$^{-1}$, and the radionuclides of Th-232 series varied between 20 and 40 Bq·kg$^{-1}$. The radioactivity level of U-238 and Th-232 series of ash and slag discharged from power plant Pécs was in the range of 200–300 Bq·kg$^{-1}$, and K-40 can be found in the ash and slag with the typical value about of 800 Bq·kg$^{-1}$. Above the waste rock piles and tailings ponds, the dose rate was 200–500 nSv h$^{-1}$.

### 2.3. Bauxite mining, alumina production

The radionuclide concentrations of bauxite ores ranged from 200 to 300 Bq·kg$^{-1}$. At the refining factories, the red mud has been produced with the activity concentration of 200–400 Bq·kg$^{-1}$ and the dose rate above the ponds of red mud was in the range of 200–400 nSv h$^{-1}$.

## 3. Conclusion

According to the first evaluation of the measurements, no extremely high values have been found at TENORM sites. Slightly higher radioactivity levels can be found at few waste rock piles of coal mines and on the tailings ponds of red mud and fly ash. The surveying program is going on, and the results are to be used for the development of the radiation protection requirements of TENORM situation and for the remedial action, too.

### Acknowledgement

Work was supported by EU Framework 5, and by Hungarian Atomic Energy Authority projects.

### Reference

[1] L. Juhasz, P. Szerbin, A. Varkonyi, Radon exhalation studies for the remediation of uranium mill tailings ponds in Hungary, Geofisica Internacional 41 (4) (2002) 483–484.

www.ics-elsevier.com

# Inaccuracies in assessing doses from radon in workplaces

C. Németh[a,b,]*, S. Tokonami[a], J. Somlai[b], T. Kovács[b], N. Kávási[b], Z. Gorjánácz[c], A. Várhegyi[c], J. Hakl[d]

[a]*Radon Research Group, National Institute of Radiological Sciences, Japan*
[b]*University of Veszprém, Hungary*
[c]*Mecsek Ore Environment Co., Pécs, Hungary*
[d]*Institute of Nuclear Research, Debrecen, Hungary*

**Abstract.** In order to assess the dose due to radon in workplaces, it is important to know the concentrations during the periods when the workers are present though the data usually come from integrating measurements over the whole time. This paper describes discussions to show the possibility of inaccuracies in dose estimation. © 2004 Elsevier B.V. All rights reserved.

*Keywords:* Radon in workplace; School; Underground site; Dose assessment

## 1. Introduction

Radon measurements, including continuous and integrating methods with various periods, were carried out in different workplaces in Hungary and Japan [1,2]. The investigated places were schools, air-conditioned office buildings, underground workplaces, etc. The results indicated that there were significant differences between the average concentrations over the whole periods and during the effective working hours. The recommended levels of radon in workplaces are mostly based on the annual whole average levels which could be fairly different from the real situation and may not be useful to predict the total exposure to radon hazard. The accurate calculation should be done based on the radon concentration measured only during the working hours.

---

\* Corresponding author. Radon Research Group, National Institute of Radiological Sciences, Anagawa 4-9-1, Inage, Chiba 263-8555, Japan. Tel.: +81 43 206 3111; fax: +81 43 206 4098.
*E-mail address:* csaba@fml.nirs.go.jp (C. Németh).

0531-5131/ © 2004 Elsevier B.V. All rights reserved.
doi:10.1016/j.ics.2004.11.121

## 2. Methods and discussion

### 2.1. Measurements at offices and schools

Long-term radon measurements with passive detectors (CR-39) were made in office buildings equipped with air-conditioning systems in Tokyo and short-term continuous measurements also were executed at the same places using AlphaGUARD device for radon and Pylon AB-5 and Pylon AEP-47 for equilibrium equivalent radon concentration (EERC). The daily average levels of radon exposure obtained from the long-term measurements are almost the same as those determined by the continuous whole-day measurements. If the data for the effective working hours are used, on the other hand, the radon concentrations are less. The results differ by a factor of more than 2. Similar measurements were done in schools and kindergartens in Hungary. The radon concentrations increase during the nights and weekends. In two typical cases, the averages were 497 Bq $m^{-3}$ (a) and 287 Bq $m^{-3}$ (b) over the whole period and during the time when the children were inside, the averages were 125 Bq $m^{-3}$ (a) and 84 Bq $m^{-3}$ (b). Consequently, the average radon concentration during the effective time is considerably lower (by a factor of more than 3) than over the whole day. Similar results were found in other schools and kindergartens as well.

### 2.2. Measurements at underground workplaces

The pattern of the change of radon activity concentration in a mine tunnel seems to be similar to the offices' ones. The ventilation works only during the working hours, so the radon level increases at nights and in the weekends, whereas it is considerably lower during the working hours.

In this experiment, track detectors were placed at different locations of a mine and some miners put on the same type of detectors during work, which were kept in a place with low radon concentration (<12 Bq $m^{-3}$) after the working hours. The average radon concentration shown by detectors placed in the mine was 3690 Bq $m^{-3}$, which represents the value over 24 h a day, while the average level coming from the detectors worn by the workers representing the working hours was only 810 Bq $m^{-3}$, which means a 4.5 time difference.

Another problem is the choosing of the correct measuring period. For example, data obtained from a 3-year-long measurement in a hospital cave showed what could be a 24-fold difference from the results of 1-month-long measurements, depending on the months selected. Even 6-month-long measurements can differ 7-fold in case of careless choice of the period.

## 3. Conclusion

The effect of the chosen measuring methods can result in considerable errors when assessing the dose due to radon at workplaces.

In the case of ordinary offices and schools, the measurements show that the radon levels during the actual working hours are significantly lower (by a factor of 2–5) than those over the whole day. Consequently, the annual dose could be overestimated if the calculation is based on the measurements over the whole period.

Other workplaces, such as underground ones, have special characteristics which should be taken into consideration.

### References

[1] S. Tokonami, et al., Radiat. Prot. Dosim. 106 (2003) 71–75.
[2] J. Somlai, et al., J. Radioanal. Nucl. Chem. 218 (1997) 61–63.

# Naturally occurring alpha emitting radionuclides in drinking water (Hungary) and assessment of dose contribution due to them

T. Kovács[a,b,*], E. Bodrogi[a], J. Somlai[a], V. Jobbágy[a], P. Dombovári[a], Cs. Németh[a,b]

[a]*University of Veszprém, Hungary*
[b]*Radon Research Group, National Institute of Radiological Sciences, Chiba, Japan*

**Abstract.** The assessment and regulation of dose coming from naturally occurring radionuclides have recently been a concern. In our work, the concentrations of gross alpha, $^{226}$Ra, $^{238}$U and $^{210}$Po were examined in Hungarian drinking waters. In the investigation, 102 drinking water samples were determined (bottled mineral, well, tap and spring waters). The dose contribution was calculated using the radionuclide concentrations and the dose conversion factors from WHO in the case of 2 L/day water consumption. In some cases, the calculated (assumed) doses were considerably higher than the limits for drinking waters. © 2004 Elsevier B.V. All rights reserved.

*Keywords:* Drinking water; Dose assessment; Alpha emitting radionuclides; Alpha spectrometry

## 1. Introduction

Drinking waters may contain radionuclides besides the numerous other contaminants. The regulation and assessment of these radionuclides are very important and the regulation of alpha emitting radionuclides is outstanding among these.

Therefore, almost all the international recommendations (WHO, UNSCEAR, IAEA) outline the assessment of alpha emitting radionuclides. In this study, investigation of 102 Hungarian drinking water samples (Hungarian bottled mineral waters [1,2], South Great Plain well waters, tap waters in Veszprém county, Balaton Highland spring waters [3]) was carried out according to the assessment process recommended by WHO.

---

* Corresponding author. Radon Research Group, National Institute of Radiological Sciences, Anagawa 4-9-1, Inage, Chiba 263-8555, Japan. Tel.: +36 88 427 681; fax: +36 88 427 681.
  *E-mail address:* csaba@fml.nirs.go.jp (T. Kovács).

0531-5131/ © 2004 Elsevier B.V. All rights reserved.
doi:10.1016/j.ics.2004.11.050

## 2. Materials and methods

*Radon activity determination*: Except for the bottled mineral waters, the Rn concentration of the samples was measured on the spot with Pylon AB-5 monitor.

*Gross Alpha–Beta activity determination*: The gross α- and β-determination was carried out using NDI ZnS(Ag)-β plastic detector of Gamma Tech. and Oxford LB-5 low level α/β gas proportional counter.

*Determination of $^{234,238}U$, $^{226}Ra$ and $^{210}Po$ isotopes by alpha spectrometry*: The uranium source was prepared by La(U)F$_3$ micro-coprecipitation using a $^{232}U$ tracer. The $^{226}Ra$ source preparation was carried out by MnO$_2$-coated polyamide discs. The $^{210}Po$ source was prepared by spontaneous deposition on nickel steel plate using $^{208}Po$ tracer. The sources were measured by Canberra alpha spectrometric chamber.

*$^{226}Ra$ determination by $^{222}Rn$ emanation procedure*: After radon in the sample was removed, it was put into the emanation chamber for 15–20 days. The growing $^{222}Rn$ was measured in Lucas-cell by EMI photo-multiplayer. The MDA was 7–12 mBq/L.

## 3. Results and discussion

### 3.1. Results of activity concentration measurements

Table 1
The activity concentrations of selected Hungarian water samples

|  | <10 | 10–100 | 100–1000 | 1000> |
|---|---|---|---|---|
| $^{222}Rn$ [Bq/L] | 17 | 47 | 0 | 0 |
| $\sum \alpha$ [mBq/L] | 5 | 79 | 15 | 3 |
| $\sum \beta$ [mBq/L] | 0 | 51 | 47 | 4 |
| $^{237,238}U+^{226}Ra+^{210}Po$ [mBq/L] | 15 | 72 | 12 | 3 |

### 3.2. Dose assessment

The dose coming from $^{234,238}U$, $^{226}Ra$ and $^{210}Po$ isotopes was assessed according to the conversion factors recommended by WHO (1993) based on 2 L/day water consumption.

## 4. Conclusion

Eleven of the 102 water samples exceeded 100 mBq/L alpha emitting radionuclide concentrations: 4 samples of bottled mineral water, 4 samples of Balaton Highland spring water and 3 samples of South Great Plain well water. Following the nuclide-specific examinations and dose assessment, it was found that only three types of bottled mineral water exceeded the level 0.1 mSv/year recommended by WHO. The mineral waters are not listed as drinking water despite the fact that the consumption of them is increasing.

## References

[1] J. Somlai, et al., J. Environ. Radioact. 62 (2002) 235–240.
[2] T. Kovács, et al., Radiat. Prot. Dosim. 108 (2004) 175–181.
[3] T. Kovács, et al., J. Radioanal. Nucl. Chem. 258 (2003) 191–194.

www.ics-elsevier.com

# Public exposure to radon and thoron progeny in Romania

## Olga Iacob*, Constantin Grecea

*Radiation Hygiene Laboratory, Institute of Public Health, 14,Victor Babes Street, Iassy 700465, Romania*

**Abstract.** The purpose of our work was to determine the activity concentrations of short-lived decay products of radon and thoron in dwellings and in free air and to evaluate, in terms of effective doses, the magnitude of the resulting annual exposure of population in eastern Romania. The activity concentrations of $^{218}$Po, $^{214}$Pb, $^{214}$Bi and $^{212}$Pb have been measured in 684 typical urban and rural houses. The average values of EEC of radon were 28 Bq m$^{-3}$ in detached houses and 9 Bq m$^{-3}$ in block of flats, with individual values ranging from 3.8 to 564 Bq m$^{-3}$. The average values of EEC of thoron were of 1.3 Bq m$^{-3}$ in detached houses and 0.6 Bq m$^{-3}$ in block of flats, with individual values ranging from 0.1 to 12.8 Bq m$^{-3}$. Outdoor average values of EEC were 4.6 Bq m$^{-3}$ for radon progeny and 0.2 Bq m$^{-3}$ for thoron progeny. The overall annual effective dose per capita was 2.0 mSv (1.65 mSv from radon progeny and 0.35 mSv from thoron progeny) with a corresponding annual collective effective dose of 10,370 manSv. © 2004 Elsevier B.V. All rights reserved.

*Keywords:* Population exposure; Radon and thoron progeny; Effective dose

## 1. Introduction

The radiation dose from inhaled radon and thoron progeny indoors and outdoors is the dominant component of population exposure to natural radiation sources accounting for about 60% in Romania [1]. The purpose of our work was to determine the activity concentrations of short-lived decay products of radon and thoron, indoors and outdoors, and to evaluate the magnitude of the resulting annual exposure of the Romanian population.

## 2. Materials and methods

The measurements of the radon and thoron short-lived decay products ($^{218}$Po, $^{214}$Pb, $^{214}$Bi and $^{212}$Pb) concentrations have been carried out in 684 typical urban and rural houses,

---

\* Corresponding author. Tel.: +40 232 410399; fax: +40 232 210399.
*E-mail address:* olga_iacob@yahoo.com (O. Iacob).

0531-5131/ © 2004 Elsevier B.V. All rights reserved.
doi:10.1016/j.ics.2004.11.019

Table 1
Indoor and outdoor average concentrations of $^{222}$Rn and $^{220}$Rn progeny and the resulting annual effective doses

| Radio nuclide | Location | EEC (Bq·m$^{-3}$) | | Annual effective dose | | | |
|---|---|---|---|---|---|---|---|
| | | | | (mSv) | | | (manSv) |
| | | Average | Range | Average | Range | Per capita | Collective |
| $^{222}$Rn progeny | INDOORS | 20.3 | | 1.20 | | 1.53 | 7930 |
| | Detached house | 28 | 3.8–564 | 1.66 | 0.23–33.3 | | |
| | Block of flats | 9 | 2.6–19.4 | 0.51 | 0.20–1.15 | | |
| | OUTDOORS | 4.6 | 1.0–7.3 | 0.09 | 0.02–0.14 | 0.12 | 593 |
| | Total $^{222}$Rn | | | 1.29 | | 1.65 | 8523 |
| $^{220}$Rn progeny | INDOORS | 1.0 | | 0.26 | | 0.33 | 1737 |
| | Detached house | 1.3 | 0.2–6.4 | 0.34 | 0.05–0.74 | | |
| | Block of flats | 0.6 | 0.1–2.8 | 0.16 | 0.03–0.74 | | |
| | OUTDOORS | 0.2 | 0.1–0.6 | 0.02 | 0.01–0.05 | 0.02 | 109 |
| | Total $^{222}$Rn | | | 0.28 | | 0.35 | 1846 |
| Total | | | | 1.57 | | 2.0 | 10,370 |

randomly selected from 8 cities and 136 villages. The method used was the active one of sucking a known volume of air through a high-efficiency filter paper and counting the deposited activity using an alpha scintillation counter [2]. Internal exposure due to inhalation of $^{222}$Rn and $^{220}$Rn daughters in indoor and outdoor air, expressed in terms of effective dose, was estimated by using the dose conversion coefficients adopted in the UNSCEAR 2000 Report [3]. An indoor occupancy factor of 0.75 and a population size of 5.2 million inhabitants (27% children and 73% adults) have been considered in all calculations.

### 3. Results and discussion

The results of our study are included in Table 1. The average value for EEC of radon was 28 Bq m$^{-3}$ in houses and 9 Bq m$^{-3}$ in flats and the average value for EEC of thoron was 1.3 Bq m$^{-3}$ in houses and 0.6 Bq m$^{-3}$ in flats. The indoor population-weighted averages of the EECs of radon and thoron were estimated at 20.3 and 1.0 Bq m$^{-3}$.

The average annual effective dose to adult was 1.20 mSv from indoor inhaled radon progeny, individual values ranging between 0.20 and 33.3 mSv. The average annual effective dose to adult due to indoor inhaled thoron progeny was 0.26 mSv, with individual values ranging between 0.03 and 1.68 mSv. The overall annual effective dose per capita, in assumption that dose received by a child is twice the adult dose, was estimated at 2.0 mSv–1.65 mSv from radon progeny and 0.35 mSv from thoron progeny—with a corresponding annual collective effective dose of 10,370 manSv.

### 4. Conclusions

The annual effective dose per capita resulting from the radon isotopes inhalation, indoors and outdoors, was estimated at 2 mSv of which 77% is due to indoor $^{222}$Rn progeny. The corresponding annual collective effective dose was calculated to be 10,370 manSv.

### References

[1] O. Iacob, E. Botezatu, Population exposure to natural radiation sources in Romania, Proc. of the 11th IRPA Congress, 23–28 May 2004 Madrid, Spain, 2004, http://www.irpa11.com/new/pdfs/6a33.pdf.
[2] Romanian Institute of Standardization, Air quality: volumetric activity determination of $^{218}$Po, $^{218}$Pb, $^{214}$Bi and $^{218}$Pb. Romanian Standard SR-13397, 1997.
[3] United Nations Scientific Committee on Effects of Atomic Radiation, Report 2000, UN, New York, 2000.

# Radon exposure in Slovenian kindergartens and schools

Janja Vaupotič*, Ivan Kobal

*Jožef Stefan Institute, Jamova 39, SI-1000 Ljubljana, Slovenia*

**Abstract.** In the Slovene Radon Project, all 730 kindergartens and play-schools and 890 elementary and high schools in Slovenia were surveyed for radon. In 45 kindergartens and 78 schools, the national radon limit of 400 Bq m$^{-3}$ was exceeded. In these buildings, the annual effective doses ranged from 0.2 to 8.8 mSv for personnel and from 0.2 to 7.2 mSv for children. As a result, 35 buildings have been successfully radon-mitigated. © 2004 Elsevier B.V. All rights reserved.

*Keywords:* Natural radioactivity; Radon; Kindergarten; School; Child; Concentration; Dose; Survey; Continuous measurement; Etched-track detectors

## 1. Introduction

In the first period of the Slovenian Radon Project, 1990–1992, instantaneous indoor air radon levels were surveyed in 730 kindergartens [1] and 890 schools [2], while in the second period, 1994–2002, attention was focused primarily on buildings with elevated radon levels. Additional monitoring, using a combination of various measuring techniques, provided data for dose estimates for personnel as well as for children, on the basis of which a decision was made whether radon-mitigation was needed or not.

## 2. Experimental

Instantaneous indoor radon concentrations were obtained and radon sources in each building identified with alpha scintillation cells produced at the Jozef Stefan Institute, Slovenia. Annual exposures were obtained with KfK etched track detectors (Karlsruhe, Germany) which were exposed all the year round. Overall average radon concentrations were distinguished from those measured during working hours only, and to which a person was actually exposed, using electret-based detectors (Genitron, Germany).

---

\* Corresponding author. Tel.: +38 614773213; fax: +38 614773811.
*E-mail address:* janja.vaupotic@ijs.si (J. Vaupotič).

0531-5131/ © 2004 Elsevier B.V. All rights reserved.
doi:10.1016/j.ics.2004.10.007

Table 1
Results of the radon survey in Slovenian kindergartens and schools

| | Kindergarten | School |
|---|---|---|
| Total number of buildings | 730 | 890 |
| Number/percentage of buildings with Rn concentration above 400 Bq m$^{-3}$ | 45/6.2 | 78/9.0 |
| Total number of children | 65,600 | 280,000 |
| Number/percentage of children in buildings with Rn concentration above 400 Bq m$^{-3}$ | 2,700/4.1 | 16,000/5.7 |
| Rn concentration in Bq m$^{-3}$: max/AM/GM | 5600/133/58 | 4680/168/82 |
| Number of children/personnel included in dosimetry | 820/267 | 2400/223 |
| Annual effective doses for children in mSv year$^{-1}$: max/AM | 6.1/2.7 | 10.4/2.1 |
| Annual effective doses for personnel in mSv year$^{-1}$: max/AM | 4.9/1.3 | 9.8/1.9 |
| Total number/percentage of mitigated buildings | 16/35.5 | 19/24.0 |
| Number/percentage of buildings for which no action has been taken | 4/9.0 | 13/17.0 |
| Number/percentage of buildings with working average Rn concentration below 400 Bq m$^{-3}$ and mitigation measures have not been requested | 25/55.5 | 46/59.0 |

## 3. Results and discussion

The data resulting from the radon survey in kindergartens and schools and a summary of the results are given in Table 1. The majority of buildings with high radon levels were found in the karstic region where cracks, fissures and underground corridors enable radon to travel larger distances [3]. Detailed radon measurements were carried out enabling annual effective doses of the personnel and children to be estimated in buildings exhibiting more than 400 Bq m$^{-3}$. On the basis of the measured radon concentration, the equilibrium factor between radon and its progeny and the duration of exposure periods, effective doses to the personnel and children were estimated, using the ICRP-65 methodology [4]. In kindergartens and in schools, 360 and 750 children, respectively, received more than 2.5 mSv year$^{-1}$, this number being 34 for kindergarten and 44 in for school personnel. As a result of the survey, 16 kindergartens and 19 schools have been successfully radon-mitigated to date.

For children in a school with an annual effective dose of more than 6 mSv, chromosome aberration analyses were performed. In some buildings with high radon levels, unattached fraction has also been measured and will be used for re-evaluation of effective doses. The results will be reported elsewhere.

Our indoor radon survey has covered more than 90% of kindergarten and school buildings and the buildings with more than 400 Bq m$^{-3}$ have been further thoroughly investigated. The buildings in which annual effective doses either for personnel or children exceeded 2.5 mSv have been successfully mitigated or are in the process of radon-mitigation. Thus, we believe that the radon problem in this environment has been adequately managed.

## References

[1] J. Vaupotič, et al., Systematic indoor radon and gamma measurements in kindergartens and play schools in Slovenia, Health Phys. 66 (1994) 550–556.
[2] J. Vaupotič, M. Šikovec, I. Kobal, Systematic indoor radon and gamma-ray measurements in Slovenian schools, Health Phys. 78 (2000) 559–562.
[3] A. Popit, J. Vaupotič, Indoor radon concentration in relation to geology in Slovenia, Environ. Geol. 42 (2002) 330–337.
[4] International Commission on Radiological Protection (ICRP), Protection against radon-222 at home and at work, ICRP Publication, vol. 65, Pergamon Press, Oxford, 1994, pp. 1–262.

www.ics-elsevier.com

# Blower door method and measurement technology in radon diagnosis

Aleš Froňka*, Ladislav Moučka

*National Radiation Protection Institute, SURO, Srobarova 48, Praha 10, 100 00, Prague, Czech Republic*

**Abstract.** The health risk and dose calculations from exposure to indoor radon concentration are based on long-term measurements and ICRP recommendations. The results of radon concentration measurements predicate more about human activities instead of the radon infiltration into the building. It means that we are not able to compare the results for different types of buildings and to characterize the radon supplies independently on indoor human activities. In our research, we strive to find out a set of methods not depending on residential habits and weather conditions, characterizing the building construction as a radon barrier. © 2004 Elsevier B.V. All rights reserved.

*Keywords:* Blower door; Radon; Ventilation; Infiltration; Pressure difference; Thermography

## 1. Introduction

For the purpose of identification and quantification of the soil radon sources and its infiltration pathways, the blower door technique [1] commonly used for energy loss studies in civil engineering was applied.

## 2. Method

The actual indoor radon concentration is a result of two competitive phenomena, the radon infiltration and ventilation. Moreover, it was proved, that at certain configuration of leakages, the radon infiltration is the increasing function of the ventilation.

The indoor–outdoor temperature difference causes the stack effect that results in outdoor–indoor pressure difference. Due to the fact that the infiltration of the radon is driven by the pressure difference $\Delta p$, the idea of the artificially produced pressure difference is

---

\* Corresponding author. Tel.: +42 241410211x173; fax: +42 241410212.
*E-mail address:* ales.fronka@suro.cz (A. Froňka).

0531-5131/ © 2004 Elsevier B.V. All rights reserved.
doi:10.1016/j.ics.2004.11.140

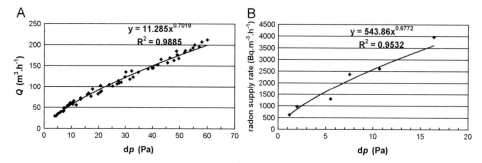

Fig. 1. (A) The blower door characteristic: the artificial ventilation rate as a function of a pressure difference; (B) the blower door test: the radon supply as a function of pressure difference.

obvious. For this reason, the technique of blower door combined with continual radon monitoring can be used.

The artificial ventilation rate $k_{BD}$ can be assessed from blower door pressure difference record. The radon supply rate can be calculated from the steady state of radon concentration or by the buildup curve analysis.

It was experimentally proved that the airflow rate $Q$ can be expressed as a power function $Q=c_{BD}(\Delta p)^{n_{BD}}$ (Fig. 1A), where $c_{BD}$ is the flow coefficient and $n_{BD}$ is a flow exponent. The $n_{BD}$ characterizes the laminar and turbulent airflow through leakages and openings. If the blower door, together with the appropriate continuous radon monitor, is used, the time behavior of the radon concentrations $a_V$ can be interpreted as a buildup curve in the model of the constant radon supply and the constant ventilation (no radon at $t=0$), $a_V=[\Phi/(\lambda+k_{BD})](1-e^{-(\lambda+k_{BD})t})$, where $\lambda$ is the radon decay constant. Analogically, we can assume for the air infiltration through the whole building envelope, that the soil air infiltration has similar characteristics and the radon supply rate can be expressed by $\Phi=c_{Rn}(\Delta p)^{n_{Rn}}$ (Fig. 1B), where the $c_{Rn}$ and $n_{Rn}$ are the soil airflow coefficient and the soil airflow exponent.

The ratio $T=\Phi/c_{soil}$, where $c_{soil}$ is soil radon concentration, may be considered to be a transfer factor quantifying the radon barrier for specific parts of the building. Other quantitative parameters, for instance, stationary concentration at certain pressure level $a_{V(\Delta p)}=\Phi/Q$, can be used for verification of the model presumptions. Several experiments were carried out to confirm the model applicability. The results indicate eligibility of this approach. However, a more extensive set of data is needed to suggest the most suitable quantitative parameters and proper reference levels as well. The unique experiment, blower door measurement combined with infrared thermography, was accomplished recently. The new approach is based on surface temperature distribution measurements within the area of leakages during the blower door test.

### Reference

[1] J. Fitzgerald, R. Nevitt, M. Blasnik, User-friendly pressure diagnostics, Home Energy Magazine (1994).

# Radon programme in the Czech Republic—experience and further research

Aleš Froňka*, Jiří Hůlka, Josef Thomas

*National Radiation Protection Institute, Prague, Czech Republic*

**Abstract.** Radon programme in the Czech republic is now incorporated in national legislation (Atomic Act, Radiation Protection Decree, Governmental Decision, etc.), and includes prevention and remedial measures, research and development. © 2004 Elsevier B.V. All rights reserved.

*Keywords:* Radon programme; Remedial; Activity index; Guidance level

## 1. Introduction

Radon programme in the Czech republic is now incorporated in national legislation (Atomic Act, Radiation Protection Decree, Governmental Decision, etc.), and includes prevention and remedial measures, research and development.

Preventive measures cover control of soil gas radon ingress, and radioactivity of building materials and of supplied water to avoid building of new houses that could exceed the indoor radon guidance levels (200 Bq/m$^3$ for indoor radon and 0.5 µGy/h for indoor gamma dose rate). The radon risk estimation of individual building site during siting of a new house and building protection according to soil radon risk category and standard technical building code are obligatory. Building materials producers are obligated to monitor natural radioactivity in produced building materials (activity index—including $K^{40}$, $Ra^{226}$ and $Th^{232}$—is used as screening level to control potential indoor gamma dose rate, limit content of $Ra^{226}$ is set to control indoor radon.) Similar regulatory system is used for public water supplies. The survey of effectiveness of preventive system was carried out. However, it was discovered that indoor radon level of 200 Bq/m$^3$ is exceeded

---

\* Corresponding author. SURO, Srobarova 48, Praha 10, 100 00, Czech Republic. Tel.: +42 241410211(173); fax: +42 241410212.

*E-mail address:* ales.fronka@suro.cz (A. Froňka).

0531-5131/ © 2004 Elsevier B.V. All rights reserved.
doi:10.1016/j.ics.2004.11.142

in some 20% of new houses. One of the reasons seems to be unexpectedly low air exchange rate in new houses. Further research of the building practice and radon concentration is under way.

The aim of governmental activities in remedial measures is to promote targeted indoor radon survey in existing buildings and to encourage owners to put into effect reasonable remedial measures (owners of family house can apply for governmental radon mitigation grants). Indoor radon survey is targeted in radon-prone areas with the help of detail geological radon prognosis maps. More than 130,000 family houses, schools and kindergartens were measured, and 20,000 of them were found to be above indoor intervention level of 400 $Bq/m^3$. Radon diagnostic methods were developed to map radon concentrations, to identify radon sources and to prepare radon mitigation project. A survey regarding long-term effectiveness of remedial measures has shown that 25% of them failed from long-term point of view. That is why it is important to monitor repeatedly the remediation effectiveness and radon expert attendance must be involved.

ELSEVIER

www.ics-elsevier.com

# Effective dose calculation using radon daughters and aerosol particles measurement in Bozkov Dolomite Cave

L. Thinova[a,*], Z. Berka[a], E. Brandejsova[a], V. Zdimal[b], D. Milka[c]

[a]*CTU, Faculty of Nuclear Sciences and Physical Engineering, Praha, Czech Republic*
[b]*Institute of Chemical Process Fundamentals, AS CR, Praha, Czech Republic*
[c]*Bozkov Dolomite Cave Direction, Bozkov, Czech Republic*

**Abstract.** Conservative methodology for estimation of a potential dose in caves employs solid state alpha track detectors. Obtained data are converted into annual effective dose in agreement with the ICRP recommendations (using the "caves factor"). The more precisely determined dose value would have a significant impact on the radon remedies or on restricting the time guides spent in the underground. To define the calculation of the effective dose in caves, the following measurements were carried out: continual radon measurement (differences in ERC between working hours and nighttime and daily and seasonal radon concentration variations); regular radon and daughters measurements; regular indoor air flow measurements to study the location of radon supply and its transfer among individual areas of the cave; natural radioactive elements content evaluation in subsoils and in water inside/outside to study radon sources in the cave; aerosol particle-size spectrum measurements for determination of free fraction (cave–apartment comparison); guides/visitors behaviour monitoring to record time spent in cave and to assign it to the continuously monitored levels of Rn concentration.
© 2004 Elsevier B.V. All rights reserved.

*Keywords:* Radon concentration; Cave; Effective dose; Aerosol campaign

## 1. Introduction

Carst caves fall among regions with exceptionally high radon concentrations, (despite the very low uranium content in limestone) that are caused by minimal airflow and negligible air exchange. The conversion factor is dependent on aerosol particle-size spectrum resolution and on distribution of radon daughters among the attached and unattached fraction [1]. With the aim to particularize the calculation of the effective dose in caves, the following

---

* Corresponding author. Tel.: +420 224 358 235; fax: +420 224 811 074.
  *E-mail address:* thinova@fjfi.cvut.cz (L. Thinova).

0531-5131/ © 2004 Elsevier B.V. All rights reserved.
doi:10.1016/j.ics.2004.11.114

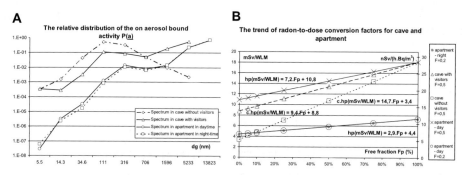

Fig. 1. (A) The results from aerosol campaign (5 days). (B) Comparison of cave with apartment environment is shown.

measurements were carried out: continual radon measurements; regular radon and daughters measurements using a sampling procedure to specify proportion of radon daughters; regular indoor air flow measurements to study the location of radon supply and its transfer among individual areas of the cave; natural radioactive elements content evaluation in subsoils and in water inside/outside to study radon sources in the cave; 5-day aerosol particle-size spectrum measurements for determination of free fraction and for comparison of aerosol spectra in apartment and cave; guides/visitors behaviour monitoring to record time spent in cave and to assign it to the continuously monitored levels of Rn concentration. All obtained data were used for the effective dose calculation.

## 2. Effective dose calculation

The entire effective dose a person receives from radon daughters was calculated as the sum of all effective doses obtained from individual sizes of aerosols (including the unattached fraction) using the following assumptions: in caves, only $^{222}$Rn and its daughters occur; the mutual ratio of radon daughters and equilibrium factor is constant; the spectrum of aerosols is constant [2] (Fig. 1).

## 3. Conclusions

Occurrence of aerosols particles with diameter 1–10 μm is caused by visitors or personnel presence. Concentration of aerosols with diameter approximately 100–200 nm seems to be stable; larger-size aerosols are produced by visitors only. Small aerosols are produced by an intensive work or movement. Time spent inside by guides must be related to the actual radon concentration at given time. Personal dosimeters are advised. Using the factors described above, the effective dose was recalculated. The resulting values are fully comparable with the "conservative estimation" values. Therefore, we have established that the methodology commonly used for dose estimation is correct. So far, we were unable to determine the ratio of attached and unattached fraction that could significantly influence the calculations.

## References

[1] ICRP, Protection against radon-222 at home and at work, ICRP Publication, vol. 65, Pergamon Press, Oxford, UK, 1993.
[2] W.C. Hihds, Aerosol Technolgy, Properties, Behavior, and Measurment of Aerosol Particles, Second edition, John Witney and Sons, New York, 1998.

# Measurement of radon daughters in water and in air using the detection unit "YAPMARE" with a YAP:Ce scintillation detector

L. Thinova[a,]*, A. Kunka[b], P. Maly[b], F. de Notaristefani[c], K. Blazek[b], T. Trojek[a], L. Moucka[d]

[a]*CTU, Faculty of Nuclear Sciences and Physical Engineering, Praha, Czech Republic*
[b]*Crytur Ltd., Turnov, Czech Republic*
[c]*INFN Sezione di Roma, Rome, Italy*
[d]*National Radiation Protection Institute, Praha, Czech Republic*

**Abstract.** With a focus on the detection of the Rn concentration in water in extreme conditions, the detection unit YAPMARE was developed by the company CRYTUR. The main part of the detection unit is a detection probe based on the Ø25×100 mm YAP:Ce detector. The measured water covers approximately 95% of the crystal surface (the detection volume is 12 ml). Measurements of $^{222}$Rn and $^{220}$Rn standards for determination of individual peaks in the spectra were conducted, including an observation of the $^{222}$Rn (and its daughters) behavior. For YAPMARE calibration, fresh water was taken from a drilled well in the Lounovice area (biotitic granodiorite–adamellite). The calibration was carried out with the assistance of radon in water monitor RADIM 4. The results of the water measurements in Lounovice were as follows: detection limit was about 50 Bq/l, and calibration coefficient was 0.0011 imp/s per 1 Bq/l. The results of the fresh water and air sample measurements have demonstrated the main *advantages* (spectrometric results of the measurement, possibility for water or air samples evaluation) and *disadvantages* (large size of scintillation crystal increases gamma ray background, calibration for each water with particular composition, and detector cleaning using HCl acid). © 2004 Elsevier B.V. All rights reserved.

*Keywords:* Radon; Water; YAP:Ce; Alpha spectrometry

## 1. Introduction

With a focus on earthquake forecasting by the detection of the Rn concentration in water, the detection unit YAPMARE was developed by the company CRYTUR. The main part of the unit is a detection probe based on the Ø25×100 mm YAP:Ce detector described by Kunka et al. [1,2]. Until May 2004, the following tests were carried out: (i) gamma and

---

\* Corresponding author. Department of Dosimetry and Application of Ionising Radiation, Czech Technical University, Brehova 7, Praha 115 19, Czech Republic. Tel.: +42 224 358 235; fax: +42 224 811 074.
 *E-mail address:* thinova@fjfi.cvut.cz (L. Thinova).

0531-5131/ © 2004 Elsevier B.V. All rights reserved.
doi:10.1016/j.ics.2004.11.083

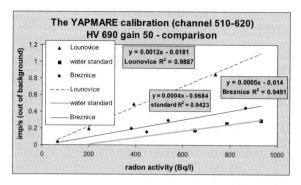

Fig. 1. The results of Rn in water measurement using the detection unit YAPMARE.

alpha energy calibration using radionuclide etalons; (ii) identification of spectrum peaks using $^{222}$Rn and $^{220}$Rn standards; (iii) a long-time monitoring for stability and background elimination; (iv) completion of water sampling methodology; (v) Rn measurements in Jachymov mines; (vi) measurements of fresh and standard Rn water; and (vii) detection-unit calibration using RADIM 4 (a monitor for water radon samples).

## 2. Materials and methods

For radon activity measurement, fresh water from a drilled well in Lounovice area (adamellite) was taken. The Rn concentration of water samples were 700–900 Bq/l. The detection unit was used in the stationary regime. The measurement of radon in water activity was started after establishment of equilibrium between radon daughters (3 h). The time measurement interval was 2 h. After each measurement, the detection unit was cleaned with HCl acid for 20 min, rinsed with water, and after that, background and energy stability were checked. The calibration of the detection unit YAPMARE was carried out with the assistance of radon in water monitor RADIM 4.

## 3. Results and conclusions

The results of the first measurements of water and air samples demonstrated the following main advantages: linearity in dependence towards gamma ray and alpha ray, availability of spectrometric results, and possibility for water or air sample evaluation. On the other hand, disadvantages were as follows: a large-size scintillation crystal increases gamma ray background in the area of $^{214}$Po, necessity of recalibration for alpha particle measurements using a new type of water, thoron influence in highly active fresh water samples, and necessity of detector cleaning by HCl. A thin plate YAP:Ce version was tested for alpha particle detection.

The results of water measurements in Lounovice were as follows: pH was 6.7, detection limit was about 50 Bq/l, and calibration coefficient was 0.0011 imp/s per 1 Bq/l (Fig. 1). The calibration coefficient and detection limit were dependent on the chemical composition of the water.

## References

[1] A. Kunka, et al., New YAP:Ce scintillation detection for determination of 222Rn in water, Radiation Measurements 38/4-6 (2004) 829–832.
[2] A. Kunka, et al., Measurement of Rn and other gamma and alpha radionuclides in water with the help of a new detection unit based on scintillation single crystal YAP:Ce (to be published in Elsevier Science "Nuclear Instruments and Measurements").

# A radon survey in some regions of Turkey

Yasemin Yarar[a,*], Tuğba Günaydı[a], Erol Kam[b]

[a]Yıldız Technical University, Physics Department, Davutpaşa Campus, 34010, Topkapı, İstanbul, Turkey
[b]Çekmece Nuclear Research And Training Centre, İstanbul, Turkey

**Abstract.** This paper presents the experimental data of a radon survey conducted in Dikili Geothermal area and Tekirdağ city, located in the west and the northwest parts of Turkey. A time-integrating passive method was applied by installing a number of CR-39 nuclear track detectors in dwellings for 3 months. Then, applying chemical etching to the exposed detectors, nuclear track numbers and corresponding indoor radon concentrations were determined. Radon concentrations of the water resources drawn from wells, municipal supplies, village fountains, and spas were also determined by using a liquid scintillation detector. Soil samples were analyzed by using a gamma-ray spectrometer connected with an HPGe detector and the natural radionuclide concentrations were obtained. Radon concentrations of the soil samples were calculated from the secular equilibrium between $^{226}$Ra–$^{222}$Rn. Risk probabilities of indoor radon inhalation were calculated.
© 2004 Elsevier B.V. All rights reserved.

*Keywords:* Indoor radon; CR-39; Dose; Cancer risk; Geothermal

## 1. Introduction

One of the important geothermal areas of Turkey is found in İzmir–Dikili. There is an active fault line in Kaynarca as indicated by the presence of the natural hot water resources in that region. These are Bademli thermal, Nebiler spas, Kocaoba thermal, and Kaynarca mud. Most of the activity found in geothermal fluids is due to the uranium decay chain. Radon, a decaying product of uranium chain, is soluble in water and is released into the atmosphere when the water or stream contacts the air. Radon and its short-lived decay products in the atmosphere are the most important contributors to human exposure from natural sources and the second largest cause of lung cancer worldwide. Therefore, radon surveys in dwellings have been made in many countries during the last decade all over the world [1]. Uranium and phosphate deposits, volcanic areas, and spas are known as having high radon concentrations.

* Corresponding author. Tel.: +90 212 449 18 60; fax: +90 212 449 15 14.
*E-mail address:* yarar@yildiz.edu.tr (Y. Yarar).

0531-5131/ © 2004 Elsevier B.V. All rights reserved.
doi:10.1016/j.ics.2004.11.011

Table 1
Annual dose equivalents and lung cancer risk probability for a lifetime from indoor radon inhalation

| Location | Average concentration (Bq/m$^3$) | Annual effective dose (mSv/year) | Probability |
|---|---|---|---|
| Bademli Village | 126±32 | 2.91 | $2.12\times10^{-4}$ |
| Dikili Town | 73±30 | 1.69 | $1.23\times10^{-4}$ |
| Kocaoba Village | 124±20 | 2.87 | $2.09\times10^{-4}$ |
| Mazılı Village | 120±23 | 2.78 | $2.02\times10^{-4}$ |
| Nebiler Village | 134±54 | 3.10 | $2.25\times10^{-4}$ |
| Tekirdağ City | 87±40 | 2.01 | $1.46\times10^{-4}$ |

Table 2
$^{226}$Ra and $^{222}$Rn concentrations of soil samples in the monitoring areas

| Location | Concentrations (Bq/kg) | |
|---|---|---|
| | Ra-226 | Rn-222 |
| Dikili Center | 82±4.59 | 81.64 |
| Bademli Village | 78±1.87 | 77.66 |
| Kaynarca area | 64±2.81 | 63.72 |
| Kocaoba Village | 81±3.72 | 80.65 |
| Nebiler Village | 287±10.33 | 285.75 |
| Tekirdağ City | 36±21 | 36.12 |

## 2. Indoor radon determination

Indoor radon monitoring was performed at 5 locations in Dikili Geothermal area and Tekirdağ city. A passive, time-integrating measuring method was performed by using CR-39 nuclear track detectors. A couple of CR-39 detector was installed in the houses for 3 months, and then chemical etching was applied to the collected detectors to make the alpha tracks on the exposed films visible. Exposed films were counted and indoor radon concentrations corresponding to net track numbers were determined. The average values for Dikili geothermal area and Tekirdağ city together with their equivalent doses and corresponding lung cancer risks caused by indoor radon inhalation, calculated according to the model given in UNCEAR 1993 and ICRP 60 taking the occupancy factor as 0.7, were given in Table 1.

## 3. Radon in water and soil samples

Radon concentrations of water samples collected from wells, municipal supplies, village fountains and spas in Dikili geothermal area were analyzed by using a liquid scintillation detector. The obtained values ranging from 29 to 3075 Bq/m$^3$ are lower than the maximum contaminant level of 11 Bq/l for radon in drinking water proposed by EPA [2]. One can find more detail about them in Ref. [3]. $^{226}$Ra concentrations of soil samples were determined by using a gamma spectrometry device connected with an HPGe detector. Using the radioactive decay equations derived from the secular equilibrium between $^{226}$Ra and $^{222}$Rn, radon concentrations of soil samples were calculated and the results were given in Table 2.

## References

[1] United Nations Scientific Committee on the Effects of Atomic Radiation, in: Sources and Effects of Ionizing Radiation: UNSCEAR 2000 Report to the General Assembly, with Scientific Annexes, vol. I, United Nations, New York, 2000, (sources).
[2] Committee on Risk Assessment of Exposure to Radon in Drinking Water, Risk Assessment of Radon in Drinking Water, National Academy Press, Washington, DC, 1999.
[3] T. Günaydı, Master of Science thesis: Determination of radon concentrations of İzmir–Dikili geothermal area. Istanbul: Yıldız Technical University, Science Institute; 2004.

www.ics-elsevier.com

# Environmental radioactivity concentrations of Tekirdağ

## Yasemin Yarar[a,*], Erol Kam[b]

[a]Yıldız Technical University, Physics Department, Davutpaşa Campus, 34010 Topkapı, İstanbul, Turkey
[b]Çekmece Nuclear Research and Training Centre, İstanbul, Turkey

**Abstract.** This paper presents the environmental radioactivity concentrations in Tekirdağ, a city in the region of Marmara in Turkey. Indoor and outdoor γ-exposure rates were measured by Eberline smart portable counter, ESP-2. Gross α- and β-counting of the water samples taken from municipal supplies, wells, and fountains were performed by using a gas-flow proportional counter, Berthold LB770-PC 10. Soil samples, collected from 40 different points in the area, were analyzed by using a γ-ray spectrometry system connected with an HPGe detector and the radioactivity concentrations of the radio nuclides of uranium and thorium series $^{40}$K and $^{137}$Cs were determined. Dose and risk calculations were made. © 2004 Elsevier B.V. All rights reserved.

*Keywords:* Environmental radioactivity; Gross alpha; Gross beta; Gamma exposure

## 1. Introduction

Naturally occurring radionuclides of terrestrial origin, also called primordial radionuclides, are present in the environment. Among the primordial radionuclides, the main contributors to external exposure are $^{40}$K and the radioactive series headed by $^{238}$U and $^{232}$Th. Relevant to the outdoor exposure. In recent years, several surveys have been performed all over the world for estimating the exposure of the populations to natural radiation [1]. This paper is also an example of such a survey conducted in Tekirdağ, a city located in the northwest of Turkey, next to Marmara Sea.

* Corresponding author. Tel.: +90 212 449 18 60; fax:+90 212 449 15 14.
*E-mail address:* yarar@yildiz.edu.tr (Y. Yarar).

0531-5131/ © 2004 Elsevier B.V. All rights reserved.
doi:10.1016/j.ics.2004.11.047

## 2. Gamma exposure dose rate measurements

Gamma exposure dose rates were measured by Eberline smart portable device, ESP-2, connected with a SPA-6 model plastic scintillation detector. Outdoor γ-dose rates in air are ranging from 30.3 to 54.3 nGy h$^{-1}$ with an arithmetical mean of 43±4 nGy h$^{-1}$. Indoor gamma doses in dwellings located in Tekirdağ city center are ranging from 30.5 to 68.8 nGy h$^{-1}$, with an arithmetic mean of 50±9 nGy h$^{-1}$.

Annual effective doses caused by γ-exposure were calculated according to the model given in UNSCEAR 1993, taking the occupancy factor of 0.7 for indoor and 0.3 for outdoor exposure. Total annual effective dose was 0.29 mSv. Taking $5 \times 10^{-2}$ Sv$^{-1}$, the fatal cancer risk for public given in ICRP 60, individual cancer risk value was obtained as $1.45 \times 10^{-5}$ deaths per year.

## 3. Radioactivity measurements in water and soil samples

Gross α- and β-counting of water samples from the region were performed by using a proportional counter with gas-flow (Berthold, LB770-PC 10). Soil samples were analyzed by using a γ-spectrometry device connected with an HPGe detector.

## 4. Results and discussion

The population-weighted averages of national surveys are given as 59 nGy h$^{-1}$ with a range from 18 to 93 nGy h$^{-1}$ for outdoor γ-exposure and as 84 nGy h$^{-1}$ with a range from 20 to 200 nGy h$^{-1}$ for indoor gamma exposure [1]. Exposure dose rates of Tekirdağ are lower than the worldwide averages. Annual equivalent doses and corresponding cancer risk values are also low.

WHO advises 0.1 Bq l$^{-1}$ for gross α- and 1.0 Bq l$^{-1}$ for gross β-activity as limit values for drinking water. Gross β-activities of all water samples are seriously under the reference value of 1.0 Bq l$^{-1}$. Gross α-activities are also significantly lower than the limit, except for Kayý, Ulaş, and Çorlu, where the activities are slightly higher than the reference. More detail: [2].

The average radioactivity concentrations of $^{226}$Ra, $^{214}$Pb, $^{214}$Bi, $^{208}$Tl, and $^{228}$Ac in Tekirdağ soil are 36.29, 26.07, 23.62, 40.19, and 37.47 Bq kg$^{-1}$, respectively. Average $^{40}$K concentration was found as 578.52 Bq kg$^{-1}$ with a range from 195 to 1466 Bq kg$^{-1}$. It is greater than the worldwide median value of 400 Bq kg$^{-1}$, ranging from 140 to 850 Bq kg$^{-1}$.

$^{137}$C is a fission product and released in a nuclear explosion or a reactor accident. The northwest of Turkey was also contaminated because of Chernobyl Reactor accident. In July 1986, $^{137}$Cs concentrations of soil of Edirne and Saray, Tekirdağ were measured as 324 and 45 Bq kg$^{-1}$, respectively [3]. In this study, average $^{137}$Cs concentration was measured as 5.17 Bq kg$^{-1}$ with, ranging from 0.11 to 30 Bq kg$^{-1}$. Sağlamtaş, Çerkezköy, Naip, and Saray are the region most highly contaminated by $^{137}$Cs. Present value of Saray measured here is 14.88 Bq kg$^{-1}$.

## References

[1] UNSCEAR 2000 report to the general assembly, with scientific annexes, New York, UN, vol. I: sources, 2000.
[2] E. Kam, Master of Science Thesis: determination of the Natural Environmental Radioactivity of Tekirdağ. Istanbul: Yıldız Technical University, Science Institute; 2004.
[3] Turkish Atomic Energy Commission, Radiation and Radioactivity Measurements in Turkey After Chernobyl, TAEK Publishing, Ankara, 1988.

www.ics-elsevier.com

# Radiological impact in an area of elevated natural radioactivity background: the case of the island of Ikaria–Aegean Sea, Greece

## Trabidou Georgia*, Florou Heleny

*ERL/INT-RP/NCSR "Demokritos", Athens, Greece*

---

**Abstract.** The levels of natural radionuclides in metallic spring water (either for spa and household use), potable water (local domestic network), soil and rock samples were measured in the island of Ikaria in the Aegean Sea, Greece. The concentrations of $^{222}$Rn and natural γ emitters were found to be significantly elevated in spring water and some soil and rock samples. The external and internal dose rates (μSv year$^{-1}$) were estimated in three critical groups according to the water use. The maximum external dose rates were found to be at the upper limits of the levels reported in the literature. Moreover, high dose rates were estimated for the workers in spa installations. © 2004 Elsevier B.V. All rights reserved.

*Keywords:* Natural radioactivity; Radon spas; Dose rate

---

## 1. Introduction

The island of Ikaria is located in the Eastern Aegean Sea, Greece. In the littoral zone around the island, there are several metallic springs and, in the sub-littoral zone, some springs bubble up from the bottom as well. They are divided into thermal bath spas (45–65 °C water temperatures) and potable spring water (20 °C water temperature).

## 2. Materials and methods

To evaluate the environmental radiation levels in the island of Ikaria, samples of soil, rock, spring water and domestic water were appropriately collected and analyzed by gamma spectrometry at the Laboratory and the activity concentrations of the collected samples of the natural radionuclides $^{238}$U, $^{226}$Ra, $^{232}$Th, $^{228}$Ra, $^{228}$Th, $^{222}$Rn and $^{40}$K were determined

---

\* Corresponding author. Tel.: +30 210 6503812; fax: +30 210 6503050.
*E-mail address:* johncats11@netonline.gr (T. Georgia).

0531-5131/ © 2004 Elsevier B.V. All rights reserved.
doi:10.1016/j.ics.2004.11.122

Table 1
Summarised results of dose rates to habitants, thermal spa workers and users of spa in the Ikaria (µSv year$^{-1}$)

| Habitants | µSv year$^{-1}$ |
|---|---|
| Dose equivalent rate due to terrestrial γ-radiation | 200–3310 |
| Effective dose equivalent due to the $^{222}$Rn intake from potable water | 30–114 |
| Effective dose equivalent due to the $^{226}$Ra intake from potable water | 25–175 |
| Effective dose equivalent due to the inhalation of $^{222}$Rn released from potable water | 0.36–85 |
| *Thermal spa workers* | |
| Effective dose equivalent due to the inhalation of $^{222}$Rn released from spa water | 5000–35,000 |
| *Spa users* | |
| Dose rate due to γ-radiation during the immersion into bath water | 0.0012–0.012 |
| Dose rate due to $^{222}$Rn radiation during the immersion into bath water | 240–1700 |
| Effective dose equivalent due to the inhalation of $^{222}$Rn released from spa water | 40–290 |

[1]. The above samples were analysed by using a high-resolution gamma spectrometry system with a HpGe detector of 20% relative efficiency to a 3″ × 3″ NaI detector.

## 3. Results and discussion

Elevated levels of natural radionuclides were detected in some of the measured samples [1] in comparison to the respective background levels in Greece [2]. In terms of pathway exposure to humans, external and internal dose rates calculations were carried out [1]. The fluctuations of external and internal dose rates for the habitants, workers and spring water users were very high (Table 1). The existence of high concentrations of natural radionuclides in the abiotic environment of the island of Ikaria resulted in some cases to overdoses in relation to the typical background (2500 µSv year$^{-1}$). In comparison to the wide Greek territory (100–1000 µSv year$^{-1}$) [2] and the reported literature, the dose equivalent rates due to terrestrial γ-radiation determined in the island of Ikaria (200–3310 µSv year$^{-1}$) were found to range in the upper limit of the value spectrum. More specific results and discussions are given below.

### 3.1. The springs as drinking water supplies for wide consumption

Considering a daily use, the impact due to the consumption of potable water of the spring named locally "Immortal Water" was high. The $^{226}$Ra ingestion resulted to the high dose rates (25–175 µSv year$^{-1}$), with the maxima exceeding the recommended limit of 100 µSv year$^{-1}$.

### 3.2. The springs used as spa

The impact of workers in spa installations was due to the waterborne $^{222}$Rn with maximum dose rate up to 35,000 µSv year$^{-1}$, which led to overexposure in terms of the 20,000 µSv year$^{-1}$ professional limit.

The impact of the spa users was due mainly to $^{222}$Rn exposure during the immersion into bath water, with a dose rate of 20–1700 µSv year$^{-1}$.

## References

[1] G. Trabidou, Radiological research study in some areas characterised by the presence of radioactive springs in Greece. PhD thesis, Dpt of Physics, Kapodistrian University of Athnes, (2004) (In Greek).
[2] M. Probonas, P. Kritidis, The exposure of the Greek population to natural gamma radiation of terrestrial origin, Radiat. Prot. Dosim. 46 (2) (1993) 123–126.

ELSEVIER

www.ics-elsevier.com

# Beyond track etch monitoring—the gap between passive and active devices has been bridged!

T. Streil*, S. Feige, V. Oeser

*SARAD GmbH, Wiesbadener Str. 20, D-01159 Dresden, Germany*

**Abstract.** Nowadays, miniaturized microprocessors enable the design of most power saving instruments with the entire performance of state-of-the-art active devices in the size of a cell phone: a new generation of instruments equipped with nuclide separation within an HV-measurement chamber, internal circular buffer, temperature and humidity sensor will be introduced to the public for the very first time. Like a track detector, this instrument is easy to handle, and a single button operation starts a measurement, indicating that the device might be operated without any additional knowledge and training. Efficiency of the detection is some magnitudes higher than track etch detectors and enables the detection of 100 Bq/m$^3$ within 3 h. Hardware characteristics such as exchangeable AA-size batteries with a standby time of more than 1 year housed in robust case completes the bodywork of a most versatile device that is definitely more than mandatory personal protective equipment. Some "killing" arguments will be presented as well as impressive examples from different fields of application, indoor and outdoor and short and long-term measurements. © 2004 Elsevier B.V. All rights reserved.

*Keywords:* Radon; Measuring technique; Time series; Cost effective; Small portable autonomous device

## 1. Introduction

A century after its discovery, radon monitoring by use of track etch detectors has definitely come out of age. It even shows some signs of aging (like a track film usually does). Claiming passive devices to be the one and only cost-effective and easy to handle method for screening measurements of radon is no longer the truth. In fact, it never had been when one takes into consideration the enormous laboratory work that is required to achieve the result! Compared to its disadvantages it definitely never had been a suitable choice for radon risk mapping at home and at workplaces but an accepted interim solution: beside its saturation exposition that will give a limit for the maximum concentration far

---

* Corresponding author. Tel.: +49 351 6580 7 12; fax: +49 351 6580 7 18.
  *E-mail address:* info@sarad.de (T. Streil).

0531-5131/ © 2004 Elsevier B.V. All rights reserved.
doi:10.1016/j.ics.2004.12.022

below the observed radon concentration at several workplaces like water facilities, the influence of humidity as well as problems to deal with decay products of both Radon-222 and Radon-220 are thought-provoking. Over the years, track etch detectors are known to give just the average value of the activity concentration and consequently seem to be responsible for the lack of any information regarding the time variation of radon and dose within the scope of thousands of measurements in the past. Why trust an average value of a track etch measurements but do not make use of the additional and valuable information that will be gained by an active device? The easiest way to lower the exposure is not to be exposed—you just need to know its variation by time! A reliable dose assessment shall neither over- nor underestimates the dose, shouldn't it?

## 2. Technical principals and system design

Nowadays, miniaturized microprocessors enable the design of most power saving instruments with the entire performance of state-of-the-art active devices in the size of a cell phone: HV-measurement chamber, internal circular buffer, temperature, humidity sensor and allocation sensor. The device is designed similar to a standard radon monitor. The gas diffuses through a membrane into the measurement chamber. Radon decay products inside the chamber, ionised after decay, is collected at the detector surface with the help of the electric field. The sensitivity of the device was determined to be 1 count or $2/(min/kBq/m^3)$. The statistical error for each 1-h value at 200 $Bq/m^3$ is ±20%. A microprocessor core, including a non-volatile memory and a real time clock, operates the device. The standard serial interface handles the data communication with a personal computer. The power supply and the high voltage generator with battery management circuit complete the system. The device is easy to handle and comfortable to wear during the work time.

## 3. Experimental results

The influence of the humidity is shown in Fig. 1A. Best performance of the device achieved by a charge collection voltage of 28 V as shown in this figure. The humidity influence using the correction algorithm is lower than ±10% in the full measuring range. The device works with 2 AA batteries or accumulators more than 3 months autonomously. Time resolutions for 3-month and 1-month operations were 3 h and 1 h, respectively. Fig. 1B shows a measurement in a house with low radon concentration.

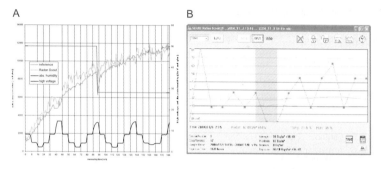

Fig. 1. (A) Humidity influence of the Radon Scout and the influence of the chamber high voltage. (B) Time series in a low radon level house the grey area shows the disturbtion by movement.

# MyRIAM: an active electronic online personal inhalation dose meter for detection of α- and β-radiation on aerosols

T. Streil*, V. Oeser

*SARAD GmbH, Wiesbadener Str. 20, D-01159 Dresden, Germany*

**Abstract.** MyRIAM is the acronym for My Radioactivity In Air Monitor, designed for personal use to detect any radioactivity in the air at the place and at the moment of the danger. The active air sampling process enables a detection limit several orders of magnitude below that of Gamma detectors. Therefore, it is the unique way to detect dangerous exposures in time. Individual protection against inhalation of long living radioactive dust (LLRD) saves human life and health. LLRD may occur in natural environment as well as in case of nuclear accident or military and terrorist attacks. But in any case, the immediate warning to the population has an important public significance. Keep in mind: it is very easy to avoid LLRD inhalation—but you have to recognise the imminent danger. The second requirement of gap-less documentation and reliable assessment of any derived LLRD exposure is building the bridge to dosimetry applications. The paper demonstrates the possibility to design small and low cost air samplers, which can be used as personal alarm dosimeters and fulfill the requirements mentioned above. Several test measurements taken by a mobilephone-size MyRIAM shall be used to demonstrate the correctness of this statement. © 2004 Elsevier B.V. All rights reserved.

*Keywords:* Inhalation dose; Long-lived α-nuclides; β-radiation; On line aerosol sampler; DU-munitions

## 1. Introduction

The increasing terrorist activity focuses on the problem of using nuclear or radioactive material by terrorists. However, the best-performed security system cannot ensure 100% protection against the multiple possibilities of terror offences. We believe

---

\* Corresponding author. Tel.: +49 351 6580 7 12; fax: +49 351 6580 7 18.
*E-mail address:* info@sarad.de (T. Streil).

0531-5131/ © 2004 Elsevier B.V. All rights reserved.
doi:10.1016/j.ics.2004.12.021

that one of the most likely scenarios is the injection of radioactive material into the ambient air by dirty bombs or simple aerosol generators. Furthermore, besides plutonium and Highly Enriched Uranium (HEU), the extremely radiotoxic and easy to acquire isotopes, radium and thorium and other radioactive isotopes from medical applications or from the radioactive waste like Sr-90, I-131, Cs-13, Co 60, etc., could also be used. Further, the inhalation dose can be determined for the public and also for the military task forces in areas where Depleted Uranium munitions were used or are in use. On the other hand it is very easy to reduce inhalation by breathing through pieces of cloth like handkerchiefs or others. The major recommendation is to detect the danger and warn the exposed people immediately. Such persons could be members of fire brigades and military task forces and police officers as well as bodyguards of politicians, or staff of public utilities.

## 2. Technical principals and system design

MyRIAM containing an internal pump with a continuous air flow of 0.25 l/min samples the nuclides on a Millipore filter with excellent spectroscopic resolution (see schema 1 and picture 1). A 1.5 $cm^2$ light protected ion-implanted silicon detector analyses the α- and β-radiation at the filter. This small detector head also contains the pre-amplification and pulse processing functions. The α- and β-radiation of the radon progeny and the long-lived α- and β-nuclides are analysed by a 60-channel spectrometer. The energy resolution of the filter spectra online analysed is in the order of 150 keV. We determine the concentration of long-lived α-nuclides, correcting the tailing effects of the radon progeny on measurements of those nuclides with use of a special algorithm. Also corrected was the influence of β-radiation from natural radon/thoron progeny on the artificial nuclides. Because of the air sampling volume of nearly 15 l/h, the system has a high efficiency. The detection limit by 2-h sampling time is an α-nuclide concentration of 0.05 $Bq/m^3$. As required, the electronic dosimeter is easy to handle and rugged enough to withstand the rough working conditions. The smallest adjustable integration time is 1 min. The following data acquired by measurement will be stored to the internal memory (capacity 300 cycles): 60 channel sum spectra, time distribution of 5 regions of interest (ROI), β-channel, α-exposure, β-exposure and average values. Free ROI set up is possible. The β-channel is fixed between 120 and 3000 keV. The system is able to warn with a time resolution of 1 min in the order of 10% of the yearly radiation limit.

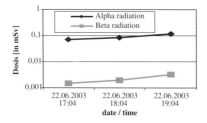

Fig. 1. Inhalation dose in a radium/uranium inhalation aerosol atmosphere.

## 3. Experimental results

Fig. 1 shows the inhalation dose in an artificially produced atmosphere with radium/uranium aerosols. The MyRIAM is continuously sampling and therefore the LLRD dose is increasing. MyRIAM will fit the requirements for personal dosimetry as well as local dosimetry or long term monitoring [1].

## Reference

[1] Proceedings of the International Conference on Physical Protection "Strengthening Global Practices for Protecting Nuclear Material", 8–13 September 2002, Salzburg, Austria, 2002, http://www.numat.at.

# Estimation of the radon dose in buildings by measuring the exhalation rate from building materials

V. Steiner[a,*], K. Kovler[b], A. Perevalov[b], H. Kelm[c]

[a] Division of Radiation and Noise, Ministry of the Environment, Jerusalem, Israel
[b] National Building Research Institute, Technion Institute of Technology, Israel
[c] Tracerlab Co., Köln, Germany

**Abstract.** The indoor radon radiation dose in buildings may be estimated by measuring the radon exhalation rate from the walls. We show that the classical accumulator method produces the same results with either passive or active radon detectors. © 2004 Elsevier B.V. All rights reserved.

*Keywords:* Radon; Exhalation; Building material; Radiation dose; Detectors; Radon sampler; Charcoal; Electret

## 1. Introduction

The radiation dose to the population, due to gamma and radon radiation from building materials, is generally not regulated. In Israel, the low NORM content in soil and building materials allows limiting the radiation dose in new buildings to less than 0.45 mSv/y [1]. The radon dose in a room with $^{222}$Rn concentration $C_{Rn}$, surface $S_R$, volume $V_R$, ventilation rate $\lambda_v$, wall density $\rho_w$, wall thickness $l$ and $^{226}$Ra concentration $C_{Ra}$ is:

$$D_{Rn} = kC_{Rn} = k\frac{1}{2}\frac{\lambda}{\lambda+\lambda_v}e_w\frac{C_{Ra}\rho_w S_R l}{V_R} \quad (1)$$

with $\lambda$ the $^{222}$Rn decay constant, $k$=0.017 mSv/y/(Bq/m$^3$) [2] and $e_w$ wall emanation, related to the exhalation rate as $E_w$=(1/2)$\lambda C_{Ra}\rho_w e_w l$. After closing a material in an accumulator [3], the radon concentration evolves as:

$$C(t) = C_0\frac{\lambda_v}{\lambda+\lambda_v}\left[1 + \frac{\lambda}{\lambda_v}\exp(-(\lambda+\lambda_v)t)\right] + C_{eq}[1 - \exp(-(\lambda+\lambda_v)t)] \quad (2)$$

with $C_0$ initial concentration, $C_{eq}$=$(ES)/(\lambda+\lambda_v)V$ equilibrium concentration, $S$ surface of the material, $V$ and $\lambda_v$ the net volume and ventilation rate of the accumulator. These parameters may be chosen to ensure a low concentration and practically a time-

---

\* Corresponding author. Postal address: Str. Knafei Nesharim 5, POB 34033, Jerusalem 95464, Israel. Tel.: +972 3 6420822; fax: +972 3 6426268.
 *E-mail address:* victors@sviva.gov.il (V. Steiner).

0531-5131/ © 2004 Elsevier B.V. All rights reserved.
doi:10.1016/j.ics.2004.11.024

Table 1
Measured exhalation rate (mBq/m² s) using the electret (EL), activated charcoal (AC), Rad7 and ERS2 detectors

| Sample | Size (cm) | EL | AC3 | Rad7 | ERS2 |
|---|---|---|---|---|---|
| Concrete block | 10, 10, 10 | 0.43 | 0.29 | 0.49 | – |
| | | 0.47 | 0.71 | 0.58 | – |
| | | 0.39 | 0.37 | 0.30 | – |
| | | 0.47 | 0.71 | 0.53 | – |
| | 12, 20, 20 | 0.37 | 0.32 | 0.31 | 0.45 |
| | | 0.77 | 0.52 | 0.66 | 0.65 |
| Pumice block | 58, 20, 21 | 0.45 | 0.52 | 0.42 | 0.38 |
| Granite tile | 2.8, 30, 28 | 0.14 | 0.30 | 0.11 | 0.25 |
| | 2.5, 24, 26 | 0.80 | – | 0.64 | 1.10 |
| Pumice wall | 20, 50, 50 | 0.68 | – | 0.52 | – |
| | | 0.83 | – | 0.82 | – |
| Concrete wall | 10, 50, 50 | 0.55 | 0.79 | 0.47 | 0.85 |
| | 10, 120, 120 | 0.41 | 0.75 | 0.50 | 0.80 |

Fig. 1. Measuring the radon exhalation rate from a massive Pumice building block using the ERS2 and activated charcoal detectors.

independent exhalation rate [4]. The exhalation rate may also be determined as $E=\lambda A/S$, with $A$ the activity of the radon source, measured by a passive detector [5].

## 2. Experimental results

In Table 1, we compare the exhalation rates measured by passive (electret and activated charcoal) and active (Rad7[1] and ERS2[2]) detectors. The ERS2 (Fig. 1), an electrostatic radon sampler coupled to an alpha-particle detector, incorporates both the setup and the exhalation rate calculation. The results are the same, within errors better than 20%.

## References

[1] Content of radioactive elements in building materials, Israeli Standard 5098, The Standards Institution of Israel, November 2002.
[2] Protection Against Radon-222 at Home and at Work, ICRP Publication 65, Annals of the ICRP, Pergamon Press, 1994.
[3] K.H. Folkerts, G. Keller, H. Muth, An experimental study on diffusion and exhalation of $^{222}$Rn and $^{220}$Rn from building materials, Radiation Protection Dosimetry 9 (1) (1984) 27.
[4] C. Samuelson, H. Pettersson, Exhalation of $^{222}$Rn from porous materials, Radiation Protection Dosimetry 7 (1–4) (1984) 95–100.
[5] K. Kovler, et al., Determination of the radon diffusion length in building materials using electrets and activated carbon, Health Physics 86 (5) (2003) 505–516.

---

[1] Durridge, Bedford, MA, USA, www.durridge.com.
[2] Tracerlab, Köln, Germany, www.tracerlab.com.

# Retrospective radon assessments in a high radon dwelling in Ireland

J.P. McLaughlin*, K. Kelleher, H. Jiménez-Nápoles, L. León-Vintró

*Department of Experimental Physics, University College Dublin, Belfield, Dublin 4, Ireland*

**Abstract.** In 2003, arising from the unusual occurrence of two cases of lung cancer in a non-smoking household in a karstic limestone region of Ireland, a medical specialist suggested that radon in the dwelling be measured. The whole-house average radon concentration was found to be 49 kBq/m$^3$, which is the highest radon level detected to date in an Irish dwelling. After remediation had taken place, measurements of surface alpha-recoil implanted $^{210}$Po activities were made on glass objects throughout the dwelling by means of a standard passive alpha-track method. Surface activities of $^{210}$Po on glass ranging up to 2500 Bq/m$^2$ were found. Using a modified Jacobi room model, it was estimated that the mean radon concentrations in different parts of the dwelling over the past 8–22 years ranged from 59 kBq/m$^3$ to as high as 212 kBq/m$^3$. Samples of dust were also taken for analysis from the surface of a 70-year-old mirror covered in a thick layer of dust. Using gamma and alpha spectrometry, it was found that the dust contained between 0.88 and 1.10 MBq/kg of $^{210}$Po/$^{210}$Pb.
© 2004 Elsevier B.V. All rights reserved.

*Keywords:* Retrospective radon assessment; $^{210}$Po alpha-recoil implantation

## 1. Introduction

To date, the highest indoor radon level detected in Ireland was found in a 200-year-old dwelling in the southwest of the country with a seasonally adjusted radon concentration of 49 kBq/m$^3$ [1]. The dwelling is in Castleisland, which is a karstic limestone region of County Kerry. The initial radon gas measurements made in the dwelling were carried out on the advice of a medical specialist, due to the unusual occurrence of two cases of lung cancer in the household. One case is that of a 52-year-old male never-smoker and the other was a 41-year-old female, who died in 1998. This latter case had quit smoking some 20 years previously before moving into the dwelling.

---

\* Corresponding author. Tel.: +353 1 716 2229; fax: +353 1 283 7275.
*E-mail address:* james.mclaughlin@ucd.ie (J.P. McLaughlin).

0531-5131/ © 2004 Elsevier B.V. All rights reserved.
doi:10.1016/j.ics.2004.10.004

Table 1
$^{210}$Po measurements made on glass objects and the estimated radon concentration for each room

| Room | Object | Object age (years) | $^{210}$Po concentration (Bq/m$^2$) | | Estimated $^{222}$Rn concentration (Bq/m$^3$) |
|---|---|---|---|---|---|
| | | | Measured | Normalised to 20 years | |
| Kitchen | picture glass | 8 | 2530 | 5319 | 212600 |
| Living room | picture glass | 15 | 1695 | 2106 | 81400 |
| Bedroom 1 | picture glass | 17 | 1418 | 1599 | 61600 |
| Bedroom 1 | mirror | 22 | 1660 | 1551 | 59500 |
| Bedroom 2 | mirror | 17 | 1460 | 1647 | 63800 |

## 2. Measurements

In February 2004, after remediation had taken place, measurements were made throughout the dwelling in order to estimate the long-term average radon concentration in the past.

The technique employed is a standard, passive method that uses CR-39 and LR-115 alpha-track detectors to measure the $^{210}$Po surface activity on suitable glass objects arising from the alpha-recoil implantation of surface-deposited radon progeny [2]. Using the measured $^{210}$Po surface activity and dwelling parameter information the average radon concentration, that the glass objects were exposed to in their lifetime, were estimated using a modified Jacobi room model (see Table 1).

In addition, samples of dust taken from the surface of a 70-year-old mirror that was located in a presently unused part of the dwelling were analysed using high-resolution gamma and alpha spectrometry. The dust was found to contain long-lived radon progeny with the following specific activities: $^{210}$Pb=0.88 MBq/kg and $^{210}$Po=1.1 MBq/kg. For comparison purposes, no detectable $^{210}$Pb activity and only 2.3 kBq/kg of $^{210}$Po were found in dust from our laboratory where the mean radon concentration is about 40 Bq/m$^3$. The very high specific, unsupported activities of $^{210}$Po and $^{210}$Pb in the mirror dust are taken as confirmation that the dust, like the glass, acquired these activities from the deposition of short-lived radon progeny.

## 3. Discussion

Using an exposure–dose conversion factor of 1 mSv per annum per 40 Bq/m$^3$, an occupant of the house would have received, prior to remediation, an annual dose of at least 1.5 Sv, which is 650 times greater than the average annual radiation dose in Ireland due to radon exposure [3]. The source of the high indoor radon concentration is not yet known but high levels of radon associated with karstic limestone formations have been found in other parts of Ireland and elsewhere.

## References

[1] C. Organo, et al., High radon concentrations in a house near castleisland County Kerry (Ireland)—identification, remediation and post-remediation, J. Radiol. Prot. 24 (2004) 107–120.
[2] J.P. McLaughlin, The application of techniques to assess radon exposure retrospectively, Radiat. Prot. Dosim. 78 (1) (1998) 1–6.
[3] J.S. Madden, et al., Radon in dwellings in selected areas of Ireland, RPII, Dublin, 1994.

# Radiation exposure from high-level radiation area and related mining and processing activities of Jos Plateau, central Nigeria

## I.I. Funtua[a,*], S.B. Elegba[b]

[a]*Centre for Energy Research and Training, Ahmadu Bello University, Zaria, Nigeria*
[b]*Nigerian Nuclear Regulatory Authority, Abuja, Nigeria*

**Abstract.** The Jos Plateau in central Nigeria is a high-radiation area, being the center of unique granitic alkaline ring complexes that are rich in accessory minerals of zircon, monazite, ilmenite, xenotime, thorite, and pyrochlore containing high concentrations of thorium and uranium. The several years of mining and processing of casseterite (tin ore) and columbite (niobium ore) in the Jos Plateau have generated large quantities of tailings that are rich in these radioactive minerals and are mostly dumped haphazardly in the environment. Radiation monitoring in the area and at some processing mills shows high levels of dose rate with values as high as 100 $\mu Sv\ h^{-1}$ for processed zircon. The in situ dose rate measurements for the public and workers showed that exposures significantly higher than the recommended values of 1 and 20 mSv $year^{-1}$, respectively. Thoron ($^{220}Rn$) and its short-lived decay products (Tn-d) exposure, long considered of low radiological significance in comparison to radon ($^{222}Rn$) and decay products (Rn-d), are the most significant in the Jos area. © 2004 Elsevier B.V. All rights reserved.

*Keywords:* Radiation exposure; Granitic alkaline rocks; Mining and processing; Tailings; Radon; Thoron

## 1. Introduction

The Jos Plateau area is the center of the so-called Younger Granites of Nigeria, which are alkaline ring complexes with zones of anomalous uranium and thorium concentrations revealed by airborne radiometric mapping. Cassiterite and columbite mining and processing have been taking place in the Jos Plateau area for several decades and associated with this activity are extensive amounts of mine tailings. There is a growing

---

\* Corresponding author. Tel.: +234 69 550397; fax: +234 69 550737.
*E-mail address:* iifuntua@yahoo.com (I.I. Funtua).

0531-5131/ © 2004 Elsevier B.V. All rights reserved.
doi:10.1016/j.ics.2004.10.006

Table 1
Range and average measured radiation dose at the processing mills ($\mu$Sv h$^{-1}$)

|  | Range | Number of measurements | Average |
| --- | --- | --- | --- |
| (1) Background in the mill premise | 5–7 | 20 | 5 |
| (2) Background in the mill shades | 5–8 | 10 | 6 |
| (3) Rapid magnetic separators | 8–10 | 8 | 8 |
| (4) High-intensity magnetic separators | 8–10 | 10 | 8 |
| (5) Processed zircon | 70–100 | 12 | 80 |
| (6) Electrostatic separator | 10–14 | 10 | 12 |
| (7) Processed monazite | 40–60 | 15 | 50 |
| (8) Unseparated tailings | 15–25 | 20 | 20 |
| (9) Tailing heaps | 20–30 | 25 | 25 |

concern over the radiological impact of these tailings since it has been established that these ores are associated with accessory minerals like zircon, monazite, xenotime, and thorite, which have high concentrations of thorium and uranium. This work reports the estimated radiological impact of the processing of cassiterite and columbite from the following mills in Jos area: the James Dung mill, Dadin Kowa, and two mills belonging to Amalgamated Tin Mines, Bukuru.

## 2. Materials and methods

An external gamma ray dose rate ($\mu$Sv h$^{-1}$) survey was performed at the mills with a portable RADOS RDS-120 Universal survey meter manufactured by RADOS Technology (Finland). The instrument has a scintillation probe and two external probes: a GM probe and an alpha pancake probe. The measurement range is from 0.05 $\mu$Sv h$^{-1}$ to 10 Sv h$^{-1}$. Measurements were taken at background areas, mill shades, magnetic separators, electrostatic separator, sorted cassiterite, monazite, zircon, unsorted tailings, and tailing heaps in the premises.

## 3. Results and discussion

Table 1 shows the dose rate at different processing points and locations at the mills, with average values ranging from 5 $\mu$Sv h$^{-1}$ for the background in the premises to 80 $\mu$Sv h$^{-1}$ for that of processed zircon. Assuming a 2000-h working year, workers in the processing mills will be exposed to an annual dose of about 2–180 mSv, far above the 20 mSv annual dose limit. The dose rates of about 25 $\mu$Sv h$^{-1}$ measured for the tailings will give an annual dose of about 50 mSv for a nonradiation worker in the vicinity of the milling plant, exceeding the 1 mSv year$^{-1}$ dose limit for the members of the public [1].

## 4. Conclusions

There are high-level radiation exposures (mainly of Th series) from the mine tailings of Jos area, Central Nigeria. Workers at the milling plants could be exposed to radiation levels far above the annual dose limit of 20 mSv, and members of the public who live near the indiscriminately dumped tailings will be exposed to doses above 1 mSv year$^{-1}$.

## Reference

[1] IAEA, International Basic Safety Standards for Protection Against Ionizing Radiation and for the Safety of Radiation Sources, Safety Series No. 115-I, IAEA, Vienna, 1994.

# Radon concentrations in caves of Parque Estadual do Alto Ribeira (PETAR), SP, Brazil: preliminary results

Simone Albergi, Brigitte R.S. Pecequilo*, Marcia P. Campos

*IPEN (Institute of Nuclear and Energetic Researches), Av. Prof. Lineu Prestes, 2242-/05508-000 São Paulo, Brazil*[1]

**Abstract.** Radon concentrations assessment in the most frequented caves of PETAR, Parque Estadual do Alto Ribeira (High Ribeira River Turistic State Park), created in 1958 and situated south of São Paulo State, Brazil are carried out with Makrofol E nuclear track detectors, installed in the most frequently visited caves. Preliminary results from October 2003 to July 2004 show radon concentrations varying from $512 \pm 86$ to $6607 \pm 179$ Bq m$^{-3}$. The complete assessment will be achieved by March 2005. © 2004 Elsevier B.V. All rights reserved.

*Keywords:* Radon; Exposure; Cave; Makrofol E; Track detector

## 1. Introduction

The most important contributors to the committed effective dose received by population due to natural sources are the short-lived decay products of radon ($^{222}$Rn). Concentrations of indoor radon and its progeny in caves vary from levels hardly higher to levels several thousand times higher than outdoor air concentrations. Prolonged exposure to such high concentration levels increases the risk of developing lung cancer and leukaemia and may also have other harmful effects.

PETAR, Parque Estadual Turístico do Alto Ribeira (High Ribeira River Turistic State Park), created in 1958, is a conservation park with an area of 35,102.8 ha, situated on the left margin of the Ribeira river, south of São Paulo State, Brazil, with more than 180 recorded caves [1]. The park has four visit centers: Santana, Caboclos, Ouro Grosso and Casa da Pedra, receiving nearly 40,000 people annually. Radon concentrations are studied at several cave galleries of Santana center and Ouro Grosso center.

* Corresponding author. Tel./fax: +55 11 38169207.
  *E-mail address:* brigitte@ipen.br (B.R.S. Pecequilo).
  [1] Work also supported by Conselho Nacional de Desenvolvimento Científico e Tecnológico-CNPq, grant 134087/03-8 and Fundação de Amparo à Pesquisa do Estado de São Paulo-Fapesp, grant 2003/08146-2.

0531-5131/ © 2004 Elsevier B.V. All rights reserved.
doi:10.1016/j.ics.2004.11.127

## 2. Methodology

The SSNTD used in this study is the polycarbonate Makrofol E. Each detector is a small plastic square of 1 cm$^2$, loaded into a diffusion chamber type KFK-FN detector, installed in the selected caves, at least 1 m away from the nearest roof. The exposure period is, at least, 3 months and will cover at least 15 months, in order to determine the long-term average levels of the indoor radon concentrations over varying seasons.

After exposure, the detectors were retrieved to the Environmental Radiometric Division, IPEN, São Paulo and processed. The exposed plastic detectors were etched for 2 h with a PEW$_{40}$ solution in a temperature-stabilized water-bath and mild stirring, at 70 °C. The track densities were read under a Zeiss/Axiolab optical microscope connected to a video camera and a personal computer.

Radon concentrations were determined considering the track densities, the exposure period and a calibration factor of $0.02874 \pm 0.00699$ tr cm$^{-2}$ Bq$^{-1}$ m$^3$ day$^{-1}$, obtained by exposing a Makrofol E detector to Pylon model RN-150 calibrated with a $^{226}$Ra source of 18 kBq.

## 3. Results and discussion

Radon concentrations covering the period from October 2003 to July 2004 are presented in Table 1.

As can be seen, all results are within the range of literature values. The highest levels represent caves far-away from the gallery entrances, which were also observed in other studies all over the world [3].

The radon concentrations of the autumn/winter period are, for some galleries, higher than the ones of the spring/summer period. However, for a possible correlation between radon levels and weather peculiarities or frequency of visitors, we are, by now, processing more information.

## References

[1] I. Karmann, J.Á. Ferrari, Carste e cavernas do Parque Estadual Turístico do Alto Ribeira (PETAR), sul do Estado de São Paulo, in: C. Schobbenhaus, D.A. Campos, E.T. Queiroz, M. Winge, M. Berbert-Born (Eds.), Sítios Geológicos e Paleontológicos do Brasil, DNPM, Brasilia, 2002, pp. 401–414.
[2] Radiation protection against radon in workplaces other than mines. International Atomic Energy Agency, Vienna, 2003 (Safety Reports Series No 33).
[3] T.A. Przylibski, Radon concentration changes in the air of two caves in Poland, J. Environ. Radioact. 45 (1999) 81–94.

Table 1
Concentrations of $^{222}$Rn in several caves of High Ribeira River Turistic State Park (PETAR)

| Center | Cave | Radon concentration (Bq/m$^3$) | | Literature values [2] |
|---|---|---|---|---|
| | | October 2003–March 2004 (Spring/Summer) | March 2004–July 2004 (Autumn/Winter) | |
| Ouro Grosso | Alambari de Baixo | 516±82 | 1327±134 | |
| Santana | Torres | 4950±159 | 4650±184 | |
| Santana | Cristo | 5811±170 | 3435±168 | |
| Santana | Descanso | 6607±179 | 6358±206 | |
| Santana | Flores 1 | 2373±122 | 2359±152 | 48–21,100 |
| Santana | Flores 2 | 3972±146 | 3452±168 | |
| Santana | Chocolate | 1957±115 | 2177±149 | |
| Santana | Plataforma | 512±86 | 1223±132 | |
| Santana | Laje Branca | 1009±96 | 2568±155 | |
| Santana | Água Suja | | 1674±140 | |
| Santana | Couto | | 1110±129 | |

# Distribution pattern of natural radionuclides in Lake Nasser bottom sediments

Ashraf E. Khater[a],*, Yasser Y. Ebaid[b], Sayed A. El-Mongy[a]

[a]*National Center for Nuclear Safety and Radiation Control, Atomic Energy Authority, Egypt*
[b]*Physics Department, Faculty of Science, Fayum, Cairo University, Egypt*

**Abstract.** In Egypt, the Nile River water is the main source of water, providing nearly 95% of water requirements. The Nile water is impounded in Lake Nasser (LN) in the south of Egypt by the High Aswan Dam (HAD). In this study, we presented the radioactivity levels in LN sediments over the time period 1992–2000 and the distribution pattern of the measured radionuclides ($^{238}$U series, $^{232}$Th series, $^{40}$K and $^{137}$Cs). In addition, the uranium concentration in water samples was measured. The distribution pattern of these radionuclides in sediments reflects the geochemical behavior and weathering processes of uranium series, thorium series and the heavy minerals in the Nile pathway and in Nasser Lake. © 2004 Published by Elsevier B.V.

*Keywords:* Natural radioactivity; Nasser Lake; Sediment; Uranium; Laser flourimetry; Polonium

## 1. Introduction

Since the construction of High Aswan Dam (HAD) in 1964, a large reservoir, Lake Nasser (LN), has been formed at the dam's upstream side. The reservoir length is calculated to be approximately 500 km; accordingly, the largest surface area and maximum storage capacity of the reservoir are estimated at 600 and 162 km$^3$, respectively [1]. This study aims at monitoring the radioactivity levels in the lake sediment and water, and to investigate the distribution pattern of different radionuclides in the lake sediments.

## 2. Experimental techniques

During four sampling trips (1992–2000), 84 bottom sediment samples and 18 surface water samples were collected from Lake Nasser. The collected sediment samples were prepared and sealed in polyethylene containers to reach secular equilibrium between

---

* Corresponding author. Present address: Physics Department, College of Science, King Saud University, P.O. Box 2455, Riyadh 11451, Kingdom of Saudi Arabia. Tel.: +966 50 241 8292; fax: +966 146 76 448.
  *E-mail address:* khater_ashraf@yahoo.com (A.E. Khater).

0531-5131/ © 2004 Published by Elsevier B.V.
doi:10.1016/j.ics.2004.11.112

Table 1
Mean specific activity of U-238 (Ra-226) series, Th-232 (Ra-228) series, K-40 and Cs-137 (Bq/kg) and activity ratio of Ra-226/Ra-228 in Nasser Lake sediments

| Ra-226/Ra-228 | Cs-137±E | K-40±E | Ra-228±E | Ra-226±E | |
|---|---|---|---|---|---|
| 0.85±0.05 (0.19, 12) | 7.6±0.8 (2.6, 11) | 309.1±12.1 (420, 12) | 19.4±1.4 (4.4, 12) | 15.7±0.6[a] (2.06, 12) | 1992 |
| 0.83±0.05 (0.27, 24) | 4.5±0.5 (2.4, 19) | 221.6±17.0 (99.8, 25) | 20.9±2.6 (12.5, 24) | 15.3±1.6 (7.9, 25) | 1998 |
| 0.96±0.07 (0.31, 20) | 5.25±0.5 (2.2, 20) | 326.2±16.4 (73.3, 20) | 24.4±2.0 (9.1, 20) | 22.0±1.8 (8.1, 20) | 1999 |
| 0.77±0.02 (0.11, 21) | 2.3±0.3 (1.3, 18) | 317.6±18.1 (82.7, 21) | 18.4±1.1 (5.2, 21) | 14.3±1.1 (4.8, 21) | 2000 |
| – | 8.9±0.6 | 310±4.3 | 24.8±0.8 | 19.1±0.4 | [4] |

[a] Mean value±standard error (standard deviation, number of samples).

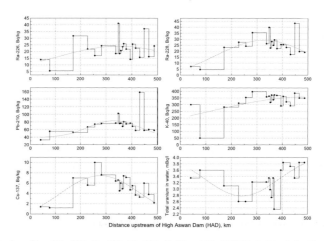

Fig. 1. Longitudinal profile of Ra-226, Ra-228, Pb-210, K-40 and Cs-137 (Bq/kg) in Lake Nasser sediments and U-238 (mBq/l) in water.

radium and thorium and their progenies [2]. Gamma spectrometer based on hyper pure germanium detector and uranium analyzer of model Sintrex UA-3 were used for radioactivity measurements of sediment and water samples [2,3].

## 3. Results and discussion

The mean specific activity of $^{238}$U ($^{226}$Ra) series, $^{232}$Th series, $^{40}$K and $^{137}$Cs in Bq/kg dry weight of sediment and activity ratio of $^{226}$Ra/$^{228}$Ra in Lake Nasser sediments are given in Table 1. Longitudinal profile of $^{226}$Ra, $^{228}$Ra, $^{40}$K and $^{137}$Cs (Bq/kg) in Lake Nasser sediments and $^{238}$U (mBq/l) in water are shown in Fig. 1.

The average concentration of uranium in the lake's water is 3.26 mBq/l, which is comparable to that in the purified Nile River water in Cairo (3.47 mBq/l) and less than that in the purified Nile water in other northern Cities (5.7–9.2 mBq/l) [5].

## References

[1] M.S. El-Manadely, et al., Lake Reserv. Res. Manage. 7 (2002) 81–86.
[2] A.J. Khater, Environ. Radioact. 71 (2004) 33–41.
[3] H. Diab, M. El-Tahawy, S. El-Mongy, Radiochim. Acta 89 (2001) 179–185.
[4] S.S. Ismail, E. Unfied, F. Grass, Radioanal. Nucl. Chem. Lett. 186 (2) (1994) 143–155.
[5] M.S. El-Tahawy, S.A.M. El-Mongy, S.Y. Omar, Isotopes Radiat. Res. 27 (2) (1995) 95–101.

# Natural radioactivity contents in tobacco

## N. Abd El-Aziz[a], A.E.M. Khater[a,*], H.A. Al-Sewaidan[b]

[a]National Center for Nuclear Safety and Radiation Control, Atomic Energy Authority, Cairo, Egypt
[b]Physics Department, College of Science, King Saud University, Riyadh, Kingdom of Saudi Arabia,
P.O. Box 7551, Nasr City, Cairo, Egypt

**Abstract.** Tobacco contains minute quantities of radioactive isotopes of uranium series and thorium series ($^{210}$Pb, $^{210}$Po and $^{226}$Ra), which are radioactive carcinogenic. Smoking of tobacco and its products increases the internal intake and radiation dose due to natural radionuclides. In a number of studies, inhalation of some naturally occurring radionuclides via smoking has been considered to be one of the most significant causes of lung cancer. In this work, Moasel tobacco samples were collected from the Saudi and Egypt markets. Natural radionuclides ($^{234}$Th, $^{226}$Ra, $^{210}$Pb, $^{214}$Bi, $^{228}$Ac, $^{40}$K and $^{210}$Po) in tobacco were measured using γ-ray spectrometer and α spectrometer. The measured data were presented and discussed. © 2004 Published by Elsevier B.V.

*Keywords:* Tobacco; Moasel; Natural radioactivity; Polonium; Smoking

## 1. Introduction

Tobacco contains minute quantities of radioactive isotopes such as uranium series and thorium series isotopes ($^{210}$Pb, $^{210}$Po and $^{226}$Ra), that are radioactive carcinogenic and could be found in smoke from burning tobacco. People who intentionally or passively inhale tobacco smoke are exposed to higher concentrations of radioactivity than non-smokers. Deposits of radioactive isotopes in the lungs of smokers, delivered to sensitive tissues for long periods of time, generating localized radiation exposures, may induce cancer both alone and synergistically with non-radioactive carcinogens. In a number of studies, inhalation of some naturally occurring radionuclides via smoking has been considered to be one of the most significant causes of lung cancer [1]. Jurak and Moasel tobacco products, smoked using hubble-bubble (water-pipe), are popular in the Middle Eastern and North African countries. It is unlike smoking cigarettes, where the filter

---

* Corresponding author. Present address: Physics Department, College of Science, King Saud University, P.O. Box 2455, Riyadh 11451, Kingdom of Saudi Arabia. Tel.: +966 50 241 8292; fax: +9661 46 76 448.
  E-mail address: khater_ashraf@yahoo.com (A.E.M. Khater).

0531-5131/ © 2004 Published by Elsevier B.V.
doi:10.1016/j.ics.2004.11.166

Table 1
Specific activity of Th-234, Ra-226, Pb-210, Ac-228 and K-40 (Bq/kg dry weight) in different types of Moasel tobacco

|  | Type K | Type S | Type Z | Average |
|---|---|---|---|---|
| Th-234 | 8.80±1.40, 3.43*(5.41–13.24) | 9.38±1.44, 3.22 (5.56–12.32) | 5.69±0.28 | 8.76±0.92, 3.18 (5.41–13.24) [12]** |
| Ra-226 | 7.54±0.89, 2.37 (3.60–9.85) | 6.32±0.76, 2.15 (4.43–9.31) | 7.68±2.46, 3.48 (5.22–10.14) | 6.98±0.56, 2.3 (3.6–10.14) [17] |
| Pb-210 | 16.42±2.21, 6.98 (5.14–25.7) | 14.44±1.34, 4.24 (10.27–22.2) | 16.59±3.96, 7.91 (5.49–22.92) | 15.62±1.21, 5.94 (5.14–25.67) [24] |
| Bi-214 | 2.42±0.42, 1.19 (1.17–4.28) | 2.00±0.40, 0.69 (1.20–2.40) | 1.25±0.04, 0.07 (1.15–1.30) | 2.02±0.26, 1.02 (1.15–4.28) [15] |
| Ac-228 | 6.49±1.0, 2.91 (1.6–10.28) | 0.82±0.21 | – | 5.78±1.19, 3.36 (0.82–10.28) [8] |
| K-40 | 617.0±28.1, 88.9 (474.6–725.8) | 795.9±38.7, 116.0 (701.0–1043.0) | 607.9±50.2, 100.3 (511.9–747.0) | 685.4±27.8, 133.1 (568.7–1043.3) [23] |
| Po-210 | 10.55±2.13, 3.69 (6.67–14.02) | 15.02±5.65, 11.30 (6.94–31.75) | – | 13.10±3.25 (6.67–31.75) [7] |
| No. of samples | 10 | 10 | 4 | 24 |

\* Mean±standard error, standard deviation (range); \*\* Number of data.

removes a minimum of 96% of the particulate phase of smoke [1]. This study aimed at assessment of radioactivity content in Moasel tobacco products to provide the necessary data to estimate the possible health effects of tobacco smoking.

## 2. Experimental techniques

Three most frequently smoked brands of Moasel tobacco were chosen. Samples were collected randomly from those available on the Saudi and Egypt markets. Samples were dried, pulverized, homogenized and transferred to polyethylene containers of 100-cm$^3$ capacity. Gamma-ray spectrometry: $^{234}$Th, $^{226}$Ra, $^{210}$Pb and $^{214}$Bi ($^{238}$U series), $^{228}$Ac ($^{232}$Th series), and $^{40}$K were measured using calibrated gamma-ray spectrometer; based on Hyper Pure Germanium detector with efficiency of 40% and full width at half maximum (FWHM) of 1.95 keV for $^{60}$Co γ-line at 1332 keV[2]. $^{210}$Pb-$^{210}$Po analysis: Polonium-210 was measured after chemical separation and plating on stainless steel disk using alpha spectrometers (CANBERRA, Mod. 7401 VR) based on passivated implanted planar silicon (PIPS) [3].

## 3. Results and discussion

Specific activity (Bq/kg) of $^{234}$Th, $^{226}$Ra, $^{210}$Pb, $^{214}$Bi, $^{228}$Ac, $^{40}$K and $^{210}$Po in 3 Moasel tobacco types and their averages are given in Table 1. Abdul-Majid (1995) reported that the average specific activity of $^{226}$Ra, $^{223}$Th and $^{40}$K in Egyptian's Moasel tobacco were 2.1, 2.8 and 471 Bq/kg, which is lower than our results. Estimation of internal dose due to inhalation of $^{234}$Th, $^{226}$Ra, $^{210}$Pb, $^{214}$Bi, $^{228}$Ac, $^{40}$K and $^{210}$Po via Moasel tobacco smoking will be performed in the near future work because several parameters are required, e.g. particulate size distribution, inhalation pattern, particle aerodynamics and the distribution factors of these radionuclides between smoke, ash and cooling water of hubble-bubble [1].

## References

[1] S. Abdul-Majid, I.I. Kutbi, M. Basabrain, J. Radioanal. Nucl. Chem., Artic. 194 (2) (1995) 371–377.
[2] M. El-Tahawy, et al., J. Nucl. Sci. 29 (1992) 361–363.
[3] A.J. Khater, Environ. Radioact. 71 (2004) 33–41.

# $^{235}$U–γ emission contribution to the 186 keV energy transition of $^{226}$Ra in environmental samples activity calculations

Y.Y. Ebaid[a,*], S.A. El-Mongy[b], K.A. Allam[b]

[a] *Faculty of science-Fayoum, Cairo University, Physics Department, Fayoum, Egypt*
[b] *NCNSRC-Atomic Energy Authority, 3 Ahmed El Zomer st, Nasr city, Cairo, Egypt*

**Abstract.** This work aims at developing a simple and reliable empirical equation to distinguish between counting rate contributions from both $^{235}$U and $^{226}$Ra to the 186 keV energy region. The method has set primarily two assumptions. First was the existence of secular equilibrium between the $^{238}$U and the $^{226}$Ra in the samples. The calculations showed that in the total count rate of the 186 keV peak consists of 58.3% of $^{226}$Ra and 41.7% of $^{235}$U at a radioactive equilibrium. Second was the absence of equilibrium, this condition needed another treatment of the obtained data. Finally, simple equations have been developed to calculate $^{226}$Ra and $^{235}$U in unenriched samples. © 2004 Published by Elsevier B.V.

*Keywords:* Environmental radioactivity; Gamma-ray spectroscopy

## 1. Introduction

Accuracy in the measurements of $^{238}$U concentrations in environmental samples by gamma spectroscopy depends on an appropriate determination of the concentrations of $^{222}$Rn decay products ($^{214}$Pb and $^{214}$Bi) in the sample under the assumption of secular equilibrium. As an inert gas, radon can leak from the sample vessel, accumulate in the void upper part of it, or leak from sample site itself. In all cases, the secular equilibrium condition is broken. Also, $^{238}$U is frequently estimated using the energy transitions of 63.3

---

\* Corresponding author. Tel.: +20 2 2873384; fax: +20 84 6370024.
*E-mail address:* yebaid@yahoo.com (Y.Y. Ebaid).

0531-5131/ © 2004 Published by Elsevier B.V.
doi:10.1016/j.ics.2004.12.020

keV (3.6%) and 92.6 (4.9%) of the $^{234}$Th (a direct daughter of $^{238}$U). Several restrictions draw attention to the use of these two lines to determine $^{238}$U [1]; interference caused by the 92.6 keV X-ray transition line from Bi, Po, U and Th, interference of the 63.3 keV line from the Th series and the significant self-absorption of both energy transitions due to their considerably low energy.

## 2. Theoretical work

The count rate (count/s) is calculated by the following form,

$$CR = A \times BR \times \xi \tag{1}$$

where BR is the branching ratio or the gamma-ray emission rate, $\xi$ is the full energy peak efficiency of the specific experimental system conditions.

If we have a sample containing both $^{226}$Ra and $^{235}$U, we have sharing of count rate at the region 186 keV. The $^{226}$Ra and $^{235}$U emits γ-transitions of 186.211 keV and 185.712 keV, respectively.

$$\text{Total count rate} = CR_{Ra} + CR_U \tag{2}$$

$$CR_{Ra}/CR_U = \frac{A_{Ra} \times BR_{Ra} \times \xi}{A_U \times BR_U \times \xi} \text{ then } CR_{Ra}/CR_U = \frac{A_{Ra} \times BR_{Ra}}{A_U \times BR_U} \tag{3}$$

Considering that the $BR_{Ra}$ and $BR_U$ are 3.59% and 57.2%, respectively, then:

$$CR_{Ra}/CR_U = 0.0359 \times A_{Ra}/0.572 \times A_U \tag{4}$$

### 2.1. At equilibrium

If we consider that the $^{226}$Ra and $^{238}$U are in equilibrium with each other and that the sample is not enriched, then the activity ratios of $^{238}$U:$^{226}$Ra would be 1:1 and, as a result, the activity ratio of $^{235}$U:$^{226}$Ra would be 0.0462:1. Then Eq. (4) will be

$$CR_U = 0.424 \times CR_T \tag{5}$$

### 2.2. At disequilibrium

In the case of disequilibrium but with a normal ratio, indicating neither enrichment nor depletion, $CR_U$ could be calculated as follows: since the activity of $^{235}$U represents about 0.0462 from that of $^{238}$U (under the condition of normal weight ratio of 0.7:99.29 for $^{235}$U:$^{238}$U), the count rate of the $^{235}$U ($CR_U$) is

$$CR_U = [0.0462 \times A(^{258}U)] \times \xi_{185.7} \times BR_{185.7}. \tag{6}$$

By substituting for the activity of $^{238}$U by (CR (at 1001 keV)/BR (at 1001 keV) $\xi$1001), then the equation will be

$$CR_{Ra} = CR_T - [0.0264264 \times \xi_{185.7} \times A(^{234m}Pa_{1001})].$$

This formula could be applied at any time with no need for the equilibrium to happen.

## 3. Conclusion

These calculations showed that $^{226}$Ra should contribute about 58.3% of the total count rate of the 186 keV peak while $^{235}$U contributes about 41.7% with the existence of equilibrium. Also, these calculations are useful in estimating the $^{226}$Ra in samples with no need for the secular equilibrium to happen between $^{226}$Ra and its respective progenies. They might also help in estimating the total uranium in the samples. Condition of equilibrium between $^{238}$U and its respective progenies is not required.

## Reference

[1] Z. Papp, Z. Dezso, S. Daroczy, Measurement of the radioactivity of $^{238}$U, $^{232}$Th, $^{226}$Ra, $^{137}$Cs and $^{40}$K in soil using direct Ge(Li) γ-ray spectroscopy, Journal of Radioanalytical and Nuclear Chemistry 222 (1–2) (1997) 171–176.

# Determination of radium isotopes in mineral water12 samples by α-spectrometry

Guogang Jia*, G. Torri, P. Innocenzi, R. Ocone, A. Di Lullo

*Italian Environmental Protection Agency and Technical Services, Via V. Brancati 48, 00144 Roma, Italy*

**Abstract.** A method for determination of low-level radium isotopes in mineral water samples by α-spectrometry has been developed. Ra-225, which is in equilibrium with its mother $^{229}$Th, was used as a yield tracer. Radium were preconcentrated from water samples by coprecipitation with BaSO$_4$ and iron (III) hydroxide, isolated from uranium, thorium and iron using a Microthene tri-octyl-phosphine oxide (TOPO) chromatography column, separated from barium in a cation-exchange resin column, electrodeposited on a stainless steel disc in a medium of oxalate, and counted by α-spectrometry. The lower limits of detection of the method are 0.12 mBq l$^{-1}$ for $^{226}$Ra and $^{224}$Ra, respectively. © 2004 Elsevier B.V. All rights reserved.

*Keywords:* $^{226}$Ra; $^{224}$Ra; Water; α-spectrometry

## 1. Introduction

Radium isotopes are the most dangerous natural radionuclides, because following their decay radon and other daughter radionuclides are produced. Moreover, the chemical and biological behaviour of radium is similar to that of other alkaline earth metals (Ca, Sr, Ba) and so they are easily incorporated into the bones of mammals and may cause an enhanced radiation dose to the public. This paper describes a new method for the determination of extremely low-level radium isotopes in mineral water samples by α-spectrometry, with emphasis on the use of $^{225}$Ra as a yield tracer, Microthene tri-octyl-phosphine oxide (TOPO) column separation of barium and radium from uranium and thorium, the cation-exchange resin separation of barium from radium using 1,2-cyclohexylene-dinitrilo-

---

* Corresponding author. Tel.: +39 06 50073219; fax: +39 06 50073287.
 *E-mail address:* jia@apat.it (G. Jia).

0531-5131/ © 2004 Elsevier B.V. All rights reserved.
doi:10.1016/j.ics.2004.09.019

tetraacetic acid monohydrate (CyDTA) as an eluant, and the preparation of carrier-free radium sources by electrodeposition.

## 2. Method

### 2.1. Preconcentration of radium in water

Ten milliliters of concentrated HCl, 10 mg of $Ba^{2+}$, 2 mg of $Fe^{3+}$ and 0.05 Bq of $^{225}Ra$ in equilibrium with its mother $^{229}Th$ as a tracer are added to 2 l of water sample. After boil by heating, 12 ml of 9 M $H_2SO_4$ are added to the solution during stirring. Then, the solution is adjusted to pH 8–9 with 30% ammonia solution to coprecipitate radium, thorium and uranium together as barium and radium sulphates and iron hydroxide. After centrifugation, the supernatant is discarded and the precipitate is dissolved with 45 ml of 8 M HCl by heating.

### 2.2. Separation of radium from thorium, uranium and barium

The obtained solution is passed through a preconditioned Microthene-TOPO column at a flow rate of 0.6–0.8 ml min$^{-1}$ and the effluent is collected. Then, the column is washed with 50 ml of 8 M HCl and the washing solution is joined with the collected effluent. In this case, all thorium and uranium remain in the column, and barium and radium come to the effluent. The $^{229}Th/^{225}Ra$ separation time is recorded.

The collected effluent is evaporated to a volume of 2–3 ml. After cooling, 1 ml of concentrated $H_2SO_4$ is added to precipitate barium and radium as sulphate. The water phase is eliminated by centrifugation and the precipitate is dissolved with 14 ml of 0.05 M CyDTA. The obtained solution is adjusted to pH 5–6, and then passed through the prepared cation-exchange column at a flow rate of 0.3–0.4 ml min$^{-1}$. The column is washed with 20 ml saturated boric acid at pH 8.5, 50 ml of 0.05 M CyDTA at pH 8.5 and 60 ml of 0.5 M HCl. The radium is eluted from the column by 40 ml of 3 M $HNO_3$ at a flow rate of 0.1 ml min$^{-1}$ and collected.

### 2.3. Electrodeposition and measurement

The collected eluent is evaporated to dryness and some $HNO_3$ and $H_2O_2$ are added to destroy any existing organic matters. Radium in the dried residue is transferred to an electrodeposition cell with 20 ml of 0.17 M $(NH_4)_2C_2O_4$ at pH 2.6 and electrodeposited on a stainless steel disc at a current density of 400 mA cm$^{-2}$ for 4 h. The electrodeposition time is recorded, which is considered as the time of $^{225}Ra/^{225}Ac$ separation or $^{225}Ac$ ingrowth. The obtained disc is counted by α-spectrometry and the time of starting and ending measurement is recorded. The chemical yield for radium is calculated from $A_{at}/A_{Ra}$, where $A_{at}$ is the measured activity of $^{217}At$, which can be converted to $^{225}Ra$ activity through the Bateman's equations, and $A_{Ra}$ is the $^{225}Ra$ activity added.

## 3. Results and discussion

Sixteen drinking water samples have been analysed with the method. The average radium yields obtained are 87.2±6.1%. The results show that the radium concentrations in the drinking waters collected in Italy have a big variation and are in the range of 0.50–60.8

mBq l$^{-1}$ for $^{226}$Ra and of ≤LLD −3.26 mBq l$^{-1}$ for $^{224}$Ra with a mean $^{224}$Ra/$^{226}$Ra ratio of 0.143±0.143. Except for one sample with a high activity, the $^{226}$Ra concentrations in most of the analysed drinking water samples are comparable with the reported data in Europe.

The high and stable yields as well as the very good resolution (34–48 keV) of the α-spectra for water samples show the wide adaptability of the method in the fields of health physics, environmental science and geochemical studies.

# Variation of terrestrial gamma radiation in Toki, Japan—comparison between gamma-ray spectrometry using Ge semiconductor and ICP-MS measurement

Y. Fujikawa[a,*], M. Fukui[a], T. Baba[b], T. Yoshimoto[a], E. Ikeda[a], M. Saito[a], H. Yamanishi[c], T. Uda[c]

[a]*Research Reactor Institute, Kyoto University, Kumatori-cho, Sennan-gun, Osaka 590-0494, Asashiro-nishi, Japan*
[b]*Kuritaz Co., Ltd., Japan*
[c]*National Institute for Fusion Science, Japan*

**Abstract.** The cause of the spatial variation in external radiation levels in the Toki area in Japan was investigated by in-situ measurements of gamma-ray spectrum and by collections and measurements of soil samples. It was found that the radionuclide contents of topsoil largely governed the radiation level while land exploitation, gardening activities, and concrete-building constructions caused the variation in radiation there. © 2004 Elsevier B.V. All rights reserved.

*Keywords:* Natural radiation; Uranium series; Thorium series; In-situ gamma-ray spectrum; Soil; ICP-MS

## 1. Introduction

The Toki area in Gifu Prefecture, Japan has a higher natural radiation level than the Japanese average because of the Th and U series radionuclide contained in the underlying geological layer there. Records from radiation monitoring stations located in the area, however, showed a complex pattern that could not be explained from the

* Corresponding author. Tel.: +81 724 51 2447; fax: +81 724 51 2620.
*E-mail address:* fujikawa@rri.kyoto-u.ac.jp (Y. Fujikawa).

0531-5131/ © 2004 Elsevier B.V. All rights reserved.
doi:10.1016/j.ics.2004.11.059

geological conditions alone. We therefore conducted a detailed research to clarify the mechanism of the variation.

## 2. Materials and methods

In-situ gamma-ray spectrum measurements were conducted with a portable Ge semiconductor detector in nine stations in the site of National Institute for Fusion Science (NIFS) as well as in Nakazawa, Kawai, and Lake Matsuno (Gifu Prefecture). Soil samples were collected, sealed in plastic vessels with epoxy resin to confine gaseous radionuclides generated from the soil, and analysed by a low-background gamma-ray spectrometry. Results of ICP-MS analyses of borate-fused soil samples were compared with those of non-destructive gamma-ray analyses to cross-check the accuracy of the U and Th series radionuclide analyses, and to detect the uranium series disequilibrium.

## 3. Results and discussion

The results of the in-situ Ge measurement (interpreted as radionuclide concentrations in soil [1]) and soil analyses are compared in Fig. 1. Stations that have names starting with I are near to the experiment facilities while those with names starting with W are located along the border of the NIFS yard. As land exploitation created flat and homogenized topsoil around the facilities, the results of in-situ Ge analyses agreed well with those of soil analyses in stations IX and IF. Concrete-building materials, fertilizers and gardening soil with high radionuclide contents increased the external radiation level in station IB. The external radiation level was lower at the stations near the border than at those near the facilities principally because the slope of the land reduced the radiation level relative to that expected from radionuclides in soil. The low Ac-228 content of organic topsoil (WA) also was the cause of the lower radiation level there. The actual radionuclide content of soil and the content estimated from in-situ Ge analysis differed significantly at the station WC probably because of nonuniform distribution of radionuclides in soil and complex geometry of the land.

Fig. 2 shows that in the locations outside NIFS, the external radiation generally agreed well with the radionuclide content of natural soil (NIFS). U-238–U-234

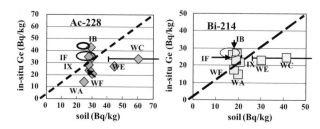

Fig. 1. Comparison between in-situ Ge analyses and soil analyses in NIFS site.

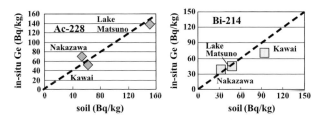

Fig. 2. Comparison between in-situ Ge analyses and soil analyses in Gifu Prefecture.

disequilibrium was found in weathered rocks from Lake Matsuno. Concerning the other soil samples, uranium series radionuclides from U-238 to Bi-214 were in radiological equilibrium.

## Reference

[1] H.L. Beck, J. DeCampo, C. Gogolack, In-situ Ge(Li) and NaI(Tl) Gamma-Ray Spectrometry. HASL-258, U.S. Atomic Energy Commission, New York, 1972.

# Spectrometry characteristics of photon fields and atmospheric radionuclide deposits monitoring in one part of Southern Bohemia

J. Kluson, L. Thinova*, T. Cechak, T. Trojek

*CTU, Faculty of Nuclear Sciences and Physical Engineering, Praha, Czech Republic*

**Abstract.** The (bio)monitoring in the neighborhood of Nuclear Power Plant (NPP) Temelin started in the year 2000—1 year before the initial power plant operation. In the years 2000, 2002, and 2004, the spectrometry characteristics of photon-spectra measurement were included in biomonitoring. The area of interest contained 29 sampled locations along eight radial profiles intersecting the area at distances from 2 to 20 km around Temelin NPP. Laboratory, as well as in situ, gamma spectrometric method enables to determine presence of natural and manmade radionuclides with very good limits of detection. In situ gamma spectrometry consists of direct air kerma rate measurements and air kerma rate calculations from photon-spectra measurement. The laboratory gamma spectrometric measurement (geometry of Marinelli containers) is used for the determination of radionuclides in the samples of pine bark, Shreber's moss, forest humus, edible mushrooms, and forest berries. The trend analysis of measurement results enables the completion of the main task of this project: to describe the influence of NPP Temelin on the radiation increase in its neighborhood. Only $^{137}$Cs of manmade radionuclides has been identified. © 2004 Elsevier B.V. All rights reserved.

*Keywords:* Gamma spectrometry; Biomonitoring; Trend analysis; Nuclear power plant; Dosimetry

## 1. Introduction

For the past 5 years, FNSPE CTU in Prague took part in monitoring the influence of Nuclear Power Plant (NPP) Temelin on the environment within 20-km radius of the plant. Using ecological principles, the changes in environment quality are indicated by biological indicator changes. We chose forest humus, surface of pine bark, Shreber's moss, edible mushrooms, and forest berries.

The year 2000 was designated as the reference year before the start of the NPP operation, and 2001 was the year of the initial operation. The biomonitoring for years 2000, 2002, and

---

* Corresponding author. Tel.: +42 224 358 235; fax: +42 224 811 074.
 *E-mail address:* thinova@fjfi.cvut.cz (L. Thinova).

0531-5131/ © 2004 Elsevier B.V. All rights reserved.
doi:10.1016/j.ics.2004.11.113

2004 also included assessment of the dosimetry and spectrometry characteristic of the photon fields.

## 2. Monitored area

Monitored area contains 29 sampled locations along eight radial profiles intersecting the area of interest (the measuring points are located 2, 5, 10, 20 km from NPP, distance of 20 km is a comparison area). The pine bark and moss were sampled at the selected sites twice yearly, forest humus once during spring months, mushrooms and berries once in growing season. The top 3 mm of tree bark was taken at reference height of 1 m. The moss samples were cut by scissors. Forest humus was sampled with respect to resolution of surface layers, according to the degree of hummification. In total, 203 samples in 2000, 222 samples in 2001, 223 samples in 2002, and 251 samples in 2003 were collected. In situ gamma spectrometry is conducted at selected 15 points in order to sufficiently cover the area of interest.

## 3. Materials and methods

In the gamma spectrometry laboratory, the samples were, after drying, enclosed in Marinelli containers (0.5l), surrounding during the measurements a coaxial HPGe detector. Processing of measured spectra in the range up to 3 MeV provided mass-related activity (Bq/kg) of naturally radioactive elements ($^{40}$K, $^{226}$Ra, and $^{232}$Th) and contaminant $^{137}$Cs (resulting from nuclear weapon tests in the fifties of last century and from Chernobyl accident fallout) using program SP DEMOS. The resulting data were used for the trend analysis. Two methods of the gamma fields in situ measurements of the dosimetric characteristics were selected (all measurements made at a height of 1 m above ground): (1) determination of air kerma rate by direct measurement (Fig. 1A); (2) air kerma rate calculation from photon spectra measurement using a portable spectrometer with scintillation detector NaI(Tl) 3″×3″ in the energy range up to 3 MeV (Fig. 1B).

## 4. Conclusions

The measured spectra represent characteristic spectra of natural background. In the calculated energy distribution of air kerma rate, it is not possible to identify (with the exception of the abovementioned $^{137}$Cs) any significant contribution of any manmade radionuclide. Based on the so-far obtained results of monitoring, as well as of the trend analysis, it can be stated that it was confirmed that JETE operation does not have any impact on the level of natural background in the measured reference points.

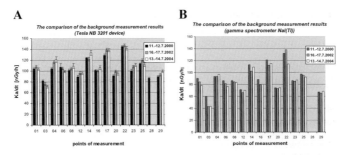

Fig. 1. Determination of air kerma rate by direct measurement and by calculation from photon spectra.

# GPS-based handheld device for measuring environmental gamma radiation and mapping contaminated areas

## J. Paridaens*

*Belgian Nuclear Research Centre, SCK-CEN, Boeretang 200, B2400 Mol, Belgium*

**Abstract.** A portable GPS-based device was developed for mapping large areas with enhanced radioactivity. It is intended for use in difficultly accessible areas, typically for foot campaigns. Standard low cost components were used, such as a wireless GPS receiver, and a wireless RS-232 device connecting a handheld NaI detector to a pocket PC (PPC). The PPC is programmed to perform simple data logging, producing a file with position and dose rate in situ, which must be post processed in the office to map the contamination. © 2004 Elsevier B.V. All rights reserved.

*Keywords:* GPS; Environmental radioactive contamination; Mapping

## 1. Introduction

Sometimes one is confronted with having to map large areas with enhanced radioactivity. Examples are mine tailings or waste rock piles, phosphate gypsum deposits, flooding zones contaminated by effluents of plants processing ores containing enhanced natural radiation, post accident sites, etc. Car borne measuring equipment is not always an option, as the terrain might only be accessible on foot. Airborne mapping with helicopters for example is fast, but expensive, not readily available and can lack the necessary detail. Hence the need arose for a portable and easily useable tool for logging radiation and location data, allowing to map the radioactivity by simply walking over the terrain with the equipment and post processing the data in the office.

## 2. Methods

The serial data output of a NaI portable gamma detector (Exploranium GR-130) was transmitted to a bluetooth® enabled pocket PC (PPC) (iPAQ 2210) using a small wireless

---

\* Tel.: +32 14 332814; fax: +32 14 321056.
*E-mail address:* jparidae@sckcen.be.

0531-5131/ © 2004 Elsevier B.V. All rights reserved.
doi:10.1016/j.ics.2004.09.028

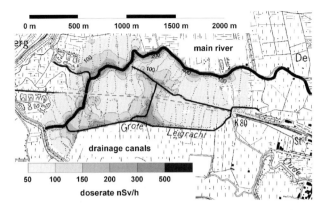

Fig. 1. An area of about 45 ha of very rough terrain was measured in a foot campaign of about 6 h and then mapped. The small crosses show the 16-km-long trajectory with about 1800 data points. The radioactivity appears to be concentrated along the main river and is also periodically transported up some of the drainage canals during periods with high rainfall.

RS-232 communication device (Brainboxes BL-521) fixed to the detector. The range of transmission is several tens of meters. A wireless GPS, the size of a computer mouse, (Navman BT 4400) transmits position data to the PPC, with a transmission range of about 10 m. This GPS is mounted on a cap on the head of the operator for optimal satellite reception. The operator carries gamma detector, GPS and PPC around over the terrain. Periodically, the position and radiation data are stored on the PPC. The time difference between position and radiation readout is never more than 1 s, which is acceptable considering the low moving speed of an operator on foot. The sampling period can be varied and has a minimum of 2 s, which allows to change the spatial density of the data points. The data logging software was home made and based on Microsoft Embedded Visual Basic.

## 3. Results

Small tests on an area with known levels of enhanced radioactivity within sharply defined perimeters were done on a former test field for studying soil to plant transfer factors. The results were satisfactory, taking the inherent precision of GPS of about 5 to 10 m into account. Next a larger area was investigated. In the flooding zone of a river with Ra-226 contamination resulting from phosphate plant effluents, about 45 ha of extremely rough terrain were measured. The data log contained about 1800 points for 16 km of trajectory and was measured in about 6 h. Fig. 1 shows the dose rate contour map resulting from post processing the data.

## 4. Conclusion

This low cost, very small and easily portable device allows mapping very large radioactively contaminated areas in reasonable times with satisfactory results. The approach could easily work with most any gamma detector with an RS-232 output, by just adapting the PPC-detector communication protocol.

www.ics-elsevier.com

# Development of a remote radiation monitoring system using unmanned helicopter

Shinichi Okuyama[a,*], Tatsuo Torii[a], Yasunori Nawa[b],
Ikuo Kinoshita[c], Akihiko Suzuki[d], Masanori Shibuya[d],
Nobuyuki Miyazaki[e]

[a] *Tsuruga Head Office, Japan Nuclear Cycle Development Institute, Japan*
[b] *NESI, Inc., Japan*
[c] *TAS Co., Ltd., Japan*
[d] *Aeronautic Operations, YAMAHA Motor Co., Ltd., Japan*
[e] *Japan Radiation Engineering Co., Ltd., Japan*

**Abstract.** Feasibility study of a remote radiation monitoring system using an autonomous unmanned helicopter, mounted the CCD cameras, the GPS sensor and a radiation detector, has been carried out as measures in a nuclear emergency and for surveying high radiation area in the environment. This system can fly for the destinations and return by the automatic operation, and the radiation data can be collected during the flight. Radiation data measured are transmitted immediately with image data to the monitoring station on the ground. It is possible to monitor these data on the map of the computer display in the real time. As a result of the flight tests, it is confirmed that the fluctuation of the dose-rate distribution on the ground is measured by this system, and it can be used for the radiation monitoring in case of a nuclear emergency. © 2004 Elsevier B.V. All rights reserved.

*Keywords:* Unmanned aeronautic vehicle; Radiation monitoring; Nuclear emergency

## 1. Introduction

In case of a nuclear emergency when an abnormal quantity of radioactive substances and/or radiation is released from a nuclear facility, emergency radiation monitoring is carried out from the sea and air as well as from the ground adjacent to the facility in order to evaluate the influence of the radiation on the surrounding environment. As a means of

---

\* Corresponding author. 2-1 Shiraki, Tsuruga-shi, Fukui-ken, 919-1279, Japan. Tel.: +81 770 39 1031; fax: +81 770 39 1554.

*E-mail address:* oku@t-hq.jnc.go.jp (S. Okuyama).

0531-5131/ © 2004 Elsevier B.V. All rights reserved.
doi:10.1016/j.ics.2004.11.154

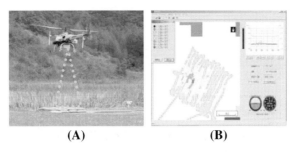

(A)          (B)

Fig. 1. (A) Measurement of radiation level above the ground put on potassic fertilizers and (B) colour map of radiation level monitored by this system.

radiation monitoring from the air, aerial monitoring using a manned helicopter is conducted. Although a manned helicopter can monitor a large area during a short time, flight at an altitude of 300 m or lower is prohibited by the Aviation Law in Japan. Therefore, it is difficult for a manned helicopter to measure the radiation profile near the ground and to measure the radiation level in a complicated terrain.

In recent years, technologies for autonomous flying of industrial unmanned helicopters have been developed and are now being applied to volcanic accidents and other disasters. In expectation of the application of unmanned helicopters to aerial radiation monitoring during a nuclear emergency, we have developed a remote radiation monitoring system and investigated its possibilities.

## 2. Remote radiation monitoring system using unmanned helicopter

### 2.1. Monitoring system

We have developed a radiation measurement system equipped with a radiation detector (plastic scintillator: 270×300×20 mm) to be loaded in an autonomous unmanned helicopter (length: 3.63 m, weight: 94 kg, usual time of flight: 60 min, other equipment: 3 CCD cameras, a GPS sensor) [1]. The radiation measurement system adopts a radiation profile mapping system, which enables radiation data to be shown at the ground station in real time by transmitting the measured data from the helicopter [2].

### 2.2. Result of the test

To perform functional tests, potassic fertilizers for gardening use were placed on the ground and measurement was conducted by flying the unmanned helicopter in the air to investigate the response characteristics of the radiation measurement system, as shown in Fig. 1. In addition, an actual flight test was conducted near to the coastline at an altitude of 30 m in order to measure the fluctuations in background radiation. We also measured the detailed fluctuation of radiation above the ground by using this system.

Accordingly, the possibility of aerial monitoring in an emergency utilizing this system was demonstrated.

## References

[1] A. Sato, Research and development and civil application of an autonomous unmanned helicopter, Proc. AHS International Forum, 2001, p. 57.
[2] S. Okuyama, et al., Study on environmental monitoring method using unmanned aerial vehicle in nuclear emergency, in: 36th Annual Meeting of Japan Health Phys. Soc., A-6, 2002 (in Japanese).

# Thermoluminescence mechanism on $SiO_2$ phosphor

Michiko Takami[a,*], Masatoshi Ohta[b], Hiroshi Yasuda[a]

[a]National Institute of Radiological Sciences, Radiation Safety Research Center, 4-9-1 Anagawa, Inage-ku, 263-8555, Chiba, Japan
[b]Niigata University, Niigata, Japan

**Abstract.** Dose-assessment for natural radiation has been investigated in regard to the high background radiation areas (HBRA) around the world. Natural $SiO_2$ crystals collected in the HBRA which exhibit TL is expected to be usable for dosimetry of the radiations from natural sources. However, the TL mechanism has not yet been well explained. In this paper, TL spectra were examined on natural $SiO_2$ and synthesized $SiO_2$ doped with $Al^{3+}$ and/or $Eu^{3+}$. As the results, it is expected that we could establish more quantitative TL dosimetry on HBRA using natural $SiO_2$ crystals by estimating quantum efficiency on each TL of $Eu^{3+}$. © 2004 Elsevier B.V. All rights reserved.

*Keywords:* $SiO_2$; Thermoluminescence; Luminescent mechanism; $Eu^{3+}$

## 1. Introduction

Many studies of dose-assessment for natural radiation have been performed in regard to the high background radiation areas (HBRA) around the world. The cosmic radiations have been measured using radiation detectors such as NaI scintillation survey meters, electronic pocket dosimeters and thermoluminescence dosimeters (TLD). In the occasion without these instruments, natural $SiO_2$ crystals collected in the HBRA is expected to be usable for dosimetry of the radiations from natural sources. The TL dosimetry using the natural $SiO_2$ crystals have already been performed in high-dose areas due to a nuclear accident like the JCO accident that happened on September 1999. The TL from a natural $SiO_2$ crystal gives historical information in relation with collected layers, not only total dose. Thus, this method can make available discussing the origin of HBRA. However, the TL mechanism has not yet been well explained; it is necessary to make the clear TL mechanism in order to establish more quantitative dosimetry.

---

* Corresponding author. Tel.: +81 43 206 3233; fax: +81 43 251 4531.
  E-mail address: mtakami@nirs.go.jp (M. Takami).

0531-5131/ © 2004 Elsevier B.V. All rights reserved.
doi:10.1016/j.ics.2004.10.017

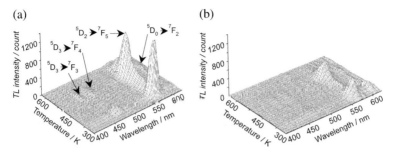

Fig. 1. TL spectrum of (a) $SiO_2$: $Eu^{3+}$ and (b) natural $SiO_2$.

## 2. Materials and methods

TL spectra was examined for synthesized $SiO_2$ doped with $Al^{3+}$ and/or $Eu^{3+}$ that are reported as the source of blue and red TL emissions from natural $SiO_2$, respectively, and the data were compared with that for natural $SiO_2$ crystals of which the impurity concentration is known. The natural $SiO_2$ was obtained from Tazawa Lake in Japan. TL and ESR spectra were measured on the $SiO_2$ sample irradiated by X-rays (about 8 and 15 kR, respectively).

## 3. Results and discussion

The TL of $SiO_2$ doped with $Eu^{3+}$ (Fig. 1(a)) had several sharp peaks ascribing to f–f transition of $Eu^{3+}$. All peaks were observed at ca. 350 and ca. 500 K. On the $SiO_2$ co-doped with $Al^{3+}$, $Eu^{3+}$, observed TL spectrum was similar to that of $Eu^{3+}$-doped sample. It was found that there are two kinds of trap, shallow trap (at ca. 350 K) and deep trap (at ca. 500 K); these TL relative to $Eu^{3+}$ strongly appeared in the TL of $SiO_2$ containing $Al^{3+}$ and $Eu^{3+}$. We have already found that an $E'_1$ center due to oxygen defects was produced in synthesized $SiO_2$, an ESR signal assigned to an Al center in the $SiO_2$ doped with $Al^{3+}$. The ESR signal of $SiO_2$ doped with $Eu^{3+}$ was combination of electron trap induced by oxygen defects and hole trap capturing a hole like Al center. As to temperature dependence, ESR signal intensity decreased at near glow peak temperature.

Thus, on $SiO_2$ doped with $Eu^{3+}$, ESR signal which diminished at near 370 K is assumed to relate to electron trap, probably Eu (from $Eu^{3+}$ to $Eu^{2+}$) and that diminished at near 500 K attributes to Eu-related center like Al-center. From these results, it is suggested that Eu substitutes Si and works as hole trap and/or electron trap that contribute to TL phenomenon. From TL spectrum of natural $SiO_2$ (Fig. 1(b)), a large number of narrow TL peaks were observed at ca. 350 and ca. 500 K and ascribed to electronic transition of $Eu^{3+}$. Particularly, the emission of $Eu^{3+}$ was observed with various colors, not only red. From this fact, it would be improper to conclude that blue and red TL of natural $SiO_2$ are attribute to $Al^{3+}$ and $Eu^{3+}$, respectively.

## 4. Conclusion

TL spectra were examined on natural $SiO_2$ and synthesized $SiO_2$ doped with $Al^{3+}$ and/or $Eu^{3+}$. The TL observed on natural $SiO_2$ can attribute to electronic transition of $Eu^{3+}$. We would like to suggest to use TL of $Eu^{3+}$ for dose-assessment using natural $SiO_2$ because the electronic transition of $Eu^{3+}$ has been well investigated. It is expected that we could establish more quantitative TL dosimetry on HBRA using natural $SiO_2$ crystals by estimating quantum efficiency on each TL of $Eu^{3+}$.

# Radon anomaly related to the 1995 Kobe earthquake in Japan

Y. Yasuoka[a,*], T. Ishii[b], S. Tokonami[c], T. Ishikawa[c], Y. Narazaki[d], M. Shinogi[a]

[a]*Kobe Pharmaceutical University, Hyogo, Japan*
[b]*University of Yamanashi, Yamanashi, Japan*
[c]*National Institute of Radiological Sciences, Chiba, Japan*
[d]*Fukuoka Institute of Health and Environmental Sciences, Fukuoka, Japan*

**Abstract.** It has been reported that several clear hydrological and geochemical anomalies were recorded before the 1995 Kobe earthquake (January 17, 1995). Remarkable changes of radon concentration in air and in groundwater were observed from November 1994 until the time of the earthquake. All of our data was monitored within the aftershock region. The increases in the atmospheric radon may reflect formation of microcracks in the granitic rocks. © 2004 Elsevier B.V. All rights reserved.

*Keywords:* Earthquake; Radon; Groundwater; Atmosphere; Preseismic

## 1. Introduction

The precursory change of the Kobe earthquake (M 7.2) that occurred on January 17, 1995, such as groundwater discharge rate, crustal strain changes, and changes of chloride ion concentration in groundwater, has been described by some researchers. Igarashi et al. [1] reported radon in groundwater at a well located in the aftershock region increased from the initial level measured in October 1994. We measured the atmospheric radon

---

\* Corresponding author. 4-19-1, Motoyamakita-machi, Higashinada-ku, Kobe, Hyogo, Japan 658-8558. Tel.: +81 78 441 7519; fax: +81 78 441 7519.
*E-mail address:* yasuoka@kobepharma-u.ac.jp (Y. Yasuoka).

0531-5131/ © 2004 Elsevier B.V. All rights reserved.
doi:10.1016/j.ics.2004.10.011

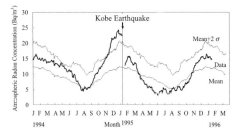

Fig. 1. Changes in atmospheric radon concentration at Kobe.

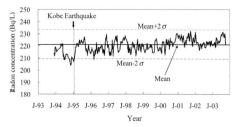

Fig. 2. Changes in radon concentration in the groundwater at Kobe.

concentration [2] and radon in the groundwater before and after the earthquake within the aftershock region and near an active fault.

## 2. Measurements and results

The measurements of atmospheric radon concentration were made continuously and automatically by the use of a flow-type ionization chamber. The daily averages of the radon concentration were calculated using the hourly data observed between April 1984 and January 1996. The mean was calculated for the radon concentration from April 1984 to March 1994. A seismic related radon anomaly which raised two or more standard deviations above the mean of radon concentration was observed for about 2 months before the earthquake (Fig. 1).

The radon in groundwater was extracted with a liquid scintillator solution and measured with a liquid scintillation counter. The groundwater samples were taken a few times every month between November 1993 and October 2003. The mean of radon concentration was calculated from the time of earthquake until October 2003. The radon data observed for about 2 months before the earthquake fell two standard deviations below the calculated mean of radon concentration (Fig. 2).

## 3. Conclusion

Remarkable changes of radon concentration in the air and in the groundwater were observed for about 2 months before the earthquake. All of our data were collected within the aftershock region. The increase and decrease of the earthquake-related changes in the radon concentration in groundwater was affected by other earthquake-related changes such as the groundwater discharge rate. The increases in the atmospheric radon may reflect the formation of microcracks in the granitic rocks.

## References

[1] G. Igarashi, S. Saeki, N. Takahata, et al., Ground-water radon anomaly before the Kobe earthquake in Japan, Science 269 (1995) 60–61.
[2] Y. Yasuoka, M. Shinogi, Anomaly in atmospheric radon concentration: a possible precursor of the 1995 Kobe, Japan, earthquake, Health Phys. 72 (5) (1997) 759–761.

# Generation of runaway electrons induced by radon progeny products in thunderstorm electric fields and the initiation of lightning discharges

Tatsuo Torii[a,*], Tatsuo Nozaki[a], Takeshi Sugita[b], Takeshi Nishijima[c], Zen-Ichiro Kawasaki[c]

[a]*Tsuruga Head Office, Japan Nuclear Cycle Development Institute, 2-1 Shiraki, Tsuruga-shi, Fukui-ken, 919-1279, Japan*
[b]*Science System Lab. Inc., Japan*
[c]*Graduate School of Engineering, Osaka University, Japan*

**Abstract.** γ-Ray dose-rate increases associated with winter thunderstorm activities have been observed in the coastal areas facing the Sea of Japan. To investigate the generation of energetic photons that originate in thunderstorm electric fields, we have calculated the behaviour of electrons and photons in electric fields with Monte Carlo method. In case of the calculation for the energetic electron emitted in the atmosphere, the electron and photon fluxes have increased greatly in the region with a strong electric field. Then, we have carried out the Monte Carlo transport calculations of the β- and γ-rays emitted by radon progeny products, as a source of energetic electrons, in thunderstorm electric fields. From the calculated results, it is confirmed that the electron flux shows notable increases in the strong electric field, while the photon flux does not fluctuate significantly. Since the radon progeny products form a large part of the energetic electrons in the atmosphere, they can serve as the source of a considerable amount of electrons. This is indicative of the role of these electrons in the initiation of a lightning discharge. © 2004 Elsevier B.V. All rights reserved.

*Keywords:* Thunderstorm; Electric field; Runaway electron; Radon progeny; Lightning

## 1. Introduction

Fluctuations in radiation intensities within or above thunderclouds have occasionally been observed by aircraft. The γ-ray dose increases are also observed by environmental radiation detectors installed in the vicinity of nuclear power stations along the coast of the Sea of Japan during thunderstorms in winter [1]. Hence, we investigated whether the

* Corresponding author. Tel.: +81 770 39 1031; fax: +81 770 39 1554.
*E-mail address:* torii@t-hq.jnc.go.jp (T. Torii).

0531-5131/ © 2004 Elsevier B.V. All rights reserved.
doi:10.1016/j.ics.2004.11.110

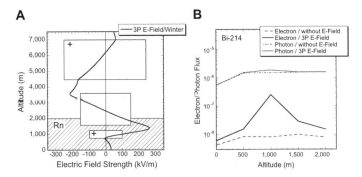

Fig. 1. Altitudinal distribution of a winter thundercloud electric field (A) and photon/electron fluxes (B).

energetic radiations are generated inside the thunderclouds, especially in the lower part. We have simulated the behaviour of energetic electrons and photons inside and under the thunderstorm electric fields using the Monte Carlo code EGS4 [2]. According to the results, the production of secondary electrons and photons increases with electric field intensity, and an electromagnetic shower is produced continuously when the electric field strength exceeds 280 $P(z)$ kV/m, where $P(z)$ is the atmospheric pressure (atm) at altitude $z$ (m). In case of the calculation for cosmic-ray electrons/photons as a source of energetic radiation, we have ascertained that these particles produce bremsstrahlung photons of up to several MeV, a part of which can reach the ground [3]. Focusing on the β/γ-rays emitting from a radon progeny, we also studied the fluctuations of these particle fluxes in a thunderstorm electric field.

## 2. Calculations and results

In view of the typical states of the atmosphere during a thunderstorm, we adopted the tripole electric field model of a winter thundercloud (see Fig. 1A) and assumed a uniform density of a radon progeny in spaces up to an altitude of 2 km, where a strong electric field should prevail and simulate the releases of β/γ-rays in $4\pi$ directions.

As the result, Fig. 1B shows the altitudinal distributions of photon/electron fluxes originating from the β/γ-rays emitted by Bi-214. Although the distributions show a slight increase in photon fluxes in the region with a strong electric field, the electron flux shows a significant increase in the same region. From the result, it can be suggested that the cause of an increase in γ-ray intensity near the ground is the cosmic ray components rather than radon progeny products. However, the increasing electron flux indicates that considerable runaway electrons are generated and that these energetic electrons play an important role in the intensive ionization of the air and, as a result, a significant growth of electric conductivity in thunderstorm electric fields. These productions may also induce the lightning discharge by these processes.

## References

[1] T. Torii, M. Takeishi, T. Hosono, Observation of gamma-ray dose increase associated with winter thunderstorm and lightning activity, J. Geophys. Res. 107 (D17) (2002) 4324.
[2] R. Nelson, H. Hirayama, D.W.O. Rogers, The EGS4 Code System, SLAC-265, Stanford Linear Accelerator Center, 1985.
[3] T. Torii, et al., Downward emission of runaway electrons and bremsstrahlung photons in thunderstorm electric fields, Geophys. Res. Lett. 31 (2004) L0511.

# Residential radon and childhood leukemia: a metaanalysis of published studies

S. Yoshinaga[a,*], S. Tokonami[a], S. Akiba[b]

[a]National Institute of Radiological Sciences, Chiba, Japan
[b]Kagoshima University, Kagoshima, Japan

**Abstract.** The carcinogenic effect of radon has not been established except for lung cancer. We conducted a metaanalysis of published epidemiological studies on residential radon and childhood leukemia. Five case-control studies were identified, with a total of 1348 cases and 2057 controls. We used a log-linear regression model to obtain a summary estimate of radon-related risk in each study, based on adjusted odds ratios corresponding to categories for radon concentration. Overall estimate of relative risk at 150 Bq/m$^3$ was 1.05 (95% CI: 0.90–1.21) based on random effect model. There is no evidence for an association between residential radon and childhood leukemia. © 2004 Elsevier B.V. All rights reserved.

*Keywords:* Metaanalysis; Radon; Childhood leukemia; Epidemiology

## 1. Introduction

There is convincing evidence from numerous studies that exposure to high levels of radon increases the risk of lung cancer. However, less attention has been paid to the risk of malignant tumors other than lung cancer associated with radon exposure. Several ecological studies showed a significant positive association between leukemia and indoor residential radon concentrations. In addition, there have been concerns from a dosimetric point of view that bone marrow doses from residential radon and its decay products might be higher than originally thought. We conducted a metaanalysis of published epidemiological studies on residential radon and childhood leukemia.

## 2. Methods

Through a literature survey using PubMed, five case-control studies [1–5] were identified (Table 1). From the five studies, adjusted odds ratios for different categories of

---

\* Corresponding author. 4-9-1, Anagawa, Inage-ku, Chiba 263-8555, Japan. Tel.: +81 43 206 3108; fax: +81 43 251 6089.
*E-mail address:* yosinaga@nirs.go.jp (S. Yoshinaga).

Table 1
Summary of the five case-control studies of residential radon and childhood leukemia

| Study | Cases/controls | Age | Leukemia type | Estimated odds ratio at 150 Bq/m$^3$ |
|---|---|---|---|---|
| Stjernfeldt et al. [1] | 7/7 | 0–13 | ALL and AML | 0.25 (95% CI: 0.01–4.28) |
| Lubin et al. [2] | 281/281 | 0–14 | ALL | 0.99 (95% CI: 0.78–1.27) |
| Steinbuch et al. [3] | 173/254 | 0–17 | AML | 1.17 (95% CI: 0.95–1.44) |
| Kaletsch et al. [4] | 82/209 | 0–14 | AL | 1.34 (95% CI: 0.28–6.56) |
| UK Childhood Cancer Study Investigators [5] | 805/1306 | 0–14 | ALL and others | 0.79 (95% CI: 0.51–1.21) |

ALL—acute lymphoblastic leukemia; AML—acute myeloid leukemia; AL—acute leukemia.

radon concentration were obtained. Mean, median, or midpoint of radon concentration in Bq/m$^3$ within categories were used as the representative values for each categories. We used a weighted log-linear regression model to obtain a summary estimate of radon-related risk in each study which corresponds to an odds ratio at 150 Bq/m$^3$, compared to reference category. Random effect model, taking into account between-study variance as well as within-study variance, was used to obtain an overall estimate of radon-related risk, using study-specific estimates of the summary risk.

## 3. Results and discussion

Each study included 7 to 805 cases of leukemia and 7 to 1306 controls, and a total of 1348 cases and 2057 controls contributed to the metaanalysis. No single study showed a statistically significant increase of odds ratios in any category. Study-specific summary odds ratios at 150 Bq/m$^3$ were estimated to be 0.25, 0.99, 1.17, 1.34, and 0.79 (Table 1). A test for homogeneity showed that there was no significant heterogeneity in radon-related risks across studies ($p=0.40$). Overall estimate of odds ratio at 150 Bq/m$^3$ was 1.05 (95% CI: 0.90–1.21) based on a random effect model.

Several ecological studies of childhood leukemia and residential radon have showed a significant positive association, while the present analysis dose not suggest an association. These significant results are likely due to chance or potential biases such as ecological fallacy.

## 4. Conclusion

Although only five studies were available in this metaanalysis, we conclude that there is no evidence for an association between residential radon and childhood leukemia.

## References

[1] M. Stjernfeldt, M. Samuelson, J. Ludvigsson, Radiation in dwellings and cancer in children, Pediatr. Hematol. Oncol. 4 (1987) 55–61.
[2] J.H. Lubin, et al., Case-control study of childhood acute lymphoblastic leukemia and residential radon exposure, J. Natl. Cancer Inst. 90 (1998) 294–300.
[3] M. Steinbuch, et al., Indoor residential radon exposure and risk of childhood acute myeloid leukemia, Br. J. Cancer 81 (1999) 900–906.
[4] U. Kaletsch, et al., Childhood cancer and residential radon exposure—results of a population-based case-control study in Lower Saxony (Germany), Radiat. Environ. Biophys. 38 (1999) 211–215.
[5] UK Childhood Cancer Study Investigators, The United Kingdom childhood cancer study of exposure to domestic sources of ionizing radiation: I. Radon gas, Br. J. Cancer 86 (2002) 1721–1726.

# New approach to dose optimization for members of the public

## Takatoshi Hattori*, Kazuo Sakai

*Low Dose Radiation Research Center, CRIEPI, 2-11-1, Iwadokita, Komae-shi, Tokyo, Japan*

**Abstract.** Using a probabilistic approach, the influence of dose distribution after the dose distribution due to artificial nuclides is added to that for NORM has been studied. On the basis of such an approach, the dose criteria of the order of 0.01 mSv/year for exemption and clearance could be set higher, for example, to the order of 0.1 mSv/year. © 2004 Elsevier B.V. All rights reserved.

*Keywords:* Optimization; Exemption; Clearance; Probability distribution; Log normal distribution; NORM

## 1. Introduction

In the case of artificial radioactive nuclides, the dose criteria of exemption and clearance that require no more endeavor for optimization are of the order of 0.01 mSv/year [1]. On the other hand, the application of such a low dose criterion to naturally occurring radioactive material (NORM) is not practicable. Recently, there have been some reports proposing to set the dose criterion for NORM to the order of 0.3 or 1 mSv/year, which are higher than that for artificial nuclides [2].

In this study, a new approach to dose optimization has been proposed aiming at the variation of total dose probabilistic distribution when the dose distribution due to artificial nuclides is added to that of natural radiation, for the purpose of discussing the dose criteria of exemption and clearance.

## 2. Materials and methods

To discuss the dose criteria of exemption and clearance, it is important to know the influences on the dose distribution after the dose distribution due to artificial nuclides is added to that for NORM. In this study, a log normal distribution with a geometric mean of

---

\* Corresponding author. Tel.: +81 3 3480 2111; fax: +81 3 3480 2493.
*E-mail address:* thattori@criepi.denken.or.jp (T. Hattori).

0531-5131/ © 2004 Elsevier B.V. All rights reserved.
doi:10.1016/j.ics.2004.09.024

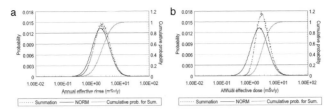

Fig. 1. Dose distribution due to NORM and total dose distribution. (a) 0.1 mSv/year for additional dose and (b) 0.5 mSv/year for additional dose.

2.0 mSv/year and a geometric standard deviation of 2.0 is assumed for the dose distribution due to NORM, according to the example of the dose distributions due to NORM in the 15 countries presented in the UNSCEAR 2000 report [3].

There are some cases in which the dose distribution for workers in nuclear facilities, whose individual doses were not limited, were similar to the log normal, which is the same as that due to NORM.

From the above reason, the dose distribution due to artificial nuclides for members of the public is treated as a log normal in the present study.

## 3. Results and discussion

The probability distribution of the summation of the dose distributions due to artificial nuclides and NORM were obtained using the Monte Carlo method. The geometric means of the additional dose distribution due to artificial nuclides were assumed to be 0.01, 0.03, 0.05, 0.1, 0.3, 0.5 and 1.0 mSv/year, and their geometric standard deviations were 10.5, 5.98, 4.61, 3.24, 1.85, 1.42 and 1.001, respectively.

Fig. 1 shows the dose distribution for NORM and the total distribution obtained by adding the additional dose distribution whose geometric means are 0.1 and 0.5 mSv/year. It is obvious from Fig. 1a that there are almost no differences between the two distributions. It can be seen from Fig. 1b that the difference between the two distributions is significant in the low-dose region, but not so marked in the high-dose region.

The dose level that is equal to cumulative probabilities of 2.5%, 50% and 97.5% in the dose distributions are selected. The 97.5% and 50% values were almost constant within the region that has an additional dose below 0.5 mSv/year. On the other hand, the 2.5% value increased gradually when the additional dose exceeded 0.1 mSv/year, but the total dose was sufficiently low at or less than 1 mSv/year within the region that has an additional dose below 0.5 mSv/year.

This indicates that an additional dose below 0.5 mSv/year is negligible for members of the public who are exposed to a relatively high dose due to NORM. In the case of members of the public who are exposed to a low dose, the dose itself is lower than the median of the dose distribution due to NORM. These results suggest the possibility of making such a low dose acceptable for members of the public.

## References

[1] IAEA, Safety Series No.115, Vienna (1996).
[2] European Commission European Commission, Radiation Protection, vol. 122, 2001 Luxemburg.
[3] UNSCEAR, Sources and Effects of Ionizing Radiation, 2000.

# A study on the necessity of boarding control for international flights

Hiroshi Yasuda*, Kazunobu Fujitaka

*Research Center for Radiation Safety, National Institute of Radiological Sciences, Japan*

**Abstract.** It is certain that part of aircraft crew are anxious about exposures to cosmic radiation. It is thus desired for experts of radiation protection to offer reliable information such as route doses and health effects of low-dose radiation, so as not to let the crew suffer from excessive anxiety. Accordingly, effective doses received on major routes of international flights from the Tokyo/Narita airport are calculated under conservative conditions. Based on the calculation, a basic framework for controlling cosmic radiation exposures of aircraft crew is proposed. © 2004 Elsevier B.V. All rights reserved.

*Keywords:* Aircraft crew; Cosmic radiation; Radiological protection; Effective dose; CARI code

## 1. Introduction

ICRP [1] recommended that there should be a requirement to include exposures to natural sources as part of occupational exposure in operation of jet aircraft. Afterwards, radiation exposures of aircraft crew have been discussed actively as an important regulation matter, particularly in Europe. In Japan, discussion on exposures of aircraft crew to cosmic radiation is now under way.

## 2. Calculation method

International flights from the Tokyo/Narita airport to major cities in the world were selected for calculation of route doses. Effective doses received in round trips to those cities were calculated using the CARI-6 code, which is issued free of charge by the Federal Aviation Administration (FAA), USA [2]. In calculation, safe side (conservative) conditions were given in keeping with radiation protection practices, except standard flight courses that are originally incorporated into the CARI-6.

---

\* Corresponding author. Postal address: 4-9-1 Anagawa, Inage-ku, Chiba 263-8555, Japan. Tel.: +81 43 206 3233; fax: +81 43 251 4531.

*E-mail address:* h_yasuda@nirs.go.jp (H. Yasuda).

0531-5131/ © 2004 Elsevier B.V. All rights reserved.
doi:10.1016/j.ics.2004.11.058

Table 1
Effective doses calculated under conservative conditions for major international flights from the Tokyo/Narita airport with boarding restriction numbers estimated for two assumed values of annual dose restrictions

| City | Flight time (h:min) | | Effective dose (µSv) | | | Boarding restriction number (N) | |
|---|---|---|---|---|---|---|---|
| | Out-ward | Home-ward | Out-ward | Home-ward | Total | Below 1 mSv year$^{-1}$ | Below 6 mSv year$^{-1}$ |
| New York | 12:20 | 14:25 | 96 | 113 | 210 | 4.8 | 28.6 |
| San Francisco | 9:00 | 11:30 | 49 | 63 | 112 | 8.9 | 53.4 |
| Honolulu | 6:35 | 8:55 | 23 | 31 | 54 | 18.5 | 111 |
| Moscow | 9:55 | 9:20 | 68 | 64 | 131 | 7.6 | 45.7 |
| London | 12:40 | 11:50 | 97 | 91 | 188 | 5.3 | 32.0 |
| Madrid | 12:25 | 12:25 | 104 | 99 | 203 | 4.9 | 29.6 |
| Frankfurt | 12:10 | 11:20 | 91 | 84 | 175 | 5.7 | 34.4 |
| Hong Kong | 5:10 | 3:45 | 15 | 11 | 26 | 38.2 | 229 |
| Bangkok | 7:00 | 5:40 | 20 | 16 | 36 | 27.5 | 165 |
| Singapore | 7:25 | 6:30 | 21 | 18 | 39 | 26.0 | 156 |
| Sydney | 9:30 | 9:40 | 31 | 31 | 62 | 16.1 | 96.5 |
| Auckland | 10:30 | 10:55 | 35 | 36 | 70 | 14.2 | 85.2 |

## 3. Results and discussion

Route doses calculated with the CARI-6 code under conservative conditions as to major international flights from the Tokyo/Narita airport are summarized in Table 1. Effective dose tends to be larger for higher latitude flights such as those to New York and London. Based on the calculation, we have estimated the maximum number ($N$) of round trips on each route corresponding to two control levels (Table 1): 1 and 6 mSv. According to average boarding time (800–900 h) of a crew, it would be difficult to keep the annual dose below 1 mSv for most cases. With the control level of 6 mSv, we need to consider limited routes at high latitude.

For aircraft crew who are assigned to various routes, a general framework to judge in advance the necessity of radiation management is given as follows:

$$\sum (n_i/N_i) > 1 \tag{1}$$

where $n_i$ is the annual boarding number of a crew on the route $i$ and $N_i$ is the boarding restriction number for the route $i$. This framework can be used desirably in a planning stage of working schedules for screening the crew who may need some management.

## Acknowledgements

The authors thank Dr. Shunji Takagi and Dr. Satoshi Iwai (Mitsubishi Research Institute) for their technical cooperation. Dr. Wallace Friedberg (Federal Aviation Administration) kindly permitted use of CARI-6 code.

## References

[1] ICRP, Recommendation of the International Commission on Radiological Protection, Publication 60, Annals of the ICRP, vol. 21(1–3), Pergamon Press, Oxford, 1991.
[2] FAA. http://www.cami.jccbi.gov/AAM-600/Radiation/radio_CARI6.htm.

# Cancer risk due to exposure to high levels of natural radon in the inhabitants of Ramsar, Iran

S.M.J. Mortazavi[a,*], M. Ghiassi-Nejad[b], M. Rezaiean[c]

[a]*Medical Physics Department, School of Medicine, Rafsanjan University of Medical Sciences (RUMS), Rafsanjan, Iran*
[b]*National Radiation Protection Department (NRPD), Iranian Nuclear Regulatory Authority (INRA), P.O. Box 14155-4494, Tehran, Iran*
[c]*Social Medicine Department, School of Medicine, Rafsanjan University of Medical Sciences (RUMS), Rafsanjan, Iran*

**Abstract.** Inhabitants of Ramsar, a city in northern Iran, are exposed to levels of natural radiation as high as 55–200 times higher than the average global dose rate. Furthermore, radon levels in some regions of Ramsar are up to 3700 Bq m$^{-3}$. To assess the association between the radon concentration and frequency of lung cancer, lung cancer patients recorded over the past 2 years in eight districts of Ramsar with different levels of radon were studied. Data from the Ramsar Health Network show that both crude lung cancer rate and adjusted lung cancer rate in one district with the highest recorded levels of external radiation and radon concentration are lower than those of the other seven districts. It can be concluded that lung cancer rate may show a negative correlation with natural radon concentration. © 2004 Elsevier B.V. All rights reserved.

*Keywords:* Radon; Lung cancer; Ramsar

## 1. Introduction

Radon has long been known to be among the main causes of lung cancer. Ramsar in Iran is famous for its high natural radiation areas and maximum radon levels in some regions of Ramsar are up to 3700 Bq m$^{-3}$. Note here that US EPA recommends some remedial actions for the houses with their indoor radon levels 200 Bq m$^{-3}$ or higher. In the

---

\* Corresponding author. Tel./fax: +98 391 822 0097.
*E-mail address:* jam23@lycos.com (S.M.J. Mortazavi).

0531-5131/ © 2004 Elsevier B.V. All rights reserved.
doi:10.1016/j.ics.2004.12.012

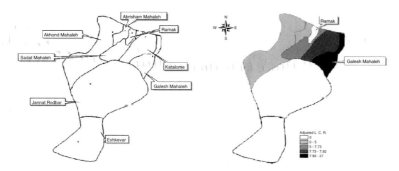

Fig. 1. Panel A. Ramsar is divided into eight health districts. Panel B. The highest lung cancer mortality rate was in Galesh Mahaleeh, where the radon levels were normal. On the other hand, the lowest lung cancer mortality rate was in Ramak, where the highest concentrations of radon in the dwellings were found.

present study, we conducted preliminary analysis of lung cancer risk among Ramsar residents.

## 2. Radon and lung cancer in Ramsar

Ramsar is divided into eight health districts (Fig. 1, Panel A) and a health center provides primary health services in each health district. Indoor radon concentration levels were previously measured in each dwelling by the Iranian Nuclear Regulatory Authority experts. A wide variety in the radon levels was observed and that is mainly due to the geological variation of the area [1,2]. The overall cancer mortality, lung cancer mortality and neonatal death rate of different districts in the years from 2000 to 2001 were collected. The radon prone areas were located in a district named Ramak. Therefore, the risk of radon-induced lung cancer was expected to be higher in this district compared to other seven districts. Our study showed that the highest lung cancer mortality rate was in Galesh Mahaleeh, where the radon levels were normal. Interestingly, the lowest lung cancer mortality rate was in Ramak, where the highest concentrations of radon in the dwellings were found (Fig. 1, Panel B).

## 3. Concluding remarks

In this study, we could not find any positive correlation between indoor radon levels and lung cancer rate in the inhabitants who lived in the dwellings with high levels of radon for many generations. Clearly, more research is needed to determine if life-long exposure to high levels of indoor radon leads to an increased risk for the development of cancer.

### References

[1] S.M.J. Mortazavi, et al., How should governments address high levels of natural radiation and radon? Lessons from the Chernobyl nuclear accident, Risk Health Saf. Environ. 13/1.2 (2002) 31–45.
[2] M. Ghiassi-nejad, et al., Very high background radiation areas of Ramsar, Iran: preliminary biological studies, Health Phys. 82 (2002) 87–93.

# Living in high natural background radiation areas in Ramsar, Iran. Is it dangerous for health?

Ali Shabestani Monfared[a,*], Farzad Jalali[b], Hossein Mozdarani[c], Mahmood Hajiahmadi[d], Hamid Samavat[e]

[a]*Department of Medical Physics, Babol University of Medical Sciences, P.O. Box 47176-41367, Babol, Iran*
[b]*Department of Internal Medicine, Babol University of Medical Sciences, Babol, Iran*
[c]*Department of Radiation Biology, Tarbiat Modares University, Tehran, Iran*
[d]*Department Social Medicine, Babol University of Medical Sciences, Babol, Iran*
[e]*Department of Medical Physics, Hamadan University of Medical Sciences, Hamadan, Iran*

**Abstract.** The main goal of the present study is to examine the health effects of high natural background radiation among the residents of Ramsar, one of the inhabited areas with highest natural background radiation. This report summarizes the data obtained from 402 residents of high background radiation area and 374 residents from an adjacent normal radiation area. Overall data showed no significant differences in the frequencies of any mental and physical disabilities as well as malignancies between residents of normal and high background radiation areas. Results also showed no significant increase in the frequency of death, abortion and mental depression among inhabitants in high background radiation area. We plan to extend our work, using more clinical and paraclinical tests on these populations in the next phase of the study.
© 2004 Elsevier B.V. All rights reserved.

*Keywords:* High background radiation; Health; Ramsar; Iran

## 1. Introduction

High background radiation has drawn researchers' attention for years. The effects of the low doses of radiation on human health are questionable and still the matter under discussion. Ramsar, a coastal city in the north part of Iran, involves the highest background radiation among the whole residential areas in the world [1]. So far the

---

* Corresponding author. Tel.: +98 911 1230475; fax: +98 111 2234367.
  *E-mail address:* monfared_ali@yahoo.com (A.S. Monfared).

0531-5131/ © 2004 Elsevier B.V. All rights reserved.
doi:10.1016/j.ics.2004.12.007

reported maximum dose is 98.5 mSv/year in Ramsar [2]. The purpose of this research is to study the health condition of the inhabitants of high background radiation (HBR) area. In this project the results of the first phase can be seen in the article and the most important point is to survey the present health condition as a final cause in radiobiological assessment of the inhabitants.

## 2. Materials and methods

From among the whole inhabitants in *Talesh Mahaleh* as HBR area and *Chaparsar* as ordinary background radiation area, 101 families (402 people) and 98 family (374 people) were selected, respectively, matching on some variables like age, sex, diet, etc. The subjects whose permanent residency was not in the study areas were excluded from the study. After explaining the aim of the project to the individuals and obtaining their consent, standard questionnaires were completed through interview and available data in local health centre.

## 3. Results

Results showed that the effective indicators on health such as economic and social features and most of hygienic indicators among the residents of two areas were statistically equal. However, the frequency of a few special diseases like cancer and cardiac disease in the HBR residents was less than that in the area with ordinary background radiation ($p$ value=0.011).

## 4. Discussion and conclusion

Lack of detrimental effect on the health of inhabitants after receiving this dose of radiation is not compatible with what is assumed by the present protection regulation. To quite a few researches, this rate can be safe for human health [3], as suggested by the present study. The lower frequency of malignancies in the HBR inhabitants is in agreement with some reports and might be due to radiation adaptive response and radiation hormesis indicating beneficial effect of low doses of radiation on health. Regarding the lower frequency of cardiac diseases, we cannot ignore the effect of other factors. We believe that more evidence and documents are required. Also by accomplishing the next phase, we hope to go into more details using clinical examinations and paraclinical tests. At any date, any comment on the risk of living in these areas should not be pessimistic because it causes to frighten the inhabitants [4].

## References

[1] UNSCEAR (United Nations Scientific Committee on the Effects of Atomic Radiation), Sources and effects of ionizing radiation, United Nations, New York, 2000.
[2] M. Sohrabi, Recent radiological studies of high level natural radiation areas of Ramsar, ICHLNR 39 (1990) 39–47.
[3] M. Ghiassi-nejad, et al., Very high background radiation areas of Ramsar, Iran. Preliminary biological studies, Health Physics 82 (1) (2002) 87–93.
[4] A.S.H. Monfared, M. Amiri, J.R. Cameron, How public fear from radiation can be reduced? Iranian Journal of Nuclear Medicine (19) (2003) 1–7.

# The need for considering social, economic, and psychological factors in warning the general public from the possible risks due to residing in HLNRAs

S.M.J. Mortazavi[a,*], A. Abbasi[b], R. Asadi[b], A. Hemmati[b]

[a]Medical Physics Department, School of Medicine, Vice Chancellor for Academic Affairs' Office, Central Building of the Rafsanjan University, Rafsanjan University of Medical Sciences (RUMS), Rafsanjan, Iran
[b]School of Medicine, Rafsanjan University of Medical Sciences (RUMS), Rafsanjan, Iran

**Abstract.** Following the discovery of X-rays and radioactivity more than 100 years ago, the need for protection against very low doses of ionizing radiation and especially different levels of natural radiation is still among the most controversial matters in radiobiology and radiation protection. According to formal reports, some areas in Ramsar, a city in northern Iran, are the inhabited areas with the highest levels of natural radiation studied so far. The people who live in these areas are usually unaware of the high levels of natural radiation in their environment. Studies performed on the residents of these areas have indicated that the effective dose of the inhabitants, in some cases, is much higher than the dose limits for occupational irradiation. Considering the new policy of ICRP regarding suggesting dose limits for exposure to natural radiation and the concentration of radon, it seems that warning the inhabitants is a must. On the other hand, considering the experiences in other countries and especially evacuation of the residents of contaminated areas after the Chernobyl accident, and according to ICRP, setting any radiation protection regulation for the inhabitants without considering social, economic, and psychological factors would waste the resources and cause harsh events. In this paper, the need for considering social, economic, and psychological factors in warning the general public from the possible risks due to residing in areas with above the normal levels of natural radioactivity in Ramsar, Iran, is discussed. © 2004 Elsevier B.V. All rights reserved.

*Keywords:* High level natural radiation areas; Ramsar; Radiation risk

## 1. Introduction

The people who live in the high level natural radiation areas (HLNRAs) of the world are of considerable interest because they and their ancestors have been exposed to

---

\* Corresponding author. Tel.: +98 391 822 0097; fax: +98 391 822 0097.
*E-mail address:* jamo23@lycos.com (S.M.J. Mortazavi).

0531-5131/ © 2004 Elsevier B.V. All rights reserved.
doi:10.1016/j.ics.2004.11.107

abnormally high radiation levels over many generations. If a radiation dose of a few hundred mSv per year is detrimental to health causing genetic abnormalities or an increased risk of cancer, it should be evident in these residents. In Ramsar, Iran, a population of about 2000 is exposed to average annual radiation levels from natural sources of 10 mGy/year, a lifetime expectancy of about 650 mSv, and the highest recorded doses are about 260 mGy/year, plus exposure to radon.

## 2. Social, economic, and psychological factors

It is well known that the possibility of being exposed to radiation causes considerable anxiety, especially among pregnant women [1]. Furthermore, after the Chernobyl accident, it has been reported that living in areas with elevated radiation levels and/or the stress and fear of living in contaminated area can lead to significant increases in nervous disorders, cardiovascular diseases, and other problems [2]. It has been shown that the most significant increase was in the suicide rate [3]. Considering these data, it has been reported that after the Chernobyl accident, there were widespread psychological reactions to the accident that were due to fear of the radiation, not due to the radiation doses. Our preliminary studies [4,5] show no significant differences between residents in HLNRAs compared to those in normal level natural radiation areas (NLNRAs) in the areas of life span, cancer incidence, or background levels of chromosomal abnormalities. Furthermore, when administered an in vitro challenge dose of 1.5 Gy of gamma rays, donor lymphocytes showed significantly reduced sensitivity to radiation as evidenced by their experiencing fewer induced chromosome aberrations among residents of HLNRAs compared to those in NLNRAs. The results obtained in our studies, along with the apparent good health of residents in HLNRAs further suggest that it is not in the public's interest to spend societal resources to relocate populations exposed to even the relatively high levels of radiation found in Ramsar and other HBRAs.

## 3. Concluding remarks

Our results suggest that exposure to elevated levels of natural radiation in HLNRAs of Ramsar does not result in increased chromosomal damage and is not detrimental to the health of residents of these areas. Relocation of the inhabitants of HLNRAs of Ramsar not only is not only unnecessary but it could lead to considerable social, economic, and psychological problems.

## References

[1] F.P. Castronovo Jr., Teratogen update: radiation and Chernobyl, Teratology 60 (2) (1999) 100–106.
[2] J. Robbins, Lessons from Chernobyl: the event, the aftermath fallout: radioactive, political, social, Thyroid 7 (2) (1997) 189–192.
[3] Z. Kamarli, A. Abdulina, Health conditions among workers who participated in the cleanup of the Chernobyl accident, World Health Stat. Q. 49 (1) (1996) 29–31.
[4] M. Ghiassi-nejad, et al., Very high background radiation areas of Ramsar, Iran: preliminary biological studies, Health Phys. 82 (2002) 87–93.
[5] S.M.J. Mortazavi, et al., How should governments address high levels of natural radiation and radon? Lessons from the Chernobyl nuclear accident, Risk Health Saf. Environ. 13/1.2 (2002) 31–45.

# Lung cancer risk due to radon exposure for 10 or 20 years

## Jing Chen*

*Radiation Protection Bureau, Health Canada, 775 Brookfield Road, Ottawa, Canada K1A 1C1*

**Abstract.** Lung cancer risks due to lifetime radon exposure are given in the BEIR VI report. However, exposures for shorter periods of time are of practical interest since exposure to elevated levels of radon may occur and end at any age. This study aims to produce practical data of lifetime relative risks due to radon exposures for 10 or 20 years, and at various radon concentrations found in most homes from 50 to 500 Bq/m$^3$. © 2004 Elsevier B.V. All rights reserved.

*Keywords:* Rn-222; Indoor radon; Lung cancer; Lifetime risk

## 1. Introduction

Radon is a naturally occurring radioactive gas. When inhaled, radon can cause mutations which lead to lung cancer. The BEIR VI report [1] outlined its preferred two risk models for exposure to radon progeny, and listed the estimated risk due to lifetime exposure. However, exposures for shorter periods of time are of practical interest because most people do not live in a house for their entire life. During the Health Canada Radon Workshop [2] held in Ottawa, March 3–4, 2004, some practitioners working in the front line of radiation protection showed strong interest in having risk tables for individuals exposed to short periods, such as 10 or 20 years. To meet this demand, lifetime relative risks for exposures of 10 or 20 years at various radon concentrations are calculated.

## 2. Methods and results

In the BEIR VI report, two preferred risk models were developed, the exposure-age-concentration model and the exposure-age-duration model. A recent publication of the U.S. EPA [3] provided a scaled version of the BEIR VI exposure-age-concentration model

---

\* Tel.: +1 613 941 5191; fax: +1 613 957 1089.
*E-mail address:* jing_chen@hc-sc.gc.ca.

0531-5131/ © 2004 Elsevier B.V. All rights reserved.
doi:10.1016/j.ics.2004.09.023

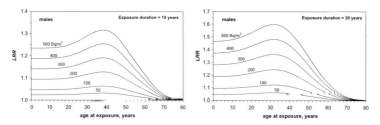

Fig. 1. Lifetime relative risks, *LRR*s, of lung cancer for males based on the EPA model.

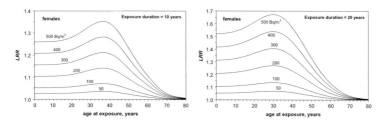

Fig. 2. Lifetime relative risks, *LRR*s, of lung cancer for females based on the EPA model.

for use in residential exposures. It yields the geometric mean of the results from the two BEIR VI preferred models. The new EPA risk model is used here.

The formulae for the calculation of lifetime risk of lung cancer are well described in the BEIR IV report [4]. In this study, we use the 1985–1989 age-specific mortality rates from the US, as used in the BEIR VI report. A lifespan of 110 years is assumed. An adjustment of lung cancer mortality rates to reflect smoking status is not applied here due to limited paper length.

As indicated in the BEIR IV report, lifetime relative risk (*LRR*) is the most suitable risk quantity to characterize individual risks of lung cancer. *LRR*s for radon exposures of 10 and 20 years are given in Figs. 1 and 2 for males and females, respectively. Age at exposure is the age exposure started.

## 3. Conclusions

Results in Figs. 1 and 2 demonstrate that the risk of developing lung cancer due to radon exposure increases with radon concentration and exposure duration. Individuals at middle ages (30–50) have higher risk when exposed to radon for several years, especially compared to later years. Therefore, individuals could lower the risk significantly by reducing radon levels earlier in their life.

## References

[1] National Research Council, Biological effects of ionizing radiation (beir) vi report, Health Effects of Exposure to Radon, National Academy Press, Washington, DC, 1999.
[2] B.L. Tracy, et al., Assessment and management of residential radon health risks: a report from the health Canada radon workshop, J. Toxicol. Environ. Health (2004) (in press).
[3] Environmental Protection Agency. EPA assessment of risks from radon in homes. Office of Radiation and Indoor Air. Washington DC, (2003).
[4] National Research Council, Biological effects of ionizing radiation (beir) iv report, Health Effects of Radon and Other Internally Deposited Alpha Emitters, National Academy Press, Washington, DC, 1988.

ELSEVIER

www.ics-elsevier.com

# Radiation doses to cardiovascular system due to absorbed radon

## Jing Chen[a],*, Richard B. Richardson[b]

[a]*Radiation Protection Bureau, Health Canada, 775 Brookfield Road, Ottawa, Canada K1A 1C1*
[b]*Radiation Biology and Health Physics Branch, Atomic Energy of Canada Limited, Chalk River, Canada*

**Abstract.** Radon is a naturally occurring radioactive gas. When it is inhaled, a significant proportion of the radon and its progeny deposited in the lung are absorbed into the blood. The absorbed radon delivers radiation doses to all tissues or organs. The thin inner layer of the coronary arteries is probably the primary target tissue of radiation-induced atherosclerosis. A Monte Carlo simulation code has been developed and used to derive the fraction of the alpha-radiation emitted by radon and progeny in blood that is absorbed in the primary and the secondary targets tissues. Age-dependent calculations have been performed for members of the public. These preliminary results indicate that young children are at a higher risk, because of their thinner intima layers and relatively higher dose fractions received in the sensitive internal layers. © 2004 Elsevier B.V. All rights reserved.

*Keywords:* $^{222}$Rn; Indoor radon; Noncancer; Cardiovascular system

## 1. Introduction

Radon is a colourless, odourless, radioactive gas that occurs naturally in the environment. In dwellings, radon, both $^{222}$Rn and $^{220}$Rn, can accumulate to relatively high concentrations. The major health effect associated with high exposure to radon is an increased risk of developing lung cancer. However, exposure to radon could increase the risk of a variety of noncancer diseases, especially coronary artery disease.

When it is inhaled, a significant proportion of radon deposited in the lung is absorbed into blood. The absorbed radon delivers radiation doses to tissues or organs other than the lung. Atherosclerosis is recognized as being caused by inflammation of inner vessel walls. Therefore, the intima, the thin inner layer of the coronary arteries containing the endothelium layer, is probably the primary target tissue of radiation-induced atheroscle-

* Corresponding author. Tel.: +1 613 941 5191; fax: 1 613 957 1089.
  *E-mail address:* jing_chen@hc-sc.gc.ca (J. Chen).

0531-5131/ © 2004 Elsevier B.V. All rights reserved.
doi:10.1016/j.ics.2004.09.027

Table 1
Dimensions of coronary artery wall layers for normal individuals based on published data [1]

| Age, gender | Diameter (mm) | | Thickness (mm) | | |
| --- | --- | --- | --- | --- | --- |
| | Lumen | External | Intima | Media | Adventitia |
| 6 months old, male/female | 1.08 | 1.44 | 0.05 | 0.06 | 0.07 |
| 25 years old, male | 2.04 | 3.47 | 0.30 | 0.21 | 0.21 |
| 25 years old, female | 1.73 | 2.95 | 0.25 | 0.18 | 0.18 |
| 70 years old, male | 2.16 | 3.77 | 0.47 | 0.18 | 0.15 |
| 70 years old, female | 1.84 | 3.20 | 0.40 | 0.16 | 0.13 |

rosis. To study radon effects on cardiovascular system, doses to the target vascular wall layers due to absorbed radon need to be determined.

## 2. Methods and results

The coronary artery wall consists of three layers around the blood lumen, the intima, media, and adventitia. A model with cylindrical symmetry is used to derive the energy deposition in the coronary artery wall from radon and its progeny in blood. Different layers of coronary artery are simulated with coaxial cylindrical shells of different thicknesses. Age-dependent dimensions of coronary artery wall layers for normal individuals (Table 1) are based on published data [1]. A Monte Carlo code has been developed which employed mass stopping power of alpha particles in human tissues [2]. It was assumed that absorbed radon is uniformly distributed in blood. Results of energies deposited by $^{222}$Rn and its progeny, $^{218}$Po and $^{214}$Po, are given in Table 2.

## 3. Discussion

It is known that only a small proportion of inhaled radon and its progeny will pass across lung into blood. The average activities of $^{222}$Rn, $^{214}$Pb, and $^{214}$Bi in blood were measured by research groups [3,4] per unit radon concentration in the atmosphere. However, the data vary significantly. There is also a tenfold uncertainty in the radon solubility in fatty human tissues, which is required to assess annual doses to organs/tissues from exposure to radon. The energy depositions per decay of $^{222}$Rn in blood showed that the highest doses from absorbed radon will be to the intima layer of newborns, because of their thinner intima layers and relatively higher dose fractions received in the sensitive internal layers. On average, females are at a higher risk than males due to smaller arterial dimensions. Because of higher solubility of radon in fat, diseased individuals with occluded lumen are expected to have higher dose to plaque and intima than that to intima of normal individuals when exposed to the same levels of radon. Later work will account for radionuclide diffusion into vascular wall and adventitial perfusion.

Table 2
Energy depositions in different coronary artery wall layers per decay of $^{222}$Rn in blood

| Age, gender | Lumen | Intima | Media | Adventitia |
| --- | --- | --- | --- | --- |
| 6 months old, male/female | 18.86 MeV | 0.309 MeV | 0.011 MeV | 0 |
| 25 years old, male | 19.02 MeV | 0.159 MeV | 0 | 0 |
| 25 years old, female | 18.99 MeV | 0.192 MeV | 0 | 0 |
| 70 years old, male | 19.03 MeV | 0.149 MeV | 0 | 0 |
| 70 years old, female | 19.00 MeV | 0.179 MeV | 0 | 0 |

## References

[1] I. Ozolanta, et al., Changes in the mechanical properties, biochemical contents and wall structure of the human coronary-arteries with age and sex, Med. Eng. Phys. 20 (1998) 523–533.
[2] International Commission on Radiation Units and Measurements, Report 49, Stopping Powers and Ranges for Protons and Alpha Particles, 1993, Bethesda, USA.
[3] G. Monchaux, Commissariat a l'Energy Atomique; Contract F14P-CT-95-0025. CEA-R-5882(E), 1999.
[4] E. Pohl, Radiological Health and Safety in Mining and Milling of Nuclear Materials, vol. 1, IAEA, Vienna, 1964, pp. 221–236.

ELSEVIER

www.ics-elsevier.com

# Award of honor

Drs Morishima, Hayata and Akiba, Chairman and Vice Chairman of the Organizing Committee and Chairman of the Programming Committee of the 6HLNRRA, proposed to the International Scientific Committee to express sincere gratitude to the founders and major contributors of the studies on high level natural radiation and radon areas. The proposal, endorsed by the committee, was approved by the International Committee of the HLNRRA on September 8, 2004, and it was decided to provide Drs Tsutomu Sugahara, Wei Luxin and Mehdi Sohrabi the Award of Honor at the 6th ICHLNRRA. Following the tradition in Japan and in many other countries, the students of those three distinguished scientists and teachers provide their words of recognition and appreciation.

**Yasuhito Sasaki**
Vice Chairman of International Scientific Committee

Professor Tsutomu Sugahara

Prof. Sugahara is a father of radiation research in Japan. He is also a father standing at the crossroads of domestic radiation research communities, connecting them to those of the world. There is a good reason for the wide and important influence he has exerted on all the fields of radiation research and it comes from his unique background training for his undergraduate and graduate studies. He was trained as a medical doctor, but after receiving the license of clinical practice, he turned his interest to physics and did his graduate study in radiation physics. After receiving a degree in physics, he started his career as a radiology doctor in a hospital. However, his background training drove him to extend his activity toward a wider field of radiation-related science. He became a professor in the Department of Fundamental Radiation Medicine at Kyoto University in 1961 where he helped many young students to become well trained researchers in the field. He has created several sub-disciplines of radiation research. He pioneered the research on sensitizer/protector to be used in clinics, fathered hyperthermia research and his contribution in basic radiation biology is also to be mentioned. However, for the 6th International Conference on High levels of Natural Radiation and

Radon Areas, his strong interest in radiation protection/risk estimation is also to be mentioned. He was quick in realizing the importance of high background radiation research and established collaborative research group between China, India, Iran and Japan.

Altogether, his interest in science is always interdisciplinary and this is the reason why he contributed so much. However, the most important source of power which has been driving him to the present accomplishment is endless curiosity about humans and nature. So, we recognize him as a giant of radiation research on whose shoulders we reside.

Thus, we would like to present this award to thank him for his great contributions in the past. Thanks also goes for his further contributions to the field.

**Ohtsura Niwa**
Kyoto Univ., Japan

Professor Luxin Wei

Prof. Luxin Wei was born in 1922, and graduated from the Medical College of Peking University, China in 1950. After graduation he worked as an assistant and lecturer in the Department of Public Health of his Alma Mater; and did advanced study on radiobiology in the Institute of Labours Hygiene and Occupational Diseases of the Academy of Medical Science, Moscow, former Soviet Union (1956–1957). Since then he has worked with the Beijing Institute of Radiation Medicine, in charge of formulation of national standards, regulations, and decrees of radiological protection, and radiation risk assessment. From 1972, he has worked with the Laboratory of Industrial Hygiene (reorganized in 2002 as the National Institute for Radiological Protection, Chinese Center for Disease Control and Prevention). Since then, he has spent almost all his working time on high background radiation research. He has been a professor of Radiation Health since 1982.

He was the member of Committee 4 of the International Commission on Radiological Protection from 1981 to 1985, and representative to UNSCEAR from 1987 to 1989. He was the chairman of the Society of Radiation Medicine and Radiological Protection, Chinese Medical Association from 1984 to 1994.

Prof. Wei is one of the principal founders of radiological protection in China and a pioneer of epidemiological study on the effect of high level natural radiation in the world. He is one of the first scientists in China to study the health effects of low-dose exposure; he has led his research group to conduct comprehensive epidemiological studies in the Yangjiang area since the early 1970s, and published a cornerstone paper in Science in 1980. In collaboration with American and Japanese colleagues, his research group conducted a series of world well-known studies on the health effects of high background radiation in the area of Yangjiang in China. He was the chairman of the fourth Conference of HLNRRA, which was held in Beijing in 1996.

Thus, we would like to present this award to thank him for his great contributions in the past. Thanks also goes for his further contributions to the field.

**Quanfu Sun**
NIRP, Chinese Center for Disease Control and Prevention

Professor Mehdi Sohrabi

Prof. Sohrabi has a Ph.D. in Health Physics from Georgia Institute of Technology (USA) in 1975 supervised by the late Prof. K. Z. Morgan, the father of health physics. He is one of the founders of health physics in Iran, established a comprehensive radiation protection (RP) infrastructure in Iran, and now is the Regional Manager at IAEA in the West Asian countries. He has held positions as Professor of Health Physics in Iran (since 1980), Member of the Iranian Academy of Sciences (since 1995); Director General of National Radiation Department (1970–1998); Heads of Authorization and Inspection Division, Radiation Dosimetry R & D and Services Division (AEOI, 1976–1979); Research Assistant and Scientist at Ga. Tech. (1972–1976), and Head of Health Physics Section, Tehran University 5 MW Research Reactor (1967–1970).

He has over 30 years' research and development experience in RP and dosimetry. His pioneering work on neutron dosimetry, alpha and radon detection, electrochemical etching of particle tracks in solids, etc. is well known and has led to some discoveries and novel dosimetry methods used worldwide. He also established major studies in HLNRAs of Ramsar, organized the 1990 conference in Ramsar, and regularized such conferences, as a founding member and Chairman of the ICHNRRAs. He has over 100 conference presentations and is the author of over 110 articles. He is one of the founders of the Ph.D. program in science and technology in Iran and was thesis advisor to over 50 M.Sc & Ph.D students. He has been editor or member of the editorial board of some international/national journals, and conference editor of the Int. J. of Nuclear Tracks. He is also a recipient of awards such as "Best Student Paper Award" (ANS, 1975), "Distinguished Ph.D. Research Award of the Sigma Xi Research Society of America" (1976), "AEOI Distinguished Director Award" (1991), "Research Medallion of the President of Iran" (1998), and "Research and Education Awards" from Presidents of four universities.

Thus, we would like to give this award to thank him for his great contributions in the past and his further contributions to this field in advance.

**Farideh Zakeri**
NRPD, Iranian Nuclear Regulatory Authority

www.ics-elsevier.com

# Concluding remarks

The inauguration of the 6th International Conference on High Levels of Natural Radiation and Radon Areas on Tuesday 7 September 2004 at the November Hall of the Kinki University in Osaka welcomed over 200 research scientists from 33 countries including Brazil, China, India and Iran, countries with high level natural radiation areas (HLNRA), and the representatives of the International Atomic Energy Agency (IAEA), the World Health Organization (WHO), and the International Committee on High Levels of Natural Radiation and Radon Areas (ICHLNRRA). The opening remarks made by the Chairman of the Organizing Committee, the Deputy Director General of the IAEA and the Chairman of International Committee emphasized the importance of studies in such areas for better understanding of the biological and health effects of low level doses from internal and external exposures to ionizing radiation.

The Conference was a highly successful event coorganized by the Kinki University (Osaka) and the National Institute of Radiological Protection (Chiba, Japan) and in cooperation with UNSCEAR, IAEA and WHO, as coordinated by the ICHLNRRA. The conference covered topical issues mainly on depleted uranium; external and internal radiation dosimetry, radiobiology, and epidemiology in HLNRA of Brazil, China, India and Iran; radon dosimetry; contribution of HLNRA studies to health risk assessment, and regulatory control. The papers presented indicated the extensive efforts made by the participating countries in achieving the main objectives of the chain of conferences towards better understanding of biological and health effects of low-level natural radiation dose to human beings. Summaries of some topical findings and shortcomings encountered in dosimetry, biological, and epidemiological studies were prepared and presented by some Rapporteurs at the conference and are published in these proceedings. It is my pleasure to provide you with some highlights of the conferences with some remarks on the conclusions of different related events held prior, during and after the conference as follows:

1. From the studies carried out in some HLNRA as presented at a one-day International Workshop on HBRA Studies on September 6 (as organized by the Health Research Foundation and Central Research Institute of Electrical Power Industry, which preceded the conference, and at the conference, it is rather difficult to conclude any confirmed radiation health effects in the study group compared to control area. There are some concerns on many existing unknowns such as lack of information on

individual doses in particular from internal exposures; on confounding factors (e.g. cigarette smoking, pesticide use, alcohol drinking, etc.) and their effects; on standardized protocols and criteria for selecting study and control area; etc. Such factors should not cast a shadow over the low risk, if any, at low-level uniform exposures of a relatively larger population in for example areas in China and India. The areas in Ramsar have low affected population but individual doses varying from low to very high might assist to obtain dose-effect related response in a well-planned study. It seems, therefore, that the experiences learned in previous studies can be applied to future studies.

2. No long term effects of depleted uranium in those individuals exposed during the war conflicts is expected due to low doses received, while some low incidence of long health determinants might be expected in those receiving higher doses. However, it seems that concrete results on health effects due to low levels of contamination to depleted uranium are yet to be experienced.

3. The internal doses of each individual in addition to external doses in the study and/or control areas should be well determined and considered in biological and epidemiological studies for obtaining dose-effect responses. For lung cancer incidence studies in residential areas, in particular, the individual absorbed dose to lung due to $^{222}$Rn and its progeny (considering the attached and unattached fractions) as well as $^{220}$Rn and its progeny specially in monazite sand areas should be well realized and retrospectively determined using proper measurement methods and conversion factors.

4. The study and control areas should have similar essential parameters in terms of environmental, sociological, and economical conditions; standards of living; confounding factors; sex and age; etc., but only differ in individual doses as the main parameter of concern. Selection of control areas in large cities with low doses but high pollutions might prove that living in HLNRA usually with low industrial pollutions has even lower life risk.

5. Standardized criteria for the selection of the HLNRA and control area: for dosimetry methods and protocols (equipment, dose units, etc.) for biological and epidemiological studies, etc. have yet to be developed as a priority by or through some international organizations or by a funded activity through the ICHLNRRA.

6. The support provided by Japan for some activities have been very instrumental in upgrading the level of understanding of radiation health effects and consolidation of results of studies in such areas. It is recommended that such a type of support be promoted further and sustained to obtain the information needed based on a standardized criteria and approaches.

7. At a lunch forum of a small group of invited participants from Brazil, China, India, and Iran held during the conference period, related to a project plan on health and psychology of residents of HLNRA, and based on a questionnaire completed by them, the psychological and sociological factors of residents in LELNRA were discussed and considered as highly important factors in such studies. In particular, interviews with individual residents would be valuable to obtain necessary information. In response, it is suggested that in the 7th International Conference in 2008, a session and/or a panel discussion be held on this issue. Moreover, it is

recommended that the studies carried out in such areas be well planned in a well consolidated manner with the co-operation of all stakeholders to avoid fragmented and ad-hoc results and to prevent any psychological effects on the public.

8. At a one-day post-conference seminar organized by the National Institute of Radiological Sciences (NIRS) in Chiba, near Tokyo, entitled "Frontiers in Radiation Protection" with over 50 local and invited participants, and by considering the findings presented also at this conference, it was concluded that the systematic studies in HLNRA are one real alternative to obtain health and biological effects at low doses to clarify further the adaptive response and bystander effects provided that all factors affecting the results be well determined in advance and considered in such studies.

9. The conference, based on the interest expressed by and verbal feedback received from the participants, seemed highly successful in meeting the objectives and standard requirements for organization of such events in an academic university environment. It can also be concluded that the organization of these conferences on a regular basis has been highly instrumental in consolidating results of studies in such areas, enhancing the awareness and importance of such studies among the international organizations and commissions, government bodies, and health professionals; and establishing a close family community of researchers working in these areas as well as fostering exchange of information and collaboration between them. In response, consolidated efforts are invited towards solving the basic response of radiation health effects at low doses as function of internal and/or external doses.

10. Many events during the conference made it memorable. The Japanese hospitable environment and personal attention by the organizers treating the participants graciously, the three major earthquakes followed by smaller ones together with typhoons while the safety of buildings in Japan against such events, the initiation of an awarding mechanism for the community, and above all sustainable organization of such conferences are worth mentioning.

11. At a formal dinner meeting of the ICHLNRRA held on 8 September 2004 at the Miyako Hotel, hosted by the Japanese organizers, the committee members and some guest members from the International Scientific Committee of the conference and the representatives of two countries, as the candidates to host the 7th International Conference in 2008, discussed and endorsed the following issues:

   ▶ The Chairman of the Committee first welcomed the members and updated the meeting on its aims and objectives and the significant progress made in this area through the organization of the chain of these conferences. In particular thanks were expressed to the conference organizers for the support, co-operation, and efforts which make the conference a success.

   ▶ The Japanese organizers briefed the meeting on status of the progress made in the organization of the conference, provided some statistics on participations, and proposed provision of "Award of Honor" to some founders and contributors to the studies on HLNRA and other related fields. The organizers further nominated three candidates based on previous consultation with and endorsement of the scientific, organizing, and program committees of this conference and the ICHLNRRA. The members further supported the proposal and unanimously endorsed the nominations.

- ▶ The venue of the 7th conference in 2008 was discussed with the representatives of Brazil Dr. L.H.S. Veiga (Institute of Radioprotection and Dosimetry, Rio de Janeiro) and India Dr. B. Rajan (Director, Regional Cancer Centre, Trivandrum) and Mr. M. V. Thampi (Bhaba Atomic Research Centre, Mumbai) who expressed high interest and commitment to host the venue. They briefed the meeting on the local and national capabilities in support of their request. In this context, Dr. Veiga withdrew her request in favor of Indian colleagues, who seemed highly authorized and committed with full charge to host the event, and congratulated and wished them success. Dr. Rajan and his colleague invited and assured the members and later the conference participants that the venue will be held during the first week of March 2008 and would be an excellent conference and at a high standard together with excellent social activities, and hospitality and entertainment unique to the Indian culture and tradition.
- ▶ The issue of establishing an International Society on Natural Radiation Environment, which was proposed and discussed at the Committee in Munich conference in 2004 was further followed up. The Committee was informed that some participants of that conference have already taken the issue further at the 7th International Symposium on Natural Radiation Environment, NRE VII held in Rhodes/Greece May 20–24, 2002 and has initiated establishing this society.
- ▶ The need to publishing a "News Letter" was discussed and proposed be established by 2008.

On behalf of the International Committee on HLNRRA and myself, I take this opportunity to express sincere thanks and appreciation to the Government of Japan, the Kinki University in Osaka, the Health Research Foundation in Kyoto, and the National Institute of Radiological Protection in Chiba for sincere and effective organization of the conference and side workshops and seminars, for high value and interest expressed to this important topic of studies of radiation health effects on man, for financial support to a number of participants (which made their participation possible) and support to studies in these areas, and treating all of us graciously and with comfort and honor.

Special thanks are due to many international organizations including IAEA, WHO, UNSCEAR for the interest and moral support to the organization of the conference. Sincere efforts made by many high officials of the above institutions which took the burden of the hard work and support are highly recognized and appreciated in particular Drs. T. Sugahara, Y. Sasaki, H. Morishima, I. Hayata, and S. Akiba and their colleagues, as well as those from other support organizations, and specially to Professor S. Akiba for his restless efforts, interest, and contributions in all aspects of the conference which made it a success.

I also wish to congratulate the Indian colleagues Dr. Rajan and Mr. Thampi for being trusted to take the burden of hosting the 7th International Conference in 2008 and wish you good health and high success in organizing a highly successful conference in March 2008 in India.

**Mehdi Sohrabi**
**ICHLNRRA**

# Author index

Abbasi, A., 440
Abd El-Aziz, N., 407
Aghamiri, S.M.R., 225
Akata, N., 307
Akiba, S., 41, 76, 97, 102, 106, 233, 236, 430
Al-Azmi, D., 350
Albergi, S., 403
Allam, K.A., 409
Al-Sewaidan, H.A., 407
Ambrosio, M., 242
Amutha, R., 329, 331
Andal Vanmathi, A.K., 344
Andrews, V.J., 8
Ånestad, K., 212
Anilkumar, V., 8
Aravindan, K.V., 8
Arnold, D., 85
Asadi, R., 440

Baba, T., 415
Babapouran, M., 169
Balajee, A., 189
Balzano, E., 242
Baradács, E., 362
Beate Ramberg, G., 212
Beitollahi, M.M., 13
Berka, Z., 381
Bhan Tiku, A., 187
Binu, V.S., 41, 102, 236
Birchall, A., 81
Blazek, K., 383
Bo, C., 313
Bodrogi, E., 371
Borisov, V.P., 354
Bovornkitti, S., 208
Brahmanandhan, G.M., 317, 319, 321, 327, 329, 331

Brandejsova, E., 381
Burkart, W., 133

Cameron, J.R., 266
Campos, M.P., 403
Carr, Z., 114, 147
Cavallo, A., 276
Cechak, T., 418
Chang, B.U., 46
Chen, C.-J., 315
Chen, D., 3
Chen, J., 311, 442, 444
Chen, M., 311
Chen, X., 309
Cheriyan, V.D., 8
Chinnaesakki, S., 339
Chittaporn, P., 72
Chruścielewski, W., 360
Csige, I., 362, 365
Czoch, I., 367

Danesi, P.R., 133
Das, B., 8
Das, S.K., 210
De Notaristefani, F., 383
Di Lullo, A., 412
Dikij, N., 274
Dombovári, P., 371

Eappen, K.P., 335, 337
Ebaid, Y.Y., 405, 409
Elegba, S.B., 401
Elisabeth, C., 118
El-Mongy, S.A., 405, 409
Esposito, A.M., 242
Ettenhuber, E., 62

Fadaei, S., 197
Fallahian, N., 13

Feige, S., 392
Fernández, P.L., 50
Frolova, A.V., 356
Froňka, A., 377, 379
Fujikawa, Y., 415
Fujimoto, K., 141
Fujinami, N., 37
Fujitaka, K., 124, 256, 434
Fukui, M., 415
Fukumoto, M., 192
Fukutsu, K., 58, 76, 251, 278
Funtua, I.I., 401
Furukawa, M., 76, 285, 307, 309

Gajendran, V., 329, 331
Gangadharan, P., 41, 102, 236
Gargioni, E., 85
Georgia, T., 390
Gerasopoulos, E., 204
Ghafourian, H., 268, 270, 272
Ghiassi-Nejad, M., 13, 201, 436
Gialanella, L., 242
Gomez, J., 50
Gorjánácz, Z., 369
Grecea, C., 373
Günaydı, T., 385
Guo, Q., 285, 307
Guryev, D.V., 181

Hajiahmadi, M., 438
Hakl, J., 369
Hanamoto, K., 249, 253
Harley, N.H., 72
Hashimoto, T., 231
Hattori, T., 432
Hayata, I., 3, 17
Hei, T.K., 21, 189
Heid, I.M., 54
Heinrich, J., 54
Heleny, F., 390
Hemmati, A., 440
Hendry, J.H., 133
Hoshi, M., 352, 356
Hou, C., 76
Huang, C.-C., 315

Hübel, K., 240
Hůlka, J., 379
Hunyadi, I., 362

Iacob, O., 373
Ichinohe, K., 177
Igari, K., 183, 262
Igari, Y., 183, 262
Iida, T., 221, 227, 287, 289, 297
Iimoto, T., 303
Ikeda, E., 415
Inagaki, M., 293, 295
Innocenzi, P., 412
Inoue, M., 227
Ioannidou, A., 204
Ishii, T., 299, 426
Ishikawa, T., 58, 76, 276, 283, 299, 301, 426
Ishimori, Y., 291, 299

Jagetia, G.C., 195
Jaikrishan, G., 8
Jalali, F., 438
Jankowski, J., 358, 360
Jayalekshmi, P., 41, 102, 236
Jia, G., 412
Jian, F., 309
Jiang, T., 3
Jigareva, T.L., 223
Jiménez-Nápoles, H., 399
Jobbágy, V., 371
Juhász, L., 362, 365, 367

Kabuto, M., 233
Kai, M., 159
Kakihara, H., 183, 262
Kalchenko, V.A., 223
Kale, R.K., 187
Kam, E., 385, 387
Kaneko, M., 162
Kariminia, A., 199
Karunakara, N., 341, 346, 348
Karuppasamy, C.V., 8
Kasianenko, A.A., 223
Kataoka, T., 249

Kataoka, Y., 299
Kato, F., 183, 260, 262
Kaul, S.C., 179
Kávási, N., 369
Kawakami, Y., 179
Kawasaki, Z.-I., 428
Kelleher, K., 399
Kelm, H., 397
Kendall, G.M., 129, 166
Khanna, D., 317, 319, 321, 323, 329, 331
Khater, A.E., 405
Khater, A.E.M., 407
Kim, C.K., 46
Kim, C.S., 46
Kim, Y.J., 46
Kinoshita, I., 422
Klimenko, N., 274
Kluson, J., 418
Kobal, I., 362, 375
Kobayashi, I., 293, 305
Kobayashi, Y., 281
Kodama, S., 155, 264
Koga, T., 37, 293, 295, 305
Kohda, A., 177
Koifman, S., 110
Koizumi, A., 251
Komori, H., 305
Komura, K., 227
Körner, S., 240
Kosako, T., 303
Kovács, T., 369, 371
Kovler, K., 397
Koya, P.K.M., 8
Kreienbrock, L., 54
Kreuzer, M., 54
Krishnan Nair, M., 41
Kubo, T., 299
Kulieva, G.A., 223
Kunka, A., 383
Kurien, C.J., 8

Lakshmi, K.S., 323, 325, 327
Lau, B.M.F., 217
Lee, H.Y., 46
Lehmann, R., 29

León-Vintró, L., 399
Leung, J.K.C., 217
Lin, D., 309
Lin, P.-H., 315
Liu, D., 192
Lloyd, D.C., 33
Lunder Jensen, C., 212
Luo, Y., 309, 311

Madhusoodhanan, M., 8
Magae, J., 179
Malathi, J., 317, 319, 321, 325, 329, 331, 344
Maleki, F., 197
Maly, P., 383
Manolopoulou, M., 204
Marenny, A.M., 352, 356
Maringer, F.-J., 365
Marsh, J.W., 81
Maruo, Y., 291
Masumura, K.-i., 25
Matarranz, J.L., 50
Matsumoto, K., 256
Matsumoto, T., 177, 179
Mayya, Y.S., 335, 337
McLaughlin, J.P., 137, 399
Medora, R., 72
Medvedeva, E., 274
Meenakshisundaram, V., 323, 325, 327, 329, 331
Merrill, R., 72
Mifune, M., 293, 295
Milka, D., 381
Minamihisamatsu, M., 3
Mitra, A., 8
Mitsunobu, F., 249
Miyazaki, N., 422
Miyazaki, T., 221
Mohagheghi, M., 106
Mohammadzadeh, S.G., 197
Mohanty, A.K., 210
Monfared, A.S., 438
Mori, M., 287
Moriizumi, J., 221, 287
Morishima, H., 3, 37, 293, 295, 305

Morozov, Yu.A., 356
Mortazavi, S.M.J., 201, 266, 436, 440
Mosavi-Jarrahi, A., 106
Motamedi, N., 106
Moučka, L., 377
Moucka, L., 383
Movafagh, A., 197
Mozdarani, H., 201, 438
Muguntha Manikandan, N., 227
Murata, Y., 227

Nagai, T., 303
Nagara, S., 221
Nair, M.K., 102, 235
Nair, R.N., 335
Nair, R.R.K., 102, 236
Nakata, Y., 231
Narayana, Y., 333, 348
Narazaki, Y., 426
Nawa, Y., 422
Nazeri, Y., 268, 270, 272
Németh, C., 283, 369
Németh, Cs., 371
Ng, F.M.F., 217
Nikezic, D., 217
Niroomand-Rad, A., 266
Nishijima, T., 428
Nitta, H., 233
Nitta, W., 299
Nohmi, T., 25
Nojiri, S., 293, 295
Nomura, T., 185
Norimura, T., 183, 260, 262
Nozaki, T., 428

Ocone, R., 412
Oeser, V., 392, 394
Oestreicher, U., 29
Oghiso, Y., 179
Ohta, M., 424
Okhrimenko, S.E., 356
Oki, Y., 297
Okuyama, S., 422
Olofsson, B., 206
Olszewski, J., 358, 360

Onyshchenko, M., 274
Ootsuyama, A., 183, 260, 262
Otsuka, K., 258

Papastefanou, C., 204
Paramesvaran, A., 344
Paridaens, J., 420
Pecequilo, B.R.S., 403
Penezev, A.V., 352, 356
Perevalov, A., 397
Piao, C.Q., 189
Pugliese, M., 242

Qiuju, G., 313
Quanfu, S., 313
Quindos, L.S., 50

Rabbani, M., 268, 270, 272
Raghu Ram Nair, K., 41
Rajan, B., 41, 102, 236
Rajashekara, K.M., 341, 348
Ramachandran, E.N., 8
Ramachandran, T.V., 335, 337, 339
Ramola, R.C., 215
Ratnikov, A.N., 223
Ravikumar, B., 8
Regner, J., 238
Repacholi, M., 114
Rezaiean, M., 436
Rho, B.H., 46
Richardson, R.B., 444
Roca, V., 242
Romano, M., 242
Ruden, L., 212

Sabbarese, C., 242
Sadeghi, S., 268, 270, 272
Saghirzadeh, M., 13
Saha, S.K., 210
Sainz, C., 50
Saito, M., 415
Sakai, K., 185, 258, 432
Saldan, I.P., 354
Samavat, H., 225, 438
Samuelsson, C., 66

Sanada, T., 299
Sarida, R., 329, 331
Sartandel, S.J., 339
Sasao, E., 287
Sato, K., 221
Savkin, M.N., 352, 356
Schaffrath Rosario, A., 54
Schmidt, P., 238
Seaward, M.R.D., 225
Sekiguchi, H., 305
Selvasekarapandian, S., 317, 319, 321, 323, 325, 327, 329, 331, 344
Sengupta, D., 210
Seshadri, M., 8
Shabestani Monfared, A., 106, 225
Shabestani-Monfared, A., 201
Shanbhag, A.A., 339
Shetty, P.K., 333
Shibuya, M., 422
Shimizu, T., 192
Shimo, M., 283
Shin Kim, Y., 285
Shinkarev, S.M., 352, 356
Shinogi, M., 299, 426
Shiraishi, K., 264
Shukla, V.K., 339
Siddappa, K., 333, 341, 346, 348
Simon, S.L., 89
Skeppström, K., 206
Skubalski, J., 358
Sohrabi, M., 169
Somashekarappa, H.M., 341, 346
Somlai, J., 369, 371
Soren, D.C., 8
Steiner, V., 397
Stephan, G., 29
Stoulos, S., 204
Strand, T., 212
Streil, T., 392, 394
Strokov, A.P., 354
Suarez Mahou, E., 50
Suda, H., 299
Sugahara, T., 3, 37, 97
Sugihara, T., 179

Sugita, T., 428
Sugiura, N., 303
Sun, Q., 76, 97, 147
Suzuki, A., 422
Suzuki, M., 21, 189, 256
Szabó, T., 365
Szalanski, P., 358
Szerbin, P., 362, 365, 367

Tachibana, A., 264
Takada, J., 245
Takahashi, H., 281, 283
Takami, M., 424
Takeda, S., 220
Takemura, T., 293, 295
Takeyasu, M., 289
Tanaka, K., 177, 179
Tanaka, Y., 293, 295
Tanizaki, Y., 249
Tao, Z., 97
Tatsuta, F., 305
Thampi, M.V., 8
Thinova, L., 381, 383, 418
Thomas, J., 379
Tokizawa, T., 221
Tokonami, S., 46, 58, 76, 151, 208, 219, 233, 251, 276, 278, 281, 283, 297, 299, 301, 307, 311, 369, 426, 430
Torii, T., 422, 428
Torri, G., 412
Toussaint, L.F., 93
Trautmannsheimer, M., 240
Trojek, T., 383, 418
Tsujimoto, T., 289
Tsuruoka, C., 256

Uda, T., 415

Várhegyi, A., 365, 369
Vaupotic, J., 362
Vaupotič, J., 375
Veiga, L.H.S., 110
Venkatesha, V.A., 195
Venoso, G., 242

Vijayan, U., 8
Vijayshankar, R., 344
Vincze, J., 365

Wadhwa, R., 179
Wanabongse, P., 208
Wang, C., 3
Wang, L., 192
Wang, W., 311
Wanitsooksumbut, W., 72
Wei, L., 3, 97
Wei, M., 309, 311
Wichmann, H.E., 54

Yamada, Y., 58, 76, 251, 278, 297, 309, 311
Yamaguchi, Y., 227
Yamanishi, H., 415
Yamaoka, K., 249, 253
Yamasaki, K., 289, 297
Yamazawa, H., 221, 287
Yarar, Y., 385, 387
Yarmoshenko, I.V., 141
Yasuda, H., 424, 434
Yasuda, N., 256
Yasuoka, Y., 299, 426
Yatabe, Y., 278
Yawata, T., 231
Yazdizadeh, B., 106
Yonehara, H., 58, 76, 281
Yonezawa, M., 264
Yoshimoto, T., 415
Yoshinaga, S., 233, 301, 430
Yu, K.N., 217
Yuan, Y., 3

Zak, A., 358
Zakeri, F., 199
Zdimal, V., 381
Zha, Y., 97
Zhang, S., 76
Zhang, W., 3
Zhao, Y.L., 189
Zhou, H., 21
Zhuo, W., 58, 76, 219, 251, 281, 283, 285, 307, 309, 311
Zou, J., 97
Zunic, Z.S., 141

# Keyword index

Absorbed dose 331
Absorbed dose rate 204
Active oxygen diseases 249
Activity 66
Activity concentration 85, 319, 323, 329
Activity index 379
Adaptive response 181, 187, 201, 264, 266
Adrenal stress hormones 199
Aerosol campaign 381
Airborne nuclides 227
Aircraft crew 434
Airflow control 287
Air–liquid interface culture 251
Airplane dose 124
Alpha 66
Alpha emitting radionuclides 371
Alpha spectrometry 371, 383
Alpha spectroscopy 358
Ambient dose 305
Angiosarcoma 192
Annual effective dose 41, 46, 169, 327, 329, 341, 344
Annual radiation dose 225
Antioxidants 249
Araxá 110
Archaeologically burnt material 231
Assessment 169
Atmosphere 426

Background radiation 106
Bacteria 268, 270
Badgastein 253
Balkan 141
$^7$Be 227
Beach sand 210
$^{215}$Bi 289

Biodosimetry 33
Biological effect 155
Biomass 268, 270, 272
Biomonitoring 418
Biosorption 268, 272
Blood glucose level 185
Blood insulin level 185
Blower door 377
Building material 339, 397
Bystander effect 21, 155
Bystander effects 256

Calculation 169
Calibration 85, 281
Cancer 97, 102, 106, 110, 155, 166
Cancer risk 147, 159, 385
Carbon ions 256
Cardiovascular system 444
CARI code 434
Case-control study 54, 233
Cave 381, 403
Cell cycle 179
Cesium 272
Chamber 276
Charcoal 397
Chemical mutagen 17
Chemiluminescence 274
Chernobyl accident 33
Childhood leukemia 233, 430
Cholangiocarcinoma 192
Chromium 268
Chromosomal aberrations 33
Chromosome aberration 3, 17, 155, 197
Chromosome aberrations 177
Chronic inflammation 274
Clastogen 17
Clearance 432

Closed can method 291
Coal fired power plant 365
Collecting plate 297
Collective effective dose 352, 356
Comet assay 258
Concentration 289, 375
Concept for institutional control 62
Continuous measurement 281, 375
Cosmic radiation 129, 434
Cosmic ray 169
Cosmic rays 124
Cost effective 392
CR-39 385
Cytogenetic analyses 29
Cytogenetic effect 223
Cytogenetics 8

Decay product 76
Deep space missions 266
Delayed mutation 262
Delayed-coincidence and time-analysis method 303
Deletion 25
Depleted uranium 133, 137, 141, 223
Depth profile study 323
Detectors 397
Dicentric and ring 3
Dicentric chromosomes 29
Diffusion in snow, frozen soil 221
Dissolved radon 208
Distance 297
Distribution 297
Diurnal variation 227, 341
DNA damage 258
Domestic 66
Dose 50, 89, 129, 362, 375, 385
Dose assessment 278, 283, 369, 371
Dose conversion factor 151
Dose rate 317, 390
Dosimetry 81, 89, 118, 151, 245, 418
Drinking water 371
DU field studies 133
DU properties 133
DU-munitions 394
Dwellings 335, 358

Earthquake 426
Effective dose 76, 238, 240, 317, 321, 323, 325, 373, 381, 434
Effective dose rate 319
Electret 397
Electric field 428
Electronic personal dosimeter 37
Elevated level 169
Emanating power 303
Enhanced natural radiation level 307
Enhanced radioactive area 141
Environmental radiation 37, 236
Environmental radiation dosimetry 41
Environmental radioactive contamination 420
Environmental radioactivity 242, 387, 409
Environmental survey meter 327
Eolian dust 307
Epidemiology 54, 66, 81, 89, 97, 118, 159, 162, 166, 233, 430
Equilibrium equivalent concentration 76
Equilibrium factor 217
Equilibrium-equivalent thoron concentration 219, 309, 311
Etched-track detectors 375
$Eu^{3+}$ 424
Exemption 432
Exhalation 397
Exhalation rate 287
Exposure 13, 81, 403
External dose 169
Extranuclear target 21

Fly ash 339

Gamma 93, 333
Gamma dose rate 309, 311
Gamma exposure 387
Gamma radiation 169, 331
Gamma ray spectrometer 321, 325, 329
Gamma rays 177
Gamma spectrometry 346, 418
Gamma-ray spectrometry 289, 350

Gamma-ray spectroscopy 409
Genetic instability 177
Genome instability 25
Geothermal 385
GIS 206
Glass 66
Global burden of disease 114
Glutathion S-transferase 197
GPS 420
Granite 293
Granitic alkaline rocks 401
Gross alpha 325, 387
Gross beta 325, 387
Ground surface radon exhalation 221
Groundwater 206, 299, 301, 426
Guarapari 110
Guidance level 379

HBRA in China 307
Health 438
Health effect 17
Health effects 118
Health impact 133
Health risks 137
Heavy ions 189
Hepatectomy 260
High background natural radiation 283
High background radiation 438
High background radiation area 155, 210, 333
High Background Radiation Area 313
High Background Radiation Areas 166
High level natural radiation areas 440
High levels of natural radiation area 37
High natural background radiation 162
High-level natural radiation 199
High-purity germanium detector 253
Himalaya 215
History of surveillance 62
Hokutolite 315
HPRT locus 256
hTERT 189
Human 192
Human lymphocyte 3

ICP-MS 415
Immune responses 199
Immunity 249
Impactor 297
Implantation 66
Indian scenario 337
Individual dose 37
Individual dose estimation 41
Indoor 58, 169
Indoor $^{221}$Rn 335
Indoor gamma dose 327
Indoor gamma radiation 344
Indoor radon 29, 208, 385, 442, 444
Infiltration 377
Influence on plants 223
Ingestion 301
Inhalation 301
Inhalation dose 394
Inhibitory damaged in pancreatic β-cells 185
In-situ gamma-ray spectrum 415
Integral counting method 299
Intercomparison 276, 297
Internal dose 169
Internal radiation exposure 133
International space station 124
Ionizing radiation 114, 262
Iran 106, 169, 197, 438
Irradiation 195

Japan 58
Jet 297

$^{40}$K 331
$^{40}$K activity 346
Kerala 8
Kindergarten 375

Large energy window 350
Laser flourimetry 405
Lifetime risk 442
Lightning 428
Linear-no-threshold hypothesis 162
Liquid scintillation counter 299

Liquid scintillation spectrometer 293
Liver 192, 260
Liver regeneration 181
Log normal distribution 432
Long-lived α-nuclides 394
Low dose 97, 147
Low dose radiation 179
Low dose rate 159, 177
Low dose rate radiation 179
Low dose-rate 258
Low dose-rate γ-irradiation 274
Low dose-rate irradiation 185
Low-dose ionizing radiation 181
Low-dose radiation 118
Low-dose rate 264
Low-level gamma spectrometry 227
LR 115 detector 217
LR-115 film 335
Luminescence dating 231
Luminescent mechanism 424
Lung cancer 54, 236, 436, 442
Lung dose 287
Lymphocytes 195

Makrofol E 403
Malformation 8
Mangiferin 195
Mapping 420
Masutomi spa 295
Measurement 169, 276
Measuring method 62
Measuring technique 392
Metaanalysis 430
Mice 260
Microarray 179
Micronuclei 195
Micronucleus 177, 251
Migration 223
Mining and processing 401
Misasa 253
Misasa, In air 293
Moasel 407
Model 352, 356

Monazite 210, 333, 348
Monazite ore 323
Mortality 106, 110
Moscow 356
Mouse 264
Mouse spleen 258
Multivariate statistics 206
Mutation 21, 183, 260

NaI(Tl) gamma ray spectrometer 317
NaI(Tl) scintillation survey meter 37
Nasser Lake 405
National regulation 62
Natural background radiation 89, 166, 337
Natural hot spring 208
Natural radiation 3, 8, 13, 97, 102, 110, 147, 159, 169, 201, 210, 356, 415
Natural radioactivity 331, 375, 390, 405, 407
Natural radionuclide 295
Network 242
Nickel 270
Non-cancer 97
Noncancer 444
Non-cancer risk 147
NORM 89, 93, 141, 339, 367, 432
Normal human fibroblasts 256
Nuclear emergency 422
Nuclear power plant 418
Nuclear weapon terrorism 245
Numerical analysis 287

Occupational exposure 129
Occupational radiation 197
Odds Ratio (OR) 236
On line aerosol sampler 394
Optically Stimulated Luminescence 305
Optically stimulated luminescence (OSL) 231
Optimization 432
Orissa 210
OSL 293

Osteoarthritis 249
Outdoor 169
Outdoor radon-223 285

p53 179, 262
Pb-212(Th)/Pb-215(U) ratio 295
$^{210}$Pb 66, 227
$^{212}$Pb 227
$^{215}$Pb 289
Pain-associated substance 249
Particle size distribution 278
Particle size measurement 72
Passive dosimeter 335
Peito hot spring 315
Personal monitor 305
Phosphogypsum 339
Pico-rad detector 293
Pleistocenic cave 204
$^{210}$Po 66
$^{210}$Po alpha-recoil implantation 399
Poços de caldas 110
Point mutation 25
Polonium 405, 407
Population 352
Population exposure 373
Powdery sample 303
Pregnancy 183
Preseismic 426
Pressure difference 377
Primordial radionuclide 321
Primordial radionuclides 317, 319
Probability distribution 432
Progeny 297, 313
Public 169

Quartz grains 231

$^{225}$Ra 412
$^{227}$Ra 348, 412
$^{227}$Ra activity 346
Radiation 17, 50, 129, 183, 187, 260, 315
β-radiation 394
Radiation dose 215, 397
γ radiation dose rate 293
Radiation exposure 62, 401

Radiation level measurement 41
Radiation monitoring 422
Radiation protection 245
Radiation risk 21, 440
Radiation-induced damages 181
Radiation-induced luminescence 231
Radioactive mineral 253
Radioactivity 315, 348, 362
Radio-adaptive response 155
Radiological protection 162, 434
Radionuclide in diet 225
Radionuclide in human body 225
Radium 93, 315
Radium-227 285
Radon 46, 50, 54, 58, 62, 66, 76, 81, 93, 151, 159, 206, 212, 215, 233, 240, 242, 251, 283, 299, 301, 309, 311, 313, 352, 354, 356, 362, 375, 377, 383, 392, 397, 401, 403, 426, 430, 436
Radon and thermal therapy 249
Radon and thoron dose 72
Radon and thoron measurement 72
Radon and thoron progeny 373
Radon chamber 278
Radon concentration 204, 238, 360, 381
Radon concentrations 212
Radon decay product 278
Radon exhalation rate 291
Radon flux 291
Radon gas measurements 350
Radon in workplace 369
Radon progeny 362, 428
Radon progeny concentration 217
Radon programme 379
Radon sampler 397
Radon spas 390
Radon survey 354
Radon surveys 212
Radon therapy 253
Radon-223 287, 293
Radon-223 flux 285
Rainwater 289
Ramsar 13, 106, 169, 201, 225, 436, 438, 440

Ratio 289
γ-Ray spectrometry 295
Recoil 66
Red-TL 231
Release 301
Remedial 379
Remedial action 367
Remediation 238, 365
Remediation measure 240
Remediation of workplace 62
Removal 268, 270, 272
Removal model 289
Residential radon 114
Retrospective 66
Retrospective radon assessment 358, 399
Risk 81
Risk communication 137
$^{221}$Rn 303
$^{223}$Rn 297, 444
$^{223}$Rn and its progeny 169
$^{223}$Rn concentration 341
Rn-223 442
Runaway electron 428
Russia 352
Ryukyu Islands of Japan 307

School 242, 369, 375
School, Child 375
Scintillation flask 303
Sediment 405
Semipalatinsk test site 245
Single-aliquot regenerative-dose (SAR) 231
$SiO_2$ 424
Size 297
Small portable autonomous device 392
Smoking 407
SMR 106
Soil 295, 415
Sources 337
South West Coast of India 346
Southwest coast of India 341
Space radiation 266

Split dose 187
SSNTD 354
Substrate 297
α-spectrometry 412
Supernova 124
Surface 66
Survey 50, 375
Survival 187

T cell receptor 183
Tailings 401
Tailings ponds 365
Tamagawa spa 295
TCR 262
Telomerase 189
TENORM 141, 365, 367
Terrestrial 169
Terrestrial gamma radiation 285
Terrestrial radiation 41
$^{233}$Th 331, 333, 348
$^{233}$Th activity 346
Thailand 208
Thermography 377
Thermoluminescence 424
Thermoluminescence dosimeter 37
Thorium 93
Thorium series 415
Thoron 46, 58, 76, 85, 93, 151, 159, 215, 219, 276, 281, 283, 309, 311, 313, 401
Thoron daughter 46
Thoron exhalation rate 219
Thoron progeny 219
Thorotrast 192
Threshold 162
Thunderstorm 428
Thyroid hormones 199
Time series 392
TLD 327, 344
Tobacco 407
Tracheal epithelial cell 251
Track detector 403
Track-etch detector 240
Transformation 189

Transgenic mouse 25
Transgenic rat 25
Translocation 3
Translocations 29
Trend analysis 418
Type II diabetes model mouse 185

$^{238}$U 331
Underground laboratory 227
Underground mine 278
Underground site 369
Underground tourist rout 360
Unmanned aeronautic vehicle 422
Uptake 270
Uranium 405
Uranium mine 291
Uranium mine disposal soil 221
Uranium mining 238
Uranium series 415

Vasoactive substance 249
Ventilation 377
Vine cellar 362

Waste rock 291
Water 383, 412
Well-type NaI(Tl) detector 350
WHO guidelines and recommendations 114
Work activities 62
Worker 362
Working place 240

Yangjiang, China 313
YAP:Ce 383